Introduction to
LINEAR ALGEBRA

Introduction to
LINEAR ALGEBRA

Roger C. McCann
Mississippi State University

HBJ

HARCOURT BRACE JOVANOVICH, PUBLISHERS

San Diego   New York   Chicago   Washington, D.C.   Atlanta
London   Sydney   Toronto

Photo Credit
Cover: Richard Williams

ISBN: 0-15-543001-7
Library of Congress Catalog Card Number: 83-82518
Printed in the United States of America

Preface

The concepts in this book are based entirely on elementary algebra and rational thinking, and they fall into two categories: (1) theory, which includes the logical development of concepts and proofs of theorems, and (2) computations, which consist of techniques to determine whether an object has a desired property, as well as techniques to create an object with a desired property. Most of these concepts have immediate and important applications. Because it is possible for a student to become so engrossed in the process that a thorough understanding of the concept is lost, this book provides numerous examples to motivate and illustrate each result.

To give the instructor flexibility in meeting the needs of students, *Introduction to Linear Algebra* includes more material than can be covered in a one-semester or one-quarter course. However, all the sections not marked "optional" in Chapters 1, 2, 4, and 5, along with Sections 6.1 and 7.1, constitute a basic course in linear algebra.

The most fundamental problem in elementary linear algebra is that of solving systems of linear equations. Almost all the other calculations are based on the ability to solve such equations. Therefore, Chapter 1 establishes the definitions, notations, and techniques needed for solving systems of linear equations. Matrices are introduced early in the chapter, and matrix arithmetic becomes a fascinating game that motivates even apathetic students.

In Chapter 2, systems of equations are used to evaluate determinants and their properties. Chapter 3 discusses vectors in two- and three-dimensional space. Although Chapter 3 may be used as an introduction to the general discussion of linear spaces in Chapter 4, its omission will not affect the continuity of the text.

Chapter 5 focuses on the linear spaces for which an inner product can be defined. Chapter 6 deals with linear transformations from one linear space into another linear space. Chapter 7 continues the focus on linear transformations by introducing eigenvalues and eigenvectors. Chapter 8 provides applications to computer graphics, analytic geometry, and differential equations. Applications to physics and approximation theory appear in various chapters. Chapter 9 concludes the text with numerical techniques for solving systems of equations and for approximating eigenvalues and eigenvectors.

Most sections include exercises that begin with routine problems and progress to more theoretical problems. Because Chapters 1 and 4 are long and contain especially important concepts, review exercises are provided at the end of these chapters. Sections marked "optional" are not required for understanding sections that are not marked "optional." The answers to the odd-numbered exercises are given at the end of the text. An Answer Manual containing answers to even-numbered exercises is available to instructors.

This book did not evolve exclusively by my own efforts. I want to thank the following reviewers, who made pertinent comments that resulted in an improved manuscript: Bruce H. Edwards (University of Florida, Gainesville), Terry L. Herdman (Virginia Polytechnic Institute and State University), Richard M. Koch (University of Oregon), Erik Schreiner (Western Michigan University), and David V. V. Wend (Montana State University). Mark Busby (Mississippi State University) also read portions of the manuscript and worked many of the exercises. Even with this assistance the book would not exist without the Harcourt Brace Jovanovich staff, including Pat Braus, Marilyn Davis, Marji James, Nancy Shehorn, and Richard Wallis; and I want to thank Mary Kitzmiller, in particular, for her careful editing of the manuscript. Most of all I would like to thank Susan McCann, who— as my wife, reviewer, and editor—provided moral support, read the manuscript, and worked most of the exercises.

Roger C. McCann

Contents

3 Vectors in 2-Space and 3-Space 100

4 Linear Spaces 144

5 Inner Product Spaces 215

6 Linear Transformations 250

7   Eigenvalues and Eigenvectors   284

8   Applications   333

9   Introduction to Numerical Linear Algebra   362

Appendix 1   381

Appendix 2   383

Answers to Odd–Numbered Exercises   385

Index   417

Introduction to
LINEAR ALGEBRA

1

Systems of Linear Equations and Matrices

As a student of linear algebra, each of you had your own particular reason for signing up for this class. Some of you had a good idea about the content of the course, while others simply knew that it was recommended or required for their particular majors. Regardless of your reason, you will find this course stimulating and challenging but easy to follow. We have compiled this book so that a student could obtain an excellent background in linear algebra even without an instructor.

Therefore, with the aid of an instructor or professor to answer questions about particular concepts that individual students may find difficult, you should be able to keep pace with the text. We want to impress on you, however, the importance of keeping up-to-date with your homework assignments. As in other mathematics courses, linear algebra builds upon previous theorems, lemmas, and definitions; therefore, you must have a good understanding of each concept before proceeding to the next.

1.1 SYSTEMS OF LINEAR EQUATIONS

A **linear equation** in the variables $x_1, x_2, \ldots, x_n$ (which are also called "unknowns") is an equation of the form

$$a_1 x_1 + a_2 x_2 + \cdots + a_n x_n = b$$

where b and the coefficients $a_1, a_2, \ldots, a_n$ are constants (usually real numbers).

A collection of linear equations, such as

$$a_{11}x_1 + a_{12}x_2 + \cdots + a_{1n}x_n = b_1$$
$$a_{21}x_1 + a_{22}x_2 + \cdots + a_{2n}x_n = b_2$$
$$\vdots \qquad \vdots \qquad \qquad \vdots \qquad \vdots$$
$$a_{m1}x_1 + a_{m2}x_2 + \cdots + a_{mn}x_n = b_m \tag{1}$$

where the subscripted a's and b's are constants, is called a **system of linear equations**. Throughout this book we shall assume that each variable has a nonzero coefficient in at least one equation in the system.

Some of you are already familiar with systems of equations and feel very comfortable with the preceding example. For those who need to refresh their memories, we can assure you that the subscripts in the example are not as difficult to decipher as they may at first appear. You will notice that the subscripts for the x's increase from left to right, beginning with 1 and increasing to some number n. If the number of variables is small, we can substitute various letters of the alphabet instead of making every variable a subscripted x. For example, when $n = 3$, we can write a system of equations as

$$a_{11}x + a_{12}y + a_{13}z = b_1$$
$$a_{21}x + a_{22}y + a_{23}z = b_2$$
$$a_{31}x + a_{32}y + a_{33}z = b_3$$
$$\vdots \qquad \vdots \qquad \vdots \qquad \vdots$$
$$a_{m1}x + a_{m2}y + a_{m3}z = b_m$$

You will notice that the *second* digit in the subscript of a also increases from left to right, whereas the *first* digit increases from top to bottom (from 1 to some number m). This is a handy way to identify any coefficient in the system. For example, if we refer to a_{23}, you will immediately realize we are referring to the coefficient of the third variable in the second equation.

In a system of linear equations all variables occur only to the *first power*, are *not multiplied* together, and are *not composed* of *trigonometric, exponential,* or *logarithmic functions.* The systems of equations

$$3x + 2y = -2$$
$$x + 4y = 1 \tag{2}$$

and

$$x_1 + 3x_2 - 2x_3 = 0$$
$$3x_1 + 6x_2 + 4x_3 = 6$$
$$-2x_1 - 9x_2 + 6x_3 = 2 \tag{3}$$

are systems of linear equations, while

$$x + 3y^2 = 7$$
$$x - y = 2$$

and

$$\sin x + \cos y - \quad z = 7$$

$$7x \quad + \quad y - \sqrt{z} = 4$$

are not systems of linear equations.

A **solution** of the system of linear equations in (1) is an ordered column of n numbers

$$\begin{bmatrix} t_1 \\ t_2 \\ \vdots \\ t_n \end{bmatrix}$$

such that we have equalities when we set $x_1 = t_1, x_2 = t_2, \ldots, x_n = t_n$ in the equations. The system will be called **consistent** if it has at least one solution and will be called **inconsistent** otherwise.

The ordered pair of numbers $\begin{bmatrix} -1 \\ 1/2 \end{bmatrix}$ is a solution of the system of equations in (2) since

$$3(-1) + 2\left(\frac{1}{2}\right) = -2$$

$$(-1) + 4\left(\frac{1}{2}\right) = 1$$

Similarly the ordered triple of numbers

$$\begin{bmatrix} 2 \\ -\dfrac{1}{3} \\ \dfrac{1}{2} \end{bmatrix}$$

is a solution of the system of equations in (3) since

$$(2) + 3\left(-\frac{1}{3}\right) - 2\left(\frac{1}{2}\right) = 0$$

$$3(2) + 6\left(-\frac{1}{3}\right) + 4\left(\frac{1}{2}\right) = 6$$

$$-2(2) - 9\left(-\frac{1}{3}\right) + 6\left(\frac{1}{2}\right) = 2$$

Before considering the general system of linear equations in (1), consider the special case in which $n = 2$ and $m = 2$. In this case the system of equations

$$a_{11}x_1 + a_{12}x_2 = b_1$$

$$a_{21}x_1 + a_{22}x_2 = b_2$$

has a nice geometrical interpretation. For ease of notation we denote x_1 by x and x_2 by y so that the system we are considering can be written as

$$a_{11}x + a_{12}y = b_1$$
$$a_{21}x + a_{22}y = b_2 \tag{4}$$

A solution of the system of equations in (4) is an ordered pair of numbers $\begin{bmatrix} x_0 \\ y_0 \end{bmatrix}$ such that

$$a_{11}x_0 + a_{12}y_0 = b_1$$
$$a_{21}x_0 + a_{22}y_0 = b_2 \tag{5}$$

You should recall that the graph of an equation of the form $ax + by = c$ is a straight line (if either a or b is nonzero). The first equation in (5) indicates that the point (x_0, y_0) is on the graph of $a_{11}x + a_{12}y = b_1$. Likewise, the second equation in (5) indicates that the point (x_0, y_0) is also on the graph of $a_{21}x + a_{22}y = b_2$. Thus, a solution of the system in (4) is a point at which the graphs (straight lines) of these equations intersect. There are three possibilities to consider:

1. The lines are parallel and do not intersect.
2. The lines are not parallel so that they intersect in precisely one point.
3. The two lines coincide.

In the first of these possibilities there are no points of intersection. Therefore, the system has no solution. The system of equations

$$x + y = 2$$
$$x + y = 4$$

is an example of an inconsistent system (see Figure 1.1a). In the second possibility there is precisely one point of intersection. Therefore, there is precisely one solution. The system of equations

$$x + y = 2$$
$$x - y = -1$$

is an example of such a system (see Figure 1.1b). The only solution is $\begin{bmatrix} 1/2 \\ 3/2 \end{bmatrix}$.

In the last possibility there are infinitely many points of intersection. Therefore, there are infinitely many solutions. The system of equations

$$x + \ y = 2$$
$$2x + 2y = 4$$

is an example of such a system (see Figure 1.1c). Every pair of numbers of the form $\begin{bmatrix} k \\ 2 - k \end{bmatrix}$, k being any number, is a solution. In particular when $k = 1$ and $k = 2$, the solutions are, respectively, $\begin{bmatrix} 1 \\ 1 \end{bmatrix}$ and $\begin{bmatrix} 2 \\ 0 \end{bmatrix}$.

The situation with $n = 2$ and $m = 2$ is typical. That is, a system of linear equations has either no solutions, precisely one solution, or infinitely many solutions. Fortunately there is a straightforward procedure to determine its solutions, if any. The basic idea of this procedure is described and illustrated by example in this section, with a detailed discussion deferred until the next section.

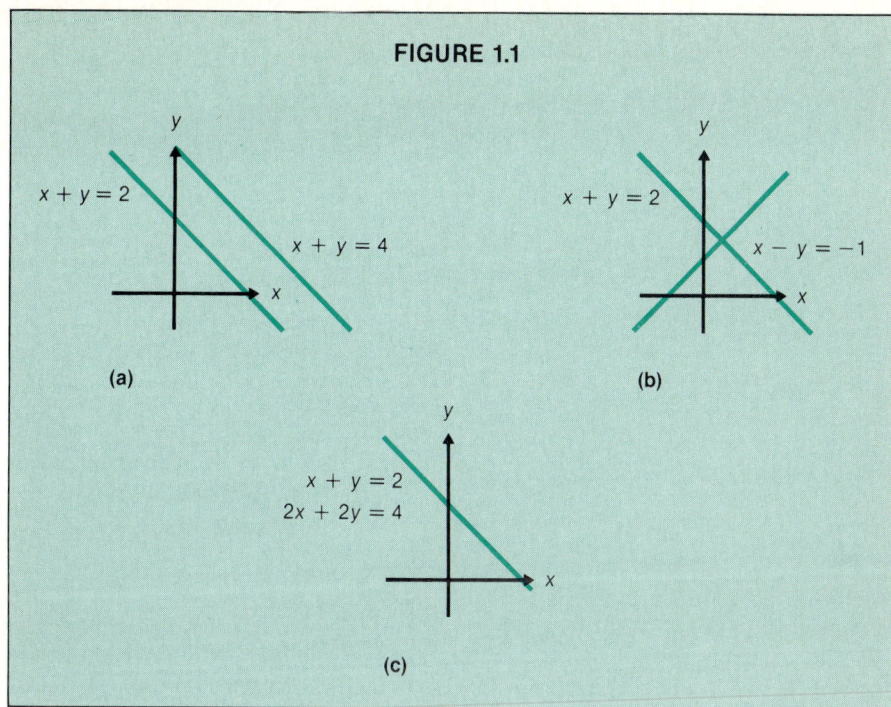

FIGURE 1.1

Two systems of linear equations are called **equivalent** if any solution of one of the systems is also a solution of the other system. That is, two systems are equivalent if they have precisely the same solutions. In order to find solutions of a given system of linear equations, we will replace it by an equivalent system that is easier to solve. This replacement procedure is continued until we obtain a system that is easy to solve. Experience has taught us, and it can be proved, that if we

1. Multiply an equation by a nonzero constant, or
2. Replace an equation by the sum of itself and a constant multiple of another equation in the system,

then we obtain an equivalent system of equations.

The following example illustrates how solutions of systems of linear equations can be found. At this point do not worry about how the operations in this example are chosen, because the next section presents a systematic procedure for choosing them.

Example 1. Consider the system

$$2x + 3y - 2z = 0$$

$$6x + 6y + 4z = 6$$

(6) $$-4x - 9y + 6z = 2$$

We begin by replacing the second equation by the sum of itself and -3 times the first equation.

$$2x + 3y - 2z = 0$$
$$-3y + 10z = 6$$
$$-4x - 9y + 6z = 2$$

Next we multiply the first equation by 2 and add the result to the third equation to obtain

$$2x + 3y - 2z = 0$$
$$-3y + 10z = 6$$
$$-3y + 2z = 2$$

Multiplying the second row by -1 and adding the result to the third equation yields

$$2x + 3y - 2z = 0$$
$$-3y + 10z = 6$$
$$-8z = -4$$

The value of z is now obtained by multiplying the third equation by $-1/8$. We find that $z = 1/2$, and this value is now inserted in the second equation to obtain $-3y + 10(1/2) = 6$, and we find that $y = -1/3$. Using these values for y and z in the first equation, we obtain $2x + 3(-1/3) - 2(1/2) = 0$. We see that $x = 1$ and thus obtain the solution

$$\begin{bmatrix} 1 \\ -\dfrac{1}{3} \\ \dfrac{1}{2} \end{bmatrix}$$

This solution can be checked by replacing x with 1, y with $-1/3$, and z with $1/2$ in the system of equations in (6).

Exercises

In Exercises 1–10 determine whether the given system of equations is a system of *linear* equations.

1. $\quad x_1 + x_2 = \sqrt{3}$
$\quad 2x_1 + 5x_2 = 0$

2. $\quad x_1 + \sqrt{x_2} = 3$
$\quad 2x_1 + 5x_2 = 0$

3. $x_1^2 + 3x_2 = 4$
$\quad x_1 - x_2 = 1$

4. $-x_1 + 4x_2 = 1.5$
$\quad x_1 + 7x_2 = 3$

5. $\quad x + 4y + 5z = 6$
$\quad x - 2y - 3z = -1.4$
$\quad -3x + 2y + z = 1$

6. $\quad x + y + 7z = 8$
$\quad -x - y + z = 0$
$\quad x + 3y + z = 1$

7.
$$x_1 + 3x_2x_3 + x_3 = 1$$
$$x_1 - x_2 + x_3 = 0$$
$$3x_1 + 5x_2 + x_3 = -9$$

8.
$$3x_1 + 6x_2 - 5x_3 = 6$$
$$3x_1 + 8x_2 - x_3 = 4$$
$$4x_1 + 8x_2 - x_2x_3 = 0$$

9.
$$-x_1 + x_2 - 6x_3 + x_4 = 5$$
$$x_1 - 2x_2 + 4x_3 = -2$$
$$x_1 + x_3 - x_4 = 3$$
$$x_2 + x_3 + 7x_4 = 8$$

10.
$$x_2 + 3x_3 + 4x_4 = 0$$
$$x_1 + 2x_2 + 5x_3 - 6x_4 = 0$$
$$-x_1 - x_2 + x_3 - x_4 = 7$$
$$\sin(x_1 + x_2 + x_3 + x_4) = 0$$

In Exercises 11–22 determine whether the ordered set of numbers is a solution of the system of linear equations.

11. $x + y = 2$
$x - y = 1$ $\begin{bmatrix} 1 \\ 1 \end{bmatrix}$

12. $x - y = 2$
$x + y = 0$ $\begin{bmatrix} 1 \\ -1 \end{bmatrix}$

13. $x + 2y = 3$
$2x - y = 1$ $\begin{bmatrix} 2 \\ 1/2 \end{bmatrix}$

14. $3x_1 + 5x_2 = 8$
$2x_1 + x_2 = -3$ $\begin{bmatrix} -1 \\ -1 \end{bmatrix}$

15. $3x - 4y = 7$
$5x + 4y = 1$ $\begin{bmatrix} 1 \\ -1 \end{bmatrix}$

16. $5x + 2y = 3/2$
$3x + 4y = -1/2$ $\begin{bmatrix} 1/2 \\ 1/2 \end{bmatrix}$

17. $x_1 + 2x_2 + 3x_3 = 0$
$2x_1 - x_2 + x_3 = 0$
$x_1 - 5x_2 - 4x_3 = 0$ $\begin{bmatrix} -1 \\ -1 \\ 1 \end{bmatrix}$

18. $3x_1 - 6x_2 + 7x_3 = 0$
$3x_1 - 2x_2 + x_3 = 1$
$4x_1 + 4x_2 - x_3 = 3$ $\begin{bmatrix} 1/2 \\ 1/4 \\ 0 \end{bmatrix}$

19. $x + 2y + z = 3$
$2x + 8y + 4z = 7$
$x + 2y - 11z = 0$ $\begin{bmatrix} 5/2 \\ 1/8 \\ 1/4 \end{bmatrix}$

20. $3x - 4y + z = 1.8$
$2x + y + 3z = 2.1$
$8x + 2y - z = .5$ $\begin{bmatrix} .1 \\ -.2 \\ .7 \end{bmatrix}$

21. $x_1 + x_2 - x_3 + x_4 = -4$
$x_1 - x_2 + x_3 - x_4 = 5$
$2x_1 + x_3 = 3$
$4x_2 + x_4 = 0$ $\begin{bmatrix} 1/2 \\ -1/2 \\ 2 \\ -2 \end{bmatrix}$

22. $x_1 - x_2 + x_3 - x_4 = 0$
$2x_1 - x_2 - 2x_3 + x_4 = 0$
$3x_1 - x_2 - x_3 - x_4 = 0$
$x_2 + x_3 - x_4 = 1$ $\begin{bmatrix} 1 \\ 1 \\ 1 \\ 1 \end{bmatrix}$

In Exercises 23–30 sketch the graphs of each of the equations. From the graphs determine whether the system of equations has no solution, precisely one solution, or infinitely many solutions.

23. $2x + 9y = 1$
$3x - 2y = 4$

24. $3x + 2y = 1$
$6x + 4y = 2$

25. $x + 5y = 2$
$2x + 10y = 3$

26. $-2x + 3y = 0$
$x - y = 2$

27. $3x + y = 7$
$-6x - 2y = -14$

28. $5x + 3y = 2$
$-5x - 3y = 1$

29. $-2x + y = 1$
$x + y = 2$

30. $3x + 3y = 1$
$x + y = 2$

In Exercises 31–36 find the solution of the given system of equations by using the method illustrated in Example 1.

31. $x + 3y = -2$
$2x + y = 1$

32. $3x + 4y = 1$
$x - 6y = 4$

33. $\begin{aligned} 2x + y + z &= -6 \\ x + 5y + 2z &= 3 \\ -3x + 2y + 3z &= -3 \end{aligned}$

34. $\begin{aligned} 3x + 2y + z &= 3 \\ 6x + 5y - 3z &= -16 \\ 2x - 2y + 3z &= 18 \end{aligned}$

35. $\begin{aligned} 2x + 4y + z &= 1 \\ -x + y + z &= 1 \\ x + y + 2z &= 1 \end{aligned}$

36. $\begin{aligned} 2x + 7y + 4z &= 6 \\ x - y + 3z &= 4 \\ 3x + 5y - z &= 2 \end{aligned}$

1.2 GAUSSIAN ELIMINATION

This section presents a systematic procedure for finding solutions of a system of linear equations:

$$a_{11}x_1 + a_{12}x_2 + \cdots + a_{1n}x_n = b_1$$
$$a_{21}x_1 + a_{22}x_2 + \cdots + a_{2n}x_n = b_2$$
$$\vdots \qquad \vdots \qquad \qquad \vdots \qquad \vdots$$
$$a_{m1}x_1 + a_{m2}x_2 + \cdots + a_{mn}x_n = b_m$$

In solving the system, you do not need to write down the plus signs and the unknowns, because they always occur in the same places. Only the coefficients of the unknowns need to be copied. Hence, we will use a notation in which the plus signs and the unknowns are omitted. We will represent the above system of equations by the rectangular array of numbers

$$\begin{bmatrix} a_{11} & a_{12} & \cdots & a_{1n} & b_1 \\ a_{21} & a_{22} & \cdots & a_{2n} & b_2 \\ \vdots & \vdots & \vdots & \vdots \\ a_{m1} & a_{m2} & \cdots & a_{mn} & b_m \end{bmatrix}$$

called the **augmented matrix** for the system.

The previous section mentions that we can obtain an equivalent system of equations if we

1. Multiply an equation by a nonzero constant, or
2. Replace an equation by the sum of itself and a constant multiple of another equation.

We can now add a third operation for obtaining an equivalent system of equations. We simply

3. Interchange the position of two equations.

Operations 1, 2, and 3 are called **elementary operations**.

Since the rows of numbers in the augmented matrix correspond to the coefficients and right sides of the equations in the system, elementary operations correspond to the following operations, called **elementary row operations**, on the rows of the augmented matrix:

1. Multiply a row by a nonzero constant.
2. Replace one row by the sum of itself and a constant multiple of another row.
3. Interchange the position of two rows.

Elementary row operations transform one augmented matrix into another augmented matrix. One of the most important properties of these operations is that they can be "undone." That is, if an augmented matrix B is obtained from an augmented matrix A by using elementary row operations, then we can also obtain the augmented matrix A from B by using elementary row operations. This property is crucial because it assures us that the system of equations corresponding to the augmented matrix A has precisely the same solutions as the system of equations corresponding to the augmented matrix B. Our goal is to use elementary row operations on a given augmented matrix until we obtain an augmented matrix in a special form called a "row-echelon form."

We say that an augmented matrix is in **row-echelon form** if the left side has a "stairlike" pattern of zero entries. The following augmented matrices are in row-echelon form:

$$\left[\begin{array}{cc|c} 1 & 2 & 5 \\ 0 & 1 & 3 \end{array}\right] \quad \left[\begin{array}{cccc|c} 4 & 2 & 5 & 0 & 2 & 5 \\ 0 & 0 & 1 & 3 & 9 & 1 \\ 0 & 0 & 0 & 0 & 2 & 7 \end{array}\right] \quad \left[\begin{array}{ccccccc|c} 6 & 1 & 2 & 2 & 1 & 3 & 1 & 0 \\ 0 & 0 & 1 & 4 & 7 & 0 & 0 & 3 \\ 0 & 0 & 0 & 0 & 0 & 4 & 0 & 9 \\ 0 & 0 & 0 & 0 & 0 & 0 & 0 & 6 \\ 0 & 0 & 0 & 0 & 0 & 0 & 0 & 0 \end{array}\right]$$

In describing the concept of row-echelon form more precisely, we shall count the **columns** (the *vertical* lists) of an augmented matrix by beginning at the left and moving to the right, while the **entries** of a column (or the *horizontal* rows) are counted from the top to the bottom. In order to determine whether or not an augmented matrix is in row-echelon form, we can check three qualifications. If the augmented matrix meets all three of the following requirements, it is in row-echelon form:

(a) The first entry of the first column is nonzero and all other entries of this column are zero.

(b) The first column that has a nonzero entry that is not in the first row has a nonzero entry in the second row, and every entry in this column below the second row equals zero.

(c) For $k \geq 3$, continue in the manner described in (b). That is, the first column that has a nonzero kth entry has every entry below the kth entry equal to zero. Moreover, this column occurs to the right of the first column that contains a nonzero $(k-1)$th entry.

In particular

$$\left[\begin{array}{ccc|c} 2 & 3 & 5 & 7 \\ 0 & 1 & 4 & -6 \\ 0 & 0 & 0 & 0 \end{array}\right] \quad \left[\begin{array}{cccc|c} -1 & 3 & 5 & 7 & 8 \\ 0 & 2 & -1 & 0 & 0 \\ 0 & 0 & 4 & 3 & 6 \\ 0 & 0 & 0 & 5 & 1 \end{array}\right]$$

are in row-echelon form, while

$$\left[\begin{array}{ccc|c} 1 & 3 & 2 & 1 \\ 0 & 2 & 4 & 2 \\ 1 & 0 & 6 & 3 \end{array}\right] \quad \left[\begin{array}{cc|c} 3 & 1 & 2 \\ 0 & 0 & 0 \\ 0 & 2 & 1 \end{array}\right]$$

are not in row-echelon form.

Using elementary row operations, we can convert any augmented matrix into one that is in row-echelon form. (An algorithm for this conversion is given later in this section.) As we shall see, it is easy to find the solutions, if any, of a system of equations whose corresponding augmented matrix is in row-echelon form.

If one row of an augmented matrix is

$$0 \ 0 \ \ldots \ 0 \,|\, a$$

where $a \neq 0$, then the associated system of equations has no solution because the corresponding equation

$$0 \cdot x_1 + 0 \cdot x_2 + \cdots + 0 \cdot x_n = a$$

from the system has no solution. If such a row does not occur when the augmented matrix is in row-echelon form, then the augmented matrix has either one solution or infinitely many solutions, as we shall see later in this section.

Before considering a general system of linear equations, reconsider the system of equations that was introduced at the end of the previous section:

$$2x + 3y - 2z = 0$$
$$6x + 6y + 4z = 6$$
$$-4x - 9y + 6z = 2$$

The augmented matrix for this system is

$$\begin{bmatrix} 2 & 3 & -2 & | & 0 \\ 6 & 6 & 4 & | & 6 \\ -4 & -9 & 6 & | & 2 \end{bmatrix}$$

In order to illustrate the relationship between elementary operations and elementary row operations, we shall find the solution of this system by using both types of operations. We shall do the calculations side by side to emphasize that the numbers in the augmented matrices are merely the numbers appearing in the system of equations.

$$2x + 3y - 2z = 0$$
$$6x + 6y + 4z = 6$$
$$-4x - 9y + 6z = 2$$

$$\begin{bmatrix} 2 & 3 & -2 & | & 0 \\ 6 & 6 & 4 & | & 6 \\ -4 & -9 & 6 & | & 2 \end{bmatrix}$$

Add -3 times the first equation to the second to obtain

$$2x + 3y - 2z = 0$$
$$-3y + 10z = 6$$
$$-4x - 9y + 6z = 2$$

Add -3 times the first row to the second to obtain

$$\begin{bmatrix} 2 & 3 & -2 & | & 0 \\ 0 & -3 & 10 & | & 6 \\ -4 & -9 & 6 & | & 2 \end{bmatrix}$$

Add 2 times the first equation to the third to obtain

$$2x + 3y - 2z = 0$$
$$-3y + 10z = 6$$
$$-3y + 2z = 2$$

Add 2 times the first row to the third to obtain

$$\begin{bmatrix} 2 & 3 & -2 & | & 0 \\ 0 & -3 & 10 & | & 6 \\ 0 & -3 & 2 & | & 2 \end{bmatrix}$$

Add -1 times the second equation to the third to obtain

$$2x + 3y - 2z = 0$$
$$-3y + 10z = 6$$
$$-8z = -4$$

Add -1 times the second row to the third to obtain

$$\begin{bmatrix} 2 & 3 & -2 & 0 \\ 0 & -3 & 10 & 6 \\ 0 & 0 & -8 & -4 \end{bmatrix}$$

Notice that the last augmented matrix is in row-echelon form. Solving the equation $-8z = -4$ for z, we find that $z = 1/2$. Inserting this value into the equation $-3y + 10z = 6$, we find that $y = -1/3$. These values for y and z are now inserted into the equation $2x + 3y - 2z = 0$, which yields $x = 1$. Thus the solution is

$$\begin{bmatrix} 1 \\ -\dfrac{1}{3} \\ \dfrac{1}{2} \end{bmatrix}$$

This process can be converted into a systematic process that transforms any augmented matrix into an augmented matrix in row-echelon form, as follows:

1. Interchange rows (if necessary) to place a nonzero number in the first entry of the first row.
2. Use the second elementary row operation as many times as necessary to place a zero as the first entry in every row below the first.
3. Hold the first row fixed, and interchange other rows (if necessary) to give the second row a nonzero entry as far left as possible.
4. Hold the first two rows fixed; use the second elementary row operation to place a zero in every row after the second row directly beneath the first nonzero entry in the second row.
5. Continue to hold the first two rows fixed, and interchange other rows (if necessary) to give the third row a nonzero entry as far left as possible.
6. Hold the first three rows fixed; use the second elementary row operation to place a zero in every row after the third row directly beneath the first nonzero entry in the third row.
7. Continue in this fashion until a matrix in row-echelon form is attained.

For ease of exposition we shall denote the elementary row operations by a shorthand notation:

1. cR_i will indicate that the ith row is multiplied by the nonzero constant c.
2. $cR_j + R_i \to R_i$ will indicate that c times the jth row is added to the ith row and the sum becomes the new ith row.
3. $R_i \leftrightarrow R_j$ will indicate that the ith and jth rows have been interchanged.

When we write two augmented matrices separated by a $\sim$ we mean that the matrix on the right is obtained from the matrix on the left by using an elementary row operation. The elementary row operation used will be indicated by the appropriate notation to the right of the second matrix. For example,

$$\begin{bmatrix} 2 & 3 & 1 \\ 5 & -2 & 4 \\ 6 & 3 & 0 \end{bmatrix} \sim \begin{bmatrix} 2 & 3 & 1 \\ 5 & -2 & 4 \\ 0 & -6 & -3 \end{bmatrix} \quad (-3)R_1 + R_3 \to R_3$$

means that we have added -3 times the first row to the third row in the matrix on the left and obtained the matrix on the right.

We will now give examples illustrating this process in the three cases in which the system of equations has (1) precisely one solution; (2) infinitely many solutions; and (3) no solutions.

Example 1.   Consider the system of equations

$$2y + 4z = 3$$

$$x - 3y + 5z = 1$$

$$3x - y - z = 1$$

The augmented matrix associated with this system is

$$\begin{bmatrix} 0 & 2 & 4 & 3 \\ 1 & -3 & 5 & 1 \\ 3 & -1 & -1 & 1 \end{bmatrix}$$

Using elementary row operations we have

$$\begin{bmatrix} 0 & 2 & 4 & 3 \\ 1 & -3 & 5 & 1 \\ 3 & -1 & -1 & 1 \end{bmatrix} \sim \begin{bmatrix} 1 & -3 & 5 & 1 \\ 0 & 2 & 4 & 3 \\ 3 & -1 & -1 & 1 \end{bmatrix} \quad R_1 \leftrightarrow R_2$$

$$\sim \begin{bmatrix} 1 & -3 & 5 & 1 \\ 0 & 2 & 4 & 3 \\ 0 & 8 & -16 & -2 \end{bmatrix} \quad (-3)R_1 + R_3 \rightarrow R_3$$

$$\sim \begin{bmatrix} 1 & -3 & 5 & 1 \\ 0 & 2 & 4 & 3 \\ 0 & 0 & -32 & -14 \end{bmatrix} \quad (-4)R_2 + R_3 \rightarrow R_3$$

The last augmented matrix corresponds to the system of equations

$$x - 3y + 5z = 1$$

$$2y + 4z = 3$$

$$- 32z = -14$$

Solving the last equation, we easily find that $z = 7/16$. Inserting this value for z in the second equation, we have $2y + 4(7/16) = 3$ or, equivalently, $y = 1/2[3 - 4(7/16)] = 5/8$. These values for y and z are now used in the first equation to yield $x - 3(5/8) + 5(7/16) = 1$ or, equivalently, $x = 1 + 3(5/8) - 5(7/16) = 11/16$. Thus the solution of the system of equations is

$$\begin{bmatrix} \dfrac{11}{16} \\ \dfrac{5}{8} \\ \dfrac{7}{16} \end{bmatrix}$$

Example 2.   Consider the system of equations

$$x_1 - 3x_2 + 2x_3 + x_4 = 2$$
$$3x_1 - 9x_2 + 10x_3 + 2x_4 = 9$$
$$2x_1 - 6x_2 + 4x_3 + 2x_4 = 4$$
$$2x_1 - 6x_2 + 8x_3 + x_4 = 7$$

The augmented matrix associated with this system is

$$\begin{bmatrix} 1 & -3 & 2 & 1 & 2 \\ 3 & -9 & 10 & 2 & 9 \\ 2 & -6 & 4 & 2 & 4 \\ 2 & -6 & 8 & 1 & 7 \end{bmatrix}$$

Using elementary row operations we have

$$\begin{bmatrix} 1 & -3 & 2 & 1 & 2 \\ 3 & -9 & 10 & 2 & 9 \\ 2 & -6 & 4 & 2 & 4 \\ 2 & -6 & 8 & 1 & 7 \end{bmatrix} \sim \begin{bmatrix} 1 & -3 & 2 & 1 & 2 \\ 0 & 0 & 4 & -1 & 3 \\ 2 & -6 & 4 & 2 & 4 \\ 2 & -6 & 8 & 1 & 7 \end{bmatrix} \quad (-3)R_1 + R_2 \rightarrow R_2$$

$$\sim \begin{bmatrix} 1 & -3 & 2 & 1 & 2 \\ 0 & 0 & 4 & -1 & 3 \\ 0 & 0 & 0 & 0 & 0 \\ 2 & -6 & 8 & 1 & 7 \end{bmatrix} \quad (-2)R_1 + R_3 \rightarrow R_3$$

$$\sim \begin{bmatrix} 1 & -3 & 2 & 1 & 2 \\ 0 & 0 & 4 & -1 & 3 \\ 0 & 0 & 0 & 0 & 0 \\ 0 & 0 & 4 & -1 & 3 \end{bmatrix} \quad (-2)R_1 + R_4 \rightarrow R_4$$

$$\sim \begin{bmatrix} 1 & -3 & 2 & 1 & 2 \\ 0 & 0 & 4 & -1 & 3 \\ 0 & 0 & 0 & 0 & 0 \\ 0 & 0 & 0 & 0 & 0 \end{bmatrix} \quad (-1)R_2 + R_4 \rightarrow R_4$$

The last augmented matrix corresponds to the system of equations

$$x_1 - 3x_2 + 2x_3 + x_4 = 2$$

(1) $$4x_3 - x_4 = 3$$

Thus we have two linear equations in four unknowns to solve. To do this we solve for two of the unknowns in terms of the other two unknowns. One way this can be done is to solve the second equation for x_4 and solve the first equation for x_1.

From the second equation, we have $x_4 = 4x_3 - 3$. Using this value for x_4 in the first equation, we find that $x_1 = 2 + 3x_2 - 2x_3 - (4x_3 - 3) = 5 + 3x_2 - 6x_3$.

We now see that x_1 and x_4 are determined in terms of x_2 and x_3, but there are no conditions to determine x_2 and x_3. This means that x_2 and x_3 can take on any values. Thus, if s and t are arbitrary numbers, then

$$\begin{bmatrix} 5 + 3s - 6t \\ s \\ t \\ 4t - 3 \end{bmatrix}$$

is a solution. In particular, if $s = 0$ and $t = 0$, then

$$\begin{bmatrix} 5 \\ 0 \\ 0 \\ -3 \end{bmatrix}$$

is a solution. Likewise if $s = 3$ and $t = 1$, then

$$\begin{bmatrix} 8 \\ 3 \\ 1 \\ 1 \end{bmatrix}$$

is a solution. Since there are infinitely many choices for s and t, there are infinitely many solutions of the system of equations in (1).

Notice that we could also find the solutions of the system of equations in (1) by solving the second equation for x_3 instead of for x_4. From this equation we obtain

$$x_3 = \frac{1}{4}(3 + x_4)$$

Using this value for x_3 in the first equation in (1) we find that

$$x_1 = 2 + 3x_2 - 2\left[\frac{1}{4}(3 + x_4)\right] - x_4 = \frac{1}{2} + 3x_2 - \frac{3}{2}x_4$$

Thus if u and v are arbitrary numbers, then

$$\begin{bmatrix} \frac{1}{2} + 3u - \frac{3}{2}v \\ u \\ \frac{3}{4} + \frac{1}{4}v \\ v \end{bmatrix}$$

is a solution. These solutions *appear* to be different from those we found when we solved for x_1 and x_4, but they are not. For example, if we set $u = 0, v = -3$ and then set $u = 3, v = 1$, we obtain the solutions

$$\begin{bmatrix} 5 \\ 0 \\ 0 \\ -3 \end{bmatrix} \quad \text{and} \quad \begin{bmatrix} 8 \\ 3 \\ 1 \\ 1 \end{bmatrix}$$

respectively. These are precisely the solutions we determined earlier. In fact every solution of the form

$$\begin{bmatrix} 5 + 3s - 6t \\ s \\ t \\ 4t - 3 \end{bmatrix}$$

also has the form

$$\begin{bmatrix} \dfrac{1}{2} + 3u - \dfrac{3}{2}v \\ u \\ \dfrac{3}{4} + \dfrac{1}{4}v \\ v \end{bmatrix}$$

and vice-versa. Traditionally the unknowns with the smaller subscripts (e.g., x_1 and x_2) are usually determined in terms of the unknowns with larger subscripts (e.g., x_5 and x_6).

The case in which the number of unknowns equals the number of equations ($n = m$) occurs frequently. The preceding process allows us to convert the original augmented matrix into one of the form

(2)
$$\begin{bmatrix} c_{11} & c_{12} & c_{13} & \cdots & c_{1n} & d_1 \\ 0 & c_{22} & c_{23} & \cdots & c_{2n} & d_2 \\ 0 & 0 & c_{33} & \cdots & c_{3n} & d_3 \\ \vdots & \vdots & \vdots & & \vdots & \vdots \\ 0 & 0 & 0 & \cdots & c_{nn} & d_n \end{bmatrix}$$

where some of the subscripted c's may be zero. As seen in the examples, such a system of equations can have exactly one solution, infinitely many solutions, or no solution.

If $c_{nn} = 0$, then the last equation has no solution if $d_n \neq 0$. If $d_n = 0$, then any number is a solution of the last equation. Hence if the system is consistent it will have infinitely many solutions. However the system may not be consistent. For example, the augmented matrix

$$\begin{bmatrix} 2 & 1 & 2 & | & 1 \\ 0 & 0 & 0 & | & 4 \\ 0 & 0 & 0 & | & 0 \end{bmatrix}$$

corresponds to a system of equations that has no solution. This is easily seen by noting that the second row of this augmented matrix corresponds to the equation $0 \cdot x_1 + 0 \cdot x_2 + 0 \cdot x_3 = 4$, which has no solution.

Example 3.   Consider the system

$$x + y + z = 3$$
$$2x + 4y + 5z = 7$$
$$x + 3y + 4z = 2$$

The augmented matrix associated with this system is

$$\begin{bmatrix} 1 & 1 & 1 & | & 3 \\ 2 & 4 & 5 & | & 7 \\ 1 & 3 & 4 & | & 2 \end{bmatrix}$$

Using elementary row operations we have

$$\begin{bmatrix} 1 & 1 & 1 & | & 3 \\ 2 & 4 & 5 & | & 7 \\ 1 & 3 & 4 & | & 2 \end{bmatrix} \sim \begin{bmatrix} 1 & 1 & 1 & | & 3 \\ 0 & 2 & 3 & | & 1 \\ 1 & 3 & 4 & | & 2 \end{bmatrix} \quad (-2)R_1 + R_2 \rightarrow R_2$$

$$\sim \begin{bmatrix} 1 & 1 & 1 & | & 3 \\ 0 & 2 & 3 & | & 1 \\ 0 & 2 & 3 & | & -1 \end{bmatrix} \quad (-1)R_1 + R_3 \rightarrow R_3$$

$$\sim \begin{bmatrix} 1 & 1 & 1 & | & 3 \\ 0 & 2 & 3 & | & 1 \\ 0 & 0 & 0 & | & -2 \end{bmatrix} \quad (-1)R_2 + R_3 \rightarrow R_3$$

The last row of the last augmented matrix corresponds to the equation $0 \cdot x + 0 \cdot y + 0 \cdot z = -2$, which has no solution because the left side is always zero while the right side is -2.

The augmented matrix in (2) corresponds to the system of equations

$$c_{11}x_1 + c_{12}x_2 + \cdots + c_{1n}x_n = d_1$$
$$c_{22}x_2 + \cdots + c_{2n}x_n = d_2$$
$$\vdots \qquad \vdots$$
$$c_{nn}x_n = d_n$$

If $c_{11} \neq 0$, $c_{22} \neq 0$, $\cdots$, $c_{nn} \neq 0$ (which occurs if and only if the original system has precisely one solution), then the solution may be found as follows.

We solve the last equation for x_n:

$$x_n = \frac{d_n}{c_{nn}}$$

When this value for x_n is used in the next-to-last equation

$$c_{n-1,n-1}x_{n-1} + c_{n-1,n}x_n = d_{n-1}$$

the value of x_{n-1} is easily determined:

$$x_{n-1} = \frac{d_{n-1} - c_{n-1,n}x_n}{c_{n-1,n-1}}$$

Continuing in this manner, we can work our way up the system of equations. At each step we determine the value of one variable until at the last step the first equation gives us the value of x_1. If one of the c_{ii} equals 0, then the system of equations has either no solution or infinitely many solutions.

The process of putting an augmented matrix into row-echelon form and then solving the associated system of equations beginning at the bottom and working upward is called **Gaussian elimination with backward substitution**. If the efficiency of a method is measured by the total number of arithmetic operations it requires, then Gaussian elimination with backward substitution is the most efficient general method for solving a system of n linear equations in n unknowns that has precisely one solution. In Example 1 we used this method to find the solution of the given system of equations.

Gaussian elimination with backward substitution can be easily programmed for a computer. For those interested, we now give an algorithm. For ease of notation we will assume that at each step the augmented matrix under consideration is written as,

$$\begin{bmatrix} a_{11} & a_{12} & \cdots & a_{1n} & b_1 \\ a_{21} & a_{22} & \cdots & a_{2n} & b_2 \\ \vdots & \vdots & & \vdots & \vdots \\ a_{n1} & a_{n2} & \cdots & a_{nn} & b_n \end{bmatrix}$$

Algorithm for Gaussian Elimination with Backward Substitution

1. INPUT the entries of the augmented matrix.
2. For $i = 1, 2, \ldots, n - 1$ do steps 3–7. (This is the elimination process.)
3. Let k be the smallest integer $i \leq k \leq n$ such that $a_{ki} \neq 0$. If no such integer k can be found, then OUTPUT "no unique solution exists" and go to step 12.
4. If $k \neq i$ perform $R_k \leftrightarrow R_i$.
5. For $j = i + 1, i + 2, \ldots, n$ do steps 6 and 7.
6. Set $\dot{p}_{kj} = a_{ji}/a_{ii}$.
7. Perform $(-p_{kj})R_i + R_j \rightarrow R_j$.
8. If $a_{nn} = 0$ then OUTPUT "no unique solution exists" and go to step 12.

9. Set $x_n = b_n/a_{nn}$. (This is the beginning of backward substitution.)

10. For $i = n - 1, n - 2 \ldots, 1$ set $x_i = (b_i - a_{i,i+1}x_{i+1} - a_{i,i+2}x_{i+2} - \cdots - a_{in}x_n)/a_{ii}$

11. OUTPUT "the unique solution is $\begin{bmatrix} x_1 \\ x_2 \\ \vdots \\ x_n \end{bmatrix}$."

12. STOP.

Exercises

In Exercise 1–8 find the augmented matrix for the given system of equations.

1. $x_1 + 3x_2 = 1$
$3x_1 + x_2 = 4$

2. $-x_1 + 4x_2 = -4$
$x_1 - x_2 = 0$

3. $5x + 3y + 2z = 0$
$-2x + 3y - 5z = 1$
$x - y + z = 6$

4. $3x_1 + x_2 - x_3 = -6$
$x_1 + 2x_2 + 5x_3 = 4$

5. $2x_1 + x_2 + 3x_3 - 4x_4 = 2$
$-x_1 + 2x_2 - x_3 + 5x_4 = 3$
$x_1 - x_2 + 4x_3 - 3x_4 = -1$

6. $x_1 + 3x_2 + 2x_3 + x_4 = -1$
$x_1 + 3x_3 - 2x_4 = 2$
$-x_1 - 2x_2 - x_4 = 0$
$2x_1 + x_2 + 5x_3 + 3x_4 = 2$

7. $6x_1 + 4x_2 - 5x_3 - 2x_4 = 1$
$2x_1 - 3x_2 + 2x_3 - 5x_4 = -1$

8. $2x_1 + x_2 + 2x_3 - x_4 = 5$
$2x_2 - 3x_3 + 5x_4 = 6$
$-x_1 + 2x_3 + 6x_4 = 1$

In Exercises 9–16 find the system of equations corresponding to the given augmented matrix.

9. $\begin{bmatrix} 1 & 2 & 3 \\ 4 & 5 & 6 \end{bmatrix}$

10. $\begin{bmatrix} -2 & 3 & 6 \\ 4 & 2 & 7 \end{bmatrix}$

11. $\begin{bmatrix} 1 & 2 & 3 & 5 \\ 6 & 2 & -1 & 7 \\ -6 & -4 & 5 & 3 \end{bmatrix}$

12. $\begin{bmatrix} 2 & 4 & 2 & -1 \\ 5 & 1 & 6 & 0 \end{bmatrix}$

13. $\begin{bmatrix} 2 & 3 & 1 & -2 & 9 \\ -3 & 4 & -1 & 5 & 5 \end{bmatrix}$

14. $\begin{bmatrix} 2 & 1 & 0 & 3 & 1 \\ 2 & 4 & 1 & 2 & 2 \\ 5 & 1 & 6 & 1 & 3 \\ 3 & 0 & 1 & 4 & 5 \end{bmatrix}$

15. $\begin{bmatrix} 3 & 2 & 4 & 0 & 8 & 5 \\ 9 & 3 & 2 & 4 & 0 & 3 \\ 2 & 2 & 3 & 2 & 5 & 3 \\ 4 & 1 & 4 & 6 & 0 & 1 \\ 5 & 4 & 3 & 2 & 1 & 0 \end{bmatrix}$

16. $\begin{bmatrix} 1 & -1 & 0 & 6 & 2 & 0 \\ 2 & 7 & 2 & 1 & 8 & 0 \\ 3 & -8 & 1 & -5 & 7 & 0 \end{bmatrix}$

In Exercise 17–24 determine which augmented matrices are in row-echelon form.

17. $\begin{bmatrix} 2 & 4 & 1 \\ 0 & 3 & 2 \end{bmatrix}$

18. $\begin{bmatrix} 6 & 5 & 1 \\ 0 & 6 & 0 \end{bmatrix}$

19. $\begin{bmatrix} 1 & 0 & 0 & 0 \\ 0 & 0 & 1 & 2 \\ 0 & 1 & 2 & 3 \end{bmatrix}$

20. $\begin{bmatrix} 2 & 4 & 6 & -1 \\ 0 & 5 & -2 & -2 \\ 0 & 0 & 0 & 3 \end{bmatrix}$

21. $\begin{bmatrix} 1 & 2 & 6 & 5 & 1 \\ 0 & 0 & 1 & 0 & 0 \\ 0 & 0 & 0 & 2 & 4 \\ 0 & 0 & 0 & 0 & 0 \end{bmatrix}$

22. $\begin{bmatrix} 2 & 1 & 0 & 0 & 0 & 1 \\ 0 & 0 & 1 & 0 & 0 & 2 \\ 0 & 0 & 0 & 0 & 5 & 4 \\ 0 & 0 & 0 & 0 & 1 & 2 \\ 0 & 0 & 0 & 0 & 0 & 0 \end{bmatrix}$

23. $\begin{bmatrix} 1 & 0 & 0 & 0 & 0 \\ 0 & 1 & 0 & 0 & 0 \\ 0 & 0 & 1 & 0 & 0 \\ 0 & 0 & 0 & 0 & 0 \\ 0 & 0 & 0 & 1 & 0 \end{bmatrix}$

24. $\begin{bmatrix} 1 & 0 & 0 & 0 & 0 \\ 0 & 1 & 0 & 1 & 0 \\ 0 & 0 & 1 & 1 & 0 \\ 0 & 0 & 0 & 0 & 0 \\ 0 & 0 & 0 & 0 & 0 \end{bmatrix}$

In Exercises 25–46 determine all of the solutions of the given system of equations by using Gaussian elimination with backward substitution.

25. $x + y = 1$
$x - y = 0$

26. $x_1 + 2x_2 = 2$
$x_1 - 4x_2 = -1$

27. $x_1 + 2x_2 + 3x_3 = 0$
$2x_1 - x_2 + x_3 = 0$
$x_1 - 5x_2 - 4x_3 = 0$

28. $3x - 4y + z = 1.8$
$2x + y + 3z = 2.1$
$8x + 2y - z = .5$

29. $2x_1 + 4x_2 - 4x_3 = 3$
$x_1 + 8x_2 + 2x_3 = 7$
$2x_1 + x_2 + x_3 = 2$

30. $3x_1 + 2x_2 + x_3 = 2$
$2x_1 + x_2 + 2x_3 = 1$
$x_1 + x_2 - x_3 = 1$

31. $x_1 - x_2 + x_3 = 0$
$2x_1 + x_2 + x_3 = 0$

32. $2x_1 + 5x_2 + x_3 = 3$
$3x_1 + 7x_2 - 3x_3 = 0$

33. $2x_1 + x_2 + x_3 = 0$
$3x_1 + 3x_2 + x_3 = 0$

34. $2x_1 + 3x_2 + 4x_3 = 4$
$x_1 - x_2 + x_3 = 1$

35 $2x_1 + 3x_2 + x_3 = 1$
$4x_1 + 6x_2 + 2x_3 = 2$
$6x_1 + 9x_2 + 3x_3 = 3$

36. $-x_1 + x_2 - x_3 - -2$
$2x_1 + x_3 = -1$
$x_1 + x_2 - x_3 = 0$

37. $x_1 + x_2 + x_3 + x_4 = 0$
$x_1 - x_2 + 2x_3 - x_4 = 2$
$x_1 - x_2 - 2x_3 + x_4 = 0$
$2x_1 + 2x_2 - 3x_3 + 3x_4 = 0$

38. $2x_1 - x_2 + x_3 - x_4 = -1$
$x_1 + 2x_2 + 3x_3 = 0$
$3x_1 - 2x_2 + x_3 - x_4 = 0$
$x_1 + x_2 + x_3 + x_4 = 7$

39. $2x_1 - x_2 + x_3 - x_4 = 0$
$3x_1 + x_2 + 2x_3 + x_4 = 2$
$x_1 + 2x_2 + x_3 + 2x_4 = 2$

40. $x_1 + x_2 + x_3 + x_4 = 2$
$2x_1 + x_2 + 3x_3 - x_4 = 1$

41. $2x_1 + x_2 + x_3 + 4x_4 + x_5 = 2$
$x_1 + x_2 + x_3 - 2x_4 + x_5 = 0$
$3x_1 - 2x_2 + x_3 + 6x_4 + x_5 = 0$
$x_1 + x_2 + x_3 + 2x_4 + x_5 = 1$
$2x_2 + x_3 - 4x_4 + x_5 = -1$

42. $x_1 + 2x_2 + x_3 + 3x_4 + 4x_5 = 3$
$2x_1 + x_2 + x_3 - 2x_5 = 0$
$x_1 + 4x_2 + x_3 - 6x_4 = 0$
$x_1 + x_2 + x_3 - 2x_5 = 0$
$x_1 + 3x_2 + x_3 + 3x_4 + 2x_5 = 3$

43. $x + y = a$
$x - y = b$

44. $2x + 3y = a$
$3x - 2y = b$

45. $x + y + z = a$
$x - y + z = b$
$2x + 2y - 2z = c$

46. $x + y + z = a$
$x + 2y + 2z = 0$
$2x + 3y + 3z = a$

In Exercises 47–48 find conditions on a, b, and c so that the system of equations is consistent.

47. $x_1 + 2x_2 + x_3 = a$
$2x_1 + x_2 - x_3 = b$
$-4x_1 + x_2 + 5x_3 = c$

48. $2x_1 - x_2 + x_3 = a$
$x_1 + x_2 - x_3 = b$
$7x_1 - 2x_2 + 2x_3 = c$

49. Find the second-degree polynomial function whose graph passes through the points $(-1, 0)$, $(0, .5)$, and $(1, 2)$.

50. Find the second-degree polynomial function whose graph passes through the points $(-1, 0)$, $(0, .5)$, and $(1, 2)$.

In Exercises 51–52 show that the given system of equations has infinitely many solutions. Find two solutions other than $x = 0$, $y = 0$, $z = 0$.

51. $x + y + z = 0$
$2x - 3y - 2z = 0$
$3x - 2y - z = 0$
$4x - y = 0$

52. $3x + 2y + 6z = 0$
$x - 2y + 5z = 0$
$2x + 4y + z = 0$
$5x + 6y + 7z = 0$

1.3 PIVOTING (OPTIONAL)

Theoretically Gaussian elimination with backward substitution yields the solution of a system of linear equations whenever the system has a unique solution. In practice this may not be the case because of errors caused by finite digit arithmetic used by computers and hand calculators. Before discussing one of the most common computational difficulties, we shall briefly discuss how numbers are handled by a computer.

The most widely used method of representing numbers in a computer is the floating-point number system. In this system a base B, usually 2, is selected along with a positive integer k. However, regardless of the numbers chosen for B and k, each nonzero number x in the floating-point system B is written in the form

$$x = \pm .a_1 a_2 \cdots a_k B^j$$

where j is an integer and each a_i is an integer such that

$$1 \le a_1 \le B - 1$$

$$0 \le a_i \le B - 1 \qquad (i = 2, 3, \cdots, k)$$

For ease of discussion we choose B to be 10 and k to be 3. Then $x = \pm .a_1 a_2 a_3 (10)^j$, where a_1 equals any integer from 1 through 9, and a_2 and a_3 are each an integer from 0 through 9. This means that we are using only the first three digits of the decimal representation of a number. For example, the numbers $4/3, 25, 3719, .5617$, and 4 are represented by $.133(10)^1, .250(10)^2, .371(10)^4, .561(10)^0$, and $.400(10)^1$, respectively, in this floating-point system.

There is nothing magical about our choice of B and k. The computational difficulty to be described arises regardless of the choices of B and k.

Consider the system of equations

$$.000100x + 2.00y = 2.00$$

$$1.00x - 2.00y = -1.00$$

which has

$$\begin{bmatrix} \dfrac{1}{1.0001} \\ \dfrac{2.0001}{2.0002} \end{bmatrix}$$

as its solution. To seven digits this solution is

$$\begin{bmatrix} .9999000 \\ .9999500 \end{bmatrix}$$

Suppose we use Gaussian elimination with backward substitution to solve this system. We multiply the first equation by $-(10)^4$ and add it to the second equation. In our floating-point system

$$-(10)^4(.0001) + 1 = -.100(10)^1 + .100(10)^1 = 0$$

and

$$-(10)^4(2.00) - 2.00 = -.200(10)^5 - .200(10)^1 = -.200(10)^5$$

and

$$-(10)^4(2.00) - 1.00 = -.200(10)^5 - .100(10)^1 = -.200(10)^5$$

After one step of the elimination process we have (in our floating-point system)

$$.100(10)^{-3}x + .200(10)^1y = .200(10)^1$$

$$-.200(10)^5y = -.200(10)^5$$

Solving the second equation for y we find that $y = 1.00$. Substituting this value for y into the first equation we find that

$$.100(10)^{-3}x + .200(10)^1 = .200(10)^1$$

Hence $x = 0$.

This answer is ridiculous! What happened? In order to discover what went wrong we need to consider the steps in Gaussian elimination. There is one case in which it is impossible to use the first equation to eliminate x from the second equation. This is when the coefficient of x in the first equation is zero. Let us consider the coefficients .0001 and 1 of x in the above system. In our floating-point system we have

$$.0001 + 1 = .100(10)^{-3} + .100(10)^1 = .100(10)^1$$

Thus, in our floating-point system .0001 acts as though it were zero when it is added to 1. In this sense the coefficient of x in the first equation acts as though it were zero.

This difficulty can be easily avoided. If the coefficient of x in the first equation were zero, we would interchange the equations. This is precisely what we do when this coefficient is very small compared to the other coefficients of x. Doing this, we find that the above system of equations becomes

$$1.00x - 2.00y = -1.00$$

$$.0001x + 2.00y = 2.00$$

If we now multiply the first equation by $-(10)^{-4}$ and add it to the second equation we obtain the system of equations (in our floating-point system)

$$.100(10)^1 x - .200(10)^1 y = -.100(10)^1$$

$$.200(10)^1 y = .200(10)^1$$

If we solve the last equation for y we find that $y = .100(10)^1$. When this value for y is used in the first equation we obtain $x = .100(10)^1$. After we interchange rows Gaussian elimination yields

$$\begin{bmatrix} .100(10)^1 \\ .100(10)^1 \end{bmatrix}$$

as the solution. This is an excellent approximation to the true solution considering that we used only three digits in our calculations.

Let us reconsider the procedure described in the preceding section. After using one row and the second row operation to place zeros at appropriate places in all lower rows, we interchanged rows so that the next lower row has a nonzero entry as far left as possible. At this stage the procedure can be modified to help prevent the difficulty illustrated above. If there is more than one row with a nonzero entry as far left as possible, we choose one whose leftmost nonzero entry has an absolute value as large as possible. The procedure remains the same except for this extra care when we interchange rows. With this modification the procedure is called **Gaussian elimination with maximum column pivoting**.

There are difficulties similar to those illustrated above that are not eliminated or even reduced by maximum column pivoting. These difficulties are caused by numbers in the calculations that are of vastly different sizes. However, there are modifications of Gaussian elimination similar to maximum column pivoting that tend to minimize these difficulties. The interested reader is referred to a numerical-analysis text for a discussion of these problems. Fortunately, the difficulties do not arise often, so Gaussian elimination with backward substitution is usually a reliable method to solve a system of linear equations.

1.4 GAUSS–JORDAN ELIMINATION

In this section we consider a modification of Gaussian elimination with backward substitution. This modification is only slightly less efficient than Gaussian elimination with backward substitution and gives a clear algorithm for finding the solutions of systems of linear equations with more than one solution. Moreover, it will be used to solve other problems later in this book.

Beginning with a system of linear equations

$$a_{11}x_1 + a_{12}x_2 + \cdots + a_{1n}x_n = b_1$$
$$a_{21}x_1 + a_{22}x_2 + \cdots + a_{2n}x_n = b_2$$
$$\vdots \qquad \vdots \qquad \qquad \vdots \qquad \vdots$$
$$a_{m1}x_1 + a_{m2}x_2 + \cdots + a_{mn}x_n = b_m$$

we use elementary row operations to reduce the corresponding augmented matrix to a matrix in row-echelon form. Instead of using backward substitution we now continue to use elementary row operations until we obtain an augmented matrix in row-echelon form with the following properties:

1. If a row does not consist entirely of zeros, then its first nonzero entry, called the **leading entry** of that row, is 1.
2. Any column containing the leading entry of some row consists entirely of zeros except for that single leading entry.

Thus whenever a "step" occurs in the row-echelon form, the entry at the step is 1 and all other entries in that column are zero. An augmented matrix in row-echelon form that satisfies these additional two properties is said to be in **reduced row-echelon form**.

For example,

$$\left[\begin{array}{ccc|c} 1 & 0 & 0 & 2 \\ 0 & 1 & 0 & 5 \\ 0 & 0 & 1 & -7 \end{array}\right] \quad \left[\begin{array}{ccc|c} 1 & 0 & 2 & 3 \\ 0 & 1 & 0 & 2 \\ 0 & 0 & 0 & 0 \\ 0 & 0 & 0 & 0 \end{array}\right] \quad \left[\begin{array}{cccc|c} 1 & 4 & 7 & 0 & 1 \\ 0 & 0 & 0 & 1 & -2 \\ 0 & 0 & 0 & 0 & 0 \end{array}\right]$$

are in reduced row-echelon form, while

$$\left[\begin{array}{ccc|c} 1 & 1 & 0 & 2 \\ 0 & 1 & 3 & 5 \\ 0 & 0 & 1 & -7 \end{array}\right] \quad \left[\begin{array}{ccc|c} 1 & 0 & 1 & 3 \\ 0 & 0 & 0 & 0 \\ 0 & 1 & 0 & 2 \\ 0 & 0 & 0 & 0 \end{array}\right] \quad \left[\begin{array}{cccc|c} 1 & 2 & 0 & 6 & 1 \\ 0 & 0 & 2 & 1 & -2 \\ 0 & 0 & 0 & 0 & 0 \end{array}\right]$$

are not in reduced row-echelon form.

Fortunately it is easy to reduce an augmented matrix to reduced row-echelon form. We will illustrate how this is done by considering the row-echelon matrices found in Section 1.2.

Example 1.   In Section 1.2 we used elementary row operations to obtain the augmented matrix

$$\left[\begin{array}{ccc|c} 2 & 3 & -2 & 0 \\ 0 & -3 & 10 & 6 \\ 0 & 0 & -8 & -4 \end{array}\right]$$

which is in row-echelon form, from the augmented matrix

$$\left[\begin{array}{rrr|r} 2 & 3 & -2 & 0 \\ 6 & 6 & 4 & 6 \\ -4 & -9 & 6 & 2 \end{array}\right] \tag{1}$$

Continuing to use elementary row operations we have

$$\left[\begin{array}{rrr|r} 2 & 3 & -2 & 0 \\ 0 & -3 & 10 & 6 \\ 0 & 0 & -8 & -4 \end{array}\right] \sim \left[\begin{array}{rrr|r} 1 & \frac{3}{2} & -1 & 0 \\ 0 & 1 & -\frac{10}{3} & -2 \\ 0 & 0 & 1 & \frac{1}{2} \end{array}\right] \quad \begin{array}{l} \frac{1}{2}R_1 \to R_1 \\ (-\frac{1}{3})R_2 \to R_2 \\ (-\frac{1}{8})R_3 \to R_3 \end{array}$$

$$\sim \left[\begin{array}{rrr|r} 1 & 0 & 4 & 3 \\ 0 & 1 & -\frac{10}{3} & -2 \\ 0 & 0 & 1 & \frac{1}{2} \end{array}\right] \quad (-\tfrac{3}{2})R_2 + R_1 \to R_1$$

$$\sim \left[\begin{array}{rrr|r} 1 & 0 & 0 & 1 \\ 0 & 1 & 0 & -\frac{1}{3} \\ 0 & 0 & 1 & \frac{1}{2} \end{array}\right] \quad \begin{array}{l} (-4)R_3 + R_1 \to R_1 \\ \frac{10}{3}R_3 + R_2 \to R_2 \end{array}$$

The last augmented matrix is in reduced row-echelon form and corresponds to the system of equations

$$\begin{aligned} x &= 1 \\ y &= -\frac{1}{3} \\ z &= \frac{1}{2} \end{aligned}$$

Thus

$$\left[\begin{array}{r} 1 \\ -\frac{1}{3} \\ \frac{1}{2} \end{array}\right]$$

is the solution of the system of equations corresponding to the augmented matrix in (1). This coincides with our findings in Section 1.2.

Example 2.   In Example 1 of Section 1.2 we used elementary row operations to obtain the augmented matrix

$$\begin{bmatrix} 1 & -3 & 5 & 1 \\ 0 & 2 & 4 & 3 \\ 0 & 0 & -32 & -14 \end{bmatrix}$$

which is in row-echelon form, from the augmented matrix

(2)
$$\begin{bmatrix} 0 & 2 & 4 & 3 \\ 1 & -3 & 5 & 1 \\ 3 & -1 & -1 & 1 \end{bmatrix}$$

Continuing to use elementary row operations we have

$$\begin{bmatrix} 1 & -3 & 5 & 1 \\ 0 & 2 & 4 & 3 \\ 0 & 0 & -32 & -14 \end{bmatrix} \sim \begin{bmatrix} 1 & -3 & 5 & 1 \\ 0 & 1 & 2 & \frac{3}{2} \\ 0 & 0 & 1 & \frac{7}{16} \end{bmatrix} \begin{array}{l} \frac{1}{2}R_2 \to R_2 \\ (-\frac{1}{32})R_3 \to R_3 \end{array}$$

$$\sim \begin{bmatrix} 1 & 0 & 11 & \frac{11}{2} \\ 0 & 1 & 2 & \frac{3}{2} \\ 0 & 0 & 1 & \frac{7}{16} \end{bmatrix} \; 3R_2 + R_1 \to R_1$$

$$\sim \begin{bmatrix} 1 & 0 & 0 & \frac{11}{16} \\ 0 & 1 & 0 & \frac{5}{8} \\ 0 & 0 & 1 & \frac{7}{16} \end{bmatrix} \begin{array}{l} (-11)R_3 + R_1 \to R_1 \\ (-2)R_3 + R_2 \to R_2 \end{array}$$

The last augmented matrix is in reduced row-echelon form and corresponds to the system of equations

$$x = \frac{11}{16}$$

$$y = \frac{5}{8}$$

$$z = \frac{7}{16}$$

Thus

$$\begin{bmatrix} \dfrac{11}{16} \\[2mm] \dfrac{5}{8} \\[2mm] 7 \\[1mm] \dfrac{7}{16} \end{bmatrix}$$

is the solution of the system of equations corresponding to the augmented matrix in (2). This coincides with our findings in Example 1 of Section 1.2.

Example 3.   In Example 2 of Section 1.2 we used elementary row operations to obtain the augmented matrix

$$\left[\begin{array}{rrrr|r} 1 & -3 & 2 & 1 & 2 \\ 0 & 0 & 4 & -1 & 3 \\ 0 & 0 & 0 & 0 & 0 \\ 0 & 0 & 0 & 0 & 0 \end{array}\right]$$

which is in row-echelon form, from the augmented matrix

$$\left[\begin{array}{rrrr|r} 1 & -3 & 2 & 1 & 2 \\ 3 & -9 & 10 & 2 & 9 \\ 2 & -6 & 4 & 2 & 4 \\ 2 & -6 & 8 & 1 & 7 \end{array}\right] \tag{3}$$

Continuing to use elementary row operations we have

$$\left[\begin{array}{rrrr|r} 1 & -3 & 2 & 1 & 2 \\ 0 & 0 & 4 & -1 & 3 \\ 0 & 0 & 0 & 0 & 0 \\ 0 & 0 & 0 & 0 & 0 \end{array}\right] \sim \left[\begin{array}{rrrr|r} 1 & -3 & 2 & 1 & 2 \\ 0 & 0 & 1 & -\frac{1}{4} & \frac{3}{4} \\ 0 & 0 & 0 & 0 & 0 \\ 0 & 0 & 0 & 0 & 0 \end{array}\right] \quad \tfrac{1}{4}R_2 \to R_2$$

$$\sim \left[\begin{array}{rrrr|r} 1 & -3 & 0 & \frac{3}{2} & \frac{1}{2} \\ 0 & 0 & 1 & -\frac{1}{4} & \frac{3}{4} \\ 0 & 0 & 0 & 0 & 0 \\ 0 & 0 & 0 & 0 & 0 \end{array}\right] \quad (-2)R_2 + R_1 \to R_1$$

The last augmented matrix is in reduced row-echelon form and corresponds to the system of equations

$$x_1 - 3x_2 + \frac{3}{2}x_4 = \frac{1}{2}$$

$$x_3 - \frac{1}{4}x_4 = \frac{3}{4}$$

so that

$$x_1 = \frac{1}{2} + 3x_2 - \frac{3}{2}x_4$$

(4)

$$x_3 = \frac{3}{4} \qquad + \frac{1}{4}x_4$$

Thus x_1 and x_3 are determined in terms of x_2 and x_4. For any choice of the unknowns x_2 and x_4 the identities in (4) enable us to determine a solution of the system of equations corresponding to the augmented matrix in (3). Setting $x_2 = u$ and $x_4 = v$ we have

$$\begin{bmatrix} \frac{1}{2} + 3u - \frac{3}{2}v \\[2mm] u \\[2mm] \frac{3}{4} + \frac{1}{4}v \\[2mm] v \end{bmatrix}$$

as a solution for every choice of the numbers u and v. Conversely every solution has this form for some choice of the numbers u and v. These solutions are the same as those found in Example 2 of Section 1.2.

The process of putting an augmented matrix in reduced row-echelon form by using elementary row operations is called **Gauss–Jordan elimination**. It needs to be emphasized that Gaussian elimination with backward substitution is the preferred method for computer implementation. However, Gauss–Jordan elimination gives us a specific method for determining the forms of the solutions of a system of linear equations with more than one solution. Later in the book we will find that Gauss–Jordan elimination is also a useful method for solving other problems.

In the remainder of this book we usually use Gaussian elimination with backward substitution to solve systems of linear equations having precisely one solution and Gauss–Jordan elimination to solve systems of linear equations having more than one solution.

Exercises

The following are Exercises 25–46 in Section 1.2, where the reader was asked to find the solutions of the given systems of equations by using Gaussian elimination with backward substitution. Now find the solutions by using Gauss–Jordon elimination and compare your calculations with those done earlier.

1. $x + y = 1$
$x - y = 0$

2. $x_1 + 2x_2 = 2$
$x_1 - 4x_2 = -1$

3. $x_1 + 2x_2 + 3x_3 = 0$
$2x_1 - x_2 + x_3 = 0$
$x_1 - 5x_2 - 4x_3 = 0$

4. $3x - 4y + z = 1.8$
$2x + y + 3z = 2.1$
$8x + 2y - z = .5$

5. $2x_1 + 4x_2 - 4x_3 = 3$
$x_1 + 8x_2 + 2x_3 = 7$
$2x_1 + x_2 + x_3 = 2$

6. $3x_1 + 2x_2 + x_3 = 2$
$2x_1 + x_2 + 2x_3 = 1$
$x_1 + x_2 - x_3 = 1$

7. $x_1 - x_2 + x_3 = 0$
$2x_1 + x_2 + x_3 = 0$

8. $2x_1 + 5x_2 + x_3 = 3$
$3x_1 + 7x_2 - 3x_3 = 0$

9. $2x_1 + x_2 + x_3 = 0$
$3x_1 + 3x_2 + x_3 = 0$

10. $2x_1 + 3x_2 + 4x_3 = 4$
$x_1 - x_2 + x_3 = 1$

11. $2x_1 + 3x_2 + x_3 = 1$
$4x_1 + 6x_2 + 2x_3 = 2$
$6x_1 + 9x_2 + 3x_3 = 3$

12. $-x_1 + x_2 - x_3 = -2$
$2x_1 + x_3 = -1$
$x_1 + x_2 - x_3 = 0$

13. $x_1 + x_2 + x_3 + x_4 = 0$
$x_1 - x_2 + 2x_3 - x_4 = 2$
$x_1 - x_2 - 2x_3 + x_4 = 0$
$2x_1 + 2x_2 - 3x_3 + 3x_4 = 0$

14. $2x_1 - x_2 + x_3 - x_4 = -1$
$x_1 + 2x_2 + 3x_3 = 0$
$3x_1 - 2x_2 + x_3 - x_4 = 0$
$x_1 + x_2 + x_3 + x_4 = 7$

15. $2x_1 - x_2 + x_3 - x_4 = 0$
$3x_1 + x_2 + 2x_3 + x_4 = 2$
$x_1 + 2x_2 + x_3 + 2x_4 = 2$

16. $x_1 + x_2 + x_3 + x_4 = 2$
$2x_1 + x_2 + 3x_3 - x_4 = 1$

17. $2x_1 + x_2 + x_3 + 4x_4 + x_5 = 2$
$x_1 + x_2 + x_3 - 2x_4 + x_5 = 0$
$3x_1 - 2x_2 + x_3 + 6x_4 + x_5 = 0$
$x_1 + x_2 + x_3 + 2x_4 + x_5 = 1$
$2x_2 + x_3 - 4x_4 + x_5 = -1$

18. $x_1 + 2x_2 + x_3 + 3x_4 + 4x_5 = 3$
$2x_1 + x_2 + x_3 - 2x_5 = 0$
$x_1 + 4x_2 + x_3 - 6x_4 = 0$
$x_1 + x_2 + x_3 - 2x_5 = 0$
$x_1 + 3x_2 + x_3 + 3x_4 + 2x_5 = 3$

19. $x + y = a$
$x - y = b$

20. $2x + 3y = a$
$3x - 2y = b$

21. $x + y + z = a$
$x - y + z = b$
$2x + 2y - 2z = c$

22. $x + y + z = a$
$x + 2y + 2z = 0$
$2x + 3y + 3z = a$

1.5 SYSTEMS OF LINEAR EQUATIONS WITH THE SAME NUMBER OF EQUATIONS AND UNKNOWNS

Throughout this section we consider a system of n linear equations in the same number of unknowns

$$a_{11}x_1 + a_{12}x_2 + \cdots + a_{1n}x_n = b_1$$
$$a_{21}x_1 + a_{22}x_2 + \cdots + a_{2n}x_n = b_2$$
$$\vdots \qquad \vdots \qquad \qquad \vdots \qquad \vdots$$
$$a_{n1}x_1 + a_{n2}x_2 + \cdots + a_{nn}x_n = b_n$$

(1)

and its corresponding augmented matrix

$$\begin{bmatrix} c_{11} & c_{12} & \cdots & c_{1n} & d_1 \\ 0 & c_{22} & \cdots & c_{2n} & d_2 \\ \vdots & \vdots & & \vdots & \vdots \\ 0 & 0 & \cdots & c_{nn} & d_n \end{bmatrix}$$

that is in reduced row-echelon form. If $c_{11} \neq 0$, $c_{22} \neq 0$, ..., and $c_{nn} \neq 0$, then we must have $c_{11} = 1, c_{22} = 1, ..., c_{nn} = 1$, and all other entries equal to zero on the left side of the augmented matrix (see Examples 1 and 2 of Section 1.4). In Section 1.2 we found that the system of equations in (1) has precisely one solution if and only if the numbers $c_{11}, c_{22}, ...,$ and c_{nn} are nonzero in any corresponding augmented matrix in row-echelon form. Combining these observations we have the following theorem:

Theorem 1. *A system of n equations in n unknowns has precisely one solution if and only if the corresponding reduced row-echelon matrix has the form*

(2)
$$\begin{bmatrix} 1 & 0 & \cdots & 0 & d_1 \\ 0 & 1 & \cdots & 0 & d_2 \\ \vdots & \vdots & & \vdots & \vdots \\ 0 & 0 & \cdots & 1 & d_n \end{bmatrix}$$

When we use elementary row operations to convert the augmented matrix

$$\begin{bmatrix} a_{11} & a_{12} & \cdots & a_{1n} & b_1 \\ a_{21} & a_{22} & \cdots & a_{2n} & b_2 \\ \vdots & \vdots & & \vdots & \vdots \\ a_{n1} & a_{n2} & \cdots & a_{nn} & b_n \end{bmatrix}$$

into one in reduced row-echelon form, the numbers $b_1, b_2, ..., b_n$ have no effect on the calculations on the left side. Thus whether the augmented matrix can be converted into one of the form in (2) depends solely on the numbers on the left side and not on those on the right. Combining this fact with Theorem 1 we have the following theorem:

Theorem 2. *If a system of n equations in n unknowns*

$$a_{11}x_1 + a_{12}x_2 + \cdots + a_{1n}x_n = b_1$$
$$a_{21}x_1 + a_{22}x_2 + \cdots + a_{2n}x_n = b_2$$
$$\vdots \qquad \vdots \qquad \qquad \vdots \qquad \vdots$$
$$a_{n1}x_1 + a_{n2}x_2 + \cdots + a_{nn}x_n = b_n$$

has precisely one solution for one particular choice of $b_1, b_2, ..., b_n$ then it has precisely one solution for every choice of $b_1, b_2, ..., b_n$.

Example 1.   In Example 1 of Section 1.4 we found that

$$2x_1 + 3x_2 - 2x_3 = 0$$
$$6x_1 + 6x_2 + 4x_3 = 6$$
$$-4x_1 - 9x_2 + 6x_3 = 2$$

has the unique solution

$$\begin{bmatrix} 1 \\ -\dfrac{1}{3} \\ \dfrac{1}{2} \end{bmatrix}$$

We now show that the system

$$
\begin{aligned}
2x_1 + 3x_2 - 2x_3 &= b_1 \\
6x_1 + 6x_2 + 4x_3 &= b_2 \\
-4x_1 - 9x_2 + 6x_3 &= b_3
\end{aligned}
\tag{3}
$$

has precisely one solution for every choice of b_1, b_2, and b_3. Using elementary row operations we have

$$
\begin{bmatrix} 2 & 3 & -2 & | & b_1 \\ 6 & 6 & 4 & | & b_2 \\ -4 & -9 & 6 & | & b_3 \end{bmatrix}
\sim
\begin{bmatrix} 2 & 3 & -2 & | & b_1 \\ 0 & -3 & 10 & | & b_2 - 3b_1 \\ 0 & -3 & 2 & | & b_3 + 2b_1 \end{bmatrix}
\quad
\begin{array}{l} (-3)R_1 + R_2 \to R_2 \\ 2R_1 + R_3 \to R_3 \end{array}
$$

$$
\sim
\begin{bmatrix} 2 & 0 & 8 & | & b_2 - 2b_1 \\ 0 & -3 & 10 & | & b_2 - 3b_1 \\ 0 & 0 & -8 & | & b_3 - b_2 + 5b_1 \end{bmatrix}
\quad
\begin{array}{l} R_2 + R_1 \to R_1 \\ (-1)R_2 + R_3 \to R_3 \end{array}
$$

$$
\sim
\begin{bmatrix} 2 & 0 & 0 & | & b_3 + 3b_1 \\ 0 & -3 & 0 & | & \frac{5}{4}b_3 - \frac{1}{4}b_2 + \frac{13}{4}b_1 \\ 0 & 0 & -8 & | & b_3 - b_2 + 5b_1 \end{bmatrix}
\quad
\begin{array}{l} R_3 + R_1 \to R_1 \\ \frac{5}{4}R_3 + R_2 \to R_2 \end{array}
$$

$$
\sim
\begin{bmatrix} 1 & 0 & 0 & | & \frac{1}{2}b_3 + \frac{3}{2}b_1 \\ 0 & 1 & 0 & | & -\frac{5}{12}b_3 + \frac{1}{12}b_2 - \frac{13}{12}b_1 \\ 0 & 0 & 1 & | & -\frac{1}{8}b_3 + \frac{1}{8}b_2 - \frac{5}{8}b_1 \end{bmatrix}
\quad
\begin{array}{l} \frac{1}{2}R_1 \to R_1 \\ (-\frac{1}{3})R_2 \to R_2 \\ (-\frac{1}{8})R_3 \to R_3 \end{array}
$$

Thus

$$\begin{bmatrix} \frac{1}{2}b_3 + \frac{3}{2}b_1 \\ -\frac{5}{12}b_3 + \frac{1}{12}b_2 - \frac{13}{12}b_1 \\ -\frac{1}{8}b_3 + \frac{1}{8}b_2 - \frac{5}{8}b_1 \end{bmatrix}$$

is the only solution of the system of equations in (3).

If the system of equations in (1) does not have a unique solution, then the corresponding augmented matrix in reduced row-echelon form cannot have the form in (2). In fact, a careful analysis of Gauss–Jordan elimination shows that in such a case there is at least one row in the reduced row-echelon matrix whose left side consists entirely of zeros. If the right sides of these rows are also zero, then the system of equations has infinitely many solutions (see Example 3 of Section 1.4). If the right side of one of these rows is nonzero, then the system of equations has no solution. By choosing $b_1, b_2, \ldots, b_n$ in appropriate fashions we can get either of these two cases to occur. We summarize this paragraph in the following theorem:

Theorem 3.   *Let*

$$a_{11}x_1 + a_{12}x_2 + \cdots + a_{1n}x_n = b_1$$

$$a_{21}x_1 + a_{22}x_2 + \cdots + a_{2n}x_n = b_2$$

$$\vdots \qquad \vdots \qquad \qquad \vdots \qquad \vdots$$

$$a_{n1}x_1 + a_{n2}x_2 + \cdots + a_{nn}x_n = b_n$$

be a system of linear equations. Then each of the following statements implies the others:

1. *The corresponding augmented matrix in reduced row-echelon form has at least one row whose left side consists entirely of zeros.*
2. *There exist numbers $b_1, b_2, \ldots, b_n$ such that the system has no solution.*
3. *There exist other numbers $b'_1, b'_2, \ldots, b'_n$ such that the system has infinitely many solutions.*
4. *For each choice of $b_1, b_2, \ldots, b_n$ either there are no solutions or there are infinitely many solutions of the system.*

Exercises

In Exercises 1–4 verify Theorem 1 for the given system of equations.

1. $2x_1 + x_2 = 4$
$\quad 4x_1 + 3x_2 = 6$

2. $3x_1 - 6x_2 = 6$
$\quad x_1 - 3x_2 = 3$

3. $\quad x_1 - 2x_2 + 3x_3 = 4$
$\quad 2x_1 - 3x_2 + 4x_3 = 0$
$\quad x_1 - \ x_2 + \ x_3 = 2$

4. $x_1 + 2x_2 + 3x_3 = 0$
$\quad x_1 - 2x_2 + 3x_3 = 1$
$\quad x_1 - 2x_2 + \ x_3 = 0$

In Exercises 5–8 show that the given system of equations has precisely one solution when $b_1 = 0, b_2 = 0$, and, if applicable, when $b_3 = 0$. Then show that the given system has precisely one solution for any choice of the b's.

5. $\quad x_1 + 3x_2 = b_1$
$\quad 2x_1 - 4x_2 = b_2$

6. $2x_1 - 3x_2 = b_1$
$\quad 5x_1 + 4x_2 = b_2$

7. $x_1 + x_2 + x_3 = b_1$
$x_1 - x_2 + x_3 = b_2$
$x_1 + x_2 - x_3 = b_3$

8. $2x_1 + 3x_2 + 4x_3 = b_1$
$x_1 + 3x_2 + 4x_3 = b_2$
$2x_1 + 3x_2 - 6x_3 = b_3$

In Exercises 9–12 show that the given system of equations has more than one solution when $b_1 = 0, b_2 = 0$, and, if applicable, when $b_3 = 0$. Then show that the given system of equations has infinitely many solutions whenever it is consistent. Finally determine for which b's the system has no solution.

9. $x_1 + 2x_2 = b_1$
$2x_1 + 4x_2 = b_2$

10. $6x_1 - 2x_2 = b_1$
$3x_1 - x_2 = b_2$

11. $2x_1 + 2x_2 + 3x_3 = b_1$
$x_1 + 3x_2 + x_3 = b_2$
$x_1 - x_2 + 2x_3 = b_3$

12. $x_1 + 4x_2 + 6x_3 = b_1$
$x_1 - 5x_2 + 3x_3 = b_2$
$2x_1 - x_2 + 9x_3 = b_3$

13. Suppose that $x_1 = 0, x_2 = 0, \ldots, x_n = 0$ is the only solution of the system of equations in (1) when $b_1 = 0, b_2 = 0, \ldots, b_n = 0$. How many solutions, if any, does the system have when $b_1 = 1, b_2 = 2, \ldots, b_n = n$?

14. Suppose that $x_1 = 0, x_2 = 0, \ldots, x_n = 0$ is not the only solution of the system of equations in (1) when $b_1 = 0, b_2 = 0, \ldots, b_n = 0$. Does the system have precisely one solution when $b_1 = 1, b_2 = 2, \ldots, b_n = n$?

1.6 MATRICES

We have already used rectangular arrays of numbers to facilitate the computation of solutions of systems of linear equations. Now we define concepts concerning rectangular arrays of numbers that may or may not represent the coefficients from a system of linear equations. When an array does represent the coefficients of a system of linear equations, these concepts will enable us to obtain information about the solutions of the system.

Rectangular arrays of numbers* are familiar to practically everyone. For example, the box score of a baseball game, the report of daily transactions on the New York Stock Exchange, trigonometric tables, and the tax schedule for income tax all involve rectangular arrays of numbers. In each of these examples the position of the numbers in the array is important. There is considerable difference between the tax due on a net income of $10,000 and that due on an income of $100,000. A ball player can make one hit in three times at bat, but he cannot make three hits in one time at bat. Arrays of numbers arise so frequently in mathematical settings that they are given a special name. The following definition gives some of the basic terminology that will be used throughout this book.

---

* By "number" we mean either a real or complex number. Except where complex numbers are explicitly indicated, there will be no loss of meaning if the reader interprets "number" as "real number" but remembers that the results also hold for complex numbers.

Definition 1.

(a) A rectangular array of numbers

$$A = \begin{bmatrix} a_{11} & a_{12} & \cdots & a_{1n} \\ a_{21} & a_{22} & \cdots & a_{2n} \\ \vdots & \vdots & & \vdots \\ a_{m1} & a_{m2} & \cdots & a_{mn} \end{bmatrix}$$

is called an $m \times n$ **matrix**.

(b) The $1 \times n$ and $m \times 1$ arrays

$$[a_{i1} \; a_{i2} \; \cdots \; a_{in}] \quad \text{and} \quad \begin{bmatrix} a_{1j} \\ a_{2j} \\ \vdots \\ a_{mj} \end{bmatrix}$$

are called the ith **row** and the jth **column**, respectively, of the matrix A.

(c) The number a_{ij} is called the ij-**component** of the matrix A.

(d) The matrix is called **square** if $m = n$.

(e) A matrix with m rows and n columns, such as A, is said to have **dimension** $m \times n$.

(f) An $n \times 1$ matrix

$$\begin{bmatrix} a_1 \\ a_2 \\ \vdots \\ a_n \end{bmatrix}$$

is called an n-**vector**, and the numbers $a_1, a_2, \ldots, a_n$ are called its **components**.

We will always denote matrices by italic capital letters, such as A, B, and C. Vectors will be denoted by boldface lowercase letters, such as **u** and **v**. The ij-component of a matrix will be denoted by the corresponding lowercase letter. For example, the ij-components of matrices A and B are a_{ij} and b_{ij}, respectively.

Example 1.   Consider the matrix

$$A = \begin{bmatrix} 1 & 2 & 3 & 4 \\ 5 & 6 & 7 & 8 \\ 9 & 0 & 1 & 2 \end{bmatrix}$$

Since A has 3 rows and 4 columns it is a 3×4 matrix. The third column of A is

$$\begin{bmatrix} 3 \\ 7 \\ 1 \end{bmatrix}$$

while the second row of A is

$$[5 \quad 6 \quad 7 \quad 8]$$

The number 7 is the 2,3-component of A because 7 is in the second row and third column of A. The 3,2-component is 0 because 0 is in the third row and second column. Thus, the ij- and ji-components of a matrix need not be the same. In fact, one component may be defined while the other is not. For example, the 3,4-component of A is 2, but A has no 4,3-component.

Two matrices A and B are said to be *equal* if and only if they have the same number of rows, the same number of columns, and the corresponding ij-components are equal.

Example 2. The matrices

$$\begin{bmatrix} 1 & 2 \\ 3 & 4 \\ 5 & 6 \end{bmatrix} \quad \text{and} \quad \begin{bmatrix} 1 & a \\ 3 & b \\ c & 6 \end{bmatrix}$$

are equal if and only if $a = 2$, $b = 4$, and $c = 5$. In particular, the matrices

$$\begin{bmatrix} 1 & 2 \\ 3 & 4 \\ 5 & 6 \end{bmatrix} \quad \text{and} \quad \begin{bmatrix} 1 & 4 \\ 3 & 2 \\ 5 & 6 \end{bmatrix}$$

are not equal even though they are both 3×2 matrices and have the same numbers as components. The corresponding 1,2 and 2,2 components are not equal.

There is an "algebra" of matrices that is similar to that of the real numbers. In fact, matrices satisfy many of the same arithmetic properties as numbers. The major exception is that the product AB of two matrices (to be defined in the next section) does not necessarily equal the product BA. In the remainder of this section we consider the addition of matrices and how to multiply a matrix by a number.

Definition 2. The sum of two $m \times n$ matrices A and B, written as $A + B$, is the $m \times n$ matrix C whose ij-component is

$$c_{ij} = a_{ij} + b_{ij}$$

where a_{ij} and b_{ij} are the ij-components of A and B, respectively.

In other words, the components of the matrix $A + B$ are the sums of the corresponding components of A and B. Notice that in order for the sum of A and B to be defined, A and B must have the same dimension.

Example 3.   Consider the matrices

$$A = \begin{bmatrix} 2 & 4 \\ 6 & 8 \\ 10 & 12 \end{bmatrix} \quad B = \begin{bmatrix} -1 & 3 \\ -2 & 5 \\ 4 & -1 \end{bmatrix} \quad C = \begin{bmatrix} -5 & 3 \\ 2 & 0 \end{bmatrix}$$

Then

$$A + B = \begin{bmatrix} 2 + (-1) & 4 + 3 \\ 6 + (-2) & 8 + 5 \\ 10 + 4 & 12 + (-1) \end{bmatrix} = \begin{bmatrix} 1 & 7 \\ 4 & 13 \\ 14 & 11 \end{bmatrix}$$

while $A + C$ and $B + C$ are not defined.

Theorem 4.   *Matrix addition is commutative and associative. That is, if A, B, and C are any $m \times n$ matrices, then*

1. *$A + B = B + A$.*
2. *$(A + B) + C = A + (B + C)$.*

Proof: Let a_{ij}, b_{ij}, and c_{ij} be the ij-components of A, B, and C, respectively. By Definition 2, the ij-components of $A + B$ and $B + A$ are $a_{ij} + b_{ij}$ and $b_{ij} + a_{ij}$, respectively. Since the addition of numbers is commutative, we have $a_{ij} + b_{ij} = b_{ij} + a_{ij}$ for $i = 1, 2, \ldots, m$ and $j = 1, 2, \ldots, n$. Therefore, the ij-components of $A + B$ and $B + A$ are equal so that $A + B = B + A$. This proves the first identity in Theorem 4.

By Definition 2, the ij-components of $(A + B) + C$ and $A + (B + C)$ are $(a_{ij} + b_{ij}) + c_{ij}$ and $a_{ij} + (b_{ij} + c_{ij})$, respectively. Since the addition of numbers is associative we have $(a_{ij} + b_{ij}) + c_{ij} = a_{ij} + (b_{ij} + c_{ij})$ for $i = 1, 2, \ldots, m$ and $j = 1, 2, \ldots, n$. Therefore, the ij-components of $(A + B) + C$ and $A + (B + C)$ are equal so that $(A + B) + C = A + (B + C)$. This proves the second identity in Theorem 4.   ∎

In light of the second identity in Theorem 4, it makes no difference whether we write $A + (B + C)$ or $(A + B) + C$. For simplicity we shall frequently drop the parentheses altogether and write merely $A + B + C$.

The $m \times n$ matrix all of whose components are zero is called a **zero matrix** and is denoted by 0 or by $0_{m \times n}$ if it is important to indicate its dimension. In the case that $n = 1$ so that $0_{m \times 1}$ is an m-vector, we will use the notation 0_m instead of $0_{m \times 1}$.

It is important to note that if A is any $m \times n$ matrix, then

$$A + 0_{m \times n} = A = 0_{m \times n} + A$$

Moreover, if a_{ij} is the ij-component of A and B is the matrix whose ij-component is $-a_{ij}$, then

$$A + B = 0_{m \times n} = B + A$$

The matrix B is called the **additive inverse** (or negative) of A and is denoted by $-A$. Thus $-A$ is the matrix obtained from A by multiplying each component of A by -1. For example, if

$$A = \begin{bmatrix} 1 & 5 & -8 \\ -7 & 4 & -2 \end{bmatrix}$$

then

$$-A = \begin{bmatrix} -1 & -5 & 8 \\ 7 & -4 & 2 \end{bmatrix}$$

If A and B are any two $m \times n$ matrices, then we shall denote $A + (-B)$ by $A - B$. The following definition describes how to multiply a matrix by a given number.

Definition 3.   Let A be an $m \times n$ matrix and c be any number. The scalar product of A and c, written as cA, is the $m \times n$ matrix B whose ij-component is

$$b_{ij} = ca_{ij}$$

In other words, the scalar product of A and c is the matrix formed by multiplying each component of A by c.

The following theorem shows that matrices have the expected distributive properties with respect to scalar multiplication.

Theorem 5.   *Let a and b be numbers and let A and B be $m \times n$ matrices. Then*

 1. $a(A + B) = aA + aB$.
 2. $(a + b)A = aA + bA$.
 3. $a(bA) = (ab)A$.

Proof: We will prove the first identity and leave the proofs of the remaining identities as exercises. Using Definitions 2 and 3 we find that the ij-components of $a(A + B)$ and $aA + aB$ are $a(a_{ij} + b_{ij})$ and $aa_{ij} + ab_{ij}$, respectively. Since multiplication of numbers is distributive we have $a(a_{ij} + b_{ij}) = aa_{ij} + ab_{ij}$ for $i = 1, 2, \ldots, m$ and $j = 1, 2, \ldots, n$. Therefore the ij-components of $a(A + B)$ and $aA + aB$ are equal so that $a(A + B) = aA + aB$. This proves the first identity in Theorem 5. ∎

Example 4.   Consider the matrices

$$A = \begin{bmatrix} 2 & 3 & 1 \\ -2 & 0 & 4 \end{bmatrix}$$

and

$$B = \begin{bmatrix} -3 & 5 & 4 \\ 2 & 1 & -1 \end{bmatrix}$$

Then

$$3(A + B) = 3\left(\begin{bmatrix} 2 & 3 & 1 \\ -2 & 0 & 4 \end{bmatrix} + \begin{bmatrix} -3 & 5 & 4 \\ 2 & 1 & -1 \end{bmatrix}\right)$$

$$= 3\begin{bmatrix} -1 & 8 & 5 \\ 0 & 1 & 3 \end{bmatrix}$$

$$= \begin{bmatrix} -3 & 24 & 15 \\ 0 & 3 & 9 \end{bmatrix}$$

and

$$3A + 3B = 3\begin{bmatrix} 2 & 3 & 1 \\ -2 & 0 & 4 \end{bmatrix} + 3\begin{bmatrix} -3 & 5 & 4 \\ 2 & 1 & -1 \end{bmatrix}$$

$$= \begin{bmatrix} 6 & 9 & 3 \\ -6 & 0 & 12 \end{bmatrix} + \begin{bmatrix} -9 & 15 & 12 \\ 6 & 3 & -3 \end{bmatrix}$$

$$= \begin{bmatrix} -3 & 24 & 15 \\ 0 & 3 & 9 \end{bmatrix}$$

so that $3(A + B) = 3A + 3B$ as is assured by the first identity in Theorem 5.

Exercises

In Exercises 1–14 perform the indicated computations, if possible, with

$$A = \begin{bmatrix} 2 & 3 \\ -1 & 0 \\ 4 & 1 \end{bmatrix}, \quad B = \begin{bmatrix} -1 & 2 \\ 2 & 1 \\ -1 & 2 \end{bmatrix}, \quad C = \begin{bmatrix} 0 & -2 \\ 3 & 1 \\ 2 & -3 \end{bmatrix}, \quad D = \begin{bmatrix} -1 & 4 & 2 \\ 2 & 0 & 1 \end{bmatrix}$$

If it is not possible, write "undefined."

1. $A + B$
2. $A - B$
3. $2A + 3B$
4. $2B + 4C$
5. $(A + B) + C$
6. $A + (B + C)$
7. $6(A + 2B)$
8. $A - D$
9. $A - B - 6D$
10. $0A$
11. $2C - B - 4A$
12. $6C - 4B - A$
13. Find a matrix E such that $(A + B) + E = 0_{3 \times 2}$.
14. Find a matrix E such that $E + D = 0_{2 \times 3}$.
15. Prove part 2 of Theorem 5.
16. Prove part 3 of Theorem 5.

In Exercises 17–22 perform the indicated operations with

$$A = \begin{bmatrix} 2 + i & 3 - i \\ 2i & -3i \end{bmatrix}, \quad B = \begin{bmatrix} 3i & 4i \\ -i & -2 + 3i \end{bmatrix}, \quad C = \begin{bmatrix} -3 - i & -3 + i \\ 2 - i & 2 + i \end{bmatrix}$$

where the components are complex numbers with $i^2 = -1$.

17. $3A + 3B$
18. $-6B + 5C$
19. $(2 + i)A - (3 + 2i)B$
20. $(3 + 7i)B - A + iC$
21. $2iA - 3B + 4iC$
22. $(2 + i)A + (2 - i)C$

1.7 Matrix Multiplication

Before defining the product AB of two matrices A and B, we anticipate the definition by noting three complications:

1. The definition is not the one that you would probably guess, but instead is more complicated. Later we shall see why it is the appropriate definition.
2. Not all matrices can be multiplied.
3. Even if the product AB is defined, it is not necessarily true that $AB = BA$ or even that BA is defined.

The product of two matrices A and B is defined whenever A has the same number of columns as B has rows. In such a case the ij-component of the product AB is the sum of the products of the corresponding components of the ith row of A and jth column of B. This can be represented pictorially as follows:

$$
\begin{bmatrix} a_{11} & a_{12} & \cdots & a_{1n} \\ \vdots & \vdots & & \vdots \\ a_{i1} & a_{i2} & \cdots & a_{in} \\ \vdots & \vdots & & \vdots \\ a_{m1} & a_{m2} & \cdots & a_{mn} \end{bmatrix}
\begin{bmatrix} b_{11} & \cdots & b_{1j} & \cdots & b_{1p} \\ \vdots & & \vdots & & \vdots \\ b_{i1} & \cdots & b_{ij} & \cdots & b_{ip} \\ \vdots & & \vdots & & \vdots \\ b_{n1} & \cdots & b_{nj} & \cdots & b_{np} \end{bmatrix}
=
\begin{bmatrix} & & \overset{j\text{th column}}{\vdots} & \\ \cdots & \sum\limits_{k=1}^{n} a_{ik}b_{kj} & \cdots & i\text{th row} \\ & & \vdots & \end{bmatrix}
$$

Usually the preceding informal definition is sufficient; however, a formal definition is sometimes preferable.

Definition 4. Let A be an $m \times n$ matrix and B be an $n \times p$ matrix. The product of A and B, written as AB, is the $m \times p$ matrix C whose ij-component is

$$c_{ij} = a_{i1}b_{1j} + a_{i2}b_{2j} + \cdots + a_{in}b_{nj}$$

$$= \sum_{k=1}^{n} a_{ik}b_{kj}$$

In particular, if A is an $n \times n$ matrix, then AA is defined and is denoted by A^2.

For example, if

$$A = \begin{bmatrix} 1 & 2 \\ 3 & 4 \\ 5 & 6 \end{bmatrix} \qquad B = \begin{bmatrix} 7 & 8 \\ 9 & 0 \end{bmatrix}$$

then

$$AB = \begin{bmatrix} 1 & 2 \\ 3 & 4 \\ 5 & 6 \end{bmatrix} \begin{bmatrix} 7 & 8 \\ 9 & 0 \end{bmatrix}$$

$$= \begin{bmatrix} (1)(7) + (2)(9) & (1)(8) + (2)(0) \\ (3)(7) + (4)(9) & (3)(8) + (4)(0) \\ (5)(7) + (6)(9) & (5)(8) + (6)(0) \end{bmatrix}$$

$$= \begin{bmatrix} 25 & 8 \\ 57 & 24 \\ 89 & 40 \end{bmatrix}$$

while BA is not defined. However, if we replace the matrix B by the matrix

$$C = \begin{bmatrix} 7 & 9 & 1 \\ 8 & 0 & 2 \end{bmatrix}$$

Then both AC and CA are defined.

$$AC = \begin{bmatrix} 1 & 2 \\ 3 & 4 \\ 5 & 6 \end{bmatrix} \begin{bmatrix} 7 & 9 & 1 \\ 8 & 0 & 2 \end{bmatrix}$$

$$= \begin{bmatrix} (1)(7) + (2)(8) & (1)(9) + (2)(0) & (1)(1) + (2)(2) \\ (3)(7) + (4)(8) & (3)(9) + (4)(0) & (3)(1) + (4)(2) \\ (5)(7) + (6)(8) & (5)(9) + (6)(0) & (5)(1) + (6)(2) \end{bmatrix}$$

$$= \begin{bmatrix} 23 & 9 & 5 \\ 53 & 27 & 11 \\ 83 & 45 & 17 \end{bmatrix}$$

$$CA = \begin{bmatrix} 7 & 9 & 1 \\ 8 & 0 & 2 \end{bmatrix} \begin{bmatrix} 1 & 2 \\ 3 & 4 \\ 5 & 6 \end{bmatrix}$$

$$= \begin{bmatrix} (7)(1) + (9)(3) + (1)(5) & (7)(2) + (9)(4) + (1)(6) \\ (8)(1) + (0)(3) + (2)(5) & (8)(2) + (0)(4) + (2)(6) \end{bmatrix}$$

$$= \begin{bmatrix} 39 & 56 \\ 18 & 28 \end{bmatrix}$$

Notice that $AC \neq CA$. In fact, AC and CA have neither the same number of rows nor the same number of columns. Even if E and D are $n \times n$ matrices, so that ED and DE are also $n \times n$ matrices, it usually happens that ED and DE are not equal. For example, let

$$D = \begin{bmatrix} 1 & 2 \\ 3 & 4 \end{bmatrix} \qquad E = \begin{bmatrix} 1 & 3 \\ 2 & 4 \end{bmatrix}$$

Then

$$DE = \begin{bmatrix} (1)(1) + (2)(2) & (1)(3) + (2)(4) \\ (3)(1) + (4)(2) & (3)(3) + (4)(4) \end{bmatrix}$$

$$= \begin{bmatrix} 5 & 11 \\ 11 & 25 \end{bmatrix}$$

$$ED = \begin{bmatrix} 1 & 3 \\ 2 & 4 \end{bmatrix} \begin{bmatrix} 1 & 2 \\ 3 & 4 \end{bmatrix}$$

$$= \begin{bmatrix} (1)(1) + (3)(3) & (1)(2) + (3)(4) \\ (2)(1) + (4)(3) & (2)(2) + (4)(4) \end{bmatrix} = \begin{bmatrix} 10 & 14 \\ 14 & 20 \end{bmatrix}$$

so that $DE \neq ED$.

Thus the order in which the matrices are multiplied is important. That is, matrix multiplication is not commutative. However, it is associative and distributive, as we see in the following theorem.

Theorem 6.　*Let A, B, and C be matrices and a be a number such that the matrix products and sums written below are defined. Then*

1. $(AB)C = A(BC)$.　(Associative property for matrix multiplication.)
2. $(A + B)C = AC + BC$.
3. $C(A + B) = CA + CB$.　(Distributive properties for matrix multiplication.)
4. $a(AB) = (aA)B = A(aB)$.

The proof of this theorem requires only the basic arithmetic properties of numbers and the definitions of addition, multiplication, and scalar multiplication for matrices. Since the proof gives little insight into the material we wish to discuss, it is omitted. However, in the exercises the reader is asked to prove this theorem in the case of all 2×2 matrices.

In light of part 1 of Theorem 6 it makes no difference whether we write $(AB)C$ or $A(BC)$. For simplicity we shall frequently drop the parentheses altogether and write merely ABC.

Example 1.　Consider the matrices

$$A = \begin{bmatrix} 1 & 2 \\ -2 & 3 \end{bmatrix} \quad B = \begin{bmatrix} -2 & 0 \\ 4 & 1 \end{bmatrix} \quad C = \begin{bmatrix} 2 & -1 & 1 \\ 3 & 0 & 6 \end{bmatrix}$$

Then

$$
\begin{aligned}
(A + B)C &= \left(\begin{bmatrix} 1 & 2 \\ -2 & 3 \end{bmatrix} + \begin{bmatrix} -2 & 0 \\ 4 & 1 \end{bmatrix} \right) \begin{bmatrix} 2 & -1 & 1 \\ 3 & 0 & 6 \end{bmatrix} \\
&= \begin{bmatrix} -1 & 2 \\ 2 & 4 \end{bmatrix} \begin{bmatrix} 2 & -1 & 1 \\ 3 & 0 & 6 \end{bmatrix} \\
&= \begin{bmatrix} (-1)(2) + (2)(3) & (-1)(-1) + (2)(0) & (-1)(1) + (2)(6) \\ (2)(2) + (4)(3) & (2)(-1) + (4)(0) & (2)(1) + (4)(6) \end{bmatrix} \\
&= \begin{bmatrix} 4 & 1 & 11 \\ 16 & -2 & 26 \end{bmatrix}
\end{aligned}
$$

and

$$
\begin{aligned}
AC + BC &= \begin{bmatrix} 1 & 2 \\ -2 & 3 \end{bmatrix} \begin{bmatrix} 2 & -1 & 1 \\ 3 & 0 & 6 \end{bmatrix} + \begin{bmatrix} -2 & 0 \\ 4 & 1 \end{bmatrix} \begin{bmatrix} 2 & -1 & 1 \\ 3 & 0 & 6 \end{bmatrix} \\
&= \begin{bmatrix} (1)(2) + (2)(3) & (1)(-1) + (2)(0) & (1)(1) + (2)(6) \\ (-2)(2) + (3)(3) & (-2)(-1) + (3)(0) & (-2)(1) + (3)(6) \end{bmatrix} \\
&\quad + \begin{bmatrix} (-2)(2) + (0)(3) & (-2)(-1) + (0)(0) & (-2)(1) + (0)(6) \\ (4)(2) + (1)(3) & (4)(-1) + (1)(0) & (4)(1) + (1)(6) \end{bmatrix}
\end{aligned}
$$

$$= \begin{bmatrix} 8 & -1 & 13 \\ 5 & 2 & 16 \end{bmatrix} + \begin{bmatrix} -4 & 2 & -2 \\ 11 & -4 & 10 \end{bmatrix}$$

$$= \begin{bmatrix} 4 & 1 & 11 \\ 16 & -2 & 26 \end{bmatrix}$$

Thus $(A + B)C = AC + BC$ as assured by part 2 of Theorem 4.

The following theorem is easily proved and is left as an exercise for the reader.

Theorem 7. *Let A be an $m \times n$ matrix. Then $A0_{n \times p} = 0_{m \times p}$. In particular, when $p = 1$, $A0_n = 0_m$.*

The theorems in Sections 1.6 and 1.7 show that matrices satisfy many of the same arithmetic properties as numbers. The major exception to this is that in general $AB \neq BA$. Another important exception is that it is possible to have $AB = 0$ without having $A = 0$ or $B = 0$. For example,

$$\begin{bmatrix} 1 & 0 \\ 0 & 0 \end{bmatrix} \begin{bmatrix} 0 & 0 \\ 0 & 1 \end{bmatrix} = \begin{bmatrix} 0 & 0 \\ 0 & 0 \end{bmatrix}$$

Also, the equation $AB = AC$ does not imply $B = C$ even if A, B, and C are not zero matrices. For example,

$$\begin{bmatrix} 1 & 1 \\ 0 & 0 \end{bmatrix} \begin{bmatrix} 1 & 0 \\ 0 & 0 \end{bmatrix} = \begin{bmatrix} 1 & 0 \\ 0 & 0 \end{bmatrix} = \begin{bmatrix} 1 & 1 \\ 0 & 0 \end{bmatrix} \begin{bmatrix} 1 & 2 \\ 0 & -2 \end{bmatrix}$$

but

$$\begin{bmatrix} 1 & 0 \\ 0 & 0 \end{bmatrix} \neq \begin{bmatrix} 1 & 2 \\ 0 & -2 \end{bmatrix}$$

We conclude this section with the observation that the jth column of AB is A times the jth column of B. This will be a useful result, and it is stated as a theorem whose proof is left as an exercise.

Theorem 8. *Let A be an $m \times n$ matrix and B be an $n \times p$ matrix. Then the jth column of AB is A times the jth column of B.*

Example 2.   Earlier in this section we found that

$$\begin{bmatrix} 1 & 2 \\ 3 & 4 \\ 5 & 6 \end{bmatrix} \begin{bmatrix} 7 & 8 \\ 9 & 0 \end{bmatrix} = \begin{bmatrix} 25 & 8 \\ 57 & 24 \\ 89 & 40 \end{bmatrix}$$

Notice that

$$\begin{bmatrix} 1 & 2 \\ 3 & 4 \\ 5 & 6 \end{bmatrix} \begin{bmatrix} 7 \\ 9 \end{bmatrix} = \begin{bmatrix} 25 \\ 57 \\ 89 \end{bmatrix}$$

is the first column of

$$\begin{bmatrix} 25 & 8 \\ 57 & 24 \\ 89 & 40 \end{bmatrix}$$

while

$$\begin{bmatrix} 1 & 2 \\ 3 & 4 \\ 5 & 6 \end{bmatrix} \begin{bmatrix} 8 \\ 0 \end{bmatrix} = \begin{bmatrix} 8 \\ 24 \\ 40 \end{bmatrix}$$

is the second column.

Exercises

In Exercises 1–14 perform the indicated computation, if possible, with

$$A = \begin{bmatrix} 2 & 3 \\ -1 & 0 \\ 4 & 1 \end{bmatrix}, \quad B = \begin{bmatrix} -1 & 2 \\ 2 & 1 \\ -1 & 2 \end{bmatrix}, \quad C = \begin{bmatrix} 0 & -2 \\ 3 & 1 \\ 2 & -3 \end{bmatrix},$$

$$D = \begin{bmatrix} -1 & 4 & 2 \\ 2 & 0 & 1 \end{bmatrix}, \quad E = \begin{bmatrix} 2 & 3 & 1 \\ 2 & 0 & 0 \end{bmatrix}$$

If the computation is not possible, write "undefined."

1. DA

2. DB

3. $D(A + B)$

4. $D(A + C)$

5. $D(2A + 3B)$

6. $D(4B - 2C)$

7. $AD + 2DB$

8. $BD + 3AC$

9. $(A + B)D$

10. $(A + 2C)D$

11. $(A + B)(D + E)$

12. $(2B - C)(3D - E)$

13. AB

14. CB

In Exercises 15–22 perform the indicated computation.

15. $\begin{bmatrix} 1 & 3 & 0 \\ -1 & 3 & 1 \\ 0 & 2 & 1 \end{bmatrix} \begin{bmatrix} 0 \\ 2 \\ 3 \end{bmatrix}$

16. $\begin{bmatrix} 1 & 2 & 3 \end{bmatrix} \begin{bmatrix} 1 & 3 & 0 \\ -1 & 3 & 1 \\ 0 & 2 & 1 \end{bmatrix}$

17. $\begin{bmatrix} 1 & 0 & 0 \\ 0 & 1 & 0 \\ 0 & 0 & 1 \end{bmatrix} \begin{bmatrix} 6 & -1 & 4 \\ 2 & 8 & -1 \\ 7 & 9 & 3 \end{bmatrix}$

18. $\begin{bmatrix} 6 & -1 & 4 \\ 2 & 8 & -1 \\ 7 & 9 & 3 \end{bmatrix} \begin{bmatrix} 1 & 0 & 0 \\ 0 & 1 & 0 \\ 0 & 0 & 1 \end{bmatrix}$

19. $\begin{bmatrix} 3 & -1 & -2 \\ -1 & 2 & 1 \\ -2 & 1 & 4 \end{bmatrix} \begin{bmatrix} 1 & 0 & 1 \\ 0 & 0 & 1 \\ 1 & 0 & 1 \end{bmatrix}$

20. $\begin{bmatrix} 3 & 1 & 0 \\ 6 & 0 & 2 \\ 0 & 1 & 3 \end{bmatrix} \begin{bmatrix} 2 & 4 & 0 \\ 0 & 2 & 4 \\ 2 & 0 & 4 \end{bmatrix}$

21. $\begin{bmatrix} a & b & 0 \\ 0 & a & b \\ a & 0 & b \end{bmatrix} \begin{bmatrix} 1 & 2 & 1 \\ 1 & 1 & 2 \\ 2 & 1 & 1 \end{bmatrix}$

where a and b are real numbers.

22. $\begin{bmatrix} 1 & 0 & 1 \\ 1 & 1 & 2 \\ 2 & 0 & 1 \end{bmatrix} \begin{bmatrix} a & 0 & b \\ 0 & 0 & 0 \\ b & 0 & a \end{bmatrix}$

where a and b are real numbers.

In Exercises 23–25 write $A = \begin{bmatrix} a & b \\ c & d \end{bmatrix}$ and then determine the numbers $a, b, c,$ and d so that the given equation is satisfied.

23. Find a matrix A such that

$$A \begin{bmatrix} 1 & 2 \\ 2 & 1 \end{bmatrix} = \begin{bmatrix} 1 & 0 \\ 0 & 1 \end{bmatrix}$$

24. Find a matrix A such that

$$\begin{bmatrix} 1 & 2 \\ 2 & 1 \end{bmatrix} A = \begin{bmatrix} 1 & 0 \\ 0 & 1 \end{bmatrix}$$

25. Find all matrices that commute with

$$B = \begin{bmatrix} 1 & 0 \\ 2 & 0 \end{bmatrix}$$

That is, find all matrices A such that $AB = BA$.

26. (a) Show that $(A + B)^2 = A^2 + AB + BA + B^2$.
(b) Find 2×2 matrices A and B such that $(A + B)^2 \neq A^2 + 2AB + B^2$.

27. Prove part 1 of Theorem 6 in the special case that A, B, and C are 2×2 matrices.

28. Find four different 2×2 matrices A satisfying $A^2 = \mathbf{0}$.

29. Find four different 2×2 matrices A satisfying $A^2 = \begin{bmatrix} 1 & 0 \\ 0 & 1 \end{bmatrix}$.

30. Find four different 2×2 matrices A satisfying $A^2 = A$.

31. Find four different pairs of 2×2 matrices A, B such that $AB = \mathbf{0}$ and $BA \neq \mathbf{0}$.

32. Prove part 2 of Theorem 6 in the special case that A, B, and C are 2×2 matrices.

33. Prove part 3 of Theorem 6 in the special case that A and B are 2×2 matrices.

34. Prove Theorem 7.   **35.** Prove Theorem 8.

In Exercises 36–39 perform the indicated operations. The components are complex numbers with $i^2 = -1$.

$$A = \begin{bmatrix} 6i & 2 - i \\ 1 + 2i & 3i \\ 2 + i & 1 + i \end{bmatrix}, \quad B = \begin{bmatrix} 3 - i & 1 + i \\ 1 - i & 3 - i \end{bmatrix}, \quad C = \begin{bmatrix} -3 + i & -1 - i \\ -1 + i & 3 + i \end{bmatrix}$$

36. AB **37.** $A(B + C)$

38. $A(2iB + 4iC)$ **39.** ABC

1.8 COMPUTATION OF A^{-1}

Consider a system of n linear equations with n unknowns

$$
\begin{aligned}
a_{11}x_1 + a_{12}x_2 + \cdots + a_{1n}x_n &= b_1 \\
a_{21}x_1 + a_{22}x_2 + \cdots + a_{2n}x_n &= b_2 \\
\vdots \qquad \vdots \qquad\qquad \vdots \quad\; &\;\; \vdots \\
a_{n1}x_1 + a_{n2}x_2 + \cdots + a_{nm}x_n &= b_n
\end{aligned}
\tag{1}
$$

and define an $n \times n$ matrix and n-vectors by

$$
A = \begin{bmatrix} a_{11} & a_{12} & \cdots & a_{1n} \\ a_{21} & a_{22} & \cdots & a_{2n} \\ \vdots & \vdots & & \vdots \\ a_{n1} & a_{n2} & \cdots & a_{nn} \end{bmatrix} \qquad \mathbf{x} = \begin{bmatrix} x_1 \\ x_2 \\ \vdots \\ x_n \end{bmatrix} \qquad \mathbf{b} = \begin{bmatrix} b_1 \\ b_2 \\ \vdots \\ b_n \end{bmatrix}
$$

Using the definition of matrix multiplication given in Section 1.7, we see that the system of equations in (1) can be rewritten in matrix notation as $A\mathbf{x} = \mathbf{b}$. The n-vector

$$
\mathbf{x} = \begin{bmatrix} x_1 \\ x_2 \\ \vdots \\ x_n \end{bmatrix}
$$

is a *solution* of $A\mathbf{x} = \mathbf{b}$ if it is a solution of the system of equations in (1). Theorem 7 showed us that $\mathbf{x} = 0_n$ is a solution of $A\mathbf{x} = 0_n$.

In the special case $n = 1$ the system of equations in (1) reduces to the single equation $a_{11}x_1 = b_1$. Assuming that $a_{11} \neq 0$ we can solve this equation by multiplying each side by a_{11}^{-1} to obtain the solution $x_1 = a_{11}^{-1}b_1$.

Notice that in finding x_1 we used the facts that $a_{11}^{-1}a_{11} = 1$ and $1 \cdot x_1 = x_1$. We can now develop a method of solving the matrix equation $A\mathbf{x} = \mathbf{b}$ that is analogous to that used to solve $a_{11}x_1 = b_1$. We begin by defining an $n \times n$ matrix I such that $I\mathbf{x} = \mathbf{x}$ for every n-vector $\mathbf{x}$. Then for suitable matrices A we define an $n \times n$ matrix A^{-1} such that $A^{-1}A = AA^{-1} = I$. If such a matrix A^{-1} exists, then, as we shall see, $\mathbf{x} = A^{-1}\mathbf{b}$ is the only solution of $A\mathbf{x} = \mathbf{b}$ just as $x_1 = a_{11}^{-1}b_1$ is the only solution of $a_{11}x_1 = b_1$.

Definition 5.   The $n \times n$ matrix

$$
I = \begin{bmatrix} 1 & 0 & 0 & \cdots & 0 \\ 0 & 1 & 0 & \cdots & 0 \\ 0 & 0 & 1 & \cdots & 0 \\ \vdots & \vdots & \vdots & \ddots & \vdots \\ 0 & 0 & 0 & \cdots & 1 \end{bmatrix}
$$

whose ij-component is 1 when $i = j$ and 0 when $i \neq j$ is called the $n \times n$ **identity matrix**. If we wish to emphasize that I is $n \times n$, we will write I_n in place of I.

It will be convenient to introduce the symbol δ_{ij}, called the **Kronecker delta**, that is defined by

$$
\delta_{ij} = \begin{cases} 1 & \text{if} \quad i = j \\ 0 & \text{if} \quad i \neq j \end{cases}
$$

Notice that the ij-component of I is δ_{ij}.

Theorem 9.   *Let A be an n × n matrix and* **x** *an n-vector. Then,*

  1. $AI_n = I_n A = A$.
  2. $I_n \mathbf{x} = \mathbf{x}$.

Proof: We shall prove the second identity of Theorem 9 and leave the first identity as an exercise. Notice that $I_n\mathbf{x}$ is an n-vector. We shall denote the ith component of $I_n\mathbf{x}$ by a_i. Then

$$a_i = \delta_{i1}x_1 + \delta_{i2}x_2 + \cdots + \delta_{in}x_n$$

where $x_1, x_2, \ldots, x_n$ are the components of **x**. Since $\delta_{ij} = 0$ except when $i = j$ and then $\delta_{ii} = 1$, we have $a_i = \delta_{ii}x_i = x_i$. Thus the ith components of $I_n\mathbf{x}$ and **x** are equal. Therefore, $I_n\mathbf{x} = \mathbf{x}$. This proves the second identity of Theorem 9.   ∎

We now return to considering the linear system of equations written in matrix notation as $A\mathbf{x} = \mathbf{b}$. Suppose that there is a matrix B such that $AB = BA = I_n$. Then

$$BA\mathbf{x} = B\mathbf{b}$$

$$I_n\mathbf{x} = B\mathbf{b}$$

$$\mathbf{x} = B\mathbf{b}$$

Therefore, if $A\mathbf{x} = \mathbf{b}$ has a solution it must be $B\mathbf{b}$. We now verify that $B\mathbf{b}$ is in fact a solution:

$$A\mathbf{x} = A(B\mathbf{b})$$

$$= (AB)\mathbf{b}$$

$$= I_n\mathbf{b}$$

$$= \mathbf{b}$$

We have proved that if there is a matrix B such that $AB = BA = I_n$, then $A\mathbf{x} = \mathbf{b}$ has $B\mathbf{b}$ as its only solution for every choice of the n-vector **b**.

Definition 6.   Let A be an $n \times n$ matrix. An $n \times n$ matrix B such that $AB = BA = I_n$ is called an **inverse** of the matrix A.

The following theorem states a basic property of $n \times n$ matrices.

Theorem 10.   *An n × n matrix has at most one inverse.*

Proof: Suppose that B and C are inverses of an $n \times n$ matrix A. Then $BA = I_n$ and $AC = I_n$ so that

$$(BA)C = I_nC = C$$

and

$$B(AC) = BI_n = B$$

By Theorem 6 we know that matrix multiplication is associative; that is, $(BA)C = B(AC)$ whenever the matrix products are defined. Hence $B = C$ so that A has at most one inverse.   ∎

Notice that Theorem 10 does not tell us that an $n \times n$ matrix A has an inverse. It merely tells us that either A has no inverse or precisely one inverse. In the case that A has an inverse we shall denote it by A^{-1} and call it the **inverse** of A.

Example 1.   Let

$$A = \begin{bmatrix} 1 & 2 \\ 3 & 4 \end{bmatrix} \qquad B = \begin{bmatrix} -2 & 1 \\ \frac{3}{2} & -\frac{1}{2} \end{bmatrix}$$

It is easily verified that $AB = BA = I_2$. Thus B is the inverse of A. Conversely, we also see that A is the inverse of B. This illustrates a general principle. Namely that if B is the inverse of A, then A is the inverse of B.

Example 2.   It needs to be emphasized that not every square matrix has an inverse. For example, consider the matrix

$$A = \begin{bmatrix} 1 & 0 \\ 0 & 0 \end{bmatrix}$$

Suppose that there is a matrix

$$B = \begin{bmatrix} b_{11} & b_{12} \\ b_{21} & b_{22} \end{bmatrix}$$

such that $I_2 = BA$. Then

$$\begin{bmatrix} 1 & 0 \\ 0 & 1 \end{bmatrix} = \begin{bmatrix} b_{11} & b_{12} \\ b_{21} & b_{22} \end{bmatrix} \begin{bmatrix} 1 & 0 \\ 0 & 0 \end{bmatrix}$$
$$= \begin{bmatrix} b_{11} & 0 \\ b_{21} & 0 \end{bmatrix}$$

This is impossible for any choice of b_{11}, b_{12}, b_{21}, and b_{22}; therefore, A has no inverse.

Much of the preceding discussion is summarized in the following theorem.

Theorem 11.   *If an $n \times n$ matrix A has an inverse, then the equation $Ax = b$ has precisely one solution for every choice of the n-vector b and that solution is $A^{-1}b$.*

Fortunately there is a relatively easy way to determine whether an $n \times n$ matrix A has an inverse and to compute A^{-1} whenever it exists. To begin we construct an $n \times 2n$ matrix, *called a* **partitioned matrix** and denoted by $[A \,|\, I_n]$, by placing A and I_n next to each other:

$$[A \,|\, I_n] = \begin{bmatrix} a_{11} & a_{12} & \cdots & a_{1n} & 1 & 0 & \cdots & 0 \\ a_{21} & a_{22} & \cdots & a_{2n} & 0 & 1 & \cdots & 0 \\ \vdots & \vdots & \ddots & \vdots & \vdots & \vdots & \ddots & \vdots \\ a_{n1} & a_{n2} & \cdots & a_{nn} & 0 & 0 & \cdots & 1 \end{bmatrix}$$

We now attempt to change the left side of this matrix into I_n using only elementary row operations. If this can be done the right side of the resulting matrix is A^{-1}. If the left side of $[A \,|\, I_n]$ cannot be changed into I_n, then A has no inverse. The validity of this method will be established in the next section.

Example 3.   Consider the matrix

$$A = \begin{bmatrix} 1 & 2 \\ 3 & 4 \end{bmatrix}$$

Then

$$\left[\begin{array}{cc|cc} 1 & 2 & 1 & 0 \\ 3 & 4 & 0 & 1 \end{array}\right] \sim \left[\begin{array}{cc|cc} 1 & 2 & 1 & 0 \\ 0 & -2 & -3 & 1 \end{array}\right] \quad (-3)R_1 + R_2 \to R_2$$

$$\sim \left[\begin{array}{cc|cc} 1 & 0 & -2 & 1 \\ 0 & -2 & -3 & 1 \end{array}\right] \quad R_2 + R_1 \to R_1$$

$$\sim \left[\begin{array}{cc|cc} 1 & 0 & -2 & 1 \\ 0 & 1 & \frac{3}{2} & -\frac{1}{2} \end{array}\right] \quad (-\tfrac{1}{2})R_2 \to R_2$$

Thus

$$A^{-1} = \begin{bmatrix} -2 & 1 \\ \frac{3}{2} & -\frac{1}{2} \end{bmatrix}$$

In Example 1 we found that this matrix is in fact the inverse of the matrix A.

The process used in this example can be formalized to one that will yield the inverse of any $n \times n$ matrix provided the inverse exists. Beginning with the partitioned matrix $[A \,|\, I_n]$:

1. Use elementary row operations to obtain a 1 as the 1,1-component.
2. By adding multiples of the first row to the remaining rows (where necessary), make the rest of the components of the first column equal to zero.
3. Without using the first row, use elementary row operations to obtain a 1 as the 2,2-component.
4. By adding multiples of the second row to the remaining rows (where necessary), make the rest of the components in the second column equal to 0.
5. Continue in this manner for each of the remaining rows, first making the k,k-component equal 1 and then making *all* other components in the kth column equal to 0.
6. The right half of the partitioned matrix is now A^{-1}. If the above process cannot be carried out, then A^{-1} does not exist.

Example 4.   Consider the matrix

$$A = \begin{bmatrix} -1 & -1 & 2 \\ 2 & 1 & -2 \\ 1 & 1 & -1 \end{bmatrix}$$

Then

$$
\begin{bmatrix}
-1 & -1 & 2 & | & 1 & 0 & 0 \\
2 & 1 & -2 & | & 0 & 1 & 0 \\
1 & 1 & -1 & | & 0 & 0 & 1
\end{bmatrix}
\sim
\begin{bmatrix}
1 & 1 & -2 & | & -1 & 0 & 0 \\
2 & 1 & -2 & | & 0 & 1 & 0 \\
1 & 1 & -1 & | & 0 & 0 & 1
\end{bmatrix}
\quad (-1)R_1 \to R_1
$$

$$
\sim
\begin{bmatrix}
1 & 1 & -2 & | & -1 & 0 & 0 \\
0 & -1 & 2 & | & 2 & 1 & 0 \\
1 & 1 & -1 & | & 0 & 0 & 1
\end{bmatrix}
\quad (-2)R_1 + R_2 \to R_2
$$

$$
\sim
\begin{bmatrix}
1 & 1 & -2 & | & -1 & 0 & 0 \\
0 & -1 & 2 & | & 2 & 1 & 0 \\
0 & 0 & 1 & | & 1 & 0 & 1
\end{bmatrix}
\quad (-1)R_1 + R_3 \to R_3
$$

$$
\sim
\begin{bmatrix}
1 & 1 & -2 & | & -1 & 0 & 0 \\
0 & 1 & -2 & | & -2 & -1 & 0 \\
0 & 0 & 1 & | & 1 & 0 & 1
\end{bmatrix}
\quad (-1)R_2 \to R_2
$$

$$
\sim
\begin{bmatrix}
1 & 0 & 0 & | & 1 & 1 & 0 \\
0 & 1 & -2 & | & -2 & -1 & 0 \\
0 & 0 & 1 & | & 1 & 0 & 1
\end{bmatrix}
\quad (-1)R_2 + R_1 \to R_1
$$

$$
\sim
\begin{bmatrix}
1 & 0 & 0 & | & 1 & 1 & 0 \\
0 & 1 & 0 & | & 0 & -1 & 2 \\
0 & 0 & 1 & | & 1 & 0 & 1
\end{bmatrix}
\quad 2R_3 + R_2 \to R_2
$$

Thus

$$
A^{-1} =
\begin{bmatrix}
1 & 1 & 0 \\
0 & -1 & 2 \\
1 & 0 & 1
\end{bmatrix}
$$

A straightforward calculation shows that

$$
AA^{-1} =
\begin{bmatrix}
-1 & -1 & 2 \\
2 & 1 & -2 \\
1 & 1 & -1
\end{bmatrix}
\begin{bmatrix}
1 & 1 & 0 \\
0 & -1 & 2 \\
1 & 0 & 1
\end{bmatrix}
=
\begin{bmatrix}
1 & 0 & 0 \\
0 & 1 & 0 \\
0 & 0 & 1
\end{bmatrix}
= I_3
$$

Similarly $A^{-1}A = I_3$ so that we have indeed found the inverse of A.

Example 5.   Consider the matrix

$$
A =
\begin{bmatrix}
1 & 2 & 3 \\
4 & 5 & 6 \\
7 & 8 & 9
\end{bmatrix}
$$

Then

$$\begin{bmatrix} 1 & 2 & 3 & | & 1 & 0 & 0 \\ 4 & 5 & 6 & | & 0 & 1 & 0 \\ 7 & 8 & 9 & | & 0 & 0 & 1 \end{bmatrix} \sim \begin{bmatrix} 1 & 2 & 3 & | & 1 & 0 & 0 \\ 0 & -3 & -6 & | & -4 & 1 & 0 \\ 7 & 8 & 9 & | & 0 & 0 & 1 \end{bmatrix} \quad (-4)R_1 + R_2 \to R_2$$

$$\sim \begin{bmatrix} 1 & 2 & 3 & | & 1 & 0 & 0 \\ 0 & -3 & -6 & | & -4 & 1 & 0 \\ 0 & -6 & -12 & | & -7 & 0 & 1 \end{bmatrix} \quad (-7)R_1 + R_3 \to R_3$$

$$\sim \begin{bmatrix} 1 & 2 & 3 & | & 1 & 0 & 0 \\ 0 & 1 & 2 & | & \frac{4}{3} & -\frac{1}{3} & 0 \\ 0 & -6 & -12 & | & -7 & 0 & 1 \end{bmatrix} \quad (-\frac{1}{3})R_2 \to R_2$$

$$\sim \begin{bmatrix} 1 & 0 & -1 & | & -\frac{5}{3} & \frac{2}{3} & 0 \\ 0 & 1 & 2 & | & \frac{4}{3} & -\frac{1}{3} & 0 \\ 0 & -6 & -12 & | & -7 & 0 & 1 \end{bmatrix} \quad (-2)R_2 + R_1 \to R_1$$

$$\sim \begin{bmatrix} 1 & 0 & -1 & | & -\frac{5}{3} & \frac{2}{3} & 0 \\ 0 & 1 & 2 & | & \frac{4}{3} & -\frac{1}{3} & 0 \\ 0 & 0 & 0 & | & 1 & -2 & 1 \end{bmatrix} \quad 6R_2 + R_3 \to R_3$$

Evidently it is not possible to transform the left side of $[A|I_3]$ into I_3 by using elementary row operations. Therefore A^{-1} does not exist.

In the special case that A is a 2×2 matrix it is easy to determine whether A has an inverse and, if it does, to compute A^{-1}. The reader is asked in the exercises to prove the following theorem.

Theorem 12. *Let*

$$A = \begin{bmatrix} a & b \\ c & d \end{bmatrix}$$

Then

    1. *If* $ad - bc = 0$, *then A has no inverse.*
    2. *If* $ad - bc \neq 0$, *then A has an inverse and*

$$A^{-1} = \frac{1}{ad - bc} \begin{bmatrix} d & -b \\ -c & a \end{bmatrix}$$

Example 6.   Consider the matrix

$$A = \begin{bmatrix} 1 & 2 \\ 3 & 6 \end{bmatrix}$$

Since $(1)(6) - (2)(3) = 0$, part 1 of Theorem 12 assures us that A has no inverse. This is also easily verified by using elementary row operations as follows:

$$[A \,|\, I_2] = \begin{bmatrix} 1 & 2 & | & 1 & 0 \\ 3 & 6 & | & 0 & 1 \end{bmatrix} \sim \begin{bmatrix} 1 & 2 & | & 1 & 0 \\ 0 & 0 & | & -3 & 1 \end{bmatrix} \quad (-3)R_1 + R_2 \to R_2$$

Evidently it is not possible to transform the left side of $[A \,|\, I_2]$ into I_2 by using elementary row operations. Hence A has no inverse, just as we concluded from part 1 of Theorem 12.

Example 7.   Consider the matrix

$$A = \begin{bmatrix} 1 & 2 \\ 3 & 4 \end{bmatrix}$$

Since $(1)(4) - (2)(3) = -2 \neq 0$, part 2 of Theorem 12 assures us that A has an inverse and that

$$A^{-1} = \frac{1}{(1)(4) - (2)(3)} \begin{bmatrix} 4 & -2 \\ -3 & 1 \end{bmatrix}$$

$$= \begin{bmatrix} -2 & 1 \\ \dfrac{3}{2} & -\dfrac{1}{2} \end{bmatrix}$$

which coincides with our findings in Examples 1 and 3.

Exercises

In Exercises 1–14, determine whether the given matrix has an inverse and find the inverse whenever possible.

1. $\begin{bmatrix} 1 & 2 \\ 3 & 4 \end{bmatrix}$

2. $\begin{bmatrix} 2 & 4 \\ 1 & 0 \end{bmatrix}$

3. $\begin{bmatrix} 1 & 2 \\ 2 & 4 \end{bmatrix}$

4. $\begin{bmatrix} 0 & 1 \\ 1 & 0 \end{bmatrix}$

5. $\begin{bmatrix} 1 & 2 & 3 \\ 0 & 2 & 4 \\ 0 & 0 & 3 \end{bmatrix}$

6. $\begin{bmatrix} 4 & 0 & 0 \\ 2 & 4 & 0 \\ 0 & 2 & 4 \end{bmatrix}$

7. $\begin{bmatrix} 1 & 2 & 3 \\ 1 & 4 & 7 \\ 2 & 4 & 9 \end{bmatrix}$

8. $\begin{bmatrix} 2 & 3 & 1 \\ 1 & 0 & 1 \\ 3 & 3 & 2 \end{bmatrix}$

9. $\begin{bmatrix} 2 & 1 & 3 \\ 1 & 1 & 2 \\ 0 & 1 & 1 \end{bmatrix}$

10. $\begin{bmatrix} 2 & 1 & -2 \\ -2 & 3 & 0 \\ 1 & 0 & 1 \end{bmatrix}$

11. $\begin{bmatrix} 0 & 1 & 1 \\ 1 & 0 & 1 \\ 1 & 1 & 0 \end{bmatrix}$

12. $\begin{bmatrix} -1 & -2 & 0 \\ 1 & 2 & 1 \\ -4 & 3 & 1 \end{bmatrix}$

13. $\begin{bmatrix} 1 & 0 & 1 & 0 \\ 0 & 1 & 0 & 1 \\ 0 & 1 & 1 & 1 \\ 1 & 1 & 1 & 0 \end{bmatrix}$

14. $\begin{bmatrix} 2 & 4 & 6 & 8 \\ 1 & 0 & 4 & 2 \\ 0 & 2 & 4 & 0 \\ 4 & 6 & 4 & 2 \end{bmatrix}$

15. Find all possible choices of a, b, and c so that

$$A = \begin{bmatrix} a & b \\ c & 0 \end{bmatrix}$$

has an inverse such that $A^{-1} = A$.

16. Find all possible values of a, b, and c so that

$$A = \begin{bmatrix} 2 & a \\ b & c \end{bmatrix}$$

has an inverse such that $A^{-1} = A$.

17. Find A^{-1} if

$$A = \begin{bmatrix} a_{11} & 0 & 0 & \cdots & 0 \\ 0 & a_{22} & 0 & \cdots & 0 \\ 0 & 0 & a_{33} & \cdots & 0 \\ \vdots & \vdots & \vdots & \ddots & \vdots \\ 0 & 0 & 0 & \cdots & a_{nn} \end{bmatrix}$$

and $a_{11} \neq 0$, $a_{22} \neq 0$, ..., $a_{nn} \neq 0$.

18. Find all values of θ for which

$$\begin{bmatrix} \cos \theta & 0 & \sin \theta \\ 0 & 1 & 0 \\ \sin \theta & 0 & \cos \theta \end{bmatrix}$$

has an inverse.

19. Find all values of θ for which

$$\begin{bmatrix} \cos \theta & 0 & \sin \theta \\ 0 & 1 & 0 \\ -\sin \theta & 0 & \cos \theta \end{bmatrix}$$

has an inverse.

20. Let S denote the set of all 2×2 matrices of the form

$$(1 - t^2)^{-1/2} \begin{bmatrix} 1 & t \\ t & 1 \end{bmatrix}$$

where t is a real number with $|t| < 1$.
(a) Show that I_2 is an element of S.
(b) Show that if A and B are elements of S, then AB is an element of S.
(c) Show that if A is an element of S, then A^{-1} exists and is also an element of S.
The set S is called the **Lorentz group**.
21. Prove part 1 of Theorem 12.
22. Prove part 2 of Theorem 12. (Hint: Multiply the matrices together.)

In Exercises 23–24 compute the inverse of the given matrix where the components are complex numbers with $i^2 = -1$.

23. $\begin{bmatrix} i & 0 \\ 1 & i \end{bmatrix}$
    24. $\begin{bmatrix} i & 1 \\ 2 & i+1 \end{bmatrix}$

1.9 NONSINGULAR MATRICES

> **Definition 7.** An $n \times n$ matrix A is called **nonsingular** if the equation
>
> $$A\mathbf{x} = \mathbf{b}$$
>
> has a solution for every choice of the n-vector $\mathbf{b}$. Otherwise the matrix A is called **singular**.

In this section we determine properties of nonsingular matrices. We begin our discussion of nonsingular matrices with a theorem that is merely a rewording of Theorem 2 in Section 1.5.

Theorem 13. *If A is a nonsingular $n \times n$ matrix, then $A\mathbf{x} = \mathbf{b}$ has precisely one solution for every n-vector $\mathbf{b}$.*

Theorem 11 in Section 1.8 tells us that if an $n \times n$ matrix A has an inverse, then $A\mathbf{x} = \mathbf{b}$ has precisely one solution for every choice of the n-vector $\mathbf{b}$. Consequently an $n \times n$ matrix that has an inverse is nonsingular. The following lemma is preparatory to proving the converse of this result: If A is nonsingular, then A has an inverse. In light of the similarities between Theorems 11 and 13, this is not an unexpected result.

Lemma 14. *If A is a nonsingular $n \times n$ matrix, then there is an $n \times n$ matrix B such that $AB = I_n$.*

Proof: Let B be the matrix whose columns are the solutions of the equations

$$A\mathbf{x} = \begin{bmatrix} 1 \\ 0 \\ 0 \\ \vdots \\ 0 \end{bmatrix}, \qquad A\mathbf{x} = \begin{bmatrix} 0 \\ 1 \\ 0 \\ \vdots \\ 0 \end{bmatrix}, \qquad \ldots, \qquad A\mathbf{x} = \begin{bmatrix} 0 \\ 0 \\ 0 \\ \vdots \\ 1 \end{bmatrix}$$

respectively. By Theorem 8 (near the end of Section 1.7) the jth column of AB is A times the jth column of B. Hence the columns of AB are

$$\begin{bmatrix} 1 \\ 0 \\ 0 \\ \vdots \\ 0 \end{bmatrix}, \begin{bmatrix} 0 \\ 1 \\ 0 \\ \vdots \\ 0 \end{bmatrix}, \cdots, \begin{bmatrix} 0 \\ 0 \\ 0 \\ \vdots \\ 1 \end{bmatrix}$$

Therefore $AB = I_n$. ∎

Theorem 15. *Let A be an $n \times n$ nonsingular matrix. If B is an $n \times n$ matrix such that $AB = I_n$, then $BA = I_n$ and A has an inverse.*

Proof: Let B be an $n \times n$ matrix such that $AB = I_n$. We first show that B is nonsingular. Let $\mathbf{b}$ be any n-vector. Then $\mathbf{c} = A\mathbf{b}$ is also an n-vector and $\mathbf{b}$ is a solution of $A\mathbf{x} = \mathbf{c}$. But

$$\mathbf{c} = I_n\mathbf{c} = (AB)\mathbf{c} = A(B\mathbf{c})$$

so that $B\mathbf{c}$ is also a solution of $A\mathbf{x} = \mathbf{c}$. By Theorem 13 we must have $B\mathbf{c} = \mathbf{b}$. We have shown that for every n-vector $\mathbf{b}$ there is an n-vector $\mathbf{c}$ such that $B\mathbf{c} = \mathbf{b}$. Hence $B\mathbf{x} = \mathbf{b}$ has a solution for every n-vector $\mathbf{b}$. Therefore, by Definition 6 the matrix B is nonsingular.

Since B is nonsingular, Lemma 14 assures us that there is an $n \times n$ matrix C such that $BC = I_n$. Then

$$A = AI_n = A(BC) = (AB)C = I_nC = C$$

so that $BC = BA$. Therefore,

$$AB = BA = I_n$$

This completes the proof. ∎

The following theorem summarizes everything that we have learned about nonsingular matrices.

Theorem 16. *Let A be an $n \times n$ matrix. Each of the following statements implies the others.*

    1. *A is nonsingular.*
    2. *$A\mathbf{x} = \mathbf{b}$ has a solution for every n-vector $\mathbf{b}$.*
    3. *$A\mathbf{x} = \mathbf{b}$ has precisely one solution for every n-vector $\mathbf{b}$.*
    4. *A^{-1} exists.*
    5. *$A^{-1}\mathbf{b}$ is the only solution of $A\mathbf{x} = \mathbf{b}$ for every n-vector $\mathbf{b}$.*
    6. *$A\mathbf{x} = \mathbf{0}$ has $\mathbf{x} = \mathbf{0}$ as its only solution.*
    7. *If one uses only elementary row operations, A can be changed into I_n.*

The equivalence of the first five statements follows from the discussions and theorems in this section. Theorems 1 and 2 of Section 1.5 assure that the last two statements are equivalent to the other five.

Example 1.   Consider the matrix

$$A = \begin{bmatrix} 1 & 2 \\ 3 & 4 \end{bmatrix}$$

Using elementary row operations we have

$$\left[\begin{array}{cc|c} 1 & 2 & b_1 \\ 3 & 4 & b_2 \end{array}\right] \sim \left[\begin{array}{cc|c} 1 & 2 & b_1 \\ 0 & -2 & b_2 - 3b_1 \end{array}\right] \quad (-3)R_1 + R_2 \to R_2$$

$$\sim \left[\begin{array}{cc|c} 1 & 2 & b_1 \\ 0 & 1 & -\frac{1}{2}b_2 + \frac{3}{2}b_1 \end{array}\right] \quad \left(-\frac{1}{2}\right)R_2 \to R_2$$

$$\sim \left[\begin{array}{cc|c} 1 & 0 & b_2 - 2b_1 \\ 0 & 1 & -\frac{1}{2}b_2 + \frac{3}{2}b_1 \end{array}\right] \quad (-2)R_2 + R_1 \to R_1$$

The last augmented matrix corresponds to the system of equations

$$x_1 = b_2 - 2b_1$$

$$x_2 = -\frac{1}{2}b_2 + \frac{3}{2}b_1$$

Thus for any 2-vector **b** the equation $A\mathbf{x} = \mathbf{b}$ has precisely one solution. In particular, when $\mathbf{b} = \mathbf{0}$ there is precisely one solution.

Notice that

$$\begin{bmatrix} x_1 \\ x_2 \end{bmatrix} = \begin{bmatrix} -2b_1 + b_2 \\ \frac{3}{2}b_1 - \frac{1}{2}b_2 \end{bmatrix}$$

$$= \begin{bmatrix} -2 & 1 \\ \frac{3}{2} & -\frac{1}{2} \end{bmatrix} \begin{bmatrix} b_1 \\ b_2 \end{bmatrix}$$

Since $\mathbf{x} = A^{-1}\mathbf{b}$ is the solution of $A\mathbf{x} = \mathbf{b}$, you may suspect that

$$\begin{bmatrix} -2 & 1 \\ \frac{3}{2} & -\frac{1}{2} \end{bmatrix}$$

is the inverse of A. In Example 7 of Section 1.8 we discovered that this matrix is indeed the inverse of A. Thus there appears to be a connection between solving $A\mathbf{x} = \mathbf{b}$ and computing A^{-1}. At the end of this section we describe this connection and use it to verify the method of computing A^{-1} described in Section 1.8.

It is sometimes convenient to have Theorem 16 stated in the following logically equivalent form.

Theorem 17. *Let A be an* $n \times n$ *matrix. Each of the following statements implies the others.*

1. *A is singular.*
2. $Ax = b$ *has no solution for some choice of the n-vector* **b**.
3. *If* $Ax = b$ *is consistent, then it has infinitely many solutions.*
4. A^{-1} *does not exist.*
5. $Ax = 0$ *has a solution other than* $x = 0$.
6. *If one uses elementary row operations, A can be changed into a matrix having at least one row that consists entirely of zeros.*

Example 2. Consider the matrix

$$A = \begin{bmatrix} 1 & 2 & 3 \\ 4 & 5 & 6 \\ 7 & 8 & 9 \end{bmatrix}$$

Using elementary row operations we have

$$\begin{bmatrix} 1 & 2 & 3 & | & b_1 \\ 4 & 5 & 6 & | & b_2 \\ 7 & 8 & 9 & | & b_3 \end{bmatrix} \sim \begin{bmatrix} 1 & 2 & 3 & | & b_1 \\ 0 & -3 & -6 & | & b_2 - 4b_1 \\ 0 & -6 & -12 & | & b_3 - 7b_1 \end{bmatrix} \quad \begin{array}{l} (-4)R_1 + R_2 \to R_2 \\ (-7)R_1 + R_3 \to R_3 \end{array}$$

$$\sim \begin{bmatrix} 1 & 2 & 3 & | & b_1 \\ 0 & 1 & 2 & | & -\frac{1}{3}b_2 + \frac{4}{3}b_1 \\ 0 & -6 & -12 & | & b_3 - 7b_1 \end{bmatrix} \quad (-\tfrac{1}{3})R_2 \to R_2$$

$$\sim \begin{bmatrix} 1 & 0 & -1 & | & -\frac{5}{3}b_1 + \frac{2}{3}b_2 \\ 0 & 1 & 2 & | & -\frac{1}{3}b_2 + \frac{4}{3}b_1 \\ 0 & 0 & 0 & | & b_3 - 2b_2 + b_1 \end{bmatrix} \quad \begin{array}{l} (-2)R_2 + R_1 \to R_1 \\ 6R_2 + R_3 \to R_3 \end{array}$$

Thus $Ax = b$ is consistent if and only if $b_3 - 2b_2 + b_1 = 0$. Since $Ax = b$ does not have a solution for every 3-vector **b**, the matrix A is singular. Whenever $b_3 - 2b_2 + b_1 = 0$ the solutions of $Ax = b$ are determined by the equations

$$x_1 \quad - \quad x_3 = -\frac{5}{3}b_1 + \frac{2}{3}b_2$$

$$x_2 + 2x_3 = -\frac{1}{3}b_2 + \frac{4}{3}b_1$$

In particular, when $b_1 = 0$, $b_2 = 0$, and $b_3 = 0$ we have

$$x_1 = x_3$$
$$x_2 = -2x_3$$

Hence for any number c

$$\begin{bmatrix} c \\ -2c \\ c \end{bmatrix}$$

is a solution of $A\mathbf{x} = \mathbf{0}$. Thus $A\mathbf{x} = \mathbf{0}$ has a solution other than $\mathbf{x} = \mathbf{0}$. When $b_1 = 1$, $b_2 = 4$, and $b_3 = 7$ we have $b_1 - 2b_2 + b_3 = 0$ and

$$\begin{aligned} x_1 \quad - \quad x_3 &= 1 \\ x_2 + 2x_3 &= 0 \end{aligned}$$

so that for every constant d

$$\begin{bmatrix} 1 + d \\ -2d \\ d \end{bmatrix}$$

is a solution of

$$A\mathbf{x} = \begin{bmatrix} 1 \\ 4 \\ 7 \end{bmatrix}$$

Thus this equation has infinitely many solutions.

Section 1.8 presented a computational procedure for finding the inverse of a nonsingular matrix. We now have two methods for finding the solution of a system of n equations in n unknowns that can be written in matrix form as $A\mathbf{x} = \mathbf{b}$ where A is a nonsingular matrix. The first is Gaussian elimination with backward substitution and the other is to compute A^{-1} and then compute $A^{-1}\mathbf{b}$. It can be shown that Gaussian elimination with backward substitution requires fewer arithmetic operations and, therefore, is the method usually used. However, if $A\mathbf{x} = \mathbf{b}$ is to be solved for several choices of the vector $\mathbf{b}$ it may be convenient to compute A^{-1} and then compute the solution $A^{-1}\mathbf{b}$. In addition, the inverse of a matrix is an important concept that facilitates the study of many problems involving matrices.

According to Definition 6, a matrix B is the inverse of a matrix A if it satisfies both of the identities $AB = I$ and $BA = I$. The following theorem shows that in fact B need only satisfy one of these identities in order to be the inverse of A.

Theorem 18.   *Let A be an $n \times n$ matrix.*

    1. *If B is an $n \times n$ matrix such that $BA = I$, then A is nonsingular and $B = A^{-1}$.*
    2. *If B is an $n \times n$ matrix such that $AB = I$, then A is nonsingular and $B = A^{-1}$.*

Proof: We will prove the first assertion and leave the other as an exercise. Suppose that B is an $n \times n$ matrix such that $BA = I$. We begin the proof by showing that A is nonsingular. This is done by showing that $A\mathbf{x} = \mathbf{0}$ has $\mathbf{x} = \mathbf{0}$ as its only solution.

Let $\mathbf{y}$ be any solution of $A\mathbf{x} = \mathbf{0}$. Then $A\mathbf{y} = \mathbf{0}$ and so we have $\mathbf{0} = B\mathbf{0} = B(A\mathbf{y}) = (BA)\mathbf{y} = I\mathbf{y} = \mathbf{y}$. Thus $\mathbf{y} = \mathbf{0}$ is the only solution of $A\mathbf{x} = \mathbf{0}$. By Theorem 16 the matrix A is nonsingular. We now multiply each side of $AB = I$ by A^{-1} to obtain

$$A^{-1}(AB) = A^{-1}I$$

On the left side of the identity we find that $A^{-1}(AB) = (A^{-1}A)B = IB = B$. On the right side we find that $A^{-1}I = A^{-1}$. Therefore, $B = A^{-1}$, which is the identity we set out to prove. $\blacksquare$

One might suppose that if AB is "very close" to I in the sense that each component of AB is "very close" to the corresponding component of I, then BA would also be "very close" to I. This is not necessarily the case. Consider the matrices*

$$A = \begin{bmatrix} 9999 & 9998 \\ 10000 & 9999 \end{bmatrix} \qquad B = \begin{bmatrix} 9999.9999 & -9997.0001 \\ -10001 & 9998 \end{bmatrix}$$

Then

$$AB = \begin{bmatrix} 1.0001 & .0001 \\ 0 & 1 \end{bmatrix} \qquad BA = \begin{bmatrix} 19998.0001 & 19995.0003 \\ -19999 & -19996 \end{bmatrix}$$

Thus AB is "close" to I while BA is not. The true inverse of A is easily found by using Theorem 12 of Section 1.8. In this case, $1/(ad - bc)$ equals 1. Therefore,

$$A^{-1} = \begin{bmatrix} 9999 & -9998 \\ -10000 & 9999 \end{bmatrix}$$

Thus, changes of approximately .01% in the components of A^{-1} have caused a dramatic change in the value of $A^{-1}A$.

This example shows that if we obtain an approximation B for the inverse of a matrix A, then it would be prudent to ascertain that AB and BA are both close to I before using B in any calculations.

We conclude this section by showing that the method for computing A^{-1} described in Section 1.8 really is valid. The essence of this method was used in the proof of Lemma 14. In that proof we showed that if B is the $n \times n$ matrix having the solutions of

(1)
$$A\mathbf{x} = \begin{bmatrix} 1 \\ 0 \\ 0 \\ \vdots \\ 0 \end{bmatrix}, \quad A\mathbf{x} = \begin{bmatrix} 0 \\ 1 \\ 0 \\ \vdots \\ 0 \end{bmatrix}, \quad \cdots, \quad A\mathbf{x} = \begin{bmatrix} 0 \\ 0 \\ 0 \\ \vdots \\ 1 \end{bmatrix}$$

* This example is taken from the article by George E. Forsythe, "Pitfalls in Computation, or Why a Math Book Isn't Enough," *American Mathematical Monthly* 77 (November 1970): 931–56.

as its columns, then $AB = I$. If one of these solutions does not exist, then A is singular (why?) and A^{-1} does not exist. Assuming that all of these solutions exist, Theorem 15 assures us that $B = A^{-1}$. Thus solving the systems of equations in (1) allows us to compute A^{-1}. If we use Gauss–Jordan elimination to solve these n systems of equations, we use elementary row operations to change each corresponding augmented matrix into a matrix of the form

$$\begin{bmatrix} 1 & 0 & 0 & \cdots & 0 & d_1 \\ 0 & 1 & 0 & \cdots & 0 & d_2 \\ 0 & 0 & 1 & \cdots & 0 & d_3 \\ \vdots & \vdots & \vdots & \ddots & \vdots & \vdots \\ 0 & 0 & 0 & \cdots & 1 & d_n \end{bmatrix}$$

In fact, exactly the same elementary row operations can be performed on each augmented matrix. Thus we use the same collection of elementary row operations n times, once for each augmented matrix. This observation leads us to a method to solve the n equations simultaneously.

We use elementary row operations to change the left side of $[A \,|\, I_n]$ into I_n. Having done this the first column on the right side is the solution of

$$A\mathbf{x} = \begin{bmatrix} 1 \\ 0 \\ 0 \\ \vdots \\ 0 \end{bmatrix}$$

The second column on the right side is the solution of

$$A\mathbf{x} = \begin{bmatrix} 0 \\ 1 \\ 0 \\ \vdots \\ 0 \end{bmatrix}$$

and so on. Thus the right side of the resulting matrix is A^{-1}.

Exercises

In Exercises 1–4:

(a) Show that given matrix A is nonsingular by showing that $A\mathbf{x} = \mathbf{0}$ has precisely one solution.
(b) Use the method in the proof of Lemma 14 to find a matrix B such that $AB = I$.
(c) Show that the matrix B found in part (b) is A^{-1}.

1. $\begin{bmatrix} 1 & 3 \\ 2 & 1 \end{bmatrix}$
   2. $\begin{bmatrix} 2 & 4 \\ 6 & 8 \end{bmatrix}$

3. $\begin{bmatrix} 2 & 0 & 4 \\ 1 & 2 & 0 \\ 0 & 4 & 2 \end{bmatrix}$

4. $\begin{bmatrix} 2 & 0 & 2 \\ 4 & 0 & 2 \\ 0 & 1 & 0 \end{bmatrix}$

In Exercises 5–8

(a) Show that the given matrix A is singular by showing that $A\mathbf{x} = \mathbf{0}$ has more than one solution.

(b) Verify that the given matrix has no inverse.

5. $\begin{bmatrix} 1 & 2 \\ 3 & 6 \end{bmatrix}$

6. $\begin{bmatrix} 4 & 2 \\ 2 & 1 \end{bmatrix}$

7. $\begin{bmatrix} 1 & 2 & 3 \\ 4 & 5 & 6 \\ 7 & 8 & 9 \end{bmatrix}$

8. $\begin{bmatrix} 1 & -2 & 4 \\ 2 & 2 & -2 \\ 3 & 0 & 2 \end{bmatrix}$

9. Let A be a nonsingular matrix and c be a nonzero constant. Show that $(cA)^{-1} = c^{-1}A^{-1}$.

10. Let A and B be $n \times n$ nonsingular matrices. Show that $(AB)^{-1} = B^{-1}A^{-1}$. (Hint: Multiply AB by $B^{-1}A^{-1}$.)

11. Let A, B, and C be nonsingular $n \times n$ matrices. What is the inverse of ABC? (Hint: See Exercise 10.)

12. Let A be a nonsingular matrix. Show that $(A^{-1})^{-1} = A$.

13. In general, the equation $AB = AC$ does not imply that $B = C$. Show that if A is nonsingular, then $AB = AC$ does imply $B = C$.

14. Let A and B be $n \times n$ matrices such that $AB = 0$. Show that if $B \neq 0$ then A is singular.

15. Find two nonsingular 2×2 matrices A and B such that $A + B$ is nonsingular. Find two other nonsingular 2×2 matrices whose sum is singular.

16. Find two singular 2×2 matrices A and B such that $A + B$ is singular. Find two singular 2×2 matrices whose sum is nonsingular.

17. Let A and B be $n \times n$ matrices. Show that if B is singular, then AB is also singular. (Hint: Show that $AB\mathbf{x} = \mathbf{0}$ has a solution other than $\mathbf{x} = \mathbf{0}$.)

18. Let A and B be $n \times n$ matrices. Show that if A is singular, then AB is also singular. (Hint: Show that $AB\mathbf{x} = \mathbf{0}$ has a solution other than $\mathbf{x} = \mathbf{0}$.)

19. Show that if $A^2 + 3A + I = 0$, then $A^{-1} = -A - 3I$.

20. Prove part 2 of Theorem 18.

1.10 MORE ON SYSTEMS OF LINEAR EQUATIONS

We now complete our discussion of the elementary properties of systems of linear equations. For simplicity of notation we write the system under consideration in the form

$$A\mathbf{x} = \mathbf{b}$$

where A is an $m \times n$ matrix, $\mathbf{x}$ is an n-vector, and $\mathbf{b}$ is an m-vector. If $\mathbf{b} = \mathbf{0}_m$ the equation is called **homogeneous**. Otherwise the equation is called **nonhomogeneous**.

A homogeneous equation is always consistent since $\mathbf{0}_n$ is a solution. This solution is called the *trivial solution*.

Theorem 19. *Let A be an m × n matrix. The homogeneous equation Ax = 0 has either*

  1. *Only the trivial solution, or*
  2. *Infinitely many solutions.*

Proof: By the above discussion $A\mathbf{x} = \mathbf{0}$ always has a trivial solution. Suppose that there is a nontrivial solution $\mathbf{z}$. If c is any nonzero constant, then $c\mathbf{z}$ is not a zero vector and

$$A(c\mathbf{z}) = c(A\mathbf{z}) = c\mathbf{0} = \mathbf{0}$$

Thus $c\mathbf{z}$ is a nontrivial solution for every choice of the nonzero constant c. This proves that if $A\mathbf{x} = \mathbf{0}$ has one nontrivial solution, then it has infinitely many nontrivial solutions.   ∎

  If the homogeneous system of equations has more unknowns n than equations m, then we always have infinitely many solutions. Before stating this as a theorem we will give an example that illustrates the validity of this statement and even indicates how it might be proved.

Example 1.   Consider the equation

$$\begin{bmatrix} 1 & 3 & 1 & 1 & 1 \\ 1 & 4 & 1 & 2 & 1 \\ 3 & 6 & 3 & 4 & 5 \end{bmatrix} \mathbf{x} = \begin{bmatrix} 0 \\ 0 \\ 0 \end{bmatrix}$$

The augmented matrix for this system is

$$\left[\begin{array}{ccccc|c} 1 & 3 & 1 & 1 & 1 & 0 \\ 1 & 4 & 1 & 2 & 1 & 0 \\ 3 & 6 & 3 & 4 & 5 & 0 \end{array}\right]$$

Using elementary row operations we have

$$\left[\begin{array}{ccccc|c} 1 & 3 & 1 & 1 & 1 & 0 \\ 1 & 4 & 1 & 2 & 1 & 0 \\ 3 & 6 & 3 & 4 & 5 & 0 \end{array}\right] \sim \left[\begin{array}{ccccc|c} 1 & 3 & 1 & 1 & 1 & 0 \\ 0 & 1 & 0 & 1 & 0 & 0 \\ 0 & -3 & 0 & 1 & 2 & 0 \end{array}\right] \quad \begin{array}{l} (-1)R_1 + R_2 \rightarrow R_2 \\ (-3)R_1 + R_3 \rightarrow R_3 \end{array}$$

$$\sim \left[\begin{array}{ccccc|c} 1 & 3 & 1 & 1 & 1 & 0 \\ 0 & 1 & 0 & 1 & 0 & 0 \\ 0 & 0 & 0 & 4 & 2 & 0 \end{array}\right] \quad 3R_2 + R_3 \rightarrow R_3$$

$$\sim \left[\begin{array}{ccccc|c} 1 & 0 & 1 & -2 & 1 & 0 \\ 0 & 1 & 0 & 1 & 0 & 0 \\ 0 & 0 & 0 & 1 & \frac{1}{2} & 0 \end{array}\right] \quad \begin{array}{l} (-3)R_2 + R_1 \rightarrow R_1 \\ \frac{1}{4}R_3 \rightarrow R_3 \end{array}$$

$$\sim \left[\begin{array}{ccccc|c} 1 & 0 & 1 & 0 & 2 & 0 \\ 0 & 1 & 0 & 0 & -\frac{1}{2} & 0 \\ 0 & 0 & 0 & 1 & \frac{1}{2} & 0 \end{array}\right] \quad \begin{array}{l} 2R_3 + R_1 \rightarrow R_1 \\ (-1)R_3 + R_2 \rightarrow R_2) \end{array} \quad (1)$$

This last matrix, which is in reduced row-echelon form, corresponds to the system of equations

$$x_1 + \quad x_3 + \quad 2x_5 = 0$$

$$x_2 - \quad \frac{1}{2}x_5 = 0$$

$$x_4 + \frac{1}{2}x_5 = 0$$

The second and third equations can be easily solved for x_2 and x_4 in terms of x_5:

(2)
$$x_2 = \frac{1}{2}x_5$$

(3)
$$x_4 = -\frac{1}{2}x_5$$

while the first equation can be solved for x_1 in terms of x_3 and x_5:

(4)
$$x_1 = -x_3 - 2x_5$$

For any choice of the unknowns x_3 and x_5 the identities in (2), (3), and (4) enable us to determine a solution. Setting $x_3 = b$ and $x_5 = c$ we have that

$$\mathbf{x} = \begin{bmatrix} -b - 2c \\ \frac{1}{2}c \\ b \\ -\frac{1}{2}c \\ c \end{bmatrix}$$

is a solution for every choice of the numbers b and c. In particular, when $b = 1$ and $c = 0$, we find that

$$\mathbf{x} = \begin{bmatrix} -1 \\ 0 \\ 1 \\ 0 \\ 0 \end{bmatrix}$$

is a solution.

Notice that a "step" does not occur in the third and fifth columns of the augmented matrix in reduced row-echelon form in (1). In equations (2), (3), and (4) we found that the unknowns corresponding to these two columns can be chosen arbitrarily and the remaining unknowns can be found in terms of the two arbitrary unknowns. Whenever there are more columns than rows in the augmented matrix, its reduced row-echelon form must have at least one column in which a "step" does not occur. This fact follows immediately from the observation that there cannot be more "steps" than rows. If a "step" does not occur in the jth column, then the variable x_j is not determined by the associated system of equations and, hence, is an independent variable. The associated system of equations determines each of the variables corresponding to a column with a "step" in terms of the independent variables that correspond to columns without a "step." This argument can be written as a proof, but it is notationally complex.

Theorem 20. *Any homogeneous system of linear equations with more unknowns than equations has infinitely many solutions.*

The following theorem describes the relationship between solutions of the homogeneous equation $A\mathbf{x} = \mathbf{0}$ and the nonhomogeneous equation $A\mathbf{x} = \mathbf{b}$.

Theorem 21. *Let A be an $m \times n$ matrix. Then*

1. *If $\mathbf{y}$ is a solution of $A\mathbf{x} = \mathbf{0}$ and $\mathbf{z}$ is a solution of $A\mathbf{x} = \mathbf{b}$, then $\mathbf{y} + \mathbf{z}$ is a solution of $A\mathbf{x} = \mathbf{b}$.*
2. *If $\mathbf{y}$ and $\mathbf{z}$ are solutions of $A\mathbf{x} = \mathbf{b}$, then $\mathbf{y} - \mathbf{z}$ is a solution of $A\mathbf{x} = \mathbf{0}$.*
3. *If $\mathbf{z}$ and $\mathbf{z}'$ are any solutions of $A\mathbf{x} = \mathbf{b}$ then there is a solution $\mathbf{y}$ of $A\mathbf{x} = \mathbf{0}$ such that $\mathbf{z}' = \mathbf{z} + \mathbf{y}$.*

Proof: We will prove the first part and leave the other two parts as exercises. Let $\mathbf{y}$ and $\mathbf{z}$ be as in part 1. Then $A(\mathbf{y} + \mathbf{z}) = A\mathbf{y} + A\mathbf{z} = \mathbf{0} + \mathbf{b} = \mathbf{b}$. ■

Example 2.   Consider the equation $A\mathbf{x} = \mathbf{b}$ where

$$A = \begin{bmatrix} 1 & 3 & 1 & 1 & 1 \\ 1 & 4 & 1 & 2 & 1 \\ 3 & 6 & 3 & 4 & 5 \end{bmatrix}, \qquad \mathbf{b} = \begin{bmatrix} 1 \\ 4 \\ 2 \end{bmatrix}$$

The corresponding augmented matrix is

$$\left[\begin{array}{ccccc|c} 1 & 3 & 1 & 1 & 1 & 1 \\ 1 & 4 & 1 & 2 & 1 & 4 \\ 3 & 6 & 3 & 4 & 5 & 2 \end{array}\right]$$

Using the same elementary row operations as we used in Example 1 to solve $A\mathbf{x} = \mathbf{0}$, we find that this augmented matrix has

$$\left[\begin{array}{ccccc|c} 1 & 0 & 1 & 0 & 2 & -4 \\ 0 & 1 & 0 & 0 & -\dfrac{1}{2} & 1 \\ 0 & 0 & 0 & 1 & \dfrac{1}{2} & 2 \end{array}\right]$$

as its reduced row-echelon form. This last augmented matrix corresponds to the system of equations

$$x_1 + \quad x_3 \quad + 2x_5 = -4$$

$$x_2 \quad\quad -\frac{1}{2}x_5 = \quad 1$$

$$x_4 + \frac{1}{2}x_5 = \quad 2$$

Setting $x_3 = b$ and $x_5 = c$ we have

$$\begin{bmatrix} x_1 \\ x_2 \\ x_3 \\ x_4 \\ x_5 \end{bmatrix} = \begin{bmatrix} -4 \\ 1 \\ 0 \\ 2 \\ 0 \end{bmatrix} + \begin{bmatrix} -b - 2c \\ \frac{1}{2}c \\ b \\ -\frac{1}{2}c \\ c \end{bmatrix}$$

is a solution of $A\mathbf{x} = \mathbf{b}$ for every choice of the numbers b and c. When $b = 0$ and $c = 0$ we find that

$$\begin{bmatrix} -4 \\ 1 \\ 0 \\ 2 \\ 0 \end{bmatrix}$$

is a solution of $A\mathbf{x} = \mathbf{b}$. In Example 1 we found that every solution of $A\mathbf{x} = \mathbf{0}$ has the form

$$\begin{bmatrix} -b - 2c \\ \frac{1}{2}c \\ b \\ -\frac{1}{2}c \\ c \end{bmatrix}$$

Hence every solution of $A\mathbf{x} = \mathbf{b}$ is the sum of a particular solution of $A\mathbf{x} = \mathbf{b}$ and a solution of $A\mathbf{x} - \mathbf{0}$.

In Theorem 17 of Section 1.9 we found a relationship between $A\mathbf{x} = \mathbf{0}$ having more than one solution and $A\mathbf{x} = \mathbf{b}$ having infinitely many solutions. In Theorem 17, A is an $n \times n$ matrix. The following theorem leads to a similar result for matrices that are not square.

Theorem 22.   *Let A be an m × n matrix. Then each of the following statements implies the others:*

     1. *Ax = 0 has a solution other than x = 0.*
     2. *Ax = 0 has infinitely many solutions.*
     3. *If Ax = b is consistent, then it has infinitely many solutions.*
     4. *If Ax = b is consistent, then it has more than one solution.*

Proof: The equivalence of the first two assertions is given in Theorem 19, while the equivalence of assertions 2 and 3 is immediate from part 1 of Theorem 21. Clearly assertion 3 implies assertion 4. From part 2 of Theorem 21 we conclude that assertion 4 implies assertion 1. Hence $(1) \Rightarrow (2) \Rightarrow (3) \Rightarrow (4) \Rightarrow (1)$. This completes the proof.   ∎

Example 3.   Consider the matrix

$$A = \begin{bmatrix} 1 & 2 & 3 \\ 2 & 4 & 6 \end{bmatrix}$$

If

$$\mathbf{b} = \begin{bmatrix} b_1 \\ b_2 \end{bmatrix}$$

is any 2-vector, then

$$\begin{bmatrix} 1 & 2 & 3 & | & b_1 \\ 2 & 4 & 6 & | & b_2 \end{bmatrix} \sim \begin{bmatrix} 1 & 2 & 3 & | & b_1 \\ 0 & 0 & 0 & | & b_2 - 2b_1 \end{bmatrix} \quad (-2)R_1 + R_2 \to R_2$$

Thus the equation $A\mathbf{x} = \mathbf{b}$ is consistent if and only if $b_2 - 2b_1 = 0$. In particular $A\mathbf{x} = \mathbf{0}$ is consistent. If $A\mathbf{x} = \mathbf{b}$ is consistent, then the second augmented matrix corresponds to the single equation $x_1 + 2x_2 + 3x_3 = b_1$, which clearly has infinitely many solutions. Therefore $A\mathbf{x} = \mathbf{b}$ has infinitely many solutions whenever it is consistent. Since $A\mathbf{x} = \mathbf{0}$ is consistent, $A\mathbf{x} = \mathbf{0}$ has infinitely many solutions.

Exercises

In Exercises 1–4 verify Theorem 19 for the given matrix.

1. $\begin{bmatrix} 1 & 2 \\ 2 & 1 \end{bmatrix}$ **2.** $\begin{bmatrix} 2 & 1 \\ 4 & 2 \end{bmatrix}$

3. $\begin{bmatrix} 1 & 4 & 7 \\ 2 & 5 & 8 \\ 3 & 6 & 9 \end{bmatrix}$ **4.** $\begin{bmatrix} 1 & 0 & 1 \\ 0 & 1 & 1 \\ 1 & 0 & 0 \end{bmatrix}$

In Exercises 5 and 6 verify Theorem 20 by finding all solutions of the given homogeneous equation.

5. $\begin{bmatrix} 1 & 2 & 3 \\ 2 & 1 & -3 \end{bmatrix} \mathbf{x} = \mathbf{0}$ **6.** $\begin{bmatrix} 1 & -1 & 1 & 4 \\ 2 & 2 & 1 & -1 \\ -1 & 1 & 4 & 1 \end{bmatrix} \mathbf{x} = \mathbf{0}$

In Exercises 7 and 8:

(a) Show that the corresponding homogeneous equation has only the trivial solution.
(b) Determine conditions on a, b, and c so that the equation is consistent.
(c) Show that the equation has precisely one solution whenever it is consistent.

7. $\begin{bmatrix} 1 & 1 \\ 2 & 1 \\ 1 & -1 \end{bmatrix} \mathbf{x} = \begin{bmatrix} a \\ b \\ c \end{bmatrix}$
  8. $\begin{bmatrix} 1 & 2 \\ 3 & 4 \\ 2 & 1 \end{bmatrix} \mathbf{x} = \begin{bmatrix} a \\ b \\ c \end{bmatrix}$

In Exercises 9 and 10:

(a) Show that the corresponding homogeneous equation has a nontrivial solution.
(b) Determine conditions on a, b, and c so that the equation is consistent.
(c) Show that the equation has infinitely many solutions whenever it is consistent.

9. $\begin{bmatrix} 1 & -2 & 3 \\ 2 & -4 & 6 \\ -3 & 6 & -9 \end{bmatrix} \mathbf{x} = \begin{bmatrix} a \\ b \\ c \end{bmatrix}$
  10. $\begin{bmatrix} 1 & 4 \\ 2 & 8 \\ -1 & -4 \end{bmatrix} \mathbf{x} = \begin{bmatrix} a \\ b \\ c \end{bmatrix}$

11. Prove part 2 of Theorem 21.
12. Prove part 3 of Theorem 21.

1.11 AN EXAMPLE FROM ANTIQUITY (OPTIONAL)

In 1770 the German dramatist and critic Gotthold Lessung became director of the library in Wolfenbuttel, a town in northern Germany, which possessed a unique collection of incunabula (books printed with movable type before 1500) and manuscripts. In 1773 he began to publish translations and commentaries on the manuscripts. In the first volume appeared a Greek epigram entitled, "A problem which Archimedes found among (some) epigrams and sent, to be solved by those in Alexandria who occupy themselves with such matters, in his letter to Eratosphenes of Cyrene."

   In an abbreviated form aided by modern notation the problem can be stated as follows:

   Compute, O friend, the host of the oxen of the Sun. Compute the number which once grazed upon the plains of the Sicilian isle Thrinacia, and which were divided according to color into four herds, one milk white, one black, one yellow, and one dappled. The number of bulls formed the majority of the animals in each herd and the relations between them were as follows:

1. White bulls $(W) = (1/2 + 1/3)$ black bulls (B) + yellow bulls (Y).
2. Black bulls $(B) = (1/4 + 1/5)$ dappled bulls (D) + yellow bulls (Y).
3. Dappled bulls $(D) = (1/6 + 1/7)$ white bulls (W) + yellow bulls (Y).

   As to the cows:

4. White cows $(w) = (1/3 + 1/4)$ black herd $(B + b)$.
5. Black cows $(b) = (1/4 + 1/5)$ dappled herd $(D + d)$.
6. Dappled cows $(d) = (1/5 + 1/6)$ yellow herd $(Y + y)$.
7. Yellow cows $(y) = (1/6 + 1/7)$ white herd $(W + w)$.

If thou canst give, O friend, the number of bulls and cows in each herd thou art not unknowing nor unskilled in numbers, but still not yet to be counted among the wise.

Consider, however, the following additional relations between the bulls of the Sun:

8. White bulls (W) + black bulls (B) = a square number.

9. Dappled bulls (D) + yellow bulls (Y) = a triangular number (a number of the form $m(m + 1)/2$ for some positive integer m).

When thou hast then computed the totals of the herds, O friend, go forth as conqueror, and rest assured that thou hast proved most skilled in the science of numbers.

The manuscipt has an appendix that gives without explanation numbers for the various types of cows and bulls with the total number of cattle being 4,031,126,560.

The first seven equations are seven linear equations in eight unknowns. Hence there are infinitely many solutions to the first seven equations. Since we are dealing with numbers of cattle we are only interested in solutions in which each number is an integer.

The first three equations can be rewritten as:

$$W - \frac{5}{6}B \qquad\qquad - Y = 0$$

$$B - \frac{9}{20}D - Y = 0$$

$$\frac{13}{42}W \qquad - D + Y = 0$$

which has

$$\begin{bmatrix} 1 & -\dfrac{5}{6} & 0 & -1 & \Big| & 0 \\[2mm] 0 & 1 & -\dfrac{9}{20} & -1 & \Big| & 0 \\[2mm] \dfrac{13}{42} & 0 & -1 & 1 & \Big| & 0 \end{bmatrix}$$

as its associated augmented matrix. Using elementary row operations it can be shown that this matrix is equivalent to:

$$\begin{bmatrix} 1 & 0 & 0 & -\dfrac{2226}{891} & \Big| & 0 \\[2mm] 0 & 1 & 0 & -\dfrac{1602}{891} & \Big| & 0 \\[2mm] 0 & 0 & 1 & -\dfrac{1580}{891} & \Big| & 0 \end{bmatrix}$$

Solving for D, B, and W in terms of Y, we find that

$$W = \frac{2226}{891} Y, \ B = \frac{1602}{891} Y, \ D = \frac{1580}{891} Y$$

Thus, in order for W, B, and D to be integers, Y must be an integer multiple of 891. Hence, $W = 2226k$, $B = 1602k$, $D = 1580k$, and $Y = 891k$, where k is a positive integer that is to be determined.

The fourth through seventh equations can now be written as

$$w - \frac{7}{12}b \qquad\qquad = \frac{7}{12}B = \frac{1869}{2}k$$

$$b - \frac{9}{20}d \qquad = \frac{9}{20}D = 711k$$

$$d - \frac{11}{30}y = \frac{11}{30}Y = \frac{3267}{10}k$$

$$-\frac{13}{42}w \qquad\qquad + \quad y = \frac{13}{42}W = 689k$$

Using Gaussian elimination with backward substitution we find that

$$w = (7{,}206{,}360/4657)k \quad b = (4{,}893{,}246/4657)k$$

$$d = (3{,}515{,}820/4657)k \quad y = (5{,}439{,}213/4657)k$$

Since 4657 is prime and we want w, b, d, and y to be positive integers, the number k must be an integer multiple of 4657. Hence $k = 4657n$ for some positive integer n. We have shown that any solution of the first seven equations is of the form

$$W = 10{,}366{,}482n \quad B = 7{,}460{,}514n \quad D = 7{,}358{,}060n \quad Y = 4{,}149{,}387n$$

$$w = 7{,}206{,}360n \quad b = 4{,}893{,}246n \quad d = 3{,}515{,}820n \quad y = 5{,}439{,}213n$$

When $n = 1$ we obtain the smallest solution satisfying the first seven equations. The total number of cattle in this case is 50,389,082. When $n = 80$ we obtain the results stated in the appendix to the manuscript.

To solve the entire problem we must choose n so that the eighth and ninth equations are satisfied. That is, we must choose n so that

$$17{,}826{,}996n = \text{a square number}$$

$$11{,}507{,}447n = \text{a triangular number}$$

Since $17{,}826{,}996 = 2^2(3)(11)(29)(4657)$ any integer n that satisfies the first of these two equations must be of the form

$$n = 3(11)(29)(4657)p^2$$
$$= 4{,}456{,}749p^2$$

for some positive integer p. If n also satisfies the second of these two equations, then there is a positive integer m such that

(1) $$\frac{m(m + 1)}{2} = (11{,}507{,}447)(4{,}456{,}749)p^2$$

If we now multiply each side of this equation by 8 and denote the coefficient of p^2 by N, we obtain $4m^2 + 4m - Np^2 = 0$ or, equivalently, $(2m + 1)^2 - Np^2 = 1$. Upon setting $x = 2m + 1$ we have the equation

$$x^2 - Np^2 = 1 \tag{2}$$

An equation of this form is called **Pell's equation**. Since neither 11,507,447 nor 4,456,749 is divisible by 2 and N is the product of these two numbers and 8, the number N is not a square. It is known that in such a case there are infinitely many pairs (x, p) of positive integers that satisfy equation (2)*. Hence there are infinitely many pairs (m, p) of positive integers that satisfy equation (1).

In 1880 Amthor determined that 206,546 digits are needed to express the total number of cattle. In fact he computed that 766 are the first three digits of this number. The number $766 \times 10^{206,543}$ is large beyond most people's imagination. In 1895 Bell published the results of four years of calculations by himself and two others. They computed the first thirty and last thirty digits of the number of cattle. Unfortunately, these results do not agree with those of Amthor. To the best of the author's knowledge a solution of this problem has not been completely computed.**

——————————Supplementary Exercises for Chapter 1——————————

Consider the following statements. If a statement is necessarily true, explain why it is true, citing appropriate theorems and definitions whenever possible. Otherwise give an example to show that the statement is not necessarily true.

In statements 1–12 the matrix A is $n \times n$ and $\mathbf{b}$ is an n-vector.

1. The equation $A\mathbf{x} = \mathbf{b}$ may have no solution.
2. The equation $A\mathbf{x} = \mathbf{b}$ may have infinitely many solutions.
3. The equation $A\mathbf{x} = \mathbf{0}$ has a solution.
4. If $A\mathbf{x} = \mathbf{b}$ has a solution for one choice of $\mathbf{b}$, then $A\mathbf{x} = \mathbf{b}$ has a solution for every choice of $\mathbf{b}$.
5. If $A\mathbf{x} = \mathbf{b}$ has one solution for one choice of $\mathbf{b}$, then $A\mathbf{x} = \mathbf{b}$ has exactly one solution for every choice of $\mathbf{b}$.
6. A is nonsingular if $A\mathbf{x} = \mathbf{0}$ has a solution other than $\mathbf{x} = \mathbf{0}$.
7. If A^{-1} exists, then A is nonsingular.
8. If A is nonsingular, then A^{-1} exists.
9. A is singular if $A\mathbf{x} = \mathbf{0}$ has a solution.
10. A is nonsingular if A can be changed into I_n by using elementary row operations.
11. If B is a $p \times q$ matrix and AB is defined, then $q = n$.
12. If B is a $p \times q$ matrix and BA is defined, then $q = n$.

---

* G. H. Hardy and E. M. Wright. *An Introduction to the Theory of Numbers.* 4th ed. London: Oxford University Press, 1965, p. 210.
** A reader interested in pursuing this problem is referred to the following papers:
  A. H. Bell. "The 'Cattle Problem' by Archimedes 251 B.C." *American Mathematical Monthly* 2 (1895): 140–41.
  M. Merriman. "The Cattle Problem of Archimedes." *Popular Science Monthly* 67 (1905): 660–65.
  No author cited. "The Cattle Problem of Archimedes." *American Mathematical Monthly* 25 (1918): 411–14.

In statements 13–17 the matrix A is $m \times n$.

13. The equation $A\mathbf{x} = \mathbf{b}$ is consistent if $m < n$.
14. The equation $A\mathbf{x} = \mathbf{b}$ is consistent if $m > n$.
15. The equation $A\mathbf{x} = \mathbf{b}$ is consistent if $m = n$.
16. If A is nonsingular, then $n = m$.
17. If A is singular, then $n = m$.

In statements 18–26 the matrices A, B, and C are $n \times n$.

18. $AB = BA$
19. $A(B + C) = AC + AB$
20. $(AB)^{-1} = A^{-1}B^{-1}$ if A and B are nonsingular.
21. $(ABC)^{-1} = C^{-1}B^{-1}A^{-1}$ if A, B, and C are nonsingular.
22. $(A + B)^{-1} = A^{-1} + B^{-1}$ if A and B are nonsingular.
23. If A and B are nonsingular, then $A + B$ is nonsingular.
24. If A and B are singular, then $A + B$ is singular.
25. If A and B are nonsingular, then AB is nonsingular.
26. If A and B are singular, then AB is singular.

2

Determinants

2.1 DETERMINANTS

Suppose we wish to solve the system of linear equations

$$a_{11}x_1 + a_{12}x_2 = b_1$$

$$a_{21}x_1 + a_{22}x_2 = b_2$$

If we multiply the first equation by a_{22}, the second equation by $-a_{12}$, and add the resulting equations, we obtain

$$(a_{11}a_{22} - a_{12}a_{21})x_1 = b_1a_{22} - b_2a_{12}$$

or

$$x_1 = \frac{b_1a_{22} - b_2a_{12}}{a_{11}a_{22} - a_{12}a_{21}} \tag{1}$$

whenever $a_{11}a_{22} - a_{12}a_{21} \neq 0$. Likewise,

$$x_2 = \frac{b_2a_{11} - b_1a_{21}}{a_{11}a_{22} - a_{12}a_{21}} \tag{2}$$

whenever $a_{11}a_{22} - a_{12}a_{21} \neq 0$.

Notice that the denominators in (1) and (2) are identical and that the numerators have the same form. Accordingly, we define, for any 2×2 matrix M a number, called the **determinant** (symbolized by det) of M, where

$$\det \begin{bmatrix} a & b \\ c & d \end{bmatrix} = ad - bc$$

We are now able to write equations (1) and (2) as

$$x_1 = \frac{\det\begin{bmatrix} b_1 & a_{12} \\ b_2 & a_{22} \end{bmatrix}}{\det\begin{bmatrix} a_{11} & a_{12} \\ a_{21} & a_{22} \end{bmatrix}} \qquad x_2 = \frac{\det\begin{bmatrix} a_{11} & b_1 \\ a_{21} & b_2 \end{bmatrix}}{\det\begin{bmatrix} a_{11} & a_{12} \\ a_{21} & a_{22} \end{bmatrix}}$$

whenever $a_{11}a_{22} - a_{12}a_{21} \neq 0$.

Next we consider the system of linear equations

$$a_{11}x_1 + a_{12}x_2 + a_{13}x_3 = b_1$$

$$a_{21}x_1 + a_{22}x_2 + a_{23}x_3 = b_2$$

$$a_{31}x_1 + a_{32}x_2 + a_{33}x_3 = b_3$$

If we solve this system for x_1, x_2, and x_3, a lengthy calculation shows that

$$x_1 = \frac{\det\begin{bmatrix} b_1 & a_{12} & a_{13} \\ b_2 & a_{22} & a_{23} \\ b_3 & a_{32} & a_{33} \end{bmatrix}}{\det\begin{bmatrix} a_{11} & a_{12} & a_{13} \\ a_{21} & a_{22} & a_{23} \\ a_{31} & a_{32} & a_{33} \end{bmatrix}}$$

$$x_2 = \frac{\det\begin{bmatrix} a_{11} & b_1 & a_{13} \\ a_{21} & b_2 & a_{23} \\ a_{31} & b_3 & a_{33} \end{bmatrix}}{\det\begin{bmatrix} a_{11} & a_{12} & a_{13} \\ a_{21} & a_{22} & a_{23} \\ a_{31} & a_{32} & a_{33} \end{bmatrix}}$$

$$x_3 = \frac{\det\begin{bmatrix} a_{11} & a_{12} & b_1 \\ a_{21} & a_{22} & b_2 \\ a_{31} & a_{32} & b_3 \end{bmatrix}}{\det\begin{bmatrix} a_{11} & a_{12} & a_{13} \\ a_{21} & a_{22} & a_{23} \\ a_{31} & a_{32} & a_{33} \end{bmatrix}}$$

where

$$(3) \quad \det\begin{bmatrix} c_{11} & c_{12} & c_{13} \\ c_{21} & c_{22} & c_{23} \\ c_{31} & c_{32} & c_{33} \end{bmatrix} = c_{11}c_{22}c_{33} + c_{13}c_{21}c_{32} + c_{12}c_{23}c_{31} - c_{11}c_{23}c_{32} - c_{12}c_{21}c_{33} - c_{13}c_{22}c_{31}$$

for any choice of the numbers c_{ij}, where $1 \leq i \leq 3$ and $1 \leq j \leq 3$.

It will be of value to determine a formula for the terms in the summation in (3). Notice that every term is of the form $c_{1j}c_{2k}c_{3l}$, where j, k, l is a permutation of $1, 2, 3$, and each term is preceded by ± 1. We use the symbol $s(j, k, l)$ called the **sign of the permutation** (j, k, l) in calculating the sign of a particular permutation. The precise value of $s(j, k, l)$ depends on how the permutation j, k, l can be obtained from $1, 2, 3$. In the next section we prove that if a permutation j, k, l of $1, 2, 3$ can be obtained by t successive interchanges of pairs of numbers and can also be obtained by r successive interchanges of pairs of numbers, then t and r are either both even or both odd. For example $2, 1, 3$ can be obtained from $1, 2, 3$ either by using one interchange or by using five successive interchanges of pairs of numbers as follows:

$$1, 2, 3 \to 2, 1, 3$$

$$1, 2, 3 \to 1, 3, 2 \to 3, 1, 2 \to 3, 2, 1 \to 2, 3, 1 \to 2, 1, 3$$

Using this fact about permutations we notice that

$$s(j, k, l) = (-1)^t$$

where t is a number of interchanges of consecutive numbers required to obtain j, k, l from $1, 2, 3$.

Since $1, 2, 3$ can be obtained from $1, 2, 3$ using no interchanges we have

$$s(1, 2, 3) = (-1)^0 = 1$$

The permutation $3, 1, 2$ can be obtained from $1, 2, 3$ in the following steps

$$1, 2, 3 \to 1, 3, 2 \to 3, 1, 2$$

Therefore, $s(3, 1, 2) = (-1)^2 = 1$. Continuing in this manner we can obtain the sign of any permutation of $1, 2, 3$: $s(1, 2, 3) = 1$; $s(3, 1, 2) = 1$; $s(2, 3, 1) = 1$; $s(1, 3, 2) = -1$; $s(3, 2, 1) = -1$; $s(2, 1, 3) = -1$. With this notation the summation in (3) can be written as

$$\sum s(j, k, l) c_{1j} c_{2k} c_{3l}$$

where the summation is over all permutations j, k, l of $1, 2, 3$. This number is called the determinant of the matrix

$$\begin{bmatrix} c_{11} & c_{12} & c_{13} \\ c_{21} & c_{22} & c_{23} \\ c_{31} & c_{32} & c_{33} \end{bmatrix}$$

We now generalize the concept of determinant to $n \times n$ matrices.

Definition 1.   Let the permutation $i_1, i_2, \ldots, i_n$ of $1, 2, \ldots, n$ be formed by t successive interchanges of pairs of numbers. Then $s(j_1, j_2, \ldots, j_n) = (-1)^t$.

As in the case when $n = 3$, if $j_1, j_2, \ldots, j_n$ can be obtained from $1, 2, \ldots, n$ by both t and r successive interchanges of pairs of numbers, then $(-1)^t = (-1)^r$ (Theorem 1 of the next section).

In order to emphasize the fact that a_{ij} is the ij-component of a matrix A, we shall write the matrix as $A = (a_{ij})$.

Definition 2.   The **determinant** of an $n \times n$ matrix $A = (a_{ij})$, denoted by det A, is given by

$$\det A = \sum s(j_1, j_2, \ldots, j_n)(a_{1j_1})(a_{2j_2}) \ldots (a_{nj_n}) \tag{4}$$

where the sum is over all permutations $j_1, j_2, \ldots, j_n$ of $1, 2, \ldots, n$.

It needs to be emphasized that det A is defined only if A is a square matrix. Each summand in the definition of det A in (4) is 1 or -1 times

$$(a_{1j_1})(a_{2j_2})\ldots(a_{nj_n})$$

Notice that a_{1j_1} is a component from the first row of A, a_{2j_1} is a component from the second row of $A, \ldots, a_{nj_n}$ is a component from the nth row of A. Thus each summand contains exactly one component from each row of A. Since $j_1, j_2, \ldots, j_n$ is a permutation of $1, 2, \ldots, n$, a similar argument shows that each summand contains exactly one component from each column of A. Therefore, each summand is 1 or -1 times a term consisting of exactly one component from each row and each column of A.

The following paragraph utilizes the symbol $n!$, which is read "n factorial." Many of you will remember from previous mathematics courses that $n!$ is the product of the first n positive integers; that is, $n! = (1)(2)\ldots(n-1)(n)$.

Since $1, 2, \ldots, n$ has $n!$ permutations, the calculation of the determinant of an $n \times n$ matrix could be a lengthy process if we used formula (4). The computation would require adding $n!$ terms, each of which is a product of n terms. Therefore, there would be $n! - 1$ additions and $(n-1)(n!)$ multiplications for a total of $n(n!) - 1$ arithmetic operations. Working at the rate of 100,000 arithmetic operations per second it would take over six years to compute the determinant of a 15×15 matrix. If n is increased to 25, it would take over 10^{14} years to evaluate the determinant. Clearly, we will need an alternate method to evaluate determinants unless n is small. Fortunately, determinants have several properties that ease their calculation. These properties and methods of evaluation are discussed later in this chapter.

Exercises

1. Verify that $\sum s(j,k,l)a_{1j}a_{2k}a_{3l} = a_{11}a_{22}a_{33} + a_{13}a_{21}a_{32} + a_{12}a_{23}a_{31} - a_{11}a_{23}a_{32} - a_{12}a_{21}a_{33} - a_{13}a_{22}a_{31}$.

In Exercises 2–5 compute the value of the determinant of the given matrix by using formula (4).

2. $\begin{bmatrix} 3 & 4 \\ 2 & 1 \end{bmatrix}$

3. $\begin{bmatrix} 2 & 4 \\ 4 & 2 \end{bmatrix}$

4. $\begin{bmatrix} 1 & 0 & 2 \\ 0 & 3 & 0 \\ 1 & 0 & 4 \end{bmatrix}$

5. $\begin{bmatrix} 1 & 6 & 2 \\ 3 & 1 & 6 \\ 2 & 0 & 1 \end{bmatrix}$

In Exercises 6–9 use determinants to find solutions of the given systems of equations.

6. $\begin{bmatrix} 1 & 2 \\ 2 & 1 \end{bmatrix} \mathbf{x} = \begin{bmatrix} 1 \\ 0 \end{bmatrix}$

7. $\begin{bmatrix} 2 & 3 \\ 4 & 1 \end{bmatrix} \mathbf{x} = \begin{bmatrix} 2 \\ 3 \end{bmatrix}$

8. $\begin{bmatrix} 1 & 2 & 0 \\ 0 & 2 & 1 \\ 2 & 0 & 1 \end{bmatrix} \mathbf{x} = \begin{bmatrix} 1 \\ 0 \\ 1 \end{bmatrix}$

9. $\begin{bmatrix} 1 & 3 & 4 \\ 6 & 2 & 2 \\ 3 & 1 & 3 \end{bmatrix} \mathbf{x} = \begin{bmatrix} 2 \\ -1 \\ 4 \end{bmatrix}$

10. In the formula for the determinant of a 5×5 matrix, there is a term $\pm a_{14}a_{25}a_{31}a_{42}a_{53}$. Is the sign plus or is it minus?

11. In the formula for the determinant of a 4×4 matrix there is a term $\pm a_{13} a_{21} a_{34} a_{42}$. Is the sign plus or is it minus?
12. Suppose that A, B, and C are matrices that are identical except for the ith row and that the ith row of C is equal to the sum of the ith row of A and B. Show that det $C =$ det $A +$ det B.
13. Find 2×2 matrices A and B such that $\det(A + B) \neq \det A + \det B$.
14. Let A be an $n \times n$ matrix. Show that $\det(cA) = c^n \det A$ for any number c.
15. Let A be an $n \times n$ matrix with one row consisting entirely of zeros. Show that det $A = 0$.
16. What is the determinant of a 1×1 matrix $A = [a_{11}]$?

2.2 PERMUTATIONS AND RUBIK'S CUBE (OPTIONAL)

The definition of determinant (Definition 2 of Section 2.1) is based on the assertion that if a permutation $x_1, x_2, \ldots, x_n$ of $1, 2, \ldots, n$ can be obtained by t and r successive interchanges of pairs of numbers, then t and r are both even or both odd; that is, $(-1)^t = (-1)^r$. This section includes a proof of this assertion and shows how it relates to Rubik's cube.

To each permutation $x_1, x_2, \ldots, x_n$ of $1, 2, \ldots, n$ we associate the nonzero number P that is the product of all numbers of the form $x_i - x_j$ where $i < j$. That is,

$$P = (x_1 - x_2)(x_1 - x_3) \ldots (x_1 - x_n)(x_2 - x_3) \ldots (x_2 - x_n) \ldots (x_{n-1} - x_n)$$
$$= \prod_{i<j} (x_i - x_j)$$

For example, if we are considering the permutation 2, 3, 4, 1 of 1, 2, 3, 4 then

$$P = (2 - 3)(2 - 4)(2 - 1)(3 - 4)(3 - 1)(4 - 1) = -12$$

Suppose that in the permutation $x_1, x_2, \ldots, x_n$ we interchange only two numbers, x_h and x_k, where $h < k$. Let P_1 denote the number associated with this new permutation of $1, 2, \ldots, n$. We will show that $P_1 = -P$ by determining what happens to the factors of P when x_h and x_k are interchanged. There are five cases to consider:

1. Any factor $x_i - x_j$, where neither i nor j equals h or k, is left unchanged and is therefore a factor of P_1.
2. If $i < h$, then $x_i - x_h$ is changed to $x_i - x_k$, and $x_i - x_k$ is changed to $x_i - x_h$. Hence $x_i - x_h$ and $x_i - x_k$ are factors of P_1.
3. If $k < i$, then $x_h - x_i$ is changed to $x_k - x_i$, and $x_k - x_i$ is changed to $x_h - x_i$. Hence $x_h - x_i$ and $x_k - x_i$ are factors of P_1.
4. If $h < i < k$, then $x_h - x_i$ is changed to $x_k - x_i$ and $x_i - x_k$ is changed to $x_i - x_h$. Since $(x_h - x_i)(x_i - x_k) = (x_k - x_i)(x_i - x_h)$, the equivalent of $(x_h - x_i)(x_i - x_k)$ is a factor of P_1.
5. When $x_h - x_k$ is changed to $x_k - x_h$, the factor equals $-(x_h - x_k)$.

Thus, each factor of P is a factor of P_1 with the sole exception that $x_h - x_k$ is replaced by $-(x_h - x_k)$; consequently $P_1 = -P$. Therefore, each time we interchange two numbers in a permutation the sign of P changes. If we successively interchange t pairs of numbers to obtain $x_1, x_2, \ldots, x_n$ from $1, 2, \ldots, n$, we have $Q = (-1)^t P$, where P and Q are the previously defined numbers associated with $1, 2, \ldots, n$ and $x_1, x_2, \ldots, x_n$, respectively. If $x_1, x_2, \ldots, x_n$ can also be obtained by r successive interchanges of pairs of numbers, then $Q = (-1)^r P$. Since $P \neq 0$, we conclude that $(-1)^t = (-1)^r$. Therefore, we have proved the following theorem.

Theorem 1. *Let* $x_1, x_2, \ldots, x_n$ *be a permutation of* $1, 2, \ldots, n$ *that can be obtained by* t *successive interchanges of numbers and also by* r *successive interchanges of numbers. Then* $(-1)^t = (-1)^r$.

Theorem 1 allows us to call a permutation **even** if it can be obtained by an even number of successive interchanges of numbers or **odd** if it can be obtained by an odd number of successive interchanges of numbers. Evidently two consecutive even or two consecutive odd permutations result in an even permutation. Similarly, two consecutive permutations, one of which is odd and the other even, result in an odd permutation.

We are now able to show that if we interchange two rows of a matrix then the determinant changes sign.

Theorem 2. *Let* A *be an* $n \times n$ *matrix and let* B *be a matrix obtained from* A *by interchanging two of its rows. Then* $\det B = -\det A$.

Proof: Suppose that rows h and k have been interchanged. Then $b_{hj} = a_{kj}$; $b_{kj} = a_{hj}$; and $b_{ij} = a_{ij}$ when $i \neq h$ and $j \neq k$. By definition,

$$\det B = \sum s(j_1, j_2, \ldots, j_n) b_{1j_1} b_{2j_2} \cdots b_{nj_n}$$

where the sum is over all permutations $j_1, j_2, \ldots, j_n$ of $1, 2, \ldots, n$. Let $l_1, l_2, \ldots, l_n$ be the permutation of $1, 2, \ldots, n$ obtained from $j_1, j_2, \ldots, j_n$ by interchanging j_h and j_k. Then $s(j_1, j_2, \ldots, j_n) = -s(l_1, l_2, \ldots, l_n)$ and

$$b_{1j_1} \cdots b_{hj_h} \cdots b_{kj_k} \cdots b_{nj_n} = b_{1j_1} \cdots b_{kj_k} \cdots b_{hj_h} \cdots b_{nj_n}$$
$$= a_{1l_1} a_{2l_2} \cdots a_{nl_n}$$

so that

$$\det B = -\sum s(l_1, l_2, \ldots, l_n) a_{1l_1} a_{2l_2} \cdots a_{nl_n}$$

where the sum is over all permutations $l_1, l_2, \ldots, l_n$ of $1, 2, \ldots, n$. Therefore $\det B = -\det A$.

Theorem 1 also gives us some understanding of Rubik's cube. When you examine this cube, you will see that it is composed of 26 smaller cubes ("cubies") plus space for a 27th cubie in the center. Notice there are three types of cubies: corner cubies (there are 8 of these); edge cubies (there are 12 of these); and face cubies (there are 6 of these). A "move" consists of rotating a single face of the cube a quarter turn. Notice that on any move:

1. The corner cubies remain corner cubies.
2. The edge cubies remain edge cubies.
3. The face cubies remain face cubies.

Thus we can view a move as a permutation of $1, 2, \ldots, 8$ (one number for each corner cubie) and also a permutation of $1, 2, \ldots, 12$ (one number for each edge cubie).

With each move four of the corners are rotated, while the other four remain fixed. If we numbered the corners in an appropriate manner, the first rotates to the second, the second to the third, the third to the fourth, and the fourth to the first. This gives us the permutation 4, 1, 2, 3, 5, 6, 7, 8 of 1, 2, 3, 4, 5, 6, 7, 8 which can be obtained as follows

$$1, 2, 3, 4, 5, 6, 7, 8 \rightarrow 1, 2, 4, 3, 5, 6, 7, 8$$

$$\rightarrow 1, 4, 2, 3, 5, 6, 7, 8$$

$$\rightarrow 4, 1, 2, 3, 5, 6, 7, 8$$

Each move corresponds to an odd permutation of $1, 2, \ldots, 8$. Likewise each move corresponds to an odd permutation of $1, 2, \ldots, 12$. Thus if we perform k moves we obtain permutations of $1, 2, \ldots, 8$ and $1, 2, \ldots, 12$ that are either both even or both odd. In terms of the cube this means that if the corners can be returned to their original positions in n moves and if the edges can be returned to their original positions in m moves, then m and n are either both even or both odd.

We are now able to show easily that not all positions of the cubies are possible. For example, suppose that the cube is in its original position except that the four corners on one face have been rotated a quarter turn. Then the corners can be returned to their original position in an odd number (one) of moves and the edges can be returned to their original position in an even (zero) number of moves. Therefore this position is impossible to attain. By examining the cube in its original position, you will discover that whenever a corner cubie is moved out of its original position, an edge cubie is also moved. Therefore, in any attempt to restore a corner position, you will also move an edge cubie.

A detailed description of the mathematical nature of Rubik's cube can be found in the writings of Singmaster* and Hofstadler.**

2.3 SOME BASIC PROPERTIES OF DETERMINANTS

Even though Definition 2 requires $n(n!) - 1$ arithmetic operations to evaluate the determinant of an $n \times n$ matrix there are some matrices whose determinants are easily evaluated by using Definition 2. In the following examples we evaluate the determinants of some such matrices.

For convenience we will repeat the definition of the determinant of an $n \times n$ matrix $A = (a_{ij})$:

$$\det A = \sum s(j_1, j_2, \ldots, j_n)(a_{1j_1})(a_{2j_2}) \ldots (a_{nj_n}) \tag{1}$$

where the sum is over all permutations of $1, 2, \ldots, n$.

Example 1.   Let

$$A = \begin{bmatrix} a_{11} & a_{12} \\ a_{21} & a_{22} \end{bmatrix}$$

---

* David Singmaster. *Notes on Rubik's "Magic Cube"*. 5th ed. London, England: Mathematical Sciences and Computing, Polytechnic of the South Bank, 1980.
** Douglas R. Hofstadler. "Metamagical Themas." *Scientific American* (March 1981): 20–39.

Then

$$\det A = s(1,2)a_{11}a_{22} + s(2,1)a_{12}a_{21}$$

$$= a_{11}a_{22} - a_{12}a_{21}$$

This formula can be remembered by using the following diagram:

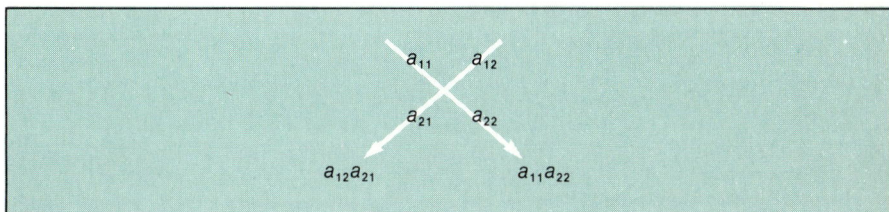

The determinant of A is the difference in the products of the numbers on the diagonals going down from left to right and from right to left. For example,

$$\det\begin{bmatrix} 1 & 2 \\ 3 & 4 \end{bmatrix} = (1)(4) - (2)(3) = -2$$

Example 2.   A method similar to that used in Example 1 for computing the determinant of a 3×3 matrix

$$A = \begin{bmatrix} a_{11} & a_{12} & a_{13} \\ a_{21} & a_{22} & a_{23} \\ a_{31} & a_{32} & a_{33} \end{bmatrix}$$

may be summarized as follows:

1. Write the first two columns of the matrix in order to the right of the third column:

$$a_{11}\ a_{12}\ a_{13}\ a_{11}\ a_{12}$$

$$a_{21}\ a_{22}\ a_{23}\ a_{21}\ a_{22}$$

$$a_{31}\ a_{32}\ a_{33}\ a_{31}\ a_{32}$$

2. In turn, multiply each of the first three components of the first row by the other two entries on its diagonal from left to right. Let S_1 denote the sum of these three products.

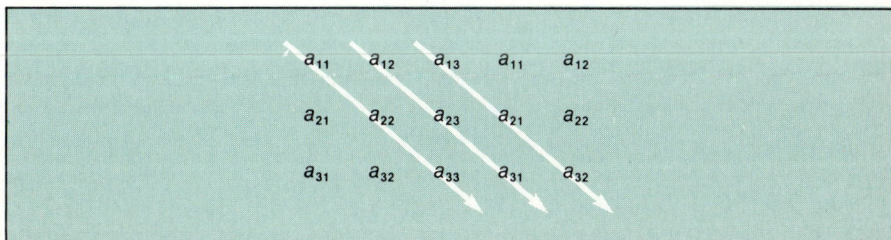

3. Similarly, multiply each of the last three components of the first row by the other two entries on its diagonal from right to left. Let S_2 denote the sum of these three products.

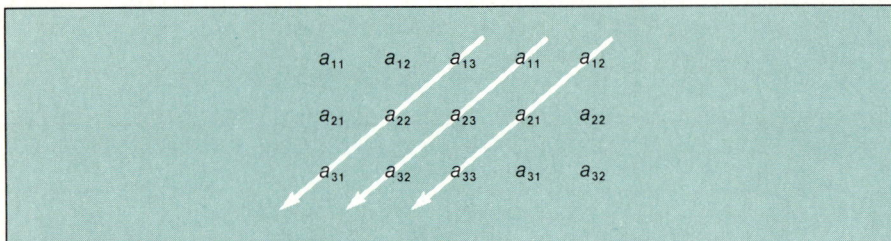

$$
\begin{array}{ccccc}
a_{11} & a_{12} & a_{13} & a_{11} & a_{12} \\
a_{21} & a_{22} & a_{23} & a_{21} & a_{22} \\
a_{31} & a_{32} & a_{33} & a_{31} & a_{32}
\end{array}
$$

4. The determinant equals S_1 minus S_2.

A short calculation shows that the preceding process computes the six terms, each with the appropriate sign, in the definition of the determinant of a 3×3 matrix. Thus, this process computes the determinant of a 3×3 matrix. **No similar method works for evaluating the determinant of an $n \times n$ matrix when $n > 3$.**

Consider the matrix

$$
A = \begin{bmatrix} 1 & 2 & -1 \\ 5 & 3 & 4 \\ -2 & 0 & 2 \end{bmatrix}
$$

Using the above procedure, we have

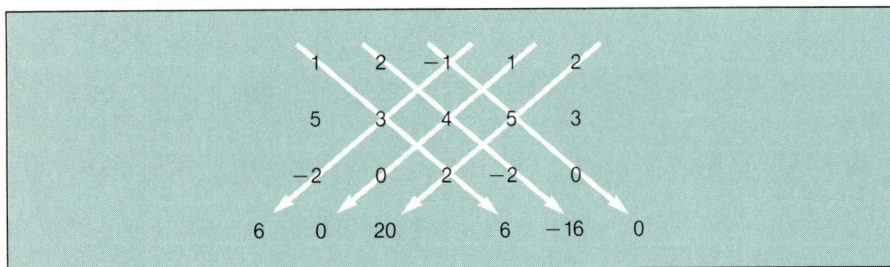

$$
\begin{array}{ccccc}
1 & 2 & -1 & 1 & 2 \\
5 & 3 & 4 & 5 & 3 \\
-2 & 0 & 2 & -2 & 0 \\
6 & 0 & 20 & 6 & -16 & 0
\end{array}
$$

so that

$$
\begin{aligned}
\det A &= [6 + (-16) + 0] - [6 + 0 + 20] \\
&= -36
\end{aligned}
$$

Example 3. An $n \times n$ matrix $A = (a_{ij})$ is called **diagonal** if $a_{ij} = 0$ whenever $i \neq j$. In order to evaluate the determinant of a diagonal matrix by using Definition 2, we must add terms of the form

$$
s(j_1, j_2, \ldots, j_n)(a_{1j_1})(a_{2j_2}) \ldots (a_{nj_n}) \tag{2}
$$

where $j_1, j_2, \ldots, j_n$ is a permutation of $1, 2, \ldots, n$. If $j_1 \neq 1$, then $a_{1j_1} = 0$ and the term in (2) is zero. If $j_2 \neq 2$, then $a_{2j_2} = 0$ and the term in (2) is zero. Continuing in this manner we see that the term in (2) is zero if $j_1 \neq 1, j_2 \neq 2, \ldots, j_n \neq n$. Thus the only summand in (1) that may be different from zero is $s(1, 2, \ldots, n)a_{11}a_{22}\ldots a_{nn}$. Since $s(1, 2, \ldots, n) = 1$ we have proved that

$$\det A = a_{11}a_{22}\ldots a_{nn}$$

whenever A is a diagonal matrix. For example,

$$\det \begin{bmatrix} 2 & 0 & 0 & 0 \\ 0 & -7 & 0 & 0 \\ 0 & 0 & 5 & 0 \\ 0 & 0 & 0 & -4 \end{bmatrix} = (2)(-7)(5)(-4) = 280$$

Example 4.   An $n \times n$ matrix $A = (a_{ij})$ is called **upper triangular** if $a_{ij} = 0$ whenever $i > j$ and **lower triangular** if $a_{ij} = 0$ whenever $i < j$. A matrix is called **triangular** if it is either upper triangular or lower triangular. An argument similar to that used in Example 3 shows that the only summand in the definition of $\det A$ that may be different from zero is $s(1, 2, \ldots, n)a_{11}a_{22}\ldots a_{nn}$. (*Hint*: In calculating the determinant of an upper triangular matrix, notice that every term except $a_{11}a_{22}\ldots a_{nn}$ contains at least one a_{ij} where $i > j$. In the case of the lower triangular matrix, every term except $a_{11}a_{22}\ldots a_{nn}$ contains at least one a_{ij} where $i < j$.) Therefore, the determinant of a triangular matrix A can also be expressed as follows:

$$\det A = a_{11}a_{22}\ldots a_{nn}$$

For example,

$$\det \begin{bmatrix} 1 & 2 & 3 & 4 \\ 0 & 5 & 6 & 7 \\ 0 & 0 & 8 & 9 \\ 0 & 0 & 0 & 2 \end{bmatrix} = (1)(5)(8)(2) = 80$$

It is sometimes convenient to consider a matrix that is obtained from a given matrix by interchanging the rows and columns of that matrix.

If A is any $m \times n$ matrix, then the **transpose** of A, denoted by A^t, is the $n \times m$ matrix whose first row is the first column of A, whose second row is the second column of $A, \ldots$, whose nth row is the nth column of A. Hence, if a_{ij} is the ij-component of A, then a_{ji} is the ij-component of A^t.

Example 5.   The transposes of

$$A = \begin{bmatrix} 1 & 2 & 3 & 4 \\ 5 & 6 & 7 & 8 \end{bmatrix} \quad B = \begin{bmatrix} 2 & 0 & 1 \\ 5 & 1 & 3 \\ 2 & 4 & 4 \end{bmatrix} \quad C = \begin{bmatrix} 2 \\ -3 \\ 4 \end{bmatrix} \quad D = \begin{bmatrix} -2 & 1 & 0 & 1 \end{bmatrix}$$

are

$$A^t = \begin{bmatrix} 1 & 5 \\ 2 & 6 \\ 3 & 7 \\ 4 & 8 \end{bmatrix} \quad B^t = \begin{bmatrix} 2 & 5 & 2 \\ 0 & 1 & 4 \\ 1 & 3 & 4 \end{bmatrix} \quad C^t = \begin{bmatrix} 2 & -3 & 4 \end{bmatrix} \quad D^t = \begin{bmatrix} -2 \\ 1 \\ 0 \\ 1 \end{bmatrix}$$

The determinant of the transpose of an $n \times n$ matrix $A = (a_{ij})$ can be determined by considering the definition of det A in (2). The term $a_{1j_1} a_{2j_2} \ldots a_{nj_n}$ is called an elementary product of det A. In Section 2.1 we noted that each term contains one component from each row and one component from each column of A. Since A^t is obtained from A by interchanging its rows and columns, we can prove that the elementary product of det A. In Section 2.1 we noted that each term contains one can also prove that the sign associated with an elementary product of det A is identical to the sign associated with the same elementary product of det A^t. In fact, the following theorem is based on such proof.

Theorem 3. *If A is any square matrix, then det $A = $ det A^t.*

Exercise 27 requires verification of this theorem for 3×3 matrices. For 2×2 matrices the proof is elementary. Let

$$A = \begin{bmatrix} a_{11} & a_{12} \\ a_{21} & a_{22} \end{bmatrix}$$

Then

$$A^t = \begin{bmatrix} a_{11} & a_{21} \\ a_{12} & a_{22} \end{bmatrix}$$

and det $A = a_{11}a_{22} - a_{12}a_{21} = $ det A^t.

Exercises

In Exercises 1–12 compute the determinant of the given matrix.

1. $\begin{bmatrix} 1 & 2 \\ 3 & 4 \end{bmatrix}$
 2. $\begin{bmatrix} -8 & 8 \\ -2 & 1 \end{bmatrix}$
 3. $\begin{bmatrix} -1 & 3 \\ -2 & -1 \end{bmatrix}$
 4. $\begin{bmatrix} 9 & -2 \\ 7 & 8 \end{bmatrix}$

5. $\begin{bmatrix} 1 & 2 & 3 \\ 4 & 5 & 6 \\ 7 & 8 & 9 \end{bmatrix}$
 6. $\begin{bmatrix} 5 & 4 & 1 \\ 2 & 3 & 0 \\ 0 & 1 & 2 \end{bmatrix}$
 7. $\begin{bmatrix} 1 & 0 & 4 \\ 4 & -5 & 7 \\ 3 & 2 & -1 \end{bmatrix}$
 8. $\begin{bmatrix} 2 & 0 & 0 \\ -1 & 6 & 5 \\ 4 & 5 & 7 \end{bmatrix}$

9. $\begin{bmatrix} 1 & 2 & 3 & -5 \\ 0 & 4 & 8 & 2 \\ 0 & 0 & 0 & 7 \\ 0 & 0 & 0 & 4 \end{bmatrix}$
 10. $\begin{bmatrix} 9 & -8 & -3 & 6 \\ 0 & 7 & 5 & 1 \\ 0 & 0 & -4 & 2 \\ 0 & 0 & 0 & 5 \end{bmatrix}$

11. $\begin{bmatrix} 6 & 0 & 0 & 0 \\ 7 & 5 & 0 & 0 \\ -8 & 2 & 4 & 0 \\ 2 & -1 & 7 & 3 \end{bmatrix}$
 12. $\begin{bmatrix} 4 & 0 & 0 & 0 \\ 5 & 1 & 0 & 0 \\ 0 & 0 & 2 & 0 \\ 0 & 7 & 0 & -2 \end{bmatrix}$

In Exercises 13–18 compute the transpose of the given matrix.

13. $\begin{bmatrix} 2 & 4 \\ 5 & -1 \\ 3 & 0 \end{bmatrix}$

14. $\begin{bmatrix} 3 & 6 & 5 \\ -1 & 7 & 4 \end{bmatrix}$

15. $\begin{bmatrix} 1 & 2 & 2 & -2 \\ 0 & 4 & 1 & 3 \\ 6 & 5 & 4 & -1 \end{bmatrix}$

16. $\begin{bmatrix} 1 & 2 & 5 \\ 3 & -2 & 1 \\ 2 & 4 & 6 \\ 7 & -8 & 0 \end{bmatrix}$

17. $[1 \ 2 \ 3 \ 4]$

18. $\begin{bmatrix} 2 \\ -6 \\ 5 \\ 1 \end{bmatrix}$

19. Show that $(A \pm B)^t = A^t \pm B^t$ for any two $m \times n$ matrices A and B.
20. (a) Show that $(AB)^t = B^t A^t$ for any 2×2 matrices A and B.
    (b) Show that $(AB)^t = B^t A^t$ for any $m \times n$ matrix A and any $n \times p$ matrix B.
21. Show that $(cA)^t = cA^t$ for any number c and any $m \times n$ matrix A.
22. Show that if A is a nonsingular $n \times n$ matrix, then $(A^t)^{-1} = (A^{-1})^t$.
23. Let A and B be nonsingular $n \times n$ matrices. Show that $(A^{-1}B^{-1})^t = (A^t B^t)^{-1}$. (Hint: Use Exercises 20 and 22.)
24. Show that $A = (A^t)^t$ for any $m \times n$ matrix A.
25. An $n \times n$ matrix A is called "symmetric" if $A^t = A$. Show that if B is any $n \times n$ matrix, then $B^t B$ and $B + B^t$ are symmetric. (Hint: Use Exercises 19, 20 and 24.)
26. Show that $(B^t A^t)^t = AB$ for any $m \times n$ matrix A and any $n \times p$ matrix B. (Hint: Use Exercises 20 and 24.)
27. Prove Theorem 3 for 3×3 matrices.
28. An $n \times n$ matrix A is called "skew-symmetric" if $A^t = -A$. Show that if B is any $n \times n$ matrix, then $B - B^t$ is skew-symmetric. (Hint: Use Exercises 19 and 24.)

2.4 EVALUATING DETERMINANTS BY USING ELEMENTARY ROW OPERATIONS

Section 2.1 indicated that generally it is impractical to use the definition of determinant (i.e., Definition 2) to compute the determinant of an $n \times n$ matrix unless n is small. Section 2.3 gave an easy way to compute the determinant of an $n \times n$ triangular matrix $A = (a_{ij})$ regardless of how large n is: det $A = a_{11}a_{22} \ldots a_{nn}$. This section shows that the problem of evaluating the determinant of a matrix can be reduced to the problem of evaluating the determinant of an appropriate triangular matrix.

We introduced the concept of determinant by considering systems of linear equations with as many unknowns as equations. In fact, we showed that the solution of such a system with either two or three unknowns can be written in terms of determinants. Previously we solved systems of linear equations by using elementary row operations. When we use this method of solution the matrix of coefficients is transformed into a triangular matrix. This leads us to investigate the effects of elementary row operations on the determinant of a matrix. These effects are described in the following theorem:

Theorem 4.   *Let A be an n × n matrix.*

   1. *If a matrix B is obtained from A by multiplying one row of A by a constant c, then det B = c det A.*
   2. *If a matrix B is obtained from A by replacing one of its rows by the sum of that row and a constant multiple of another row, then det B = det A.*
   3. *If a matrix B is obtained from A by interchanging two of its rows, then det B = − det A.*

The proof of the first part follows directly from the definition of a determinant and is left as an exercise. Part 2 is more difficult to prove; Exercise 18 asks for the proof in the special case when $n = 2$. Part 3 of Theorem 4 is precisely Theorem 2 (in Section 2.2), which has already been proved.

Example 1.   Let

$$A = \begin{bmatrix} 5 & 2 \\ 3 & 6 \end{bmatrix}$$

By part 1 of Theorem 4

$$\det \begin{bmatrix} 5 & 2 \\ 3 & 6 \end{bmatrix} = 3 \det \begin{bmatrix} 5 & 2 \\ 1 & 2 \end{bmatrix}$$

This can easily be checked by computing the determinants:

$$\det \begin{bmatrix} 5 & 2 \\ 3 & 6 \end{bmatrix} = (5)(6) - (2)(3) = 24 \qquad (1)$$

$$3 \det \begin{bmatrix} 5 & 2 \\ 1 & 2 \end{bmatrix} = 3[(5)(2) - (2)(1)] = 3(8) = 24$$

If we add -2 times the second row of A to the first row we obtain the matrix

$$B = \begin{bmatrix} -1 & -10 \\ 3 & 6 \end{bmatrix}$$

By part 2 of Theorem 4

$$\det B = \det A$$

This can also be checked by computing the determinants:

$$\det B = \begin{bmatrix} -1 & -10 \\ 3 & 6 \end{bmatrix} = (-1)(6) - (-10)(3) = 24$$

while det A was found to equal 24 in (1).
   If we interchange the rows of A, then

$$\det \begin{bmatrix} 3 & 6 \\ 5 & 2 \end{bmatrix} = (3)(2) - (6)(5) = -24 = -\det A$$

as is assured by part 3 of Theorem 4.

   The most practical general method of evaluating the determinant of an $n \times n$ matrix A is to use elementary row operations until the determinant of A is equal to a constant times the determinant of a triangular matrix, which is easily evaluated. We will illustrate this process with three examples. In order to indicate which elementary row operations are used we will use the same notation as in Chapter 1. We will also indicate which part of Theorem 4 is used by [1], [2], or [3].

Example 2.

$$\det \begin{bmatrix} 1 & 2 & 3 \\ 4 & 5 & 6 \\ 7 & 8 & 9 \end{bmatrix} = \det \begin{bmatrix} 1 & 2 & 3 \\ 0 & -3 & -6 \\ 0 & -6 & -12 \end{bmatrix} \quad \begin{array}{l} (-4)R_1 + R_2 \to R_2 \quad [2] \\ (-7)R_1 + R_3 \to R_3 \quad [2] \end{array}$$

$$= \det \begin{bmatrix} 1 & 2 & 3 \\ 0 & -3 & -6 \\ 0 & 0 & 0 \end{bmatrix} \quad (-2)R_2 + R_3 \to R_3 \quad [2]$$

$$= (1)(-3)(0) = 0$$

Example 3.

$$\det \begin{bmatrix} 2 & 2 & 5 & 1 \\ 2 & 1 & 1 & 3 \\ 1 & 0 & 1 & 0 \\ 0 & 1 & 3 & 2 \end{bmatrix} = -\det \begin{bmatrix} 1 & 0 & 1 & 0 \\ 2 & 1 & 1 & 3 \\ 2 & 2 & 5 & 1 \\ 0 & 1 & 3 & 1 \end{bmatrix} \quad R_1 \leftrightarrow R_3 \quad [3]$$

$$= -\det \begin{bmatrix} 1 & 0 & 1 & 0 \\ 0 & 1 & -1 & 3 \\ 0 & 2 & 3 & 1 \\ 0 & 1 & 3 & 2 \end{bmatrix} \quad \begin{array}{l} (-2)R_1 + R_2 \to R_2 \quad [2] \\ (-2)R_1 + R_3 \to R_3 \quad [2] \end{array}$$

$$= -\det \begin{bmatrix} 1 & 0 & 1 & 0 \\ 0 & 1 & -1 & 3 \\ 0 & 0 & 5 & -5 \\ 0 & 0 & 4 & -1 \end{bmatrix} \quad \begin{array}{l} (-2)R_2 + R_3 \to R_3 \quad [2] \\ (-1)R_2 + R_4 \to R_4 \quad [2] \end{array}$$

$$= -5 \det \begin{bmatrix} 1 & 0 & 1 & 0 \\ 0 & 1 & -1 & 3 \\ 0 & 0 & 1 & -1 \\ 0 & 0 & 4 & -1 \end{bmatrix} \quad [1]$$

$$= -5 \det \begin{bmatrix} 1 & 0 & 1 & 0 \\ 0 & 1 & -1 & 3 \\ 0 & 0 & 1 & -1 \\ 0 & 0 & 0 & 3 \end{bmatrix} \quad (-4)R_3 + R_4 \to R_4 \quad [2]$$

$$= -5(1)(1)(1)(3) = -15$$

Example 4.

$$\det \begin{bmatrix} 2 & 4 & -6 & 8 \\ 5 & 6 & 7 & 8 \\ 1 & 2 & -3 & -4 \\ 4 & 8 & 7 & 8 \end{bmatrix} = 2 \det \begin{bmatrix} 1 & 2 & -3 & 4 \\ 5 & 6 & 7 & 8 \\ 1 & 2 & -3 & -4 \\ 4 & 8 & 7 & 8 \end{bmatrix} \quad [1]$$

$$= 2 \det \begin{bmatrix} 1 & 2 & -3 & 4 \\ 0 & -4 & 22 & -12 \\ 0 & 0 & 0 & -8 \\ 0 & 0 & 19 & -8 \end{bmatrix} \quad \begin{matrix} (-5)R_1 + R_2 \to R_2 & [2] \\ (-1)R_1 + R_3 \to R_3 & [2] \\ (-4)R_1 + R_4 \to R_4 & [2] \end{matrix}$$

$$= -2 \det \begin{bmatrix} 1 & 2 & -3 & 4 \\ 0 & -4 & 22 & -12 \\ 0 & 0 & 19 & -8 \\ 0 & 0 & 0 & -8 \end{bmatrix} \quad R_3 \leftrightarrow R_4 \quad [3]$$

$$= -2(1)(-4)(19)(-8) = -1216$$

--- Exercises ---

In Exercises 1–10 compute the determinant of the given matrix by using elementary row operations.

1. $\begin{bmatrix} 1 & 2 & 3 \\ 3 & 2 & 4 \\ 4 & 1 & 5 \end{bmatrix}$

2. $\begin{bmatrix} 2 & 6 & 7 \\ 2 & 4 & 5 \\ 4 & 5 & 3 \end{bmatrix}$

3. $\begin{bmatrix} 4 & 8 & -4 \\ 1 & 2 & -1 \\ 2 & 0 & 3 \end{bmatrix}$

4. $\begin{bmatrix} 9 & 1 & 0 \\ 6 & 1 & 2 \\ 3 & 1 & -1 \end{bmatrix}$

5. $\begin{bmatrix} 1 & 0 & 1 & 0 \\ 0 & 1 & 0 & 1 \\ 1 & 0 & 0 & 1 \\ 0 & 1 & 1 & 0 \end{bmatrix}$

6. $\begin{bmatrix} 1 & 0 & 0 & 1 \\ 0 & 1 & 1 & 0 \\ 1 & 0 & 1 & 0 \\ 0 & 1 & 0 & 1 \end{bmatrix}$

7. $\begin{bmatrix} 1 & 2 & 3 & 5 \\ 5 & 6 & 7 & 8 \\ 4 & 3 & 2 & 1 \\ -1 & 2 & -3 & 4 \end{bmatrix}$

8. $\begin{bmatrix} 1 & 2 & 3 & 4 \\ 5 & 6 & 7 & 8 \\ 4 & 3 & 2 & 6 \\ 1 & -2 & 3 & -4 \end{bmatrix}$

9. $\begin{bmatrix} 1 & 0 & 2 & 0 \\ 2 & 1 & 1 & 0 \\ 3 & 1 & 3 & 1 \\ 0 & 1 & 0 & 0 \end{bmatrix}$

10. $\begin{bmatrix} 3 & 4 & 5 & 6 \\ 2 & 3 & 4 & 5 \\ 1 & 1 & 1 & 1 \\ 1 & 1 & 1 & 2 \end{bmatrix}$

11. Let A be an $n \times n$ matrix with two identical rows. Show that det $A = 0$. (Hint: Use part 3 of Theorem 4).
12. Let A be an $n \times n$ matrix. Use part 1 of Theorem 4 to show that $\det(cA) = c^n(\det A)$ for any number c.
13. An $n \times n$ matrix A is called skew symmetric if $A^t = -A$. Show that if A is skew-symmetric and n is odd, then det $A = 0$. [Hint: Show that $\det A = (-1)^n(\det A)$ for any n.]
14. Let A be an $n \times n$ matrix with two identical columns. Show that det $A = 0$. (Hint: Consider A^t and use Exercise 11.)
15. Let one row of an $n \times n$ matrix A be the sum of two other rows of A. Show that det $A = 0$.
16. Let one column of an $n \times n$ matrix A be the sum of two other columns of A. Show that det $A = 0$. (Hint: Consider A^t and use Exercise 15.)
17. Prove part 1 of Theorem 4.
18. Prove part 2 of Theorem 4 in the case $n = 2$.

2.5 MORE BASIC PROPERTIES OF DETERMINANTS

This section contains proofs of two of the most important properties of determinants:

1. An $n \times n$ matrix A is nonsingular if and only if det $A \neq 0$.
2. If A and B are $n \times n$ matrices, then $\det AB = (\det A)(\det B)$.

Unfortunately, it is not easy to establish either of these results. To begin we define a class of $n \times n$ matrices that are easily obtained from I_n and then note their effect when multiplied by another $n \times n$ matrix.

Any matrix that can be obtained from I_n by using only one elementary row operation is called an **elementary matrix**. The effect of multiplying an elementary matrix by another $n \times n$ matrix can be easily described.

Let E be the elementary matrix obtained from I_n by multiplying its ith row by a nonzero constant c. Then EB is the matrix obtained from B by multiplying its ith row by c. For example, if we multiply the second row of I_3 by c, then we have

$$\begin{bmatrix} 1 & 0 & 0 \\ 0 & c & 0 \\ 0 & 0 & 1 \end{bmatrix} \begin{bmatrix} b_{11} & b_{12} & b_{13} \\ b_{21} & b_{22} & b_{23} \\ b_{31} & b_{32} & b_{33} \end{bmatrix} = \begin{bmatrix} b_{11} & b_{12} & b_{13} \\ cb_{21} & cb_{22} & cb_{23} \\ b_{31} & b_{32} & b_{33} \end{bmatrix}$$

Let E be the elementary matrix obtained from I_n by multiplying its jth row by a nonzero constant c and adding it to the ith row ($i \neq j$). Then EB is the matrix obtained from B by multiplying its jth row by c and adding it to the ith row. For example, if we multiply the second row of I_3 by c and add it to the first row, then we have

$$\begin{bmatrix} 1 & c & 0 \\ 0 & 1 & 0 \\ 0 & 0 & 1 \end{bmatrix} \begin{bmatrix} b_{11} & b_{12} & b_{13} \\ b_{21} & b_{22} & b_{23} \\ b_{31} & b_{32} & b_{33} \end{bmatrix} = \begin{bmatrix} b_{11} + cb_{21} & b_{12} + cb_{22} & b_{13} + cb_{23} \\ b_{21} & b_{22} & b_{23} \\ b_{31} & b_{32} & b_{33} \end{bmatrix}$$

Let E be the elementary matrix obtained from I_n by interchanging its ith and jth rows. Then EB is the matrix obtained from B by interchanging its ith and jth rows.

For example, if the second and third rows of I_3 are interchanged, then we have

$$\begin{bmatrix} 1 & 0 & 0 \\ 0 & 0 & 1 \\ 0 & 1 & 0 \end{bmatrix} \begin{bmatrix} b_{11} & b_{12} & b_{13} \\ b_{21} & b_{22} & b_{23} \\ b_{31} & b_{32} & b_{33} \end{bmatrix} = \begin{bmatrix} b_{11} & b_{12} & b_{13} \\ b_{31} & b_{32} & b_{33} \\ b_{21} & b_{22} & b_{23} \end{bmatrix}$$

The following three theorems describe the fundamental properties of elementary matrices.

Theorem 5.   *Elementary matrices are nonsingular, and the inverse of an elementary matrix is another elementary matrix.*

Proof: Let E be the elementary matrix obtained from I_n by multiplying its ith row by a nonzero constant c. Let F be the elementary matrix obtained from I_n by multiplying its ith row by c^{-1}. Then EF is the matrix obtained from F by multiplying its ith row by c. Therefore $EF = I_n$. By Theorem 18 of Section 1.9 the matrix E is nonsingular and $F = E^{-1}$. The proof for the remaining two types of elementary matrices is left as an exercise. ∎

Theorem 6.

1. *Let E be the elementary matrix obtained by multiplying one of the rows of I_n by the nonzero constant c; then det E = c.*
2. *Let E be the elementary matrix obtained by replacing a row of I_n by the sum of that row and a constant multiple of another row; then det E = 1.*
3. *Let E be the elementary matrix obtained by interchanging two rows of I_n; then det E = −1.*

Proof: Since I_n is a diagonal matrix with a one as each component on its diagonal, we have det $I_n = 1$. Replacing A by I_n in Theorem 4 of Section 2.4 yields the desired results. ∎

Theorem 7.   *Let E be an elementary $n \times n$ matrix and A be an $n \times n$ matrix. Then det EA = det E det A.*

Proof: We shall prove Theorem 7 for the case where E is obtained by multiplying one row of I_n by a nonzero constant c. Exercise 14 asks for proof of the other two cases. Since EA is a matrix obtained from A by multiplying one row by c, part 1 of Theorem 4 assures us that det $EA = c$ det A. From part 1 of Theorem 6 we have det $E = c$. Therefore det $EA = c$ det $A = $ det E det A, which is what we wished to prove. ∎

In learning to solve the matrix equation $A\mathbf{x} = \mathbf{b}$ when A is an $n \times n$ matrix, we found two distinct cases: A was singular or A was nonsingular. The following theorem shows that the determinant of A may be used to distinguish these cases.

Theorem 8.   *Let A be an $n \times n$ matrix. Then A is nonsingular if and only if det $A \neq 0$.*

Proof: By part 7 of Theorem 16 and part 6 of Theorem 17 of Section 1.9 we can use elementary row operations to change A into I_n if A is nonsingular, or into a matrix that has at least one row consisting entirely of zeros if A is singular. Thus there are elementary matrices $E_1, E_2, \ldots, E_k$ such that

(1) $$E_1 E_2 \ldots E_k A = B$$

where $B = I_n$ if A is nonsingular, or B has one row consisting entirely of zeros if A is singular. The determinant of an $n \times n$ matrix with one row consisting entirely of zeros is zero (part 1 of Theorem 4 with $c = 0$). Therefore $\det B \neq 0$ if A is nonsingular and $\det B = 0$ if A is singular. By Theorem 5 each of the E_i is nonsingular and E_i^{-1} is another elementary matrix. Multiplying each side of the equation in (1) by E_1^{-1}, then by E_2^{-1}, $\ldots$, then by E_k^{-1} we find that

$$A = E_k^{-1} \ldots E_2^{-1} E_1^{-1} B$$

Repeated application of Theorem 7 yields

$$\det A = (\det E_k^{-1}) \ldots (\det E_2^{-1})(\det E_1^{-1})(\det B)$$

Since each E_i^{-1} is an elementary matrix, Theorem 6 assures us that $\det E_i^{-1} \neq 0$ for each i. Therefore $\det A \neq 0$ if and only if $\det B \neq 0$. Since $\det B \neq 0$ if and only if A is nonsingular, we conclude that A is nonsingular if and only if $\det A \neq 0$.

Example 1.   Consider the matrix

$$A = \begin{bmatrix} 1 & 2 \\ 3 & 4 \end{bmatrix}$$

Since $\det A = (1)(4) - (2)(3) = -2 \neq 0$, the matrix A is nonsingular. This was also shown in Example 3 of Section 1.8, where we computed A^{-1}, and in Example 1 of Section 1.9, where we showed that the equation $A\mathbf{x} = \mathbf{b}$ has a solution for every 2-vector $\mathbf{b}$.

It is sometimes convenient to have Theorem 8 stated in the following logically equivalent form.

Theorem 9.   *Let A be an $n \times n$ matrix. Then A is singular if and only if $\det A = 0$.*

Example 2.   In Example 2 of Section 2.4 we found that

$$\det \begin{bmatrix} 1 & 2 & 3 \\ 4 & 5 & 6 \\ 7 & 8 & 9 \end{bmatrix} = 0$$

Hence, by Theorem 9, the matrix

$$A = \begin{bmatrix} 1 & 2 & 3 \\ 4 & 5 & 6 \\ 7 & 8 & 9 \end{bmatrix}$$

is singular. This was also shown in Example 5 of Section 1.8, where we showed that A does not have an inverse, and in Example 2 of Section 1.9, where we showed that the equation $A\mathbf{x} = \mathbf{b}$ does not have a solution for every 3-vector $\mathbf{b}$.

Theorem 10.   _If A and B are n × n matrices, then det AB = (det A)(det B)._

Proof: We first consider the case in which A is nonsingular. By Theorem 16 of Section 1.9 the matrix A can be changed into I_n by using elementary row operations. Thus there exist elementary matrices $E_1, E_2, \ldots, E_k$ such that

$$E_1 E_2 \ldots E_k A = I_n$$

By Theorem 5 the inverse of each of the E_i exists so that

$$A = E_k^{-1} \ldots E_2^{-1} E_1^{-1} I_n$$
$$= E_k^{-1} \ldots E_2^{-1} E_1^{-1}$$

Since the inverse of an elementary matrix is another elementary matrix (Theorem 5) we can use Theorem 7 repeatedly to obtain

$$\det AB = \det (E_k^{-1} \ldots E_2^{-1} E_1^{-1} B)$$
$$= (\det E_k^{-1}) \ldots (\det E_2^{-1})(\det E_1^{-1})(\det B)$$
$$= (\det E_k^{-1} \ldots E_2^{-1} E_1^{-1})(\det B)$$
$$= (\det A)(\det B)$$

We next consider the case in which A is singular. Then by Exercise 18 of Section 1.9 the matrix AB is also singular. By Theorem 9 we have

$$\det AB = 0 = 0 \cdot \det B = (\det A)(\det B)$$

We have shown that for any $n \times n$ matrices A and B that det $AB =$ (det A)(det B).   ∎

Example 3.   Consider the matrices

$$A = \begin{bmatrix} 2 & 1 \\ -2 & 3 \end{bmatrix} \quad \text{and} \quad B = \begin{bmatrix} 4 & -2 \\ 5 & 1 \end{bmatrix}$$

Then

$$AB = \begin{bmatrix} 2 & 1 \\ -2 & 3 \end{bmatrix} \begin{bmatrix} 4 & -2 \\ 5 & 1 \end{bmatrix} = \begin{bmatrix} 13 & -3 \\ 7 & 7 \end{bmatrix}$$

and

$$\det A = (2)(3) - (1)(-2) = 8$$
$$\det B = (4)(1) - (-2)(5) = 14$$
$$\det AB = (13)(7) - (-3)(7) = 112$$

Thus det $AB = $ (det A)(det B) as is assured by Theorem 10.

─────────────────── **Exercises** ───────────────────

In Exercises 1–6 use determinants to determine whether the given matrix is nonsingular.

1. $\begin{bmatrix} 2 & 3 \\ 1 & 2 \end{bmatrix}$

2. $\begin{bmatrix} 2 & 4 \\ 4 & 8 \end{bmatrix}$

3. $\begin{bmatrix} 2 & 3 & 4 \\ 2 & 5 & 1 \\ 2 & 4 & 0 \end{bmatrix}$ **4.** $\begin{bmatrix} 2 & 3 & 5 \\ -2 & 1 & 4 \\ 6 & 5 & 6 \end{bmatrix}$

5. $\begin{bmatrix} 1 & 2 & 3 \\ 4 & 5 & 6 \\ 5 & 7 & 9 \end{bmatrix}$ **6.** $\begin{bmatrix} -3 & 1 & 1 \\ 1 & 4 & 1 \\ 1 & 2 & 5 \end{bmatrix}$

In Exercises 7–10 use determinants to determine whether the given equation has a unique solution.

7. $\begin{bmatrix} 1 & 2 & 4 \\ 2 & 5 & 1 \\ 0 & 4 & 1 \end{bmatrix} \mathbf{x} = \begin{bmatrix} 2 \\ 7 \\ 4 \end{bmatrix}$ **8.** $\begin{bmatrix} -2 & 4 & 1 \\ 3 & 2 & 1 \\ -5 & 2 & 0 \end{bmatrix} \mathbf{x} = \begin{bmatrix} 1 \\ 5 \\ 7 \end{bmatrix}$

9. $\begin{bmatrix} 7 & 4 & 2 \\ 6 & -2 & 3 \\ 1 & 6 & -1 \end{bmatrix} \mathbf{x} = \begin{bmatrix} -3 \\ 6 \\ 4 \end{bmatrix}$ **10.** $\begin{bmatrix} 4 & 1 & 1 \\ 2 & 5 & 1 \\ 1 & 0 & 3 \end{bmatrix} \mathbf{x} = \begin{bmatrix} -1 \\ 2 \\ 3 \end{bmatrix}$

11. Let A be a nonsingular matrix. Show that det $A^{-1} = (\det A)^{-1}$.

12. Any $n \times n$ matrix A is called "orthogonal" if $A^{-1} = A^t$. Show that for an orthogonal matrix A, det $A = \pm 1$.

13. Give an example to show that in general det $(A + B) \neq \det A + \det B$.

14. Complete the proof of Theorem 7.

2.6 COFACTOR EXPANSIONS

This section considers another method for evaluating determinants that is not only useful for hand calculations, but also has important theoretical applications.

Definition 3.   Let A be an $n \times n$ matrix. The $(n-1) \times (n-1)$ matrix M_{ij} obtained by deleting the ith row and jth column from A is called the **minor** of a_{ij}. The number $(-1)^{i+j} \det M_{ij}$, denoted by C_{ij}, is called the **cofactor** of a_{ij}.

Example 1.   Consider the matrix

$$A = \begin{bmatrix} 2 & -3 & 1 \\ 4 & -2 & 5 \\ 1 & -1 & 3 \end{bmatrix}$$

The minors of A are

$$M_{11} = \begin{bmatrix} -2 & 5 \\ -1 & 3 \end{bmatrix} \quad M_{12} = \begin{bmatrix} 4 & 5 \\ 1 & 3 \end{bmatrix} \quad M_{13} = \begin{bmatrix} 4 & -2 \\ 1 & -1 \end{bmatrix}$$

$$M_{21} = \begin{bmatrix} -3 & 1 \\ -1 & 3 \end{bmatrix} \quad M_{22} = \begin{bmatrix} 2 & 1 \\ 1 & 3 \end{bmatrix} \quad M_{23} = \begin{bmatrix} 2 & -3 \\ 1 & -1 \end{bmatrix}$$

$$M_{31} = \begin{bmatrix} -3 & 1 \\ -2 & 5 \end{bmatrix} \quad M_{32} = \begin{bmatrix} 2 & 1 \\ 4 & 5 \end{bmatrix} \quad M_{33} = \begin{bmatrix} 2 & -3 \\ 4 & -2 \end{bmatrix}$$

The cofactors of A are

$$C_{11} = (-1)^{1+1} \det M_{11} \quad C_{12} = (-1)^{1+2} \det M_{12} \quad C_{13} = (-1)^{1+3} \det M_{13}$$
$$= -1 \qquad\qquad\qquad = -7 \qquad\qquad\qquad = -2$$

$$C_{21} = (-1)^{2+1} \det M_{21} \quad C_{22} = (-1)^{2+2} \det M_{22} \quad C_{23} = (-1)^{2+3} \det M_{23}$$
$$= 8 \qquad\qquad\qquad = 5 \qquad\qquad\qquad = -1$$

$$C_{31} = (-1)^{3+1} \det M_{31} \quad C_{32} = (-1)^{3+2} \det M_{32} \quad C_{33} = (-1)^{3+3} \det M_{33}$$
$$= -13 \qquad\qquad\qquad = -6 \qquad\qquad\qquad = 8$$

From the definition of determinant of an $n \times n$ matrix $A = (a_{ij})$ (Definition 2 in Section 2.1) we see that each summand contains one component from each row and each column of A. If we focus our attention on the first row, we can write the determinant of A as

$$\det A = A_{11}a_{11} + A_{12}a_{12} + \cdots + A_{1n}a_{1n} \tag{1}$$

where $A_{11}a_{11}, A_{12}a_{12}, \ldots, A_{1n}a_{1n}$ are obtained by grouping all terms which contain $a_{11}, a_{12}, \ldots, a_{1n}$, respectively. We want to find a simple way to express the numbers $A_{11}, A_{12}, \ldots, A_{1n}$.

We begin by considering the case $n = 2$. Then

$$\det \begin{bmatrix} a_{11} & a_{12} \\ a_{21} & a_{22} \end{bmatrix} = a_{11}a_{22} - a_{12}a_{21}$$

Since the determinant of a 1×1 matrix $[b_{11}]$ is simply b_{11}, we have

$$C_{11} = (-1)^{1+1} a_{22} = a_{22}$$
$$C_{12} = (-1)^{1+2} a_{21} = -a_{21}$$

so that

$$\det A = C_{11}a_{11} + C_{12}a_{12}$$

When $n = 3$ we have (see Section 2.1)

$$\det \begin{bmatrix} a_{11} & a_{12} & a_{13} \\ a_{21} & a_{22} & a_{23} \\ a_{31} & a_{32} & a_{33} \end{bmatrix} = a_{11}a_{22}a_{33} + a_{13}a_{21}a_{32}$$
$$+ a_{12}a_{23}a_{31} - a_{11}a_{23}a_{32}$$
$$- a_{12}a_{21}a_{33} - a_{13}a_{22}a_{31}$$
$$= a_{11}(a_{22}a_{33} - a_{23}a_{32})$$
$$+ a_{12}(a_{23}a_{31} - a_{21}a_{33})$$
$$+ a_{13}(a_{21}a_{32} - a_{22}a_{31})$$

Notice that

$$C_{11} = (-1)^{1+1} \det \begin{bmatrix} a_{22} & a_{23} \\ a_{32} & a_{33} \end{bmatrix} = a_{22}a_{33} - a_{23}a_{32}$$

$$C_{12} = (-1)^{1+2} \det \begin{bmatrix} a_{21} & a_{23} \\ a_{31} & a_{33} \end{bmatrix} = a_{23}a_{31} - a_{21}a_{33}$$

$$C_{13} = (-1)^{1+3} \det \begin{bmatrix} a_{21} & a_{22} \\ a_{31} & a_{32} \end{bmatrix} = a_{21}a_{32} - a_{22}a_{31}$$

so that

$$\det A = C_{11}a_{11} + C_{12}a_{12} + C_{13}a_{13}$$

The cases $n = 2$ and $n = 3$ lead us to conjecture that for any n the coefficient A_{1j} of a_{1j} in equation (1) equals C_{1j}. The following theorem, which is given without proof, assures us that our conjecture is true. Moreover, it does not matter which row or column we use to obtain an expression for the determinant of A.

Theorem 11. *The determinant of an $n \times n$ matrix $A = (a_{ij})$ is the sum of the products of the components in any row (or column) by their cofactors. That is, for each i and j, $1 \le i \le n$ and $1 \le j \le n$,*

(2) $$\det A = C_{i1}a_{i1} + C_{i2}a_{i2} + \cdots + C_{in}a_{in}$$

and

(3) $$\det A = C_{1j}a_{1j} + C_{2j}a_{2j} + \cdots + C_{nj}a_{nj}$$

The summation in (2) is called the **cofactor expansion of det A along the ith row**. Similarly, the summation in (3) is called the **cofactor expansion of det A along the jth column**.

Example 2.   Consider the matrix

$$A = \begin{bmatrix} 2 & -3 & 1 \\ 4 & -2 & 5 \\ 1 & -1 & 3 \end{bmatrix}$$

The cofactor expansion of det A along the second row is

$$\det A = 4C_{21} + (-2)C_{22} + 5C_{23}$$

Using Example 1 we find

$$\det A = 4(8) - 2(5) + 5(-1) = 17$$

The cofactor expansion of det A along the third column is

$$\det A = 1C_{13} + 5C_{23} + 3C_{33}$$
$$= 1(-2) + 5(-1) + 3(8) = 17$$

Example 3.   It is frequently convenient to use both elementary row operations and cofactor expansions when the determinant of an $n \times n$ matrix, $n \ge 4$, is evaluated by using hand calculations. For example,

$$\det\begin{bmatrix} -2 & 6 & -4 & 1 \\ 1 & -2 & 3 & 4 \\ 3 & 1 & 2 & 3 \\ -2 & 1 & -1 & 1 \end{bmatrix} = \det\begin{bmatrix} 0 & 2 & 2 & 9 \\ 1 & -2 & 3 & 4 \\ 0 & 7 & -7 & -9 \\ 0 & -3 & 5 & 9 \end{bmatrix}$$

$$\begin{array}{l} 2R_2 + R_1 \rightarrow R_1 \\ (-3)R_2 + R_3 \rightarrow R_3 \\ 2R_2 + R_4 \rightarrow R_4 \end{array}$$

$$= -\det\begin{bmatrix} 2 & 2 & 9 \\ 7 & -7 & -9 \\ -3 & 5 & 9 \end{bmatrix}$$

Cofactor expansion along the first column. (Note the new matrix is M_{21}.)

$$= -\det\begin{bmatrix} 2 & 2 & 9 \\ 9 & -5 & 0 \\ -5 & 3 & 0 \end{bmatrix}$$

$$\begin{array}{l} R_1 + R_2 \rightarrow R_2 \\ (-1)R_1 + R_3 \rightarrow R_3 \end{array}$$

$$= -9\det\begin{bmatrix} 9 & -5 \\ -5 & 3 \end{bmatrix}$$

Cofactor expansion along the third column.

$$= (-9)(\overset{.}{2}7 - 25) = -18$$

Exercises

In Exercises 1–6 find the cofactor of the given component of the matrix

$$\begin{bmatrix} 1 & 0 & 4 & 2 \\ 3 & 5 & 9 & -2 \\ 7 & -1 & -3 & 8 \\ -5 & -8 & 6 & -4 \end{bmatrix}$$

1. 4   **2.** −3
3. 3   **4.** 8
5. 7   **6.** 9

In Exercises 7–14 use cofactor expansions to evaluate the determinant of the given matrix.

7. $\begin{bmatrix} 3 & 4 \\ 2 & 5 \end{bmatrix}$

8. $\begin{bmatrix} 6 & 7 \\ -2 & -3 \end{bmatrix}$

9. $\begin{bmatrix} 2 & 1 & 3 \\ -6 & 0 & 4 \\ 7 & 0 & 9 \end{bmatrix}$

10. $\begin{bmatrix} 7 & 8 & 8 \\ 2 & 0 & -1 \\ 0 & 7 & 0 \end{bmatrix}$

11. $\begin{bmatrix} 2 & 7 & 5 \\ 1 & 4 & 2 \\ 8 & 5 & 3 \end{bmatrix}$

12. $\begin{bmatrix} -2 & -2 & 3 \\ 2 & -6 & -4 \\ -3 & 4 & -5 \end{bmatrix}$

13. $\begin{bmatrix} 3 & 2 & 0 & 0 & 4 \\ 1 & 5 & 0 & 0 & 5 \\ 6 & 3 & 5 & 5 & 4 \\ 0 & 2 & 0 & 0 & 0 \\ 1 & 3 & 7 & 0 & 1 \end{bmatrix}$

14. $\begin{bmatrix} 3 & 1 & 0 & 0 & 0 \\ 1 & 3 & 1 & 0 & 0 \\ 0 & 1 & 3 & 1 & 0 \\ 0 & 1 & 3 & 1 & 0 \\ 0 & 0 & 0 & 3 & 3 \end{bmatrix}$

In Exercises 15–18 evaluate the determinant of the given matrix.

15. $\begin{bmatrix} 1 & 2 & 3 & 4 \\ 5 & -2 & 7 & 4 \\ 0 & 1 & 0 & 3 \\ 3 & 4 & 5 & 6 \end{bmatrix}$
16. $\begin{bmatrix} -2 & -4 & -6 & 8 \\ 3 & 1 & 2 & 0 \\ 0 & -4 & 2 & 1 \\ 5 & 0 & 2 & 3 \end{bmatrix}$

17. $\begin{bmatrix} 1 & -1 & 1 & -1 & 1 \\ -1 & 1 & 1 & -1 & 1 \\ 1 & -1 & -1 & 1 & 1 \\ 1 & 1 & 1 & 1 & 1 \\ -1 & 1 & 1 & 1 & -1 \end{bmatrix}$
18. $\begin{bmatrix} 1 & 2 & 1 & 0 & 1 \\ 2 & 1 & 1 & 1 & 1 \\ 0 & 1 & 2 & 1 & 1 \\ 1 & 2 & 1 & 1 & 1 \\ 2 & 0 & 1 & 2 & 1 \end{bmatrix}$

19. Let $a_1, a_2, \ldots, a_n$ be numbers that are not necessarily distinct and consider the matrix

$$A_n = \begin{bmatrix} 1 & 1 & \cdots & 1 \\ a_1 & a_2 & \cdots & a_n \\ a_1^2 & a_2^2 & \cdots & a_n^2 \\ \vdots & \vdots & & \vdots \\ a_1^{n-1} & a_2^{n-1} & \cdots & a_n^{n-1} \end{bmatrix}$$

Show that

(1)
$$\det A_n = \prod_{1 \leq i < j \leq n} (a_j - a_i)$$

where the symbol Π stands for "product." For example,

$$\prod_{1 \leq i < j \leq 3} (a_j - a_i) = (a_3 - a_2)(a_3 - a_1)(a_2 - a_1)$$

The identity in (1) can be proved by induction as follows:

(a) Show that the identity in (1) holds when $n = 2$.
(b) Now assume that

$$\det A_k = \prod_{1 \leq i < j \leq k} (a_j - a_i)$$

where k is an integer greater than or equal to 2. We want to use this identity to establish the identity in (1) when $n = k + 1$. Consider the function

$$F(x) = \det \begin{bmatrix} 1 & 1 & & 1 & 1 \\ a_1 & a_2 & \cdots & a_k & x \\ a_1^2 & a_2^2 & \cdots & a_k^2 & x^2 \\ \vdots & \vdots & & \vdots & \vdots \\ a_1^k & a_2^k & \cdots & a_k^k & x^k \end{bmatrix}$$

Show that $F(x)$ is a polynomial of degree k with det A_k as the coefficient of x^k.

(c) Show that $F(a_i) = 0$ for $i = 1, 2,..., k$.

(d) Show that $F(x) = (\det A_k) \prod_{i=1}^{k} (x - a_i)$.

(e) Show that det $A_{k+1} = F(a_{k+1})$.

(f) Show that det $A_{k+1} = \prod_{1 \leqslant i < j \leqslant k+1} (a_j - a_i)$.

This completes the proof that the identity in (1) holds for all n. The determinant of A_n is known as the **Vandermonde* determinant**.

2.7 CRAMER'S RULE

This section presents alternate methods of computing the inverse of a nonsingular matrix A and of solving the equation $A\mathbf{x} = \mathbf{b}$. Even though these methods are not computationally efficient, they are of theoretic interest because they allow us to obtain properties of A^{-1} and $A^{-1}\mathbf{b}$ without actually computing A^{-1} or $A^{-1}\mathbf{b}$.

Definition 4. If $A = (a_{ij})$ is an $n \times n$ matrix and C_{ij} is the cofactor of a_{ij}, then the matrix

$$\begin{bmatrix} C_{11} & C_{12} & \cdots & C_{1n} \\ C_{21} & C_{22} & \cdots & C_{2n} \\ \vdots & \vdots & & \vdots \\ C_{n1} & C_{n2} & \cdots & C_{nn} \end{bmatrix}$$

is called the **matrix of cofactors** of A. The transpose of this matrix is called the **adjoint** of A and is denoted by adj A.

Example 1. In Example 1 of Section 2.6 we found that the cofactors of

$$A = \begin{bmatrix} 2 & -3 & 1 \\ 4 & -2 & 5 \\ 1 & -1 & 3 \end{bmatrix}$$

are

$$C_{11} = -1 \qquad C_{12} = -7 \qquad C_{13} = -2$$
$$C_{21} = 8 \qquad C_{22} = 5 \qquad C_{23} = -1$$
$$C_{31} = -13 \qquad C_{32} = -6 \qquad C_{33} = 8$$

so that the matrix of cofactors of A is

$$\begin{bmatrix} -1 & -7 & -2 \\ 8 & 5 & -1 \\ -13 & -6 & 8 \end{bmatrix}$$

---

* Named for A. T. Vandermonde (1735–1796), a French mathematician.

Taking the transpose of this matrix we obtain the adjoint of A

$$\text{adj } A = \begin{bmatrix} -1 & 8 & -13 \\ -7 & 5 & -6 \\ -2 & -1 & 8 \end{bmatrix}$$

Example 2. Let

$$A = \begin{bmatrix} a & b \\ c & d \end{bmatrix}$$

The cofactors of A are

$$C_{11} = d \qquad C_{12} = -c$$
$$C_{21} = -b \qquad C_{22} = a$$

so that the adjoint of A is given by

$$\text{adj } A = \begin{bmatrix} d & -c \\ -b & a \end{bmatrix}^t$$

$$= \begin{bmatrix} d & -b \\ -c & a \end{bmatrix}$$

In Theorem 12 of Section 1.8 we found that if A is nonsingular, then

$$A^{-1} = \frac{1}{ad - bc} \begin{bmatrix} d & -b \\ -c & a \end{bmatrix}$$

Since det $A = ad - bc$, we have

(1)
$$A^{-1} = \frac{1}{\det A} \text{ adj } A$$

The following theorems show that the formula in equation (1) holds whenever A is any nonsingular matrix.

Theorem 12. *If A is a nonsingular $n \times n$ matrix, then*

$$A^{-1} = \frac{1}{\det A} \text{ adj } A$$

Proof: We begin by showing that A adj $A = (\det A)I_n$. This is done by considering the product A adj A:

$$A \text{ adj } A = \begin{bmatrix} a_{11} & a_{12} & \cdots & a_{1n} \\ a_{21} & a_{22} & \cdots & a_{2n} \\ \vdots & \vdots & & \vdots \\ a_{n1} & a_{n2} & \cdots & a_{nn} \end{bmatrix} \begin{bmatrix} C_{11} & C_{21} & \cdots & C_{n1} \\ C_{12} & C_{22} & \cdots & C_{n2} \\ \vdots & \vdots & & \vdots \\ C_{1n} & C_{2n} & \cdots & C_{nn} \end{bmatrix}$$

The *ij*-component of A adj A is

$$a_{i1}C_{j1} + a_{i2}C_{j2} + \cdots + a_{in}C_{jn} \tag{2}$$

If $i = j$, then this component is the cofactor expansion along the *i*th row for det A. The reader is asked to show in Exercise 13 that this component is zero whenever $i \neq j$. Therefore,

$$A \text{ adj } A = \begin{bmatrix} \det A & 0 & 0 & \cdots & 0 \\ 0 & \det A & 0 & \cdots & 0 \\ 0 & 0 & \det A & \cdots & 0 \\ \vdots & \vdots & \vdots & \ddots & \vdots \\ 0 & 0 & 0 & \cdots & \det A \end{bmatrix} = (\det A)I_n$$

Multiplying each side of this equation by A^{-1} yields

$$\text{adj } A = A^{-1}(\det A)I_n = (\det A)A^{-1}I_n = (\det A)A^{-1}$$

Since A is nonsingular we have det $A \neq 0$ (Theorem 8 of Section 2.5) so that

$$\frac{1}{\det A} \text{ adj } A = A^{-1}$$

This completes the proof. ∎

Example 3. Consider the matrix

$$A = \begin{bmatrix} 2 & -3 & 1 \\ 4 & -2 & 5 \\ 1 & -1 & 3 \end{bmatrix}$$

In Example 2 of the previous section and Example 1 of this section we found that det $A = 17$ and that

$$\text{adj } A = \begin{bmatrix} -1 & 8 & -13 \\ -7 & 5 & -6 \\ -2 & -1 & 8 \end{bmatrix}$$

Therefore

$$A^{-1} = \frac{1}{17} \begin{bmatrix} -1 & 8 & -13 \\ -7 & 5 & -6 \\ -2 & -1 & 8 \end{bmatrix}$$

Let A be an $n \times n$ nonsingular matrix. The solution $\mathbf{x}$ of the matrix equation $A\mathbf{x} = \mathbf{b}$, which represents a system of n linear equations in n unknowns, can be written as

$$\mathbf{x} = A^{-1}\mathbf{b}$$

In light of Theorem 12 this solution may be rewritten as

$$\mathbf{x} = \frac{1}{\det A}(\text{adj } A)\mathbf{b}$$

We will use this formula to obtain a formula for the components of the solution **x**. The jth component, x_j, of **x** is $(\det A)^{-1}$ times the jth component of $(\text{adj } A)\mathbf{b}$. If C_{ij} is the cofactor of a_{ij}, then

$$(\text{adj } A)\mathbf{b} = \begin{bmatrix} C_{11} & C_{21} & \cdots & C_{n1} \\ C_{12} & C_{22} & \cdots & C_{n2} \\ \vdots & \vdots & & \vdots \\ C_{1n} & C_{2n} & \cdots & C_{nn} \end{bmatrix} \begin{bmatrix} b_1 \\ b_2 \\ \vdots \\ b_n \end{bmatrix}$$

Therefore, the jth component, $[(\text{adj } A)\mathbf{b}]_j$, of $(\text{adj } A)\mathbf{b}$ can be computed as follows:

(3) $$[(\text{adj } A)\mathbf{b}]_j = C_{1j}b_1 + C_{2j}b_2 + \cdots + C_{nj}b_n$$

We now consider the $n \times n$ matrix obtained from A by replacing the jth column of A by **b**:

$$A_j = \begin{matrix} & & & \overset{\displaystyle j\text{th}}{\underset{\displaystyle \text{column}}{\ }} & \\ \begin{bmatrix} a_{11} & a_{12} & \cdots & b_1 & \cdots & a_{1n} \\ a_{21} & a_{22} & \cdots & b_2 & \cdots & a_{2n} \\ \vdots & \vdots & & \vdots & & \vdots \\ a_{n1} & a_{n2} & \cdots & b_n & \cdots & a_{nn} \end{bmatrix} \end{matrix}$$

If we compute the determinant of A_j by using the cofactor expansion along the jth column (which is **b**), we find that

$$\det A_j = C_{1j}b_1 + C_{2j}b_2 + \cdots + C_{nj}b_n$$

Using equation (3) we conclude that the jth component, x_j, of the solution **x** of $A\mathbf{x} = \mathbf{b}$ is given by

$$x_j = \frac{\det A_j}{\det A}$$

We have proved the following theorem that is known as **Cramer's rule**.

Theorem 13. *Let A be an $n \times n$ nonsingular matrix. The components of the solutions **x** of $A\mathbf{x} = \mathbf{b}$ are given by*

$$x_1 = \frac{\det A_1}{\det A}, \qquad x_2 = \frac{\det A_2}{\det A}, \ldots, x_n = \frac{\det A_n}{\det A}$$

*where A_j is the matrix obtained from A by replacing the jth column of A by **b**.*

Example 4.   Consider the system of equations

$$2x_1 + x_2 - x_3 = 4$$
$$x_1 - 2x_2 + x_3 = 1$$
$$3x_1 - 3x_2 + 2x_3 = 0$$

Then

$$A = \begin{bmatrix} 2 & 1 & -1 \\ 1 & -2 & 1 \\ 3 & -3 & 2 \end{bmatrix} \qquad \mathbf{x} = \begin{bmatrix} x_1 \\ x_2 \\ x_3 \end{bmatrix} \qquad \mathbf{b} = \begin{bmatrix} 4 \\ 1 \\ 0 \end{bmatrix}$$

We know that A is nonsingular (because det $A = -4$). Therefore, we can use Cramer's rule. Since

$$A_1 = \begin{bmatrix} 4 & 1 & -1 \\ 1 & -2 & 1 \\ 0 & -3 & 2 \end{bmatrix} \qquad A_2 = \begin{bmatrix} 2 & 4 & -1 \\ 1 & 1 & 1 \\ 3 & 0 & 2 \end{bmatrix} \qquad A_3 = \begin{bmatrix} 2 & 1 & 4 \\ 1 & -2 & 1 \\ 3 & -3 & 0 \end{bmatrix}$$

then

$$x_1 = \frac{\det A_1}{\det A} = \frac{3}{4} \qquad x_2 = \frac{\det A_2}{\det A} = -\frac{11}{4} \qquad x_3 = \frac{\det A_3}{\det A} = -\frac{21}{4}$$

For $n \times n$ nonsingular matrices A with $n \geq 3$, the methods of computing A^{-1} given in Section 1.8 and of obtaining the solution of $A\mathbf{x} = \mathbf{b}$ given in Section 1.2 are more efficient than those given in Theorems 12 and 13 in this section. Nonetheless, these theorems have theoretical importance because they can be used to obtain properties of A^{-1} or the solution of $A\mathbf{x} = \mathbf{b}$ without actually computing A^{-1} or the solution. For example, suppose that the components of A and $\mathbf{b}$ are integers. Then det A_j is an integer for every j because the components of A_j are integers. From Theorem 13 we conclude that the components of the solution of $A\mathbf{x} = \mathbf{b}$ are integers whenever det $A = \pm 1$.

We cannot overemphasize the fact that Theorems 12 and 13 are theoretically useful, even though they are not efficient for many computations.

Exercises

In Exercises 1–4 compute the inverse of the given matrix by using Theorem 12.

1. $\begin{bmatrix} 1 & 2 \\ 4 & 3 \end{bmatrix}$

2. $\begin{bmatrix} 2 & 9 \\ -7 & 4 \end{bmatrix}$

3. $\begin{bmatrix} 1 & 2 & -1 \\ 2 & 3 & 4 \\ 0 & 1 & 5 \end{bmatrix}$

4. $\begin{bmatrix} 3 & 1 & 1 \\ 1 & 4 & 2 \\ 1 & 2 & 5 \end{bmatrix}$

5. Compute the inverse of the following matrix by using Theorem 12 and then compute the inverse by using the method in Section 1.8.

$$\begin{bmatrix} 1 & 1 & 1 & 1 \\ 1 & 2 & 1 & 1 \\ 1 & 1 & 2 & 1 \\ 1 & 1 & 1 & 2 \end{bmatrix}$$

In Exercises 6–9 find the solution of the given equations by using Cramer's rule.

6. $\begin{bmatrix} 1 & 2 \\ 4 & 3 \end{bmatrix} \mathbf{x} = \begin{bmatrix} 1 \\ 2 \end{bmatrix}$

7. $\begin{bmatrix} 2 & 9 \\ -7 & 4 \end{bmatrix} \mathbf{x} = \begin{bmatrix} 0 \\ 1 \end{bmatrix}$

8. $\begin{bmatrix} 1 & 2 & -1 \\ 2 & 3 & 4 \\ 0 & 1 & 5 \end{bmatrix} \mathbf{x} = \begin{bmatrix} -1 \\ 2 \\ 0 \end{bmatrix}$
   9. $\begin{bmatrix} 3 & 1 & 1 \\ 1 & 4 & 2 \\ 1 & 2 & 5 \end{bmatrix} \mathbf{x} = \begin{bmatrix} 0 \\ -1 \\ 0 \end{bmatrix}$

10. Find the solution of the following equation by using Cramer's rule and then find the solution by using Gaussian elimination.

$$\begin{bmatrix} 1 & 1 & 1 & 1 \\ 1 & 2 & 1 & 1 \\ 1 & 1 & 2 & 1 \\ 1 & 1 & 1 & 2 \end{bmatrix} \mathbf{x} = \begin{bmatrix} 1 \\ 2 \\ 3 \\ 4 \end{bmatrix}$$

11. Let A be an $n \times n$ matrix whose components are integers. Show that if det $A = 1$, then the components of A^{-1} are integers.

12. Show that if A is nonsingular, then

$$(\text{adj } A)^{-1} = \text{adj } A^{-1}$$

[Hint: Show that $(\text{adj } A)^{-1} = (\det A)^{-1}A$. Replace A by A^{-1} in Theorem 12 and use Exercise 11 of Section 2.5 to show that $A = \det A$ adj A^{-1}.]

13. Show that the ij-component of A adj A given in formula (2) is zero whenever $i \neq j$. This may be done as follows:

(a) Let B be the matrix formed by replacing the jth row of A by the ith row so that the ith and jth rows of B are identical. Show that the sum given by formula (2) is the cofactor expansion for det B along the jth row. That is, show that det B is the ij-component of A adj A.

(b) Show that the ij-component of A adj A is zero by showing that det $B = 0$.

3

Vectors in 2-Space
and 3-Space

3.1 VECTORS IN 2-SPACE

Many quantities of importance in the physical sciences cannot be completely described by a single number representing the magnitude of the quantity under consideration. For example, suppose that we apply a force to an object. In order to describe the effect of this force, we must do more than give the magnitude of the force. We must also give the direction in which the force acts.

Displacement is another example of a quantity that is not completely described by its magnitude. If a person starts at the intersection of two city streets and walks one block, we do not know where the person is unless we know in which direction he walked.

Intuitively a vector is a quantity having **magnitude** or **length** as well as **direction**. In elementary situations, these quantities can be represented in either two-dimensional space (which we call **2-space**) or three-dimensional space (**3-space**). This chapter begins by discussing two ways in which the concept of vectors in 2-space can be made precise and continues with the same type of discussion for vectors in 3-space.

Let $A = (a_1, a_2)$ and $B = (b_1, b_2)$ be two points in a plane (2-space) described by Cartesian coordinates. The **directed line segment** from A to B, denoted by $\overrightarrow{AB}$, is the straight line segment extending from A to B (see Figure 3.1). The point A of the directed line segment $\overrightarrow{AB}$ is called the **initial point** of $\overrightarrow{AB}$, and the point B is called the **terminal point** of $\overrightarrow{AB}$. The **direction** of $\overrightarrow{AB}$ is the angle measured in radians that

FIGURE 3.1

FIGURE 3.2

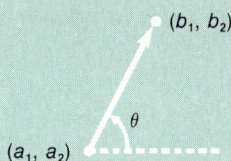

the directed line segment makes with a ray emanating to the right of A along the line $y = a_2$ (see Figure 3.2). It is traditional to restrict θ so that $0 \leq \theta < 2\pi$.

Notice that the directed line segment $\overrightarrow{BA}$ consists of the same line segment as $\overrightarrow{AB}$ but has the opposite direction (see Figure 3.3). By **opposite direction** we mean that the directions of the two directed line segments differ by either π or $-\pi$. That is, $\theta - \phi = \pi$ or $\theta - \phi = -\pi$.

Since $A = (a_1, a_2)$ and $B = (b_1, b_2)$, then the direction θ of $\overrightarrow{AB}$ (see Figure 3.4) satisfies the following equation:

$$\tan \theta = \frac{b_2 - a_2}{b_1 - a_1}$$

whenever $b_1 \neq a_1$. If $b_1 = a_1$ and $b_2 > a_2$, then $\overrightarrow{AB}$ is a directed line segment in the direction of the positive vertical axis. Hence, in such a case the direction of $\overrightarrow{AB}$ is $\pi/2$. Similarly if $b_1 = a_1$ and $b_2 < a_2$, then $\overrightarrow{AB}$ is a directed line segment in the direction of the negative vertical axis and has direction $3\pi/2$. If $b_1 = a_1$ and $b_2 = a_2$, then $A = B$ and we shall assign no direction to $\overrightarrow{AB}$.

The **length** of $\overrightarrow{AB}$ is the length of the line segment from A to B. That is, if $A = (a_1, a_2)$ and $B = (b_1, b_2)$, then the length of $\overrightarrow{AB}$, denoted by $|\overrightarrow{AB}|$, is given by

$$|\overrightarrow{AB}| = \sqrt{(b_1 - a_1)^2 + (b_2 - a_2)^2}$$

or

$$|\overrightarrow{AB}| = \sqrt{(a_1 - b_1)^2 + (a_2 - b_2)^2}$$

FIGURE 3.3

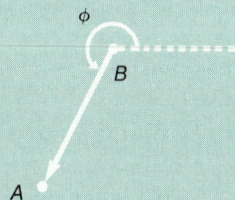

FIGURE 3.4

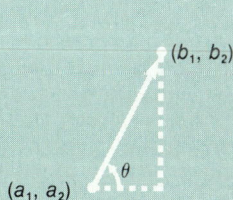

FIGURE 3.5

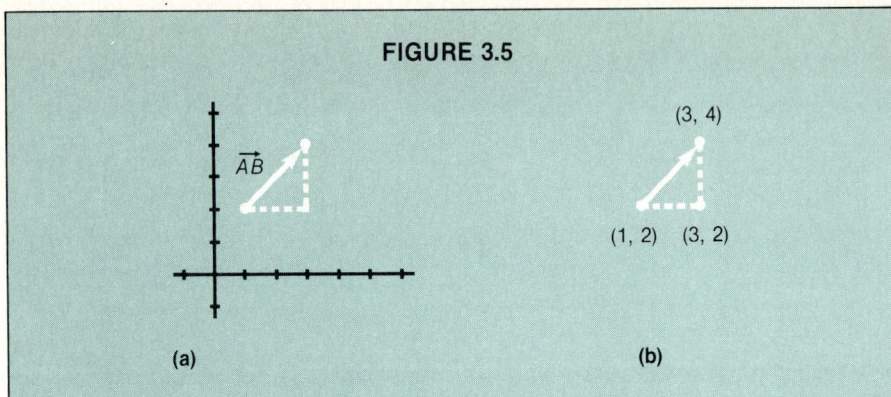

(a)   (b)

Example 1. Find the direction and length of the directed line segment $\overrightarrow{AB}$ where $A = (1, 2)$ and $B = (3, 4)$, as in Figure 3.5.

Since $\tan \theta = 1$ and $0 \le \theta \le \pi/2$, the direction of $\overrightarrow{AB}$ is $\pi/4$. Moreover,

$$|\overrightarrow{AB}| = \sqrt{(1-3)^2 + (2-4)^2}$$
$$= 2\sqrt{2}$$

Example 2. Find the direction and length of the directed line segment $\overrightarrow{AB}$ where $A = (-2, 1)$ and $B = (1, 1 - 3\sqrt{3})$, as in Figure 3.6.

Since $\tan \theta = -\sqrt{3}$ and $3\pi/2 \le \theta < 2\pi$, the direction of $\overrightarrow{AB}$ is $5\pi/3$. Moreover,

$$|AB| = \sqrt{[1-(-2)]^2 + [(1 - 3\sqrt{3}) - 1]^2}$$
$$= 6$$

The trouble with describing a directed line segment by its initial end $A = (a_1, a_2)$ and its terminal end $B = (b_1, b_2)$ is that the two crucial pieces of information, direction and length, are "hidden." Rather than writing the terminal end as (b_1, b_2) it is usually more convenient to write it as $(a_1 + h, a_2 + k)$ where $h = b_1 - a_1$ and

FIGURE 3.6

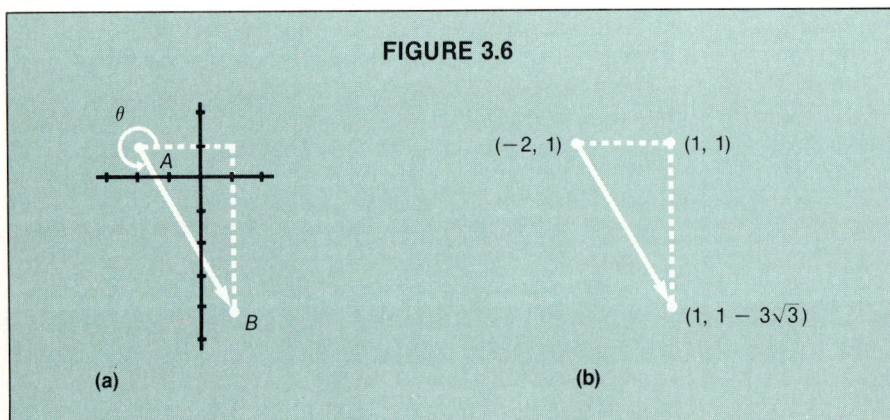

(a)   (b)

$k = b_2 - a_2$. From this form the direction and length are more apparent. Thus a directed line segment can be described by two ordered pairs of numbers:

1. (a_1, a_2), indicating the **initial point** of the directed line segment, and
2. (h, k), indicating the **displacement** of the terminal end from the initial end.

When directed line segments arise in practice, the initial end is usually known. In such a case the directed line segment is completely determined by the displacement pair of numbers (h, k), called the **components** of the vector. We shall call the 2-vector

$$\begin{bmatrix} h \\ k \end{bmatrix}$$

the **vector** corresponding to the displacement pair of numbers (h, k). Thus each displacement pair of numbers, and hence each directed line segment in a plane, is associated with a 2-vector. Since directed line segments with the same length and same direction determine the same displacement pair of numbers, these directed line segments are associated with the same 2-vector. For example, all of the directed line segments in Figure 3.7 are associated with the same 2-vector.

Frequently we shall refer to 2-vectors simply as **vectors**. Moreover, since 2-vectors can be represented geometrically by directed line segments in a plane, we shall also refer to them as **vectors in 2-space** whenever we wish to emphasize their geometric interpretation.

The idea of having two or more distinct items identified with a single item is not uncommon in mathematics or elsewhere. For example, in mechanics it is common to treat an object as though it were a single point, its center of mass. Likewise, the symbols 1/2, 2/4, 18/36, and 207/414 are all different; yet when considered as rational numbers, they represent the number 1/2. Similarly, a vector can be thought of as being represented by any directed line segment that has a given length and given direction. Any such directed line segment is called a **representation** of the vector. In particular, any directed line segment with the same length and direction as the directed line segment from the origin to the point (h, k) is a representation of the vector

$$\begin{bmatrix} h \\ k \end{bmatrix}$$

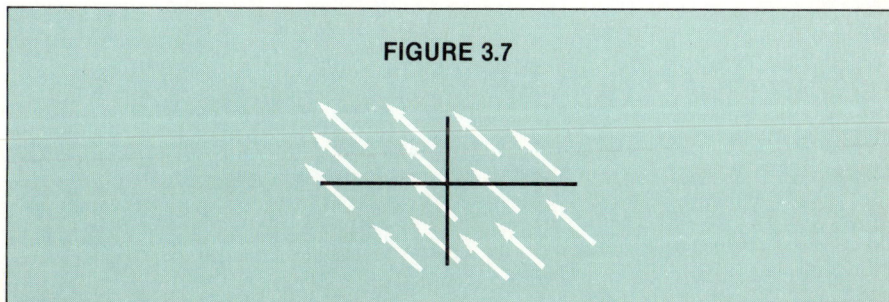

FIGURE 3.7

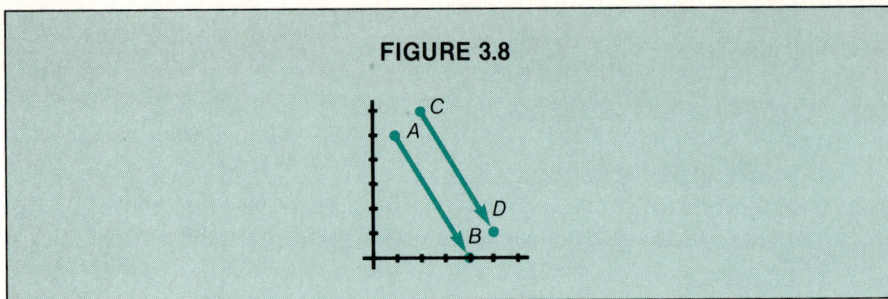

FIGURE 3.8

Example 3.   Consider the directed line segments $\overrightarrow{AB}$ and $\overrightarrow{CD}$ where $A = (1, 5)$, $B = (4, 0)$, $C = (2, 6)$, and $D = (5, 1)$, as in Figure 3.8. These directed line segments have the same length and direction. The displacement pair of numbers is $(3, -5)$ for each directed line segment. Therefore they are both representations of the vector

$$\begin{bmatrix} 3 \\ -5 \end{bmatrix}$$

─────────── Exercises ───────────

In Exercises 1–10 sketch the directed line segment $\overrightarrow{AB}$; determine its length; and find the vector having $\overrightarrow{AB}$ as a representation.

1. $A = (2, 4)$; $B = (5, 7)$.

2. $A = (-2, 1)$; $B = (-5, -2)$.

3. $A = (2, -\sqrt{3})$; $B = (3, 1 - \sqrt{3})$

4. $A = (5, 6)$; $B = (7, 6 + 2\sqrt{3})$.

5. $A = (2, 5)$; $B = (2, 8)$.

6. $A = (3, -2)$; $B = (7, -2)$.

7. $A = (2, 8)$; $B = (2, 5)$.

8. $A = (7, -2)$; $B = (3, -2)$.

9. $A = (1, 2)$; $B = (7, 11)$.

10. $A = (-2, 5)$; $B = (6, 8)$.

In Exercises 11–22 determine the direction and the length of any representation of the given vector.

11. $\begin{bmatrix} 2 \\ -2 \end{bmatrix}$

12. $\begin{bmatrix} 3 \\ 1 \end{bmatrix}$

13. $\begin{bmatrix} -\sqrt{7} \\ -\sqrt{7} \end{bmatrix}$

14. $\begin{bmatrix} \cos \dfrac{\pi}{5} \\ \sin \dfrac{\pi}{5} \end{bmatrix}$

15. $\begin{bmatrix} \sin \dfrac{\pi}{7} \\ \cos \dfrac{\pi}{7} \end{bmatrix}$

16. $\begin{bmatrix} \cos 1 \\ \sin 1 \end{bmatrix}$

17. $\begin{bmatrix} \sin .5 \\ \cos .5 \end{bmatrix}$

18. $\begin{bmatrix} 4 \\ 3 \end{bmatrix}$

19. $\begin{bmatrix} 3 \\ 4 \end{bmatrix}$

20. $\begin{bmatrix} 2 \\ -5 \end{bmatrix}$

21. $\begin{bmatrix} -3 \\ -2 \end{bmatrix}$

22. $\begin{bmatrix} -7 \\ 10 \end{bmatrix}$

3.2 SUMS, SCALAR PRODUCTS, AND NORMS OF VECTORS IN 2-SPACE

Since vectors in 2-space can be considered to be objects represented by directed line segments, it is natural that the "length" of a vector is a concept of interest. Anticipating future terminology and notation, we shall use the word "norm" instead of the word "length" and define the **norm** of a vector $\mathbf{v}$, denoted by $\|\mathbf{v}\|$, to be the length of any representation of $\mathbf{v}$. That is, if $\mathbf{v} = \begin{bmatrix} a \\ b \end{bmatrix}$ (as in Figure 3.9), then

$$\|\mathbf{v}\| = \sqrt{a^2 + b^2}$$

A vector with norm 1 is called a **unit vector**.

Example 1.   Let $\mathbf{v}$ be the vector represented by the directed line segment $\overrightarrow{AB}$ where $A = (-1, 4)$ and $B = (2, -3)$, as in Figure 3.10. Then

$$\|\mathbf{v}\| = \sqrt{[2 - (-1)]^2 + [(-3) - 4]^2} = \sqrt{58}$$

Example 2.   The norm of the vector $\mathbf{v} = \begin{bmatrix} 2 \\ -5 \end{bmatrix}$ is given by

$$\|\mathbf{v}\| = \sqrt{2^2 + (-5)^2} = \sqrt{29}$$

Definitions 2 and 3 in Section 1.6 demonstrated how to add two $m \times n$ matrices and how to multiply a matrix by a scalar. Therefore, since vectors in 2-space are

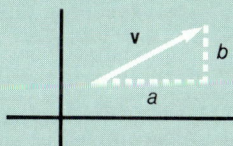

FIGURE 3.9

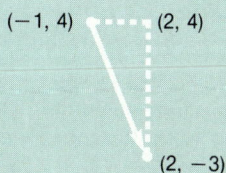

FIGURE 3.10

2×1 matrices, we have already defined the way to add them and to multiply them by a scalar. For convenience we repeat those definitions, except we adapt them to vectors in 2-space.

Definition 1. Let $\mathbf{u} = \begin{bmatrix} a \\ b \end{bmatrix}$ and $\mathbf{v} = \begin{bmatrix} c \\ d \end{bmatrix}$, and let k be any real number. The sum of $\mathbf{u}$ and $\mathbf{v}$, denoted by $\mathbf{u} + \mathbf{v}$, is defined by

$$\mathbf{u} + \mathbf{v} = \begin{bmatrix} a + c \\ b + d \end{bmatrix}$$

and the scalar product of $\mathbf{u}$ by k, denoted by $k\mathbf{u}$, is defined by

$$k\mathbf{u} = \begin{bmatrix} ka \\ kb \end{bmatrix}$$

Example 3.   Let

$$\mathbf{u} = \begin{bmatrix} 4 \\ -6 \end{bmatrix} \text{ and } \mathbf{v} = \begin{bmatrix} 2 \\ 7 \end{bmatrix}$$

Then

$$\mathbf{u} + \mathbf{v} = \begin{bmatrix} 4 \\ -6 \end{bmatrix} + \begin{bmatrix} 2 \\ 7 \end{bmatrix} = \begin{bmatrix} 6 \\ 1 \end{bmatrix}$$

and

$$\frac{7}{2}\mathbf{u} = \frac{7}{2}\begin{bmatrix} 4 \\ -6 \end{bmatrix} = \begin{bmatrix} \left(\frac{7}{2}\right)4 \\ \left(\frac{7}{2}\right)(-6) \end{bmatrix} = \begin{bmatrix} 14 \\ -21 \end{bmatrix}$$

The geometric interpretation of the sum of two vectors

$$\mathbf{u} = \begin{bmatrix} a \\ b \end{bmatrix} \quad \text{and} \quad \mathbf{v} = \begin{bmatrix} c \\ d \end{bmatrix}$$

is simple. The vectors $\mathbf{u}$ and $\mathbf{u} + \mathbf{v}$ are represented by the directed line segments with their initial ends at $(0, 0)$ and their terminal ends at (a, b) and $(a + c, b + d)$, respectively. Notice that (c, d) is the displacement pair of numbers for the directed line segment from (a, b) to $(a + c, b + d)$. Hence this directed line segment is a representation of the vector $\mathbf{v}$, as shown in Figure 3.11.

In fact, we can make an analogous argument for any representation of $\mathbf{u}$. Take any representation of $\mathbf{v}$ having its initial point at the terminal point of the representation of $\mathbf{u}$. A representation for the sum $\mathbf{u} + \mathbf{v}$ is the directed line segment from the initial point of the representation of $\mathbf{u}$ to the terminal point of the representation of $\mathbf{v}$ (see Figure 3.12).

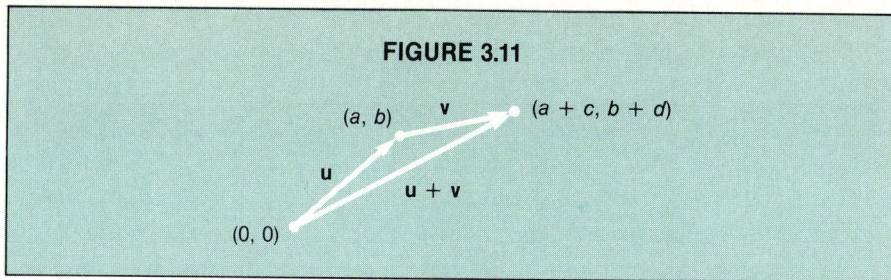

FIGURE 3.11

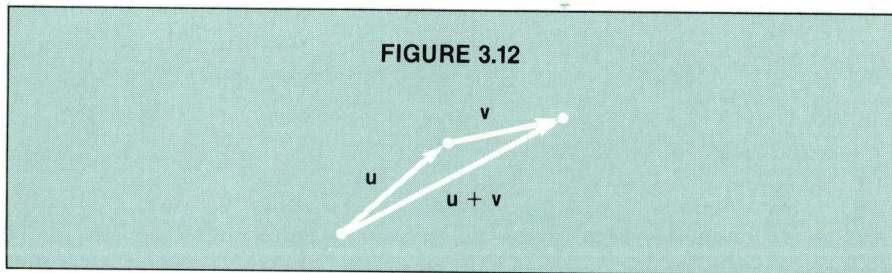

FIGURE 3.12

Since the shortest path between two points in a plane is a line segment connecting the two points, we see from Figure 3.12 that

$$\|\mathbf{u} + \mathbf{v}\| \le \|\mathbf{u}\| + \|\mathbf{v}\|$$

This inequality is called the **triangle inequality**.

The geometric interpretation of scalar multiplication is also simple. Notice that if $\mathbf{v} = \begin{bmatrix} a \\ b \end{bmatrix}$ is any vector and k is any real number, then

$$\|k\mathbf{v}\| = \left\| \begin{bmatrix} ka \\ kb \end{bmatrix} \right\|$$
$$= \sqrt{(ka)^2 + (kb)^2}$$
$$= \sqrt{k^2(a^2 + b^2)}$$
$$= |k|\sqrt{a^2 + b^2}$$
$$= |k|\,\|\mathbf{v}\|$$

Thus, multiplying a nonzero vector $\mathbf{v}$ by k produces a longer vector if $|k| > 1$, a shorter vector if $|k| < 1$, or a vector the same length if $|k| = 1$. The direction of $k\mathbf{v}$ is the same as that of $\mathbf{v}$ if $k > 0$ and the opposite of $\mathbf{v}$ if $k < 0$. If $k = 0$, then $k\mathbf{v}$ is the zero vector $\mathbf{0} = \begin{bmatrix} 0 \\ 0 \end{bmatrix}$ and has no direction.

If $\mathbf{v} = \begin{bmatrix} a \\ b \end{bmatrix}$ is any vector, then

$$\mathbf{v} + \mathbf{0} = \begin{bmatrix} a \\ b \end{bmatrix} + \begin{bmatrix} 0 \\ 0 \end{bmatrix} = \begin{bmatrix} a + 0 \\ b + 0 \end{bmatrix} = \begin{bmatrix} a \\ b \end{bmatrix} = \mathbf{v}$$

and

$$0 + v = \begin{bmatrix} 0 \\ 0 \end{bmatrix} + \begin{bmatrix} a \\ b \end{bmatrix} = \begin{bmatrix} 0 + a \\ 0 + b \end{bmatrix} = \begin{bmatrix} a \\ b \end{bmatrix} = v$$

so that

$$v + 0 = 0 + v = v$$

for every vector **v**. Also

$$v + (-1)v = \begin{bmatrix} a \\ b \end{bmatrix} + (-1)\begin{bmatrix} a \\ b \end{bmatrix} = \begin{bmatrix} a \\ b \end{bmatrix} + \begin{bmatrix} -a \\ -b \end{bmatrix} = \begin{bmatrix} a - a \\ b - b \end{bmatrix} = 0$$

The vector $(-1)v$ is called the **additive inverse** (or **negative**) of **v** and is denoted by $-v$.

Section 1.6 defined the additive inverse (negative) of an $m \times n$ matrix and the difference between two $m \times n$ matrices. When these definitions are applied to vectors in 2-space, we find that the additive inverse, $-v$, of a vector **v** is $(-1)v$ and that the difference, $u - v$, between two vectors is $u + (-1)v$. Notice that if **v** is represented by a directed line segment $\overrightarrow{AB}$, then $-v$ is represented by the directed line segment $\overrightarrow{BA}$.

There are two ways of geometrically interpreting the difference $u - v$ between two vectors:

1. Noticing that $u - v = u + (-v)$, place the initial end of a representation of $-v$ at the terminal end of a representation of **u**. A representation of $u - v$ is the directed line segment from the initial end of the representation for **u** to the terminal end of the representation of $-v$ (see Figure 3.13a).

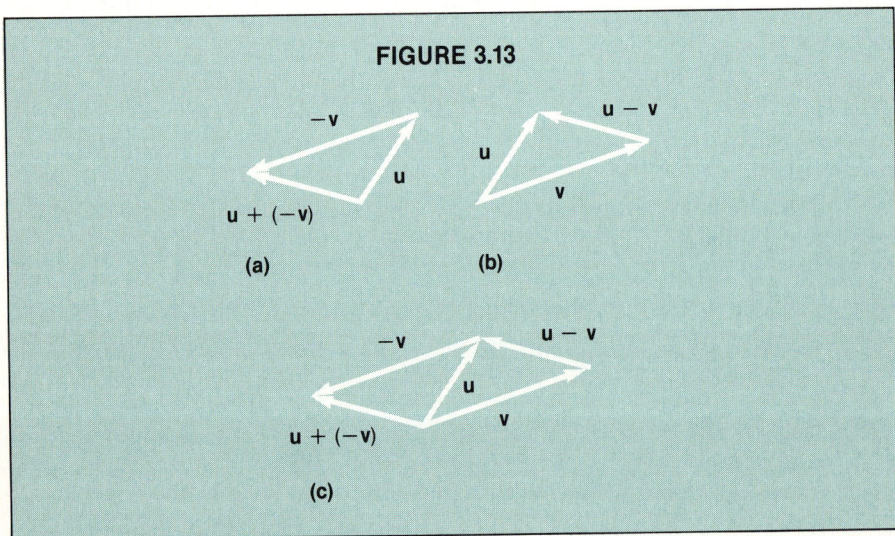

FIGURE 3.13

(a)

(b)

(c)

2. The identity $\mathbf{u} = \mathbf{v} + (\mathbf{u} - \mathbf{v})$ shows that if we take any representations of $\mathbf{u}$ and $\mathbf{v}$ having the same initial point, then a representation of $\mathbf{u} - \mathbf{v}$ is the directed line segment from the terminal end of $\mathbf{v}$ to the terminal end of $\mathbf{u}$ (see Figure 3.13b).

When we combine the two triangles in Figures 3.13a and 3.13b, we obtain the parallelogram in Figure 3.13c. Thus the first method interprets $\mathbf{u} - \mathbf{v}$ as one side of a parallelogram, while the second method interprets $\mathbf{u} - \mathbf{v}$ as the opposite side of the parallelogram.

The following theorem adapts the results from Section 1.6 to vectors in 2-space.

Theorem 1.   *Let* $\mathbf{u}$, $\mathbf{v}$, *and* $\mathbf{w}$ *be vectors in 2-space and* k, k' *be arbitrary numbers. Then*

1. $\mathbf{u} + \mathbf{v} = \mathbf{v} + \mathbf{u}$
2. $(\mathbf{u} + \mathbf{v}) + \mathbf{w} = \mathbf{u} + (\mathbf{v} + \mathbf{w})$
3. $k(\mathbf{u} + \mathbf{v}) = k\mathbf{u} + k\mathbf{v}$
4. $(k + k')\mathbf{u} = k\mathbf{u} + k'\mathbf{u}$
5. $\mathbf{u} + \mathbf{0} = \mathbf{u}$
6. $(kk')\mathbf{u} = k(k'\mathbf{u})$

Proof: We shall prove part 1 and leave the proofs of the remaining parts as exercises. Let

$$\mathbf{u} = \begin{bmatrix} a \\ b \end{bmatrix} \qquad \text{and} \qquad \mathbf{v} = \begin{bmatrix} c \\ d \end{bmatrix}$$

Then

$$\mathbf{u} + \mathbf{v} = \begin{bmatrix} a \\ b \end{bmatrix} + \begin{bmatrix} c \\ d \end{bmatrix} = \begin{bmatrix} a + c \\ b + d \end{bmatrix} = \begin{bmatrix} c + a \\ d + b \end{bmatrix}$$

and

$$\mathbf{v} + \mathbf{u} = \begin{bmatrix} c \\ d \end{bmatrix} + \begin{bmatrix} a \\ b \end{bmatrix} = \begin{bmatrix} c + a \\ d + b \end{bmatrix}$$

so that

$$\mathbf{u} + \mathbf{v} = \mathbf{v} + \mathbf{u} \quad \blacksquare$$

--------------------------------- Exercises ---------------------------------

In Exercises 1–6 determine the norm of the given vectors.

1. $\begin{bmatrix} -2 \\ 4 \end{bmatrix}$ 　　　　**2.** $\begin{bmatrix} 2 \\ -8 \end{bmatrix}$ 　　　　**3.** $\begin{bmatrix} -3 \\ -5 \end{bmatrix}$

4. $\begin{bmatrix} 8 \\ 2 \end{bmatrix}$ 　　　　**5.** $\begin{bmatrix} 7 \\ 11 \end{bmatrix}$ 　　　　**6.** $\begin{bmatrix} -7 \\ 11 \end{bmatrix}$

In Exercises 7–14 compute the indicated vector where:

$$\mathbf{u} = \begin{bmatrix} 1 \\ 2 \end{bmatrix} \qquad \mathbf{v} = \begin{bmatrix} -4 \\ 3 \end{bmatrix} \qquad \mathbf{w} = \begin{bmatrix} 5 \\ -3 \end{bmatrix}$$

7. $3\mathbf{u} + 2\mathbf{v}$

8. $2\mathbf{u} - 4\mathbf{v}$

9. $\dfrac{1}{\|\mathbf{u}\|}\mathbf{u} + \dfrac{1}{\|\mathbf{v}\|}\mathbf{v}$

10. $\dfrac{1}{\|\mathbf{u}\| + \|\mathbf{v}\|}(\mathbf{u} + \mathbf{v})$

11. $\dfrac{1}{\|\mathbf{u} + \mathbf{v}\|}(\mathbf{u} + \mathbf{v})$

12. $4\mathbf{u} + \mathbf{v} - \mathbf{w}$

13. $-2\mathbf{u} + 3\mathbf{v} + \mathbf{w}$

14. $\|\mathbf{v}\|\mathbf{u} - \|\mathbf{w}\|\mathbf{v} + \|\mathbf{u}\|\mathbf{w}$

In Exercises 15–20 find the vector $\mathbf{v}$ having the given norm and given direction.

15. $\|\mathbf{v}\| = 1; \theta = \dfrac{\pi}{6}.$

16. $\|\mathbf{v}\| = 4; \theta = \pi.$

17. $\|\mathbf{v}\| = 3; \theta = \dfrac{7\pi}{4}.$

18. $\|\mathbf{v}\| = 1; \theta = \dfrac{3}{4}\pi.$

19. $\|\mathbf{v}\| = 2; \theta = \dfrac{3}{2}\pi.$

20. $\|\mathbf{v}\| = 3; \theta = \dfrac{2}{3}\pi.$

21. Find a unit vector having the direction of $\begin{bmatrix} 3 \\ -4 \end{bmatrix}$.

22. Find a unit vector having the direction of $\begin{bmatrix} -3 \\ 6 \end{bmatrix}$.

23. Let $\mathbf{u}$ be a nonzero vector. Show that $\dfrac{1}{\|\mathbf{u}\|}\mathbf{u}$ is a unit vector.

24. Let $\mathbf{u} = \begin{bmatrix} a \\ b \end{bmatrix}$ and $\mathbf{v} = \begin{bmatrix} c \\ d \end{bmatrix}$. Show that

$$\|\mathbf{u}\|^2 + \|\mathbf{v}\|^2 = \|\mathbf{u} + \mathbf{v}\|^2 \text{ if and only if } ac + bd = 0.$$

25. Show that $\|\mathbf{u}\| - \|\mathbf{v}\| \le \|\mathbf{u} + \mathbf{v}\|$ for any two vectors $\mathbf{u}$ and $\mathbf{v}$. [Hint: $\mathbf{u} = (\mathbf{u} + \mathbf{v}) - \mathbf{v}$.]
26. Prove part 2 of Theorem 1.
27. Prove part 3 of Theorem 1.
28. Prove part 4 of Theorem 1.
29. Prove part 5 of Theorem 1.
30. Prove part 6 of Theorem 1.

3.3 INNER PRODUCT AND ORTHOGONALITY OF VECTORS IN 2-SPACE

This section discusses a type of vector multiplication that will give us an easy means of determining the angle between two vectors. This type of vector multiplication produces a number, not another vector.

Definition 2. If $\mathbf{u} = \begin{bmatrix} u_1 \\ u_2 \end{bmatrix}$ and $\mathbf{v} = \begin{bmatrix} v_1 \\ v_2 \end{bmatrix}$ are vectors in 2-space, then the **inner product**, or **dot product**, of $\mathbf{u}$ and $\mathbf{v}$ (denoted by $\mathbf{u} \cdot \mathbf{v}$) is defined by

$$\mathbf{u} \cdot \mathbf{v} = u_1 v_1 + u_2 v_2$$

We emphasize that the inner product $\mathbf{u} \cdot \mathbf{v}$ of two vectors $\mathbf{u}$ and $\mathbf{v}$ is a *number*, not another vector.

Example 1.   Let

$$\mathbf{u} = \begin{bmatrix} -3 \\ 4 \end{bmatrix} \quad \text{and} \quad \mathbf{v} = \begin{bmatrix} 5 \\ 2 \end{bmatrix}$$

Then

$$\mathbf{u} \cdot \mathbf{v} = (-3)(5) + (4)(2) = -7$$

The following theorem lists the basic properties of the inner product of two vectors.

Theorem 2.   *Let* $\mathbf{u}$, $\mathbf{v}$, *and* $\mathbf{w}$ *be vectors in 2-space. Then:*

1. $\mathbf{u} \cdot \mathbf{v} = \mathbf{v} \cdot \mathbf{u}$
2. $\mathbf{u} \cdot \mathbf{u} = \|\mathbf{u}\|^2$
3. $\mathbf{u} \cdot \mathbf{u} > 0$ if $\mathbf{u} \neq \mathbf{0}$
4. $\mathbf{u} \cdot \mathbf{0} = 0$
5. $(\mathbf{u} \pm \mathbf{v}) \cdot \mathbf{w} = \mathbf{u} \cdot \mathbf{w} \pm \mathbf{v} \cdot \mathbf{w}$
6. *If* k *is any number, then* $(k\mathbf{u}) \cdot \mathbf{v} = k(\mathbf{u} \cdot \mathbf{v}) = \mathbf{u} \cdot (k\mathbf{v})$.

Proof: We shall prove part 1 and leave the proofs of the remaining parts as exercises. Let

$$\mathbf{u} = \begin{bmatrix} u_1 \\ u_2 \end{bmatrix} \quad \text{and} \quad \mathbf{v} = \begin{bmatrix} v_1 \\ v_2 \end{bmatrix}$$

Then

$$\mathbf{u} \cdot \mathbf{v} = u_1 v_1 + u_2 v_2 = v_1 u_1 + v_2 u_2 = \mathbf{v} \cdot \mathbf{u} \quad \blacksquare$$

In order to describe the geometrical significance of the inner product we need to define the angle between two nonzero vectors. The angle θ between any two directed line segments with the same initial point and nonzero lengths can always be chosen to be in the interval $[0, \pi]$. Some typical cases are illustrated in Figure 3.14.

Definition 3.   The angle between two nonzero vectors $\mathbf{u}$ and $\mathbf{v}$ in 2-space is the angle $\theta, 0 \leq \theta \leq \pi$, between any representations of these vectors having the same initial point.

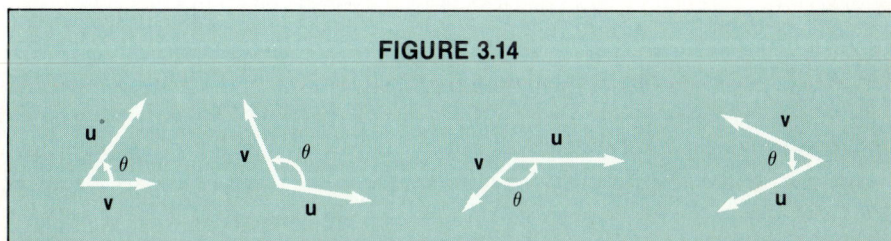

FIGURE 3.14

FIGURE 3.15

FIGURE 3.16

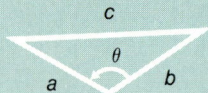

Let **u** and **v** be any two nonzero vectors with representations as indicated in Figure 3.15. The angle between these vectors can be determined by using the Law of Cosines, which states that for any given triangle, such as the one in Figure 3.16, $c^2 = a^2 + b^2 - 2ab \cos \theta$. Noticing that the lengths of the sides of the triangle in Figure 3.15 are $\|\mathbf{u}\|$, $\|\mathbf{v}\|$, and $\|\mathbf{u} - \mathbf{v}\|$, we use the Law of Cosines to obtain

$$\|\mathbf{u} - \mathbf{v}\|^2 = \|\mathbf{u}\|^2 + \|\mathbf{v}\|^2 - 2\|\mathbf{u}\| \, \|\mathbf{v}\| \cos \theta \tag{1}$$

Example 2.   Consider the vectors

$$\mathbf{u} = \begin{bmatrix} 1 \\ 3 \end{bmatrix} \qquad \mathbf{v} = \begin{bmatrix} 2 \\ -4 \end{bmatrix}$$

Since

$$\mathbf{u} - \mathbf{v} = \begin{bmatrix} 1 \\ 3 \end{bmatrix} - \begin{bmatrix} 2 \\ -4 \end{bmatrix} = \begin{bmatrix} -1 \\ 7 \end{bmatrix}$$

the identity in (1) for these vectors can be written as

$$\left\| \begin{bmatrix} -1 \\ 7 \end{bmatrix} \right\|^2 = \left\| \begin{bmatrix} 1 \\ 3 \end{bmatrix} \right\|^2 + \left\| \begin{bmatrix} 2 \\ -4 \end{bmatrix} \right\|^2 - 2 \left\| \begin{bmatrix} 1 \\ 3 \end{bmatrix} \right\| \left\| \begin{bmatrix} 2 \\ -4 \end{bmatrix} \right\| \cos \theta$$

$$50 = 10 + 20 - 2\sqrt{10}\sqrt{20} \cos \theta$$

so that the angle θ between **u** and **v** satisfies

$$\cos \theta = -\frac{1}{\sqrt{2}}$$

Hence $\theta = 3\pi/4$ radians. Notice that

$$\mathbf{u} \cdot \mathbf{v} = (1)(2) + (3)(-4) = -10$$

$$\|\mathbf{u}\| = \sqrt{10}$$

$$\|\mathbf{v}\| = \sqrt{20}$$

and that

$$\cos\theta = -\frac{1}{\sqrt{2}} = -\frac{10}{\sqrt{10}\sqrt{20}} = \frac{\mathbf{u} \cdot \mathbf{v}}{\|\mathbf{u}\|\,\|\mathbf{v}\|}$$

In fact, the identity

$$\cos\theta = \frac{\mathbf{u} \cdot \mathbf{v}}{\|\mathbf{u}\|\,\|\mathbf{v}\|}$$

where θ is the angle between $\mathbf{u}$ and $\mathbf{v}$, holds for any two nonzero vectors $\mathbf{u}$ and $\mathbf{v}$. The proof of this follows almost immediately from the following important identity.

Theorem 3. *Let $\mathbf{u}$ and $\mathbf{v}$ be any two vectors in 2-space. Then*

$$\|\mathbf{u} - \mathbf{v}\|^2 = \|\mathbf{u}\|^2 + \|\mathbf{v}\|^2 - 2\mathbf{u} \cdot \mathbf{v}$$

Proof:

$$\|\mathbf{u} - \mathbf{v}\|^2 = (\mathbf{u} - \mathbf{v}) \cdot (\mathbf{u} - \mathbf{v})$$

$$= \mathbf{u} \cdot \mathbf{u} + \mathbf{v} \cdot \mathbf{v} - 2\mathbf{u} \cdot \mathbf{v}$$

$$= \|\mathbf{u}\|^2 + \|\mathbf{v}\|^2 - 2\mathbf{u} \cdot \mathbf{v} \quad \blacksquare$$

Combining this theorem with the identity in (1), we arrive at the following theorem.

Theorem 4. *Let $\mathbf{u}$ and $\mathbf{v}$ be any two vectors in 2-space. Then*

$$\mathbf{u} \cdot \mathbf{v} = \|\mathbf{u}\|\,\|\mathbf{v}\|\,\cos\theta$$

where θ is the angle between $\mathbf{u}$ and $\mathbf{v}$.

Example 3. Consider the vectors

$$\mathbf{u} = \begin{bmatrix} 3 \\ 7 \end{bmatrix} \quad \text{and} \quad \mathbf{v} = \begin{bmatrix} -2 \\ 1 \end{bmatrix}$$

Then $\mathbf{u} \cdot \mathbf{v} = 1$, $\|\mathbf{u}\| = \sqrt{58}$, and $\|\mathbf{v}\| = \sqrt{5}$, so that $1 = \sqrt{58}\sqrt{5}\cos\theta$ where θ is the angle between $\mathbf{u}$ and $\mathbf{v}$. Therefore

$$\theta = \cos^{-1}\frac{1}{\sqrt{58}\sqrt{5}} \approx 1.51 \text{ radians}$$

which is the angle between $\mathbf{u}$ and $\mathbf{v}$.

If the angle between two nonzero vectors is $\pi/2$, it is natural to think of these vectors as being perpendicular. Once again we anticipate the terminology that is to come and use another word to describe such a situation.

Definition 4. Let θ denote the angle between two nonzero vectors **u** and **v** in 2-space. Then **u** and **v** are said to be **orthogonal** if $\theta = \pi/2$ and **parallel** if $\theta = 0$ or $\theta = \pi$. The zero vector is orthogonal to every vector.

From Theorem 4 we obtain the following characterization of orthogonal vectors.

Theorem 5. *Let **u** and **v** be vectors in 2-space. Then **u** and **v** are orthogonal if and only if* $\mathbf{u} \cdot \mathbf{v} = 0$.

Example 4. Most elementary geometry courses prove that the diagonals of a square intersect at right angles. That is, the vectors indicated in Figure 3.17 are orthogonal. The inner product gives us a means to prove this geometric relationship. In fact, the following proof shows that the diagonals of any **rhombus** (a parallelogram with all sides of equal length) intersect at right angles.

Let A, B, C, D be the vertices of a rhombus and let r be the length of each side (see Figure 3.18a). For simplicity let **u** and **v** denote the vectors represented by $\overrightarrow{AB}$ and $\overrightarrow{AD}$, respectively. Since the figure is a rhombus, $\overrightarrow{DC}$ and $\overrightarrow{BC}$ are also representations of **u** and **v**, respectively. Let $\mathbf{w}_1$ and $\mathbf{w}_2$ denote the vectors represented by $\overrightarrow{AC}$ and $\overrightarrow{DB}$

FIGURE 3.17

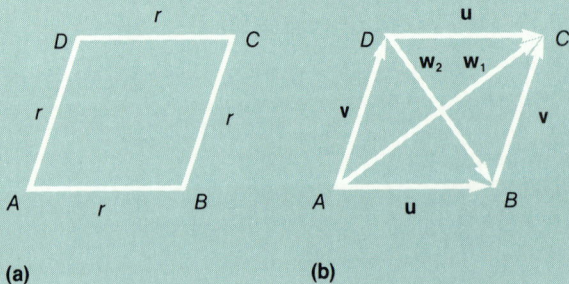

FIGURE 3.18

(a)   (b)

respectively (see Figure 3.18b). Using the triangle with A, B, C as its vertices, we obtain

$$\mathbf{w}_1 = \mathbf{u} + \mathbf{v}$$

Similarly, using the triangle with A, B, D as its vertices, we obtain

$$\mathbf{v} = \mathbf{u} + \mathbf{w}_2$$

so that

$$\mathbf{w}_2 = \mathbf{v} - \mathbf{u}$$

Now

$$\begin{aligned}
\mathbf{w}_1 \cdot \mathbf{w}_2 &= (\mathbf{v} + \mathbf{u}) \cdot (\mathbf{v} - \mathbf{u}) \\
&= \mathbf{v} \cdot (\mathbf{v} - \mathbf{u}) + \mathbf{u} \cdot (\mathbf{v} - \mathbf{u}) \\
&= \mathbf{v} \cdot \mathbf{v} - \mathbf{u} \cdot \mathbf{v} + \mathbf{u} \cdot \mathbf{v} - \mathbf{u} \cdot \mathbf{u} \\
&= \mathbf{v} \cdot \mathbf{v} - \mathbf{u} \cdot \mathbf{u} \\
&= \|\mathbf{v}\|^2 - \|\mathbf{u}\|^2
\end{aligned}$$

Recalling that the norm of a vector is the length of any of its representations and that our geometric figure is a rhombus with each side having length r, we conclude that $\|\mathbf{v}\| = r = \|\mathbf{u}\|$. Therefore $\|\mathbf{v}\|^2 = \|\mathbf{u}\|^2$ and $\mathbf{w}_1 \cdot \mathbf{w}_2 = 0$. By Theorem 5 the vectors $\mathbf{w}_1$ and $\mathbf{w}_2$ are orthogonal. Hence the diagonals of any rhombus intersect at right angles.

In working with a force that causes an object to move, a physicist often decomposes the force into two components: one in the direction of the motion and the other perpendicular to this direction. This practice can be easily described in terms of vectors. Given two nonzero vectors $\mathbf{u}$ and $\mathbf{v}$, we can obtain vectors $\mathbf{w}_1$, parallel to $\mathbf{v}$, and $\mathbf{w}_2$, orthogonal to $\mathbf{v}$, such that

$$\mathbf{u} = \mathbf{w}_1 + \mathbf{w}_2$$

The vectors $\mathbf{w}_1$ and $\mathbf{w}_2$ can be found as follows. Choose representations of $\mathbf{u}$ and $\mathbf{v}$ having the same initial point P, let L_1 denote the line passing through the initial and terminal points of $\mathbf{v}$, and let L_2 denote the line perpendicular to L_1 passing through the terminal point Q_1 of $\mathbf{u}$ (see Figure 3.19a). Let Q_2 denote the point of intersection of L_1 and L_2. From Figure 3.19b we see that

1. $\overrightarrow{PQ_2}$ is a representation of a vector $\mathbf{w}_1$ that is a scalar multiple of $\mathbf{v}$.
2. $\overrightarrow{Q_2Q_1}$ is a representation of a vector $\mathbf{w}_2$ that is orthogonal to $\mathbf{v}$.
3. $\mathbf{u} = \mathbf{w}_1 + \mathbf{w}_2$.

From Figure 3.19b we also see that

$$\|\mathbf{w}_1\| = \|\mathbf{u}\| \cos \theta$$

so that $\mathbf{w}_1$ has the same direction as $\mathbf{v}$ and length $\|\mathbf{u}\| \cos \theta$. Since $\dfrac{1}{\|\mathbf{v}\|} \mathbf{v}$ is a unit

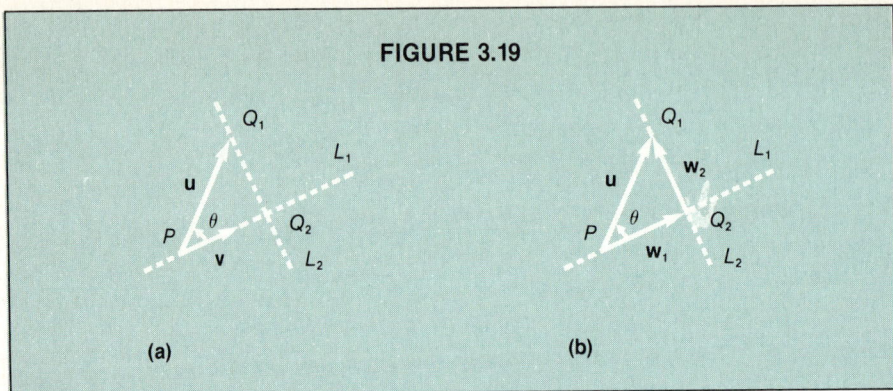

FIGURE 3.19

(a)   (b)

vector with the same direction as **v**, the vector

$$(\|\mathbf{u}\| \cos \theta)\left(\frac{1}{\|\mathbf{v}\|}\mathbf{v}\right) = \left(\frac{\|\mathbf{u}\|}{\|\mathbf{v}\|}\cos \theta\right)\mathbf{v}$$

$$= \left(\frac{\|\mathbf{u}\|}{\|\mathbf{v}\|}\frac{\mathbf{u}\cdot\mathbf{v}}{\|\mathbf{u}\|\,\|\mathbf{v}\|}\right)\mathbf{v} \qquad \text{(Theorem 4)}$$

$$= \left(\frac{\mathbf{u}\cdot\mathbf{v}}{\|\mathbf{v}\|^2}\right)\mathbf{v}$$

$$= \frac{\mathbf{u}\cdot\mathbf{v}}{\mathbf{v}\cdot\mathbf{v}}\mathbf{v}$$

has length $\|\mathbf{u}\| \cos \theta$ and the same direction as **v**. That is,

$$\mathbf{w}_1 = \frac{\mathbf{u}\cdot\mathbf{v}}{\mathbf{v}\cdot\mathbf{v}}\mathbf{v} \qquad (2)$$

Since $\mathbf{u} = \mathbf{w}_1 + \mathbf{w}_2$, we have

$$\mathbf{w}_2 = \mathbf{u} - \mathbf{w}_1 = \mathbf{u} - \frac{\mathbf{u}\cdot\mathbf{v}}{\mathbf{v}\cdot\mathbf{v}}\mathbf{v} \qquad (3)$$

In determining the vectors $\mathbf{w}_1$ and $\mathbf{w}_2$, we have relied heavily on Figure 3.19b. In fact, the vectors $\mathbf{w}_1$ and $\mathbf{w}_2$ given in (2) and (3) are orthogonal with $\mathbf{u} = \mathbf{w}_1 + \mathbf{w}_2$ regardless of the choice of the nonzero vectors **u** and **v**. The reader is asked to verify this in the exercises.

The vector $\mathbf{w}_1$ is called the **orthogonal projection of u onto v**, while $\mathbf{w}_2$ is called the **component of u orthogonal to v**.

Example 5.   Let

$$\mathbf{u} = \begin{bmatrix} 6 \\ 1 \end{bmatrix} \quad \text{and} \quad \mathbf{v} = \begin{bmatrix} -3 \\ 4 \end{bmatrix}$$

Then

$$\mathbf{u}\cdot\mathbf{v} = (6)(-3) + (1)(4) = -14$$

$$\mathbf{v}\cdot\mathbf{v} = (-3)(-3) + (4)(4) = 25$$

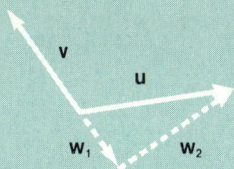

FIGURE 3.20

The orthogonal projection of **u** onto **v** is

$$\mathbf{w}_1 = \frac{\mathbf{u} \cdot \mathbf{v}}{\mathbf{v} \cdot \mathbf{v}} \mathbf{v} = -\frac{14}{25} \begin{bmatrix} -3 \\ 4 \end{bmatrix} = \begin{bmatrix} \dfrac{42}{25} \\ -\dfrac{56}{25} \end{bmatrix}$$

The component of **u** orthogonal to **v** is

$$\mathbf{w}_2 = \mathbf{u} - \frac{\mathbf{u} \cdot \mathbf{v}}{\mathbf{v} \cdot \mathbf{v}} \mathbf{v} = \begin{bmatrix} 6 \\ 1 \end{bmatrix} - \begin{bmatrix} \dfrac{42}{25} \\ -\dfrac{56}{25} \end{bmatrix} = \begin{bmatrix} \dfrac{108}{25} \\ \dfrac{81}{25} \end{bmatrix}$$

The vectors **u**, **v**, **w**$_1$, and **w**$_2$ are illustrated in Figure 3.20.

--- **Exercises** ---

In Exercises 1–6 compute **u** · **v**.

1. $\mathbf{u} = \begin{bmatrix} 2 \\ 1 \end{bmatrix}$ $\mathbf{v} = \begin{bmatrix} 5 \\ -3 \end{bmatrix}$ 2. $\mathbf{u} = \begin{bmatrix} 5 \\ 4 \end{bmatrix}$ $\mathbf{v} = \begin{bmatrix} 2 \\ -1 \end{bmatrix}$

3. $\mathbf{u} = \begin{bmatrix} 3 \\ -2 \end{bmatrix}$ $\mathbf{v} = \begin{bmatrix} 2 \\ 6 \end{bmatrix}$ 4. $\mathbf{u} = \begin{bmatrix} 3 \\ -2 \end{bmatrix}$ $\mathbf{v} = \begin{bmatrix} 3 \\ 5 \end{bmatrix}$

5. $\mathbf{u} = \begin{bmatrix} -2 \\ 7 \end{bmatrix}$ $\mathbf{v} = \begin{bmatrix} 7 \\ 2 \end{bmatrix}$ 6. $\mathbf{u} = \begin{bmatrix} 1 \\ 6 \end{bmatrix}$ $\mathbf{v} = \begin{bmatrix} 5 \\ -2 \end{bmatrix}$

In Exercises 7–14 find the angle between **u** and **v**.

7. $\mathbf{u} = \begin{bmatrix} 1 \\ 3 \end{bmatrix}$ $\mathbf{v} = \begin{bmatrix} -1 \\ 2 \end{bmatrix}$ 8. $\mathbf{u} = \begin{bmatrix} 2 \\ 3 \end{bmatrix}$ $\mathbf{v} = \begin{bmatrix} 5 \\ 1 \end{bmatrix}$

9. $\mathbf{u} = \begin{bmatrix} 2 \\ 4 \end{bmatrix}$ $\mathbf{v} = \begin{bmatrix} -3 \\ -6 \end{bmatrix}$ 10. $\mathbf{u} = \begin{bmatrix} 0 \\ 2 \end{bmatrix}$ $\mathbf{v} = \begin{bmatrix} 3 \\ 1 \end{bmatrix}$

11. $\mathbf{u} = \begin{bmatrix} 3 \\ 2 \end{bmatrix}$ $\mathbf{v} = \begin{bmatrix} 3\sqrt{3} \\ -2 \end{bmatrix}$ **12.** $\mathbf{u} = \begin{bmatrix} 4 \\ 2 \end{bmatrix}$ $\mathbf{v} = \begin{bmatrix} 1 \\ 3 \end{bmatrix}$

13. $\mathbf{u} = \begin{bmatrix} 1 \\ 1 \end{bmatrix}$ $\mathbf{v} = \begin{bmatrix} -2 \\ 3 \end{bmatrix}$ **14.** $\mathbf{u} = \begin{bmatrix} 2 \\ 3 \end{bmatrix}$ $\mathbf{v} = \begin{bmatrix} 1 \\ 1 \end{bmatrix}$

In Exercises 15–18 find the orthogonal projection of $\mathbf{u}$ onto $\mathbf{v}$ and the component of $\mathbf{u}$ orthogonal to $\mathbf{v}$.

15. $\mathbf{u} = \begin{bmatrix} 1 \\ -1 \end{bmatrix}$ $\mathbf{v} = \begin{bmatrix} 2 \\ 1 \end{bmatrix}$ **16.** $\mathbf{u} = \begin{bmatrix} 7 \\ 5 \end{bmatrix}$ $\mathbf{v} = \begin{bmatrix} 2 \\ 8 \end{bmatrix}$

17. $\mathbf{u} = \begin{bmatrix} 3 \\ 2 \end{bmatrix}$ $\mathbf{v} = \begin{bmatrix} 3 \\ -2 \end{bmatrix}$ **18.** $\mathbf{u} = \begin{bmatrix} -2 \\ 1 \end{bmatrix}$ $\mathbf{v} = \begin{bmatrix} -2 \\ -3 \end{bmatrix}$

In Exercises 19–22 compute the indicated number or vector where $\mathbf{u} = \begin{bmatrix} 4 \\ 5 \end{bmatrix}$ and $\mathbf{v} = \begin{bmatrix} 3 \\ -4 \end{bmatrix}$.

19. $\|(\mathbf{u} \cdot \mathbf{v})\mathbf{v}\|$ **20.** $(\mathbf{v} \cdot \mathbf{v})\mathbf{u} - (\mathbf{u} \cdot \mathbf{u})\mathbf{v}$

21. $\dfrac{1}{\|\mathbf{u} + \mathbf{v}\|}(\mathbf{u} \cdot \mathbf{v})\mathbf{u}$ **22.** $\|\mathbf{v} - (\mathbf{v} \cdot \mathbf{u})\mathbf{v}\|$

23. For what value of k is $\mathbf{u} = \begin{bmatrix} 5 \\ 2 \end{bmatrix}$ orthogonal to $\mathbf{v} = \begin{bmatrix} k \\ 3 \end{bmatrix}$?

24. Find two unit vectors orthogonal to $\begin{bmatrix} 7 \\ 4 \end{bmatrix}$.

25. Explain why each of the following expressions is not defined.

      (a) $(\mathbf{u} \cdot \mathbf{v}) \cdot \|\mathbf{v}\|$
      (b) $(\mathbf{u} \cdot \mathbf{v}) \cdot \mathbf{w}$
      (c) $\mathbf{u} - (\mathbf{v} \cdot \mathbf{w})$

26. Show that $\|\mathbf{u}\|^2 + \|\mathbf{v}\|^2 = \|\mathbf{u} + \mathbf{v}\|^2$ if and only if $\mathbf{u}$ and $\mathbf{v}$ are orthogonal.
27. Show that $\|\mathbf{u} + \mathbf{v}\|^2 + \|\mathbf{u} - \mathbf{v}\|^2 = 2(\|\mathbf{u}\|^2 + \|\mathbf{v}\|^2)$.
28. Show that $\|\mathbf{u} + \mathbf{v}\|^2 - \|\mathbf{u} - \mathbf{v}\|^2 = 4\mathbf{u} \cdot \mathbf{v}$.
29. Prove part 2 of Theorem 2.
30. Prove part 3 of Theorem 2.
31. Prove part 4 of Theorem 2.
32. Prove part 5 of Theorem 2.
33. Prove part 6 of Theorem 2.
34. Show that the vectors $\mathbf{w}_1$ and $\mathbf{w}_2$ in equations (2) and (3) are orthogonal. (Hint: Show $\mathbf{w}_1 \cdot \mathbf{w}_2 = 0$.)

3.4 VECTORS IN 3-SPACE

In our discussion of vectors in 2-space we repeatedly used the fact that points in the plane can be represented by ordered pairs of real numbers. Analogously points in 3-space can be represented by ordered triples of real numbers. We begin with the xy-plane and introduce a third coordinate axis perpendicular to the xy-plane, passing

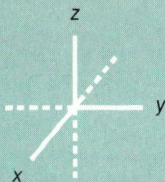

FIGURE 3.21

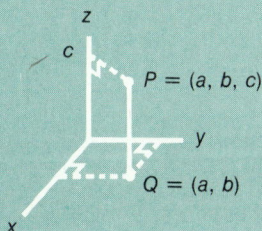

FIGURE 3.22

through the origin, and directed upward. This third axis is commonly called the z-axis. Figure 3.21 shows the traditional way the z-axis is positioned in relation to the xy-plane. In this figure the positive axes are indicated by solid half lines, and the negative axes by dotted half lines. Usually the units of length used for measuring distances are the same along the three axes.

Let P be any point in 3-space; pass a line through P perpendicular to the xy-plane. Let $Q = (a, b)$ denote the point of intersection between the line and the xy-plane (see Figure 3.22). Pass a line through P perpendicular to and intersecting the z-axis. Let c denote the coordinate on the z-axis that corresponds to this point of intersection (see Figure 3.22). The point P is represented by the ordered triple (a, b, c). Geometrically $|c|$ represents how far above (if $c > 0$) or below (if $c < 0$) the xy-plane the point P is located. If $c = 0$ the point P lies in the xy-plane.

If $A = (a_1, b_1, c_1)$ and $B = (a_2, b_2, c_2)$ are any two points in 3-space, then the distance between A and B is the length $|\overrightarrow{AB}|$ of the directed line segment from A to B. The Pythagorean theorem allows us to obtain a formula for $|\overrightarrow{AB}|$. To find this formula we construct a rectangular box having A and B as diagonally opposite corners and having its edges parallel to the coordinate axes (see Figure 3.23). Notice that $|\overrightarrow{AC}| = |b_2 - b_1|$, $|\overrightarrow{CD}| = |a_2 - a_1|$, and $|\overrightarrow{DB}| = |c_2 - c_1|$. Using these identities and applying the Pythagorean theorem to the right triangles ACD and ADB, we obtain

$$|\overrightarrow{AD}|^2 = |\overrightarrow{CD}|^2 + |\overrightarrow{AC}|^2$$
$$= (a_2 - a_1)^2 + (b_2 - b_1)^2$$

FIGURE 3.23

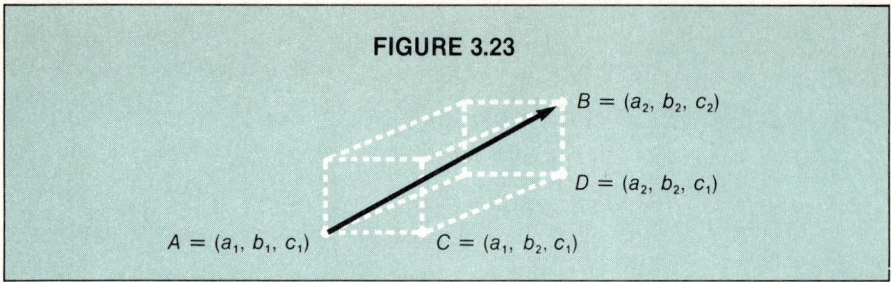

and

$$|\overrightarrow{AB}|^2 = |\overrightarrow{AD}|^2 + |\overrightarrow{DB}|^2$$
$$= (a_2 - a_1)^2 + (b_2 - b_1)^2 + (c_2 - c_1)^2$$

Therefore, we have proved the following theorem.

Theorem 6. *The distance between the points $A = (a_1, b_1, c_1)$ and $B = (a_2, b_2, c_2)$ is given by*

$$|\overrightarrow{AB}| = \sqrt{(a_2 - a_1)^2 + (b_2 - b_1)^2 + (c_2 - c_1)^2}$$

Example 1. The distance between $A = (2, -4, 5)$ and $B = -3, 0, -1)$ is

$$|\overrightarrow{AB}| = \sqrt{[2 - (-3)]^2 + [-4 - 0]^2 + [5 - (-1)]^2}$$
$$= \sqrt{25 + 16 + 36} = \sqrt{77}$$

Let $A = (a_1, b_1, c_1)$ and $B = (a_2, b_2, c_2)$ be two points in 3-space. Just as in the case for directed line segments in 2-space, it is usually convenient to write the terminal end of $\overrightarrow{AB}$ as $(a_1 + h, b_1 + k, c_1 + m)$ where $h = a_2 - a_1$; $k = b_2 - b_1$; and $m = c_2 - c_1$. Thus a directed line segment in 3-space can be described by two ordered triples of numbers:

1. (a_1, b_1, c_1), indicating the initial end of the directed line segment, and
2. (h, k, m), indicating the displacement of the terminal end from the initial end.

We shall call the 3-vector

$$\begin{bmatrix} h \\ k \\ m \end{bmatrix}$$

the vector corresponding to the displacement triple of numbers (h, k, m). Thus each displacement triple of numbers, and hence each directed line segment in 3-space, is associated with a 3-vector. Moreover, all directed line segments with the same displacement triple of numbers are associated with the same 3-vector.

Frequently in this chapter we shall refer to 3-vectors simply as vectors, or as vectors in 3-space whenever we wish to emphasize their geometric properties. We shall always be careful to use the word *vector* in a context where it is clear whether we mean a 2-vector or a 3-vector.

If $A = (a_1, b_1, c_1)$ and $B = (a_2, b_2, c_2)$ are two points in 3-space, then the directed line segment $\overrightarrow{AB}$ is a representation of the vector

$$\mathbf{v} = \begin{bmatrix} a_2 - a_1 \\ b_2 - b_1 \\ c_2 - c_1 \end{bmatrix}$$

The **norm** of a vector $\mathbf{v}$, denoted by $\|\mathbf{v}\|$, is the length of any representation of $\mathbf{v}$. Thus if

$$\mathbf{v} = \begin{bmatrix} a \\ b \\ c \end{bmatrix}$$

then

$$\|\mathbf{v}\| = \sqrt{a^2 + b^2 + c^2}$$

Since vectors in 3-space are 3×1 matrices, Section 1.6 has already defined the way to add them and to multiply them by a scalar. For convenience, these definitions, adapted to vectors in 3-space, are now repeated.

Definition 5.   Let

$$\mathbf{u} = \begin{bmatrix} a_1 \\ b_1 \\ c_1 \end{bmatrix} \qquad \text{and} \qquad \mathbf{v} = \begin{bmatrix} a_2 \\ b_2 \\ c_2 \end{bmatrix}$$

be two vectors in 3-space and let k be any real number. The **sum** of $\mathbf{u}$ and $\mathbf{v}$, denoted by $\mathbf{u} + \mathbf{v}$, is defined by

$$\mathbf{u} + \mathbf{v} = \begin{bmatrix} a_1 + a_2 \\ b_1 + b_2 \\ c_1 + c_2 \end{bmatrix}$$

and the **scalar product** of $\mathbf{u}$ by k, denoted by $k\mathbf{u}$, is defined by

$$k\mathbf{u} = \begin{bmatrix} ka_1 \\ kb_1 \\ kc_1 \end{bmatrix}$$

Example 2.   Let

$$\mathbf{u} = \begin{bmatrix} 2 \\ -7 \\ 11 \end{bmatrix} \qquad \text{and} \qquad \mathbf{v} = \begin{bmatrix} -7 \\ -16 \\ 23 \end{bmatrix}$$

Then

$$\mathbf{u} + \mathbf{v} = \begin{bmatrix} 2 \\ -7 \\ 11 \end{bmatrix} + \begin{bmatrix} -7 \\ -16 \\ 23 \end{bmatrix} = \begin{bmatrix} 2 + (-7) \\ -7 + (-16) \\ 11 + 23 \end{bmatrix} = \begin{bmatrix} -5 \\ -23 \\ 34 \end{bmatrix}$$

and

$$\sqrt{2}\mathbf{u} = \sqrt{2}\begin{bmatrix} 2 \\ -7 \\ 11 \end{bmatrix} = \begin{bmatrix} 2\sqrt{2} \\ -7\sqrt{2} \\ 11\sqrt{2} \end{bmatrix}$$

The geometric interpretation of the sum of two vectors **u** and **v** in 3-space is the same as that of the sum of two vectors in 2-space. Take any representation of **u** and the representation of **v** with its initial point at the terminal point of the representation of **u**. A representation for the sum **u** + **v** is the directed line segment from the initial point of the representation of **u** to the terminal point of the representation of **v** (see Figure 3.24.)

The geometric interpretation of scalar multiplication is also the same as before. Multiplying a nonzero vector **v** by k produces a longer vector if $|k| > 1$, a shorter vector if $|k| < 1$, or a vector of the same length if $|k| = 1$. The direction of $k\mathbf{v}$ is the same as that of **v** if $k > 0$ and the opposite of **v** if $k < 0$. If $k = 0$, then $k\mathbf{v}$ is the zero vector, which has no direction.

In analogy to the case for 2-vectors, the additive inverse, $-\mathbf{v}$, of a vector in 3-space is obtained by multiplying **v** by -1. The difference, $\mathbf{u} - \mathbf{v}$, between two vectors in 3-space is obtained by adding **u** and $-\mathbf{v}$. That is, $-\mathbf{v} = (-1)\mathbf{v}$ and $\mathbf{u} - \mathbf{v} = \mathbf{u} + (-\mathbf{v})$.

The geometric interpretation of the difference between two vectors **u** and **v** in 3-space is the same as its interpretation in 2-space. Take any representations of **u** and **v** with the same initial point. The directed line segment from the terminal end of the representation of **v** to the terminal end of the representation of **u** is a representation of $\mathbf{u} - \mathbf{v}$ (see Figure 3.25).

The properties of addition and scalar multiplication of vectors in 3-space are analogous to those in 2-space.

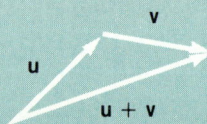

FIGURE 3.24

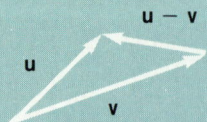

FIGURE 3.25

Theorem 7. *Let* **u**, **v**, *and* **w** *be arbitrary vectors in 3-space and* k, k' *be real numbers. Then*

    1. $\mathbf{u} + \mathbf{v} = \mathbf{v} + \mathbf{u}$
    2. $(\mathbf{u} + \mathbf{v}) + \mathbf{w} = \mathbf{u} + (\mathbf{v} + \mathbf{w})$
    3. $k(\mathbf{u} \pm \mathbf{v}) = k\mathbf{u} \pm k\mathbf{v}$
    4. $(k + k')\mathbf{u} = k\mathbf{u} + k'\mathbf{u}$
    5. $\mathbf{u} + \mathbf{0} = \mathbf{u}$
    6. $\|\mathbf{u} + \mathbf{v}\| \le \|\mathbf{u}\| + \|\mathbf{v}\|$

The proofs of the parts of this theorem are similar to the proofs of the analogous parts of Theorem 1 for vectors in 2-space and will be omitted.

──────── Exercises ────────

In Exercises 1–4 sketch the directed line segment $\overrightarrow{AB}$ and determine its length.

1. $A = (1, 2, 3)$; $B = (2, 1, 4)$.   **2.** $A = (-1, -2, 3)$; $B = (-2, 1, 5)$.
3. $A = (2, 4, -1)$; $B = (5, 4, 2)$.   **4.** $A = (2, -1, 2)$; $B = (0, 2, 4)$.

In Exercises 5–8 determine the norm of the given vector

5. $\begin{bmatrix} 2 \\ 3 \\ 1 \end{bmatrix}$ **6.** $\begin{bmatrix} 3 \\ 2 \\ -3 \end{bmatrix}$ **7.** $\begin{bmatrix} -1 \\ -2 \\ 3 \end{bmatrix}$ **8.** $\begin{bmatrix} 3 \\ -5 \\ 7 \end{bmatrix}$

In Exercises 9–16 determine the indicated vector where

$$\mathbf{u} = \begin{bmatrix} 1 \\ -2 \\ 3 \end{bmatrix} \qquad \mathbf{v} = \begin{bmatrix} 4 \\ 6 \\ -7 \end{bmatrix} \qquad \mathbf{w} = \begin{bmatrix} 2 \\ 0 \\ 1 \end{bmatrix}$$

9. $3\mathbf{u} + 2\mathbf{v} - 2\mathbf{w}$ **10.** $4\mathbf{u} - 3\mathbf{v} + 2\mathbf{w}$

11. $\dfrac{1}{\|\mathbf{u}\|}\mathbf{u} - \dfrac{1}{\|\mathbf{v}\|}\mathbf{v} + \dfrac{1}{\|\mathbf{w}\|}\mathbf{w}$ **12.** $\dfrac{1}{\|\mathbf{u} + \mathbf{v}\|}\mathbf{w}$

13. $\dfrac{1}{\|\mathbf{u} + \mathbf{v} + \mathbf{w}\|}(\mathbf{u} + \mathbf{v} + \mathbf{w})$ **14.** $\|\mathbf{u} - \mathbf{v}\|\mathbf{v} + \|\mathbf{u} + \mathbf{v}\|\mathbf{v} + \mathbf{w}$

15. $\left\|\dfrac{1}{\|\mathbf{w}\|}\mathbf{v}\right\|$ **16.** $(\|\mathbf{u}\| - \|\mathbf{v}\|)\mathbf{w}$

17. Show that $\|\mathbf{u}\| - \|\mathbf{v}\| \le \|\mathbf{u} + \mathbf{v}\|$. [Hint: $\mathbf{u} = (\mathbf{u} + \mathbf{v}) - \mathbf{v}$.]

3.5 INNER PRODUCT IN 3-SPACE

Proceeding by analogy with the discussion of vectors in 2-space, we define the inner product of two vectors in 3-space.

    Definition 6.   If

$$\mathbf{u} = \begin{bmatrix} a_1 \\ b_1 \\ c_1 \end{bmatrix} \qquad \text{and} \qquad \mathbf{v} = \begin{bmatrix} a_2 \\ b_2 \\ c_2 \end{bmatrix}$$

are two vectors in 3-space, then the **inner product**, or **dot product**, of **u** and **v** (denoted by **u · v**) is defined by

$$\mathbf{u} \cdot \mathbf{v} = a_1 a_2 + b_1 b_2 + c_1 c_2$$

We emphasize that, just as in 2-space, the inner product of two vectors is a *number*, not another vector.

Example 1.   If

$$\mathbf{u} = \begin{bmatrix} 6 \\ -4 \\ 8 \end{bmatrix} \quad \text{and} \quad \mathbf{v} = \begin{bmatrix} -5 \\ 1 \\ 3 \end{bmatrix}$$

then

$$\mathbf{u} \cdot \mathbf{v} = (6)(-5) + (-4)(1) + (8)(3) = -10$$

The following theorem lists the basic properties of the inner product of two vectors. Its proof is analogous to that of Theorem 2 and will be omitted.

Theorem 8.   *Let* **u**, **v**, *and* **w** *be arbitrary vectors in 3-space. Then*

1. $\mathbf{u} \cdot \mathbf{v} = \mathbf{v} \cdot \mathbf{u}$
2. $\mathbf{u} \cdot \mathbf{u} = \|\mathbf{u}\|^2$
3. $\mathbf{u} \cdot \mathbf{u} > 0$ if $\mathbf{u} \neq 0$
4. $\mathbf{u} \cdot \mathbf{0} = 0$
5. $(\mathbf{u} \pm \mathbf{v}) \cdot \mathbf{w} = \mathbf{u} \cdot \mathbf{w} \pm \mathbf{v} \cdot \mathbf{w}$
6. *If k is any number, then* $(k\mathbf{u}) \cdot \mathbf{v} = k(\mathbf{u} \cdot \mathbf{v}) = \mathbf{u} \cdot (k\mathbf{v})$.

The geometric interpretation of the inner product of vectors in 3-space is precisely the same as for vectors in 2-space. Let **u** and **v** be any nonzero vectors in 3-space. Let A be any point in 3-space and let $\overrightarrow{AB}$ and $\overrightarrow{AC}$ be representations of **u** and **v**, respectively (see Figure 3.26). The three points A, B, and C determine a plane. (Section 3.7 shows how to determine this plane.) Thus $\overrightarrow{AB}$ and $\overrightarrow{AC}$ are directed line segments in a plane. The angle between **u** and **v** is defined to be the angle between

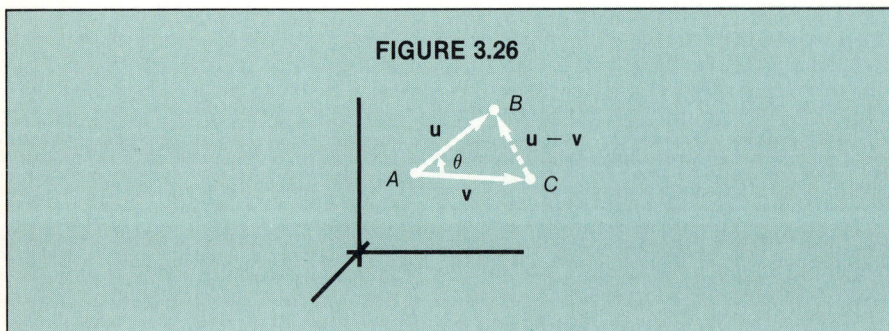

FIGURE 3.26

$\overrightarrow{AB}$ and $\overrightarrow{AC}$. Using the Law of Cosines, we see that

(1)
$$\|\mathbf{u} - \mathbf{v}\|^2 = \|\mathbf{u}\|^2 + \|\mathbf{v}\|^2 - 2\|\mathbf{u}\|\,\|\mathbf{v}\| \cos \theta$$

The identity in the following theorem can be established precisely as we established the same identity for vectors in 2-space (see the proof of Theorem 3 in Section 3.3).

Theorem 9. *Let* $\mathbf{u}$ *and* $\mathbf{v}$ *be any two vectors in 3-space. Then*

$$\|\mathbf{u} - \mathbf{v}\|^2 = \|\mathbf{u}\|^2 + \|\mathbf{v}\|^2 - 2\mathbf{u} \cdot \mathbf{v}$$

Combining the identities in (1) and Theorem 9, we obtain the following theorem.

Theorem 10. *Let* $\mathbf{u}$ *and* $\mathbf{v}$ *be any two vectors in 3-space. If* θ *is the angle between* $\mathbf{u}$ *and* $\mathbf{v}$, *then*

$$\mathbf{u} \cdot \mathbf{v} = \|\mathbf{u}\|\,\|\mathbf{v}\| \cos \theta$$

This is a powerful theorem. It is usually difficult to compute directly the angle between two vectors $\mathbf{u}$ and $\mathbf{v}$. Theorem 10 relates this angle to three quantities ($\|\mathbf{u}\|$, $\|\mathbf{v}\|$, and $\mathbf{u} \cdot \mathbf{v}$) that are easy to compute. However, in practice it is usually $\mathbf{u} \cdot \mathbf{v}$, rather than the angle between $\mathbf{u}$ and $\mathbf{v}$, that is needed.

Example 2. By Theorem 10, the angle θ between

$$\mathbf{u} = \begin{bmatrix} 1 \\ 2 \\ 3 \end{bmatrix} \quad \text{and} \quad \mathbf{v} = \begin{bmatrix} 2 \\ -7 \\ 9 \end{bmatrix}$$

satisfies

$$\cos \theta = \frac{\mathbf{u} \cdot \mathbf{v}}{\|\mathbf{u}\|\,\|\mathbf{v}\|} = \frac{(1)(2) + (2)(-7) + (3)(9)}{\sqrt{1^2 + 2^2 + 3^2}\,\sqrt{2^2 + (-7)^2 + 9^2}} = \frac{15}{\sqrt{1876}}$$

so that $\theta \simeq 1.217$ radians.

Example 3. If a constant force of f pounds moves an object a distance d feet along a straight line and if the force is acting in the direction of the motion of the object, then physicists define the work W done by the force by $W = fd$.

Suppose that a force denoted by a vector $\mathbf{F}$ causes an object to move from a point A to a point B (see Figure 3.27). Then the component of $\mathbf{F}$ in the direction of the motion is $\cos \theta\, \mathbf{F}$. Let $\mathbf{D}$ denote the vector, called the displacement vector, having

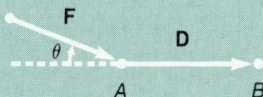

FIGURE 3.27

AB as a representation. Then the work done by $\mathbf{F}$ is given by

$$W = \|\mathbf{F} \cos \theta\| \|\mathbf{D}\|$$
$$= \|\mathbf{F}\| \|\mathbf{D}\| \cos \theta$$

Using Theorem 10 we conclude that

$$W = \mathbf{F} \cdot \mathbf{D}$$

That is, the work is the inner product of the force vector and the displacement vector.

Definition 7.   Let θ be the angle between two nonzero vectors $\mathbf{u}$ and $\mathbf{v}$ in 3-space. Then $\mathbf{u}$ and $\mathbf{v}$ are said to be **orthogonal** if $\theta = \pi/2$ and **parallel** if $\theta = 0$ or $\theta = \pi$. The zero vector is orthogonal to every vector.

From Theorem 10 we obtain the following characterization of orthogonal vectors.

Theorem 11.   *Let $\mathbf{u}$ and $\mathbf{v}$ be vectors in 3-space. Then $\mathbf{u}$ and $\mathbf{v}$ are orthogonal if and only if* $\mathbf{u} \cdot \mathbf{v} = 0$.

Often it is convenient to write a vector

$$\mathbf{v} = \begin{bmatrix} a \\ b \\ c \end{bmatrix}$$

as

$$\mathbf{v} = a\mathbf{i} + b\mathbf{j} + c\mathbf{k}$$

where

$$\mathbf{i} = \begin{bmatrix} 1 \\ 0 \\ 0 \end{bmatrix} \qquad \mathbf{j} = \begin{bmatrix} 0 \\ 1 \\ 0 \end{bmatrix} \qquad \mathbf{k} = \begin{bmatrix} 0 \\ 0 \\ 1 \end{bmatrix}$$

We can easily verify that $\mathbf{i} \cdot \mathbf{j} = 0$, $\mathbf{i} \cdot \mathbf{k} = 0$, and $\mathbf{j} \cdot \mathbf{k} = 0$; therefore, each of $\mathbf{i}$, $\mathbf{j}$, and $\mathbf{k}$ is orthogonal to the other two. Each of these vectors has a norm of 1. Their directions along the coordinate axes are shown in Figure 3.28.

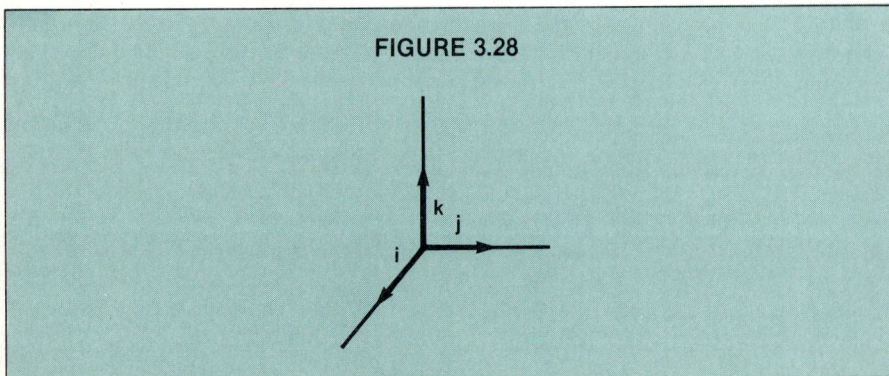

FIGURE 3.28

─────────────────────── **Exercises** ───────────────────────

In Exercises 1–6 compute $\mathbf{u} \cdot \mathbf{v}$.

1. $\mathbf{u} = \begin{bmatrix} 1 \\ 5 \\ 4 \end{bmatrix} \quad \mathbf{v} = \begin{bmatrix} 2 \\ -2 \\ 1 \end{bmatrix}$
 2. $\mathbf{u} = \begin{bmatrix} -1 \\ 3 \\ 6 \end{bmatrix} \quad \mathbf{v} = \begin{bmatrix} 3 \\ 2 \\ 7 \end{bmatrix}$

3. $\mathbf{u} = \begin{bmatrix} -2 \\ 3 \\ 7 \end{bmatrix} \quad \mathbf{v} = \begin{bmatrix} 6 \\ -1 \\ 0 \end{bmatrix}$
 4. $\mathbf{u} = \begin{bmatrix} 6 \\ 8 \\ -2 \end{bmatrix} \quad \mathbf{v} = \begin{bmatrix} 3 \\ 2 \\ 7 \end{bmatrix}$

5. $\mathbf{u} = \mathbf{i} - \mathbf{j} + 2\mathbf{k}; \mathbf{v} = 2\mathbf{i} + 3\mathbf{k}.$
 6. $\mathbf{u} = 3\mathbf{i} - 5\mathbf{j} + 4\mathbf{k}; \mathbf{v} = 2\mathbf{i} - \mathbf{j} + 7\mathbf{k}.$

In Exercises 7–10 find the angle between the given vectors.

7. $\mathbf{u} = \begin{bmatrix} 1 \\ 2 \\ 3 \end{bmatrix} \quad \mathbf{v} = \begin{bmatrix} 1 \\ -3 \\ 1 \end{bmatrix}$
 8. $\mathbf{u} = \begin{bmatrix} 1 \\ 0 \\ 1 \end{bmatrix} \quad \mathbf{v} = \begin{bmatrix} -3 \\ -3 \\ 5 \end{bmatrix}$

9. $\mathbf{u} = 2\mathbf{i} + 4\mathbf{j} - 6\mathbf{k}; \mathbf{v} = \mathbf{i} - 5\mathbf{j} + 2\mathbf{k}.$
10. $\mathbf{u} = -5\mathbf{i} + 2\mathbf{j} + \mathbf{k}; \mathbf{v} = 3\mathbf{i} + 7\mathbf{k}.$

In Exercises 11–18 compute the indicated vector or number where $\mathbf{u} = 3\mathbf{i} + 2\mathbf{j} - \mathbf{k}$ and $\mathbf{v} = 2\mathbf{i} - \mathbf{j} + 3\mathbf{k}.$

11. $2\mathbf{u} + 4\mathbf{v}$
 12. $3\mathbf{u} - 5\mathbf{v}$

13. $\mathbf{u} \cdot \mathbf{v}$
 14. $\mathbf{u} \cdot \mathbf{v} - \|\mathbf{v}\|$

15. $\|\mathbf{u}\| + \|\mathbf{v}\|$
 16. $\|\|\mathbf{u}\|\mathbf{v}\|$

17. $\dfrac{1}{\|\mathbf{u}\|}\mathbf{v} + \dfrac{1}{\|\mathbf{v}\|}\mathbf{u}$
 18. $\dfrac{1}{\|\mathbf{u} + \mathbf{v}\|}(\mathbf{u} + \mathbf{v})$

19. For what value of c is $\mathbf{u} = 3\mathbf{i} + 4\mathbf{j} + \mathbf{k}$ orthogonal to $\mathbf{v} = 2\mathbf{i} - \mathbf{j} + c\mathbf{k}$?
20. Is there a value of c such that $\mathbf{u} = 2\mathbf{i} + 3\mathbf{j} + c\mathbf{k}$ and $\mathbf{v} = -3\mathbf{i} + 4\mathbf{j} + c\mathbf{k}$ are orthogonal?
21. Find all vectors that are orthogonal to

$$\mathbf{v} = \begin{bmatrix} 1 \\ 2 \\ 3 \end{bmatrix}$$

22. Show that $\|\mathbf{u}\|^2 + \|\mathbf{v}\|^2 = \|\mathbf{u} + \mathbf{v}\|^2$ if and only if $\mathbf{u}$ and $\mathbf{v}$ are orthogonal.
23. Show that $\|\mathbf{u} + \mathbf{v}\|^2 + \|\mathbf{u} - \mathbf{v}\|^2 = 2\|\mathbf{u}\|^2 + 2\|\mathbf{v}\|^2.$
24. Show that $\|\mathbf{u} + \mathbf{v}\|^2 - \|\mathbf{u} - \mathbf{v}\|^2 = 4\mathbf{u} \cdot \mathbf{v}.$

3.6 CROSS PRODUCT

In many applications of vectors in physics and engineering it is necessary to construct a vector that is orthogonal to two given vectors. We now define a type of vector multiplication that yields a vector orthogonal to the two vectors being multiplied.

Definition 8. Let

$$\mathbf{u} = \begin{bmatrix} a_1 \\ b_1 \\ c_1 \end{bmatrix} \qquad \text{and} \qquad \mathbf{v} = \begin{bmatrix} a_2 \\ b_2 \\ c_2 \end{bmatrix}$$

be vectors in 3-space. The **cross product** of **u** and **v**, denoted by **u** × **v**, is the vector defined by

$$\mathbf{u} \times \mathbf{v} = \begin{bmatrix} b_1c_2 - c_1b_2 \\ c_1a_2 - a_1c_2 \\ a_1b_2 - b_1a_2 \end{bmatrix}$$

Example 1. If

$$\mathbf{u} = \begin{bmatrix} 2 \\ 5 \\ 3 \end{bmatrix} \qquad \text{and} \qquad \mathbf{v} = \begin{bmatrix} 4 \\ -2 \\ 1 \end{bmatrix}$$

then

$$\mathbf{u} \times \mathbf{v} = \begin{bmatrix} (5)(1) - (3)(-2) \\ (3)(4) - (2)(1) \\ (2)(-2) - (5)(4) \end{bmatrix} = \begin{bmatrix} 11 \\ 10 \\ -24 \end{bmatrix}$$

A mnemonic method for remembering the formula for **u** × **v** can be written in terms of a "determinant." To help in remembering the formula, we shall use the vectors **i**, **j**, and **k** (as in Section 3.5 following Theorem 11). Then

$$\mathbf{u} \times \mathbf{v} = (b_1c_2 - c_1b_2)\mathbf{i} + (c_1a_2 - a_1c_2)\mathbf{j} + (a_1b_2 - b_1a_2)\mathbf{k}$$

$$= \det\begin{bmatrix} b_1 & c_1 \\ b_2 & c_2 \end{bmatrix}\mathbf{i} - \det\begin{bmatrix} a_1 & c_1 \\ a_2 & c_2 \end{bmatrix}\mathbf{j} + \det\begin{bmatrix} a_1 & b_1 \\ a_2 & b_2 \end{bmatrix}\mathbf{k} \tag{1}$$

If we treat **i**, **j**, and **k** strictly as symbols and recall the cofactor expansions in Section 2.6, we see that the expression for **u** × **v** in (1) is the same as

$$\det\begin{bmatrix} \mathbf{i} & \mathbf{j} & \mathbf{k} \\ a_1 & b_1 & c_1 \\ a_2 & b_2 & c_2 \end{bmatrix} \tag{2}$$

Thus

$$\mathbf{u} \times \mathbf{v} = \det\begin{bmatrix} \mathbf{i} & \mathbf{j} & \mathbf{k} \\ a_1 & b_1 & c_1 \\ a_2 & b_2 & c_2 \end{bmatrix} \tag{3}$$

We emphasize that the "determinants" in (2) and (3) are not determinants as defined in Chapter 2, because **i**, **j**, and **k** are not numbers. Nonetheless, the formula in (3) gives a convenient way to remember the formula for **u** × **v**.

Example 2.   Consider the vectors **u** and **v** from Example 1. Then

$$\mathbf{u} \times \mathbf{v} = \det \begin{bmatrix} \mathbf{i} & \mathbf{j} & \mathbf{k} \\ 2 & 5 & 3 \\ 4 & -2 & 1 \end{bmatrix}$$

$$= \det \begin{bmatrix} 5 & 3 \\ -2 & 1 \end{bmatrix} \mathbf{i} - \det \begin{bmatrix} 2 & 3 \\ 4 & 1 \end{bmatrix} \mathbf{j} + \det \begin{bmatrix} 2 & 5 \\ 4 & -2 \end{bmatrix} \mathbf{k}$$

$$= 11\mathbf{i} + 10\mathbf{j} - 24\mathbf{k}$$

$$= \begin{bmatrix} 11 \\ 10 \\ -24 \end{bmatrix}$$

Example 3.   Recall from Section 3.5 that

$$\mathbf{i} = \begin{bmatrix} 1 \\ 0 \\ 0 \end{bmatrix} \qquad \mathbf{j} = \begin{bmatrix} 0 \\ 1 \\ 0 \end{bmatrix} \qquad \mathbf{k} = \begin{bmatrix} 0 \\ 0 \\ 1 \end{bmatrix}$$

Then

$$\mathbf{i} \times \mathbf{j} = \begin{bmatrix} \mathbf{i} & \mathbf{j} & \mathbf{k} \\ 1 & 0 & 0 \\ 0 & 1 & 0 \end{bmatrix}$$

$$= \det \begin{bmatrix} 0 & 0 \\ 1 & 0 \end{bmatrix} \mathbf{i} - \det \begin{bmatrix} 1 & 0 \\ 0 & 0 \end{bmatrix} \mathbf{j} + \det \begin{bmatrix} 1 & 0 \\ 0 & 1 \end{bmatrix} \mathbf{k}$$

$$= \mathbf{k}$$

Similar calculations show that

$$\mathbf{i} \times \mathbf{i} = 0 \qquad \mathbf{j} \times \mathbf{i} = -\mathbf{k} \qquad \mathbf{k} \times \mathbf{i} = \mathbf{j}$$

$$\mathbf{i} \times \mathbf{j} = \mathbf{k} \qquad \mathbf{j} \times \mathbf{j} = 0 \qquad \mathbf{k} \times \mathbf{j} = -\mathbf{i}$$

$$\mathbf{i} \times \mathbf{k} = -\mathbf{j} \qquad \mathbf{j} \times \mathbf{k} = \mathbf{i} \qquad \mathbf{k} \times \mathbf{k} = 0$$

Figure 3.29 is useful for remembering these identities. The cross product of two consecutive vectors going clockwise is the next vector. The cross product of two consecutive vectors going counterclockwise is the negative of the next vector.

One of the most important properties of **u** × **v** is that it is orthogonal to both **u** and **v**. This may be verified by direct computation. If the dot product of **u** and **u** × **v**

FIGURE 3.29

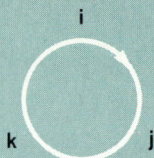

---

equals zero, then **u** and **u** × **v** are orthogonal. We shall use vectors **u** and **v** from Definition 8 to demonstrate the procedure.

$$\mathbf{u} \cdot (\mathbf{u} \times \mathbf{v}) = \begin{bmatrix} a_1 \\ b_1 \\ c_1 \end{bmatrix} \cdot \begin{bmatrix} b_1 c_2 - c_1 b_2 \\ c_1 a_2 - a_1 c_2 \\ a_1 b_2 - b_1 a_2 \end{bmatrix}$$

$$= a_1(b_1 c_2 - c_1 b_2) + b_1(c_1 a_2 - a_1 c_2) + c_1(a_1 b_2 - b_1 a_2)$$

$$= a_1 b_1 c_2 - a_1 c_1 b_2 + b_1 c_1 a_2 - b_1 a_1 c_2 + c_1 a_1 b_2 - c_1 b_1 a_2$$

$$= 0$$

A similar calculation shows that **v** and **u** × **v** are orthogonal. These and other properties of the cross product are listed in the following theorem.

Theorem 12. *Let* **u**, **v**, *and* **w** *be arbitrary vectors in 3-space and let k be any number. Then*

1. $\mathbf{u} \times \mathbf{0} = \mathbf{0} \times \mathbf{u} = \mathbf{0}$
2. $\mathbf{u} \times \mathbf{v} = -\mathbf{v} \times \mathbf{u}$
3. $(k\mathbf{u}) \times \mathbf{v} = k(\mathbf{u} \times \mathbf{v}) = \mathbf{u} \times (k\mathbf{v})$
4. $\mathbf{u} \times (\mathbf{v} + \mathbf{w}) = (\mathbf{u} \times \mathbf{v}) + (\mathbf{u} \times \mathbf{w})$
5. $(\mathbf{u} \times \mathbf{v}) \cdot \mathbf{w} = \mathbf{u} \cdot (\mathbf{v} \times \mathbf{w})$
6. $\mathbf{u} \cdot (\mathbf{u} \times \mathbf{v}) = \mathbf{v} \cdot (\mathbf{u} \times \mathbf{v}) = 0$
7. $\|\mathbf{u} \times \mathbf{v}\|^2 = \|\mathbf{u}\|^2 \|\mathbf{v}\|^2 - (\mathbf{u} \cdot \mathbf{v})^2$ *(Lagrange's identity)*

The proofs of the identities in this theorem involve only straightforward calculations and will be left as exercises. The reader is warned that in general

$$\mathbf{u} \times (\mathbf{v} \times \mathbf{w}) \neq (\mathbf{u} \times \mathbf{v}) \times \mathbf{w}$$

For example, using the vectors **i**, **j**, and **k** of Example 3, we find that

$$\mathbf{i} \times (\mathbf{i} \times \mathbf{j}) = \mathbf{i} \times \mathbf{k} = -\mathbf{j}$$

and

$$(\mathbf{i} \times \mathbf{i}) \times \mathbf{j} = \mathbf{0} \times \mathbf{j} = \mathbf{0}$$

so that

$$\mathbf{i} \times (\mathbf{i} \times \mathbf{j}) \neq (\mathbf{i} \times \mathbf{i}) \times \mathbf{j}$$

We have seen that **u** × **v** is orthogonal to both **u** and **v**. Let $\overrightarrow{AB}$ and $\overrightarrow{AC}$ be representations of **u** and **v**, respectively, having a point A as their common initial

FIGURE 3.30

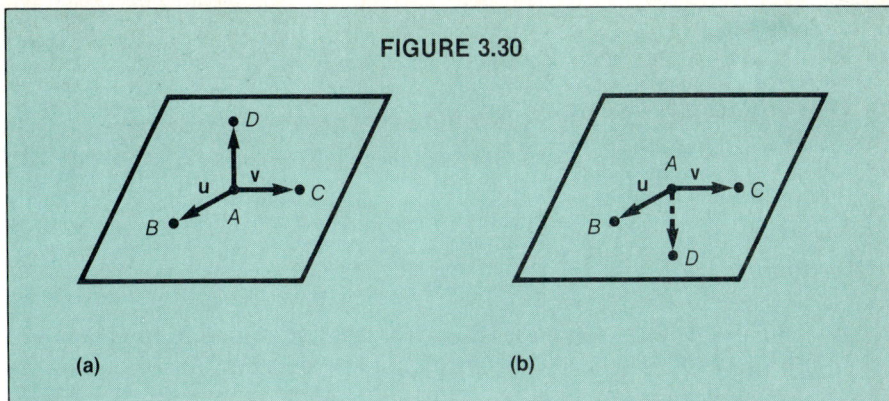

(a)   (b)

FIGURE 3.31

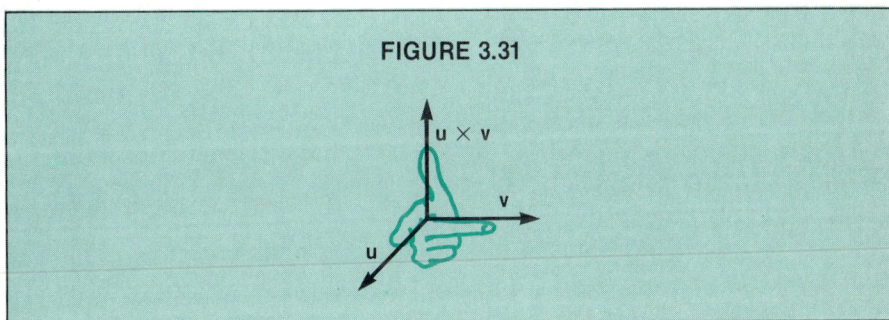

point. The three points A, B, and C determine a plane. (Section 3.7 shows how to find this plane.) Therefore, a representation $\overrightarrow{AD}$ of $\mathbf{u} \times \mathbf{v}$ is a line segment perpendicular to this plane. There are only two directions in which this direct line segment can point (see Figure 3.30).

The direction of $\mathbf{u} \times \mathbf{v}$ can be determined by the following rule, called the **right-hand rule**: If the right hand is placed so that the index finger points in the direction of $\mathbf{u}$ and the middle finger points in the direction of $\mathbf{v}$, then the thumb points in the direction of $\mathbf{u} \times \mathbf{v}$ (see Figure 3.31).

A geometric interpretation of $\|\mathbf{u} \times \mathbf{v}\|$ can be obtained by using Lagrange's identity from part 7 of Theorem 12 and the identity $\mathbf{u} \cdot \mathbf{v} = \|\mathbf{u}\| \|\mathbf{v}\| \cos \theta$ from Theorem 10. Then

$$\|\mathbf{u} \times \mathbf{v}\|^2 = \|\mathbf{u}\|^2 \|\mathbf{v}\|^2 - (\mathbf{u} \cdot \mathbf{v})^2$$
$$= \|\mathbf{u}\|^2 \|\mathbf{v}\|^2 - \|\mathbf{u}\|^2 \|\mathbf{v}\|^2 \cos^2 \theta$$
$$= \|\mathbf{u}\|^2 \|\mathbf{v}\|^2 (1 - \cos^2 \theta)$$
$$= \|\mathbf{u}\|^2 \|\mathbf{v}\|^2 \sin^2 \theta$$

Since $0 \leq \theta \leq \pi$ we have

$$\|\mathbf{u} \times \mathbf{v}\| = \|\mathbf{u}\| \|\mathbf{v}\| \sin \theta$$

This identity has an important geometric interpretation. The vectors $\mathbf{u}$ and $\mathbf{v}$ can be considered to be adjacent sides of a parallelogram (see Figure 3.32). The area of a

FIGURE 3.32

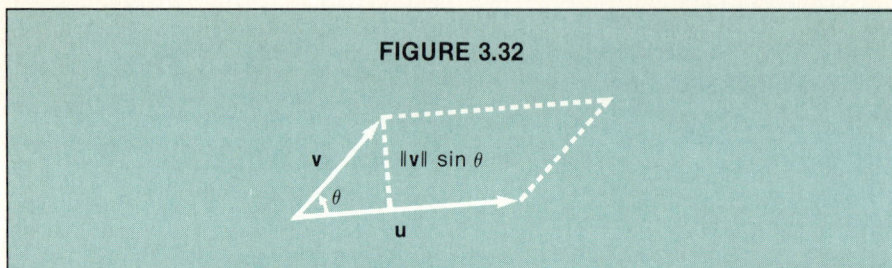

parallelogram is the product of the length of a base times its altitude. In Figure 3.32, $\|\mathbf{u}\|$ is the length of a base and $\|\mathbf{v}\|\sin\theta$ is its altitude. Therefore $\|\mathbf{u}\|\,\|\mathbf{v}\|\sin\theta$ is the area of the parallelogram determined by $\mathbf{u}$ and $\mathbf{v}$. We have proved the following theorem.

Theorem 13. *Let $\mathbf{u}$ and $\mathbf{v}$ be nonzero vectors in 3-space; let θ denote the angle between $\mathbf{u}$ and $\mathbf{v}$; and let A denote the area of the parallelogram determined by $\mathbf{u}$ and $\mathbf{v}$. Then*

$$\|\mathbf{u}\times\mathbf{v}\| = \|\mathbf{u}\|\,\|\mathbf{v}\|\sin\theta = A$$

Thus $\mathbf{u}\times\mathbf{v}$ is a vector that is orthogonal to both $\mathbf{u}$ and $\mathbf{v}$; its direction is determined by the right hand rule; and its length is equal to the area of the parallelogram determined by $\mathbf{u}$ and $\mathbf{v}$.

Example 4. Consider the quadrangle with the points $A = (2,\ 4,\ 6)$, $B = (3,\ 6,\ 9)$, $C = (-1,\ 4,\ 6)$, and $D = (-2,\ 2,\ 3)$ as vertices. Straightforward calculations show that both $\overrightarrow{AB}$ and $\overrightarrow{DC}$ are representations of the vector

$$\begin{bmatrix} 1 \\ 2 \\ 3 \end{bmatrix}$$

while both $\overrightarrow{AD}$ and $\overrightarrow{BC}$ are representations of the vector

$$\begin{bmatrix} -4 \\ -2 \\ -3 \end{bmatrix}$$

Therefore, the quadrangle is a parallelogram. By Theorem 13 the area of the parallelogram is given by

$$\|\mathbf{u}\times\mathbf{v}\| = \left\| \det \begin{bmatrix} \mathbf{i} & \mathbf{j} & \mathbf{k} \\ 1 & 2 & 3 \\ -4 & -2 & -3 \end{bmatrix} \right\|$$

$$= \left\| \det \begin{bmatrix} 2 & 3 \\ -2 & -3 \end{bmatrix}\mathbf{i} - \det \begin{bmatrix} 1 & 3 \\ -4 & -3 \end{bmatrix}\mathbf{j} + \det \begin{bmatrix} 1 & 2 \\ -4 & -2 \end{bmatrix}\mathbf{k} \right\|$$

$$= \|-9\mathbf{j} + 6\mathbf{k}\| = \left\| \begin{bmatrix} 0 \\ -9 \\ 6 \end{bmatrix} \right\| = 3\sqrt{13}$$

From Theorem 13 we obtain the following characterization of parallel vectors.

Theorem 14. *Two nonzero vectors* **u** *and* **v** *in 3-space are parallel if and only if* **u** × **v** = 0.

The proof of this theorem is left as an exercise.

──────────────── **Exercises** ────────────────

In Exercises 1–8 compute **u** · **v**.

1. $\mathbf{u} = \begin{bmatrix} 1 \\ 2 \\ 3 \end{bmatrix}$ $\mathbf{v} = \begin{bmatrix} 2 \\ 0 \\ -1 \end{bmatrix}$

2. $\mathbf{u} = \begin{bmatrix} 2 \\ 1 \\ 2 \end{bmatrix}$ $\mathbf{v} = \begin{bmatrix} 1 \\ 6 \\ 0 \end{bmatrix}$

3. $\mathbf{u} = \begin{bmatrix} 3 \\ -2 \\ -4 \end{bmatrix}$ $\mathbf{v} = \begin{bmatrix} -7 \\ -1 \\ -4 \end{bmatrix}$

4. $\mathbf{u} = \begin{bmatrix} 3 \\ 2 \\ 0 \end{bmatrix}$ $\mathbf{v} = \begin{bmatrix} -3 \\ 4 \\ -1 \end{bmatrix}$

5. $\mathbf{u} = -\mathbf{i} + 3\mathbf{j} + 4\mathbf{k}; \mathbf{v} = 2\mathbf{i} - 5\mathbf{j} + \mathbf{k}$.
6. $\mathbf{u} = 5\mathbf{i} + 3\mathbf{k}; \mathbf{v} = \mathbf{j} - \mathbf{k}$.
7. $\mathbf{u} = -\mathbf{i} + 4\mathbf{j} + 3\mathbf{k}; \mathbf{v} = 2\mathbf{i} - 7\mathbf{j} + 5\mathbf{k}$.
8. $\mathbf{u} = 2\mathbf{i} + 3\mathbf{j}; \mathbf{v} = 3\mathbf{i} + 2\mathbf{k}$.

In Exercises 9–16 compute the indicated number or vector where $\mathbf{u} = 2\mathbf{i} - \mathbf{j} + 3\mathbf{k}$ and $\mathbf{v} = \mathbf{i} + 2\mathbf{j} - \mathbf{k}$.

9. $(\mathbf{u} \times \mathbf{v}) - (\mathbf{v} \times \mathbf{u})$
10. $\mathbf{u} \times (\mathbf{u} + \mathbf{v})$
11. $(\mathbf{u} \times \mathbf{v}) \cdot \mathbf{u}$
12. $[(\mathbf{v} \times \mathbf{u}) + \mathbf{u}] \cdot \mathbf{v}$
13. $\|\mathbf{u} \times \mathbf{v}\| - \mathbf{u} \cdot \mathbf{v}$
14. $\|\mathbf{u} \times \mathbf{v}\| (\mathbf{u} \cdot \mathbf{v})(\mathbf{u} - \mathbf{v})$
15. $\mathbf{i} \times (\mathbf{u} \times \mathbf{j})$
16. $\mathbf{u} \times (\mathbf{i} \times \mathbf{v})$
17. Compute the area of the parallelogram having vertices $(1, 2, -1)$, $(2, 3, 0)$, $(2, 4, 0)$, and $(1, 3, -1)$.
18. Compute the area of the triangle with vertices $(0, 2, 7)$, $(2, -5, 3)$, and $(1, 1, 1)$.
19. Let $\mathbf{u} = \mathbf{i} - 2\mathbf{j} + \mathbf{k}$ and $\mathbf{w} = -2\mathbf{i} + \mathbf{j} + 4\mathbf{k}$. Find all vectors **v** such that $\mathbf{u} \times \mathbf{v} = \mathbf{w}$.
20. Let $\mathbf{u} = u_1\mathbf{i} + u_2\mathbf{j} + u_3\mathbf{k}; \mathbf{v} = v_1\mathbf{i} + v_2\mathbf{j} + v_3\mathbf{k}$; and $\mathbf{w} = w_1\mathbf{i} + w_2\mathbf{j} + w_3\mathbf{k}$. Show that

$$\mathbf{u} \cdot (\mathbf{v} \times \mathbf{w}) = \det \begin{bmatrix} u_1 & u_2 & u_3 \\ v_1 & v_2 & v_3 \\ w_1 & w_2 & w_3 \end{bmatrix}$$

21. Let $\mathbf{u} = \mathbf{i} + 2\mathbf{j} + 3\mathbf{k}$. Find vectors **v** and **w** that are both orthogonal to **u** and to each other.

3.7 PLANES IN 3-SPACE

In 2-space a line is uniquely determined by a point on the line and the slope of the line. In a similar fashion a plane in 3-space is uniquely determined by a point in the plane and by the inclination of the plane. The most common way to describe the inclination of a plane is to specify a vector **N** perpendicular to the plane. Such a vector is called a **normal** vector to the plane.

Let $A = (x_0, y_0, z_0)$ be a point in 3-space and let $\mathbf{N} = a\mathbf{i} + b\mathbf{j} + c\mathbf{k}$ be a nonzero vector. It is easy to find an equation for the plane that contains the point A and has the normal vector **N**. Indeed, if $B = (x, y, z)$ is any point in the plane that contains point A, then the vector $\mathbf{v}$ represented by $\overrightarrow{AB}$ is orthogonal to $\mathbf{N}$. That is,

$$
\begin{aligned}
0 &= \mathbf{v} \cdot \mathbf{N} \\
&= [(x - x_0)\mathbf{i} + (y - y_0)\mathbf{j} + (z - z_0)\mathbf{k}] \cdot (a\mathbf{i} + b\mathbf{j} + c\mathbf{k}) \\
&= a(x - x_0) + b(y - y_0) + c(z - z_0)
\end{aligned}
\tag{1}
$$

is an equation, called the **point-normal equation**, for the plane containing the point (x_0, y_0, z_0) with the normal vector $a\mathbf{i} + b\mathbf{j} + c\mathbf{k}$.

The importance of the normal vector $\mathbf{N}$ is found in its *direction* and not simply in $\mathbf{N}$ itself. For example, if c_1 is any nonzero number, then the point-normal form of an equation for the plane passing through (x_0, y_0, z_0) having the vector $c_1\mathbf{N} = c_1 a\mathbf{i} + c_1 b\mathbf{j} + c_1 c\mathbf{k}$ as a normal vector is

$$
c_1 a(x - x_0) + c_1 b(y - y_0) + c_1 c(z - z_0) = 0
$$

Since $c_1 \neq 0$, this equation is clearly equivalent to the equation in (1). Thus, if our calculations yield $\mathbf{N} = 4\mathbf{i} + 16\mathbf{j} - 12\mathbf{k}$, we can obtain the same plane by using $\mathbf{N}_1 = \mathbf{i} + 4\mathbf{j} - 3\mathbf{k}$ and the arithmetic is easier.

Example 1.   Let $A = (2, 6, 1)$ be a point in 3-space and let $\mathbf{N} = 7\mathbf{i} + 2\mathbf{j} - 3\mathbf{k}$. Then the point-normal form of an equation for the plane passing through A and having $\mathbf{N}$ as a normal is

$$
7(x - 2) + 2(y - 6) - 3(z - 1) = 0 \tag{2}
$$

If we multiply out and collect terms in equation (1), we have

$$
ax + by + cz + d = 0 \tag{3}
$$

where $d = -(ax_0 + by_0 + cz_0)$. This equation is called the **general form** of the equation of a plane. We have just shown that every plane is the graph of an equation of the form in (3). The following theorem shows that the converse is also true, provided at least one of a, b, and c is nonzero.

Theorem 15.   *If $a, b, c,$ and d are numbers with at least one of $a, b,$ and c nonzero, then the graph of the equation*

$$
ax + by + cz + d = 0
$$

is a plane having the vector $\mathbf{N} = a\mathbf{i} + b\mathbf{j} + c\mathbf{k}$ as a normal vector.

Proof: By hypothesis, at least one of the coefficients $a, b,$ and c is nonzero. Suppose that $a \neq 0$. Then the equation $ax + by + cz + d = 0$ can be rewritten as

$$
a\left(x + \frac{d}{a}\right) + by + cz = 0
$$

which is the point-normal form of an equation of a plane passing through the point $(-d/a, 0, 0)$ having $\mathbf{N} = a\mathbf{i} + b\mathbf{j} + c\mathbf{k}$ as a normal vector.

If $a = 0$, then either $b \neq 0$ or $c \neq 0$. In either case a slight modification of the above argument completes the proof. ∎

The planes determined by the equations $z = 0$, $y = 0$, and $x = 0$ are called the xy-plane, the xz-plane, and the yz-plane, respectively (see Figure 3.33).

FIGURE 3.33

xy-plane   xz-plane   yz-plane

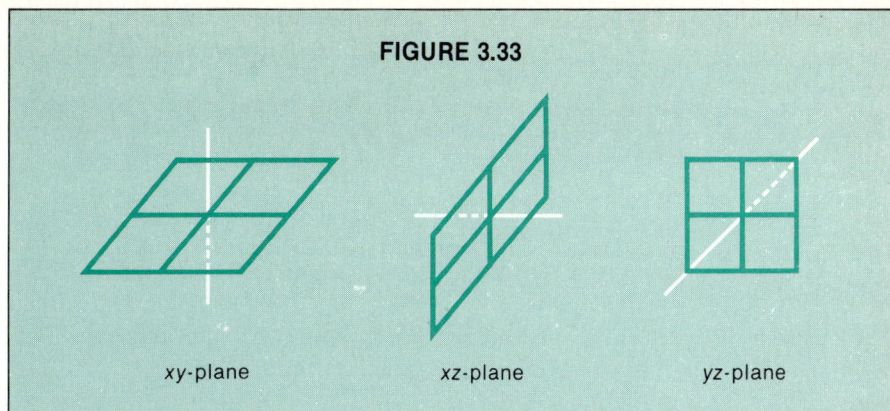

The cross product gives us an easy means of finding an equation of the plane containing three given points that do not lie on a line.

Example 2.   We shall use the cross product to find an equation for the plane containing the three points $A = (1, 2, 3)$, $B = (-3, 1, 0)$, and $C = (5, 4, -2)$. Let $\mathbf{u}$ and $\mathbf{v}$ be the vectors having $\overrightarrow{AB}$ and $\overrightarrow{AC}$, respectively, as representations. Since $\mathbf{u} \times \mathbf{v}$ is orthogonal to both $\mathbf{u}$ and $\mathbf{v}$, we can choose a normal vector $\mathbf{N}$ to be $\mathbf{u} \times \mathbf{v}$. Evidently $\mathbf{u} = -4\mathbf{i} - \mathbf{j} - 3\mathbf{k}$, $\mathbf{v} = 4\mathbf{i} + 2\mathbf{j} - 5\mathbf{k}$, and

$$\mathbf{N} = \mathbf{u} \times \mathbf{v} = \det \begin{bmatrix} \mathbf{i} & \mathbf{j} & \mathbf{k} \\ -4 & -1 & -3 \\ 4 & 2 & -5 \end{bmatrix}$$

$$= \det \begin{bmatrix} -1 & -3 \\ 2 & -5 \end{bmatrix} \mathbf{i} - \det \begin{bmatrix} -4 & -3 \\ 4 & -5 \end{bmatrix} \mathbf{j} + \det \begin{bmatrix} -4 & -1 \\ 4 & 2 \end{bmatrix} \mathbf{k}$$

$$= 11\mathbf{i} - 32\mathbf{j} - 4\mathbf{k}$$

$$= \begin{bmatrix} 11 \\ -32 \\ -4 \end{bmatrix}$$

Using the point-normal form of the equation for a plane passing through the point A and having $\mathbf{u} \times \mathbf{v}$ as a normal vector, we find that

$$11(x - 1) - 32(y - 2) - 4(z - 3) = 0$$

is an equation for the plane passing through A, B, and C. Elementary algebra allows us to rewrite this equation as

$$11x - 32y - 4z + 65 = 0$$

which is the general form for an equation for the plane passing through A, B, and C.

In Section 1.1 we saw that the solutions of a system of two linear equations in two unknowns

$$a_{11}x + a_{12}y = k_1$$

$$a_{21}x + a_{22}y = k_2$$

correspond to points of intersection of the lines determined by $a_{11}x + a_{12}y = k_1$ and $a_{21}x + a_{22}y = k_2$. Similarly, the solutions of three linear equations in three unknowns

$$a_1 x + b_1 y + c_1 z = d_1$$
$$a_2 x + b_2 y + c_2 z = d_2 \qquad\qquad (4)$$
$$a_3 x + b_3 y + c_3 z = d_3$$

correspond to points of intersection of the planes determined by $a_1 x + b_1 y + c_1 z = d_1$, $a_2 x + b_2 y + c_2 z = d_2$, and $a_3 x + b_3 y + c_3 z = d_3$. In Chapter 1 we found that a system of linear equations such as that in (4) has no solution, exactly one solution, or infinitely many solutions. Some geometric interpretations of these three cases are illustrated in Figure 3.34.

FIGURE 3.34

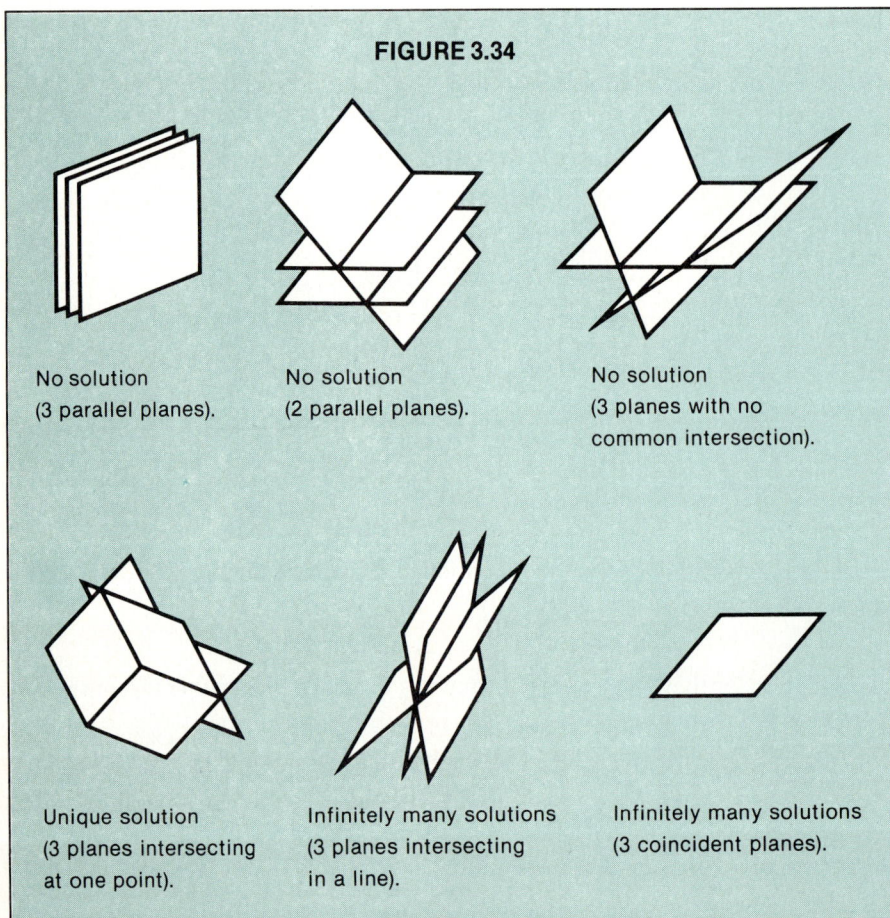

No solution
(3 parallel planes).

No solution
(2 parallel planes).

No solution
(3 planes with no
common intersection).

Unique solution
(3 planes intersecting
at one point).

Infinitely many solutions
(3 planes intersecting
in a line).

Infinitely many solutions
(3 coincident planes).

Definition 9. Two planes determined by

$$a_1 x + b_1 y + c_1 z + d_1 = 0$$

$$a_2 x + b_2 y + c_2 z + d_2 = 0$$

are **parallel** if their normal vectors $a_1 \mathbf{i} + b_1 \mathbf{j} + c_1 \mathbf{k}$ and $a_2 \mathbf{i} + b_2 \mathbf{j} + c_2 \mathbf{k}$ are parallel and are **orthogonal** if their normal vectors $a_1 \mathbf{i} + b_1 \mathbf{j} + c_1 \mathbf{k}$ and $a_2 \mathbf{i} + b_2 \mathbf{j} + c_2 \mathbf{k}$ are orthogonal.

Example 3.   Intuitively we expect the xy-plane, the xz-plane, and the yz-plane to be orthogonal to each other. Since the vectors $\mathbf{i}, \mathbf{j}$, and $\mathbf{k}$ are normal vectors for these three planes, it is easily verified that the planes are in fact orthogonal to each other.

We conclude this section by finding a formula for the distance from a point A to a plane having $\mathbf{N}$ as a normal vector. Let B be any point in the plane and let C be the point of intersection between the plane and the line perpendicular to the plane passing through A (see Figure 3.35).

The vector $\mathbf{u}$ represented by the directed line segment $\overrightarrow{CA}$ is parallel to $\mathbf{N}$. (In Figure 3.35 these vectors have been drawn in the same direction. In fact, they may have opposite directions.) Therefore $\mathbf{u} = k\mathbf{N}$ for some constant k. Let $\mathbf{v}$ be the vector having the directed line segment $\overrightarrow{BA}$ as a representation and let θ be the angle between $\mathbf{v}$ and $k\mathbf{N}$ (see Figure 3.36). Then the length D of $k\mathbf{N}$ is given by

$$D = |\,\|\mathbf{v}\| \cos \theta\,|$$

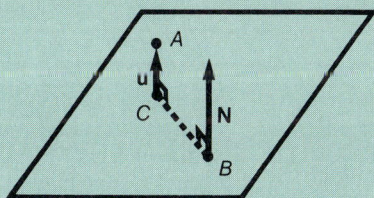

FIGURE 3.35

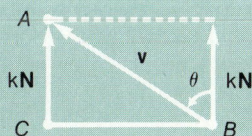

FIGURE 3.36

The absolute value has been introduced to allow for the case $\pi/2 < \theta \leq \pi$. This occurs when $\mathbf{u}$ and $\mathbf{N}$ have opposite directions. By Theorem 10 of Section 3.5,

$$\cos \theta = \frac{\mathbf{v} \cdot (k\mathbf{N})}{\|\mathbf{v}\| \|k\mathbf{N}\|} = \frac{k(\mathbf{v} \cdot \mathbf{N})}{|k| \|\mathbf{v}\| \|\mathbf{N}\|}$$

so that

$$D = \frac{|\mathbf{v} \cdot \mathbf{N}|}{\|\mathbf{N}\|} \tag{5}$$

where $\mathbf{v}$ is the vector with $\overrightarrow{BA}$ as a representation.

If the plane is determined by the equation $ax + by + cz + d = 0$, then there is a simpler formula for determining the distance from a point $A = (x_1, y_1, z_1)$ to the plane. To derive this formula we begin by noticing that $\mathbf{N} = a\mathbf{i} + b\mathbf{j} + c\mathbf{k}$ is a normal vector for the plane and, if $a \neq 0$, the point $(-d/a, 0, 0)$ is on the plane. In such a case the vector

$$\mathbf{v} = \left(x_1 + \frac{d}{a}\right)\mathbf{i} + y_1\mathbf{j} + z_1\mathbf{k}$$

has the directed line segment $\overrightarrow{BA}$ as a representation. Then

$$\mathbf{v} \cdot \mathbf{N} = \left(x_1 + \frac{d}{a}\right)(a) + (y_1)(b) + (z_1)(c)$$

$$= ax_1 + by_1 + cz_1 + d$$

and

$$\|\mathbf{N}\| = \sqrt{a^2 + b^2 + c^2}$$

The distance D from the point A to the plane is now determined directly from formula (5):

$$D = \frac{|ax_1 + by_1 + cz_1 + d|}{\sqrt{a^2 + b^2 + c^2}} \tag{6}$$

It is left as an exercise to show that this formula also holds when $a = 0$. Accepting this as being done, we have proved that the distance from a point $A = (x_1, y_1, z_1)$ to the plane determined by $ax + by + cz + d = 0$ is given by the formula in (6). Thus it is not necessary to find a point B on the plane or to find the vector having $\overrightarrow{BA}$ as a representation in order to find the distance from A to the plane.

Example 4.   Consider the point $A = (1, 2, 3)$ and the plane determined by $4x + 6y - 2z - 4 = 0$. Using formula (6), we can find the distance D from the point A to the plane in the following way:

$$D = \frac{|(4)(1) + (6)(2) + (-2)(3) + (-4)|}{\sqrt{4^2 + 6^2 + (-2)^2}}$$

$$= \frac{3}{\sqrt{14}}$$

Exercises

In Exercises 1–4 find the point-normal equation and the general form of the equation of the plane passing through P with the given normal vector.

1. $P = (3, 4, 1)$ $\mathbf{N} = \begin{bmatrix} -1 \\ 3 \\ 2 \end{bmatrix}$

2. $P = (-2, 3, -6)$ $\mathbf{N} = \begin{bmatrix} 5 \\ 8 \\ 9 \end{bmatrix}$

3. $P = (-2, -3, 0)$; $\mathbf{N} = 5\mathbf{i} + 3\mathbf{j} + 2\mathbf{k}$.
4. $P = (7, 8, -5)$; $\mathbf{N} = 4\mathbf{i} + 7\mathbf{j} - 5\mathbf{k}$.

In Exercises 5 and 6 find an equation for the plane passing through the given points.

5. $(1, 0, 5)$, $(2, 3, -1)$, $(5, -7, 2)$.
6. $(-2, 3, 4)$, $(6, 8, 2)$, $(3, -2, -3)$.

In Exercises 7 and 8 find a point-normal form of the plane with the general equation.

7. $6x + 4y + 5z + 2 = 0$.
8. $2x - 4y - 5z + 7 = 0$.

In Exercises 9–12 find the distance from the given point P to the plane whose general equation is given.

9. $P = (-5, 2, 1)$; $3x + 2y - z + 1 = 0$.
10. $P = (3, -1, 4)$; $-5x - 4y + 2z + 5 = 0$.
11. $P = (0, 0, 0)$; $3x - 2y + 5z + 4 = 0$.
12. $P = (0, 0, 0)$; $5x + 4y - 3z + 2 = 0$.

In Exercises 13 and 14 find the distance between the given planes.

13. $2x + y - 2z + 7 = 0$; $2x + y - 2z - 13 = 0$.
14. $-4x + 5y + 8z - 4 = 0$; $4x - 5y - 8z + 10 = 0$.

In Exercises 15 and 16 find an equation for the plane that passes through the point P and that is parallel to the plane whose general equation is given.

15. $P = (2, 3, -5)$; $3x - 7y + 2z + 1 = 0$.
16. $P = (-6, 4, 1)$; $-2x + 5y + 3z + 6 = 0$.

In Exercises 17 and 18 find the equations for all planes that are orthogonal to the plane whose general equation is given.

17. $2x + 3y + z + 4 = 0$.
18. $3x - 5y + 6z + 2 = 0$.
19. Show that formula (6) for the distance from a point (x_1, y_1, z_1) to a plane determined by $ax + by + cz + d = 0$ holds when $b \neq 0$. Show that it holds when $c \neq 0$.

3.8 LINES IN 3-SPACE

In a plane, a line is uniquely determined by specifying two distinct points on the line. The same is true in 3-space. Let $A = (a_1, b_1, c_1)$ and $B = (a_2, b_2, c_2)$ be two distinct points on a line L in 3-space. The directed line segment $\overrightarrow{AB}$ is a representation of the vector

$$\mathbf{v} = (a_2 - a_1)\mathbf{i} + (b_2 - b_1)\mathbf{j} + (c_2 - c_1)\mathbf{k}$$

Let $C = (x, y, z)$ be any point on the line. Then the directed line segment $\overrightarrow{AC}$ is a representation of the vector

$$\mathbf{w} = (x - a_1)\mathbf{i} + (y - b_1)\mathbf{j} + (z - c_1)\mathbf{k}$$

that is parallel to the vector $\mathbf{v}$. Since these vectors are parallel and $\mathbf{v}$ is not the zero vector, there is a number t such that

$$\mathbf{w} = t\mathbf{v}$$

That is,

$$(x - a_1)\mathbf{i} + (y - b_1)\mathbf{j} + (z - c_1)\mathbf{k} = t(a_2 - a_1)\mathbf{i} + t(b_2 - b_1)\mathbf{j} + t(c_2 - c_1)\mathbf{k}$$

so that

$$\begin{aligned} x &= a_1 + t(a_2 - a_1) \\ y &= b_1 + t(b_2 - b_1) \\ z &= c_1 + t(c_2 - c_1) \end{aligned} \tag{1}$$

These equations are called **parametric equations** for the line L because L is traced out by $x\mathbf{i} + y\mathbf{j} + z\mathbf{k}$ as the parameter t varies over the interval $(-\infty, \infty)$.

In the equations in (1) the numbers a_1, b_1, c_1 are the components of a point on the line, and the numbers $a_2 - a_1, b_2 - b_1, c_2 - c_1$ are the components of a vector that is parallel to the line. Therefore, if (x_0, y_0, z_0) is any point in 3-space and if $\mathbf{v} = a\mathbf{i} + b\mathbf{j} + c\mathbf{k}$ is any nonzero vector, then

$$\begin{aligned} x &= x_0 + ta \\ y &= y_0 + tb \\ z &= z_0 + tb \end{aligned} \tag{2}$$

are parametric equations for the line passing through the point (x_0, y_0, z_0) parallel to the vector $\mathbf{v} = a\mathbf{i} + b\mathbf{j} + c\mathbf{k}$.

Example 1. Parametric equations for the line L passing through the points $(1, -2, 4)$ and $(-5, 3, 2)$ can be found by setting $(a_1, b_1, c_1) = (1, -2, 4)$ and $(a_2, b_2, c_2) = (-5, 3, 2)$ and using the equations in (1). Doing this we find that

$$\begin{aligned} x &= 1 - 6t \\ y &= -2 + 5t \\ z &= 4 - 2t \end{aligned}$$

are parametric equations for L.

If we interchange the roles of the two points so that $(a_1, b_1, c_1) = (-5, 3, 2)$ and $(a_2, b_2, c_2) = (1, -2, 4)$, we find that

$$\begin{aligned} x &= -5 + 6t \\ y &= 3 - 5t \\ z &= 2 + 2t \end{aligned}$$

are also parametric equations for L. Thus a given line can be determined by more than one set of parametric equations.

Example 2.   Parametric equations for the line L passing through the point $(3, -5, 4)$ parallel to the vector $\mathbf{v} = -\mathbf{i} - 7\mathbf{j} + 3\mathbf{k}$ can be found by setting $(x_0, y_0, z_0) = (3, -5, 4)$, setting $a\mathbf{i} + b\mathbf{j} + c\mathbf{k} = -\mathbf{i} - 7\mathbf{j} + 3\mathbf{k}$, and using the equation in (2). Doing this we find that

$$x = \quad 3 - t$$
$$y = -5 - 7t$$
$$z = \quad 4 + 3t$$

are parametric equations for L.

In elementary algebra we found that a line in a plane (2-space) is uniquely determined by a point on the line and the slope of the line. The slope of the line determines the direction of the line. To determine the direction of a line in 3-space we specified a direction vector, that is denoted by $\mathbf{v}$ in the preceding discussion. Using a vector to determine the direction of a line works in 2- or 3- (or even higher) dimensional space, whereas the slope of the line determines its direction only in 2-space.

Consider two planes determined by the equations

(3)
$$a_1 x + b_1 y + c_1 z = d_1$$
$$a_2 x + b_2 y + c_2 z = d_2$$

If these planes are not parallel, then the planes intersect in a line. Parametric equations for this line can be found by finding the solutions of the system of equations in (3).

Example 3.   Consider the planes determined by the equations

$$x + 2y + 3z = 2$$
$$2x + 3y + 2z = 4$$

Solving this system of equations we find that

$$x = 2 + 5z$$
$$y = -4z$$

Setting $z = t$ we obtain parametric equations for the line of intersection of the two planes

$$x = 2 + 5t$$
$$y = \quad -4t$$
$$z = \quad t$$

These equations show that the line of intersection of the two planes has the same direction as the vector $5\mathbf{i} - 4\mathbf{j} + \mathbf{k}$. Since the line lies in each plane, the vector

$5\mathbf{i} - 4\mathbf{j} + \mathbf{k}$ is orthogonal to the normal vectors of each plane. The reader should verify that this is the case.

By setting $t = 0$ in these equations, we find that $(2, 0, 0)$ is a point on the line of intersection of the planes. The reader should also verify that $(2, 0, 0)$ is a point on each plane.

In some problems it is convenient to have equations for a line that do not depend on a parameter. If we solve each of the equations in (2) for t we obtain

$$t = \frac{x - x_0}{a} \qquad t = \frac{y - y_0}{b} \qquad t = \frac{z - z_0}{c}$$

provided a, b and c are nonzero. It is now easy to eliminate the parameter t. If a, b, and c are nonzero, we obtain

$$\frac{x - x_0}{a} = \frac{y - y_0}{b} = \frac{z - z_0}{c} \tag{4}$$

which are called **symmetric equations** for the line passing through the point (x_0, y_0, z_0) parallel to the vector $\mathbf{v} = a\mathbf{i} + b\mathbf{j} + c\mathbf{k}$. If one or more of a, b, and c are zero, then the symmetric equations are obtained by eliminating t where possible. For example, if $a = 0$, $b \neq 0$, and $c \neq 0$, then the equations in (2) can be rewritten as

$$x = x_0 \qquad t = \frac{y - y_0}{b} \qquad t = \frac{z - z_0}{c}$$

or equivalently

$$x = x_0 \qquad \frac{y - y_0}{b} = \frac{z - z_0}{c} \tag{5}$$

Example 4.   In order to find symmetric equations for the line L passing through the points $A = (1, -2, 4)$ and $B = (-5, 3, 2)$, we need to determine a vector that is parallel to this line. The vector $\mathbf{v} = -6\mathbf{i} + 5\mathbf{j} - 2\mathbf{k}$, having $\overrightarrow{AB}$ as a representation, is such a vector. If we use the point A and the vector $\mathbf{v}$ in equation (4), we find that

$$\frac{x - 1}{-6} = \frac{y + 2}{5} = \frac{z - 4}{-2}$$

are symmetric equations for the line L. If we use the point B in place of the point A, we find that

$$\frac{x + 5}{-6} = \frac{y - 3}{5} = \frac{z - 2}{-2}$$

are also symmetric equations for the line L. Thus a given line can be determined by more than one set of symmetric equations.

In Example 1 we found parametric equations for the line L that was determined by the above sets of symmetric equations.

Example 5.   **Let** L denote the line passing through the point $(3, 7, 2)$ parallel to the vector $\mathbf{v} = 6\mathbf{j} - 9\mathbf{k}$. From equations (5) we find that

$$x = 3 \qquad \frac{y - 7}{6} = \frac{z - 2}{-9}$$

are symmetric equations for L.

─────────────────────── **Exercises** ───────────────────────

In Exercises 1–6 find the parametric and symmetric equations for the line through the two given points.

1. $(2, 1, 3);\quad (-3, 0, 4)$.
2. $(1, 5, 7);\quad (2, 6, -3)$.
3. $(5, 3, -1);\quad (3, 6, 8)$.
4. $(0, 7, 6);\quad (3, 9, 2)$.
5. $(-2, -1, -4);\quad (-8, -9, -5)$.
6. $(8, 2, 4);\quad (-7, -1, -2)$.

In Exercises 7 and 8 find parametric and symmetric equations for the line that passes through the point P and that is parallel to the vector $\mathbf{v}$.

7. $P = (2, 3, 1);\quad \mathbf{v} = 3\mathbf{i} - 2\mathbf{j} + \mathbf{k}$.
8. $P = (-3, 5, 4);\quad \mathbf{v} = 5\mathbf{i} + 3\mathbf{j} - 6\mathbf{k}$.

In Exercises 9 and 10 find parametric equations for the line of intersection between the planes whose general equations are given.

9. $2x + 3y + 5z + 4 = 0;\ 2x - 5y + 6z - 2 = 0$.
10. $3x - 5y + 4z - 2 = 0;\ 4x + 7y + 2z + 1 = 0$.

In Exercises 11 and 12 determine whether or not the given lines intersect.

11. $x = 1 + 2t,\ y = 2 + 3t,\ z = 3 + 4t$; and $x = 2 + 3t,\ y = 5 + 4t,\ z = -1 + 2t$.
12. $x = 3 + 2t,\ y = 3 + 2t,\ z = 2 + t$; and $x = 1 - 4t,\ y = -2 - 6t,\ z = -2 - 4t$.

In Exercises 13 and 14 find symmetric equations for the line that passes through the point $(1, 2, 3)$ and that is parallel to the line whose symmetric equations are given.

13. $\dfrac{x + 2}{1} = \dfrac{y + 3}{2} = \dfrac{z - 1}{3}$.

14. $\dfrac{x - 4}{2} = \dfrac{y + 3}{5} = \dfrac{z - 2}{7}$.

In Exercises 15 and 16 find the distance between the given lines.

15. $x = 3 + t,\ y = 1 - 2t,\ z = 2 + 2t$; and $x = 7 + t,\ y = 1 - 2t,\ z = -3 + 2t$.
16. $x = 2 - t,\ y = 3 + 4t,\ z = -3 + t$; and $x = 5 - t,\ y = 4 + 4t,\ z = 2 + t$.

In Exercises 17 and 18 find parametric equations for the lines through the given point that are perpendicular to the given line.

17. $(7, 1, 5)$ and $x = 2 + t,\ y = -4 + t,\ z = 2t$.
18. $(-3, 0, 2)$ and $x = 4t,\ y = 3 - 2t,\ z = 1 + t$.

4

Linear Spaces

4.1 R^n

In the previous chapter we studied vectors in 2-space and 3-space. There is really no need to restrict our attention to these two dimensions. It is just as easy to consider ordered quadruples of numbers

$$\begin{bmatrix} a_1 \\ a_2 \\ a_3 \\ a_4 \end{bmatrix}$$

or even ordered n-tuples of numbers,

$$\begin{bmatrix} a_1 \\ a_2 \\ \vdots \\ a_n \end{bmatrix}$$

where n is any positive integer. Ordered n-tuples of numbers arise naturally in many applications. For example, if a company requires n raw materials to make a product, then an ordered n-tuple can be used to represent the per unit costs of these materials. This section defines and discusses the basic arithmetic properties of the set of all ordered n-tuples of real numbers.

Definition 1. R^n will denote the set of all n-vectors with real components. That is, $\mathbf{u}$ is an element of R^n if and only if there are real numbers $u_1, u_2, \ldots, u_n$ such that

$$\mathbf{u} = \begin{bmatrix} u_1 \\ u_2 \\ \vdots \\ u_n \end{bmatrix}$$

The arithmetic operations of addition and scalar multiplication on R^n are defined as in Definitions 2 and 3 in Section 1.6. Specifically, if

$$\mathbf{u} = \begin{bmatrix} u_1 \\ u_2 \\ \vdots \\ u_n \end{bmatrix} \qquad \mathbf{v} = \begin{bmatrix} v_1 \\ v_2 \\ \vdots \\ v_n \end{bmatrix}$$

are two elements of R^n and a is any scalar, then

$$\mathbf{u} + \mathbf{v} = \begin{bmatrix} u_1 + v_1 \\ u_2 + v_2 \\ \vdots \\ u_n + v_n \end{bmatrix} \qquad \text{and} \qquad a\mathbf{u} = \begin{bmatrix} au_1 \\ au_2 \\ \vdots \\ au_n \end{bmatrix}$$

For example

$$\mathbf{u} = \begin{bmatrix} 2 \\ 6 \\ -4 \end{bmatrix} \qquad \text{and} \qquad \mathbf{v} = \begin{bmatrix} -1 \\ 3 \\ 5 \end{bmatrix}$$

are elements of R^3. Using the above definitions for addition and scalar multiplication, we find that

$$\mathbf{u} + \mathbf{v} = \begin{bmatrix} 2 \\ 6 \\ -4 \end{bmatrix} + \begin{bmatrix} -1 \\ 3 \\ 5 \end{bmatrix} = \begin{bmatrix} 2 + (-1) \\ 6 + 3 \\ (-4) + 5 \end{bmatrix} = \begin{bmatrix} 1 \\ 9 \\ 1 \end{bmatrix}$$

and

$$3\mathbf{u} = \begin{bmatrix} 3(2) \\ 3(6) \\ 3(-4) \end{bmatrix} = \begin{bmatrix} 6 \\ 18 \\ -12 \end{bmatrix}$$

The zero element of R^n and the inverse of $\mathbf{u}$ (i.e., $-\mathbf{u}$) are

$$\begin{bmatrix} 0 \\ 0 \\ \vdots \\ 0 \end{bmatrix} \qquad \text{and} \qquad \begin{bmatrix} -u_1 \\ -u_2 \\ \vdots \\ -u_n \end{bmatrix}$$

respectively.

For every positive integer n the set R^n satisfies the same arithmetic properties as the set of all vectors in 2-space or 3-space. These properties are listed in the following theorem. Since these properties are easy to establish, the proof of this theorem is left as an exercise.

Theorem 1. *If $\mathbf{u}$, $\mathbf{v}$, and $\mathbf{w}$ are any vectors in R^n and if a and b are any real numbers, then*

1. $\mathbf{u} + \mathbf{v}$ *is an n-vector.*
2. $\mathbf{u} + \mathbf{v} = \mathbf{v} + \mathbf{u}$.
3. $(\mathbf{u} + \mathbf{v}) + \mathbf{w} = \mathbf{u} + (\mathbf{v} + \mathbf{w})$.
4. *There is a zero element $\mathbf{0}$ in R^n such that $\mathbf{u} + \mathbf{0} = \mathbf{u}$ for every $\mathbf{u}$ in R^n.*
5. *There is an n-vector $-\mathbf{u}$ such that $\mathbf{u} + (-\mathbf{u}) = \mathbf{0}$.*
6. $a\mathbf{u}$ *is an n-vector.*
7. $(ab)\mathbf{u} = a(b\mathbf{u})$.
8. $(a + b)\mathbf{u} = a\mathbf{u} + b\mathbf{u}$.
9. $a(\mathbf{u} + \mathbf{v}) = a\mathbf{u} + a\mathbf{v}$.
10. $1\mathbf{u} = \mathbf{u}$

Exercises

In Exercises 1–10 the vectors $\mathbf{u}$, $\mathbf{v}$, and $\mathbf{w}$ are as follows:

$$\mathbf{u} = \begin{bmatrix} 1 \\ 2 \\ 0 \\ 4 \\ 6 \end{bmatrix} \qquad \mathbf{v} = \begin{bmatrix} 2 \\ -4 \\ 3 \\ -1 \\ -7 \end{bmatrix} \qquad \mathbf{w} = \begin{bmatrix} 1 \\ 8 \\ 2 \\ 0 \\ 0 \end{bmatrix}$$

1. Find $\mathbf{u} + 7\mathbf{v}$.
2. Find $8\mathbf{u} - 7\mathbf{v} + 5\mathbf{w}$.
3. Find $-2\mathbf{u} + 3\mathbf{v} + 4\mathbf{w}$.
4. Find $7\mathbf{u} + \mathbf{v} - (8\mathbf{w} + \mathbf{u})$.
5. Find $2(5\mathbf{u} - 4\mathbf{w}) + 3\mathbf{v}$.
6. Find $-3(2\mathbf{v} + 4\mathbf{w}) - (2\mathbf{u} + 3\mathbf{v} - \mathbf{w})$
7. Find the vector $\mathbf{x}$ such that $3\mathbf{u} + \mathbf{x} = \mathbf{v}$.
8. Find the vector $\mathbf{x}$ such that $\mathbf{u} - 2\mathbf{v} + \mathbf{x} = 3\mathbf{x} + \mathbf{w}$.
9. Are there numbers a, b, and c other than $a = 0, b = 0, c = 0$ such that $a\mathbf{u} + b\mathbf{v} + c\mathbf{w} = \mathbf{0}$?
10. Can numbers a, b, and c be found to satisfy the following equation?

$$a\mathbf{u} + b\mathbf{v} + c\mathbf{w} = \begin{bmatrix} 3 \\ 32 \\ -3 \\ 9 \\ 19 \end{bmatrix}$$

11. Show that for every two vectors $\mathbf{u}$ and $\mathbf{v}$ in R^n, $-(\mathbf{u} + \mathbf{v}) = -\mathbf{u} - \mathbf{v}$.
12. Show that for every two vectors $\mathbf{u}$ and $\mathbf{v}$ in R^n, $-(\mathbf{u} - \mathbf{v}) = -\mathbf{u} + \mathbf{v}$.
13. Let $\mathbf{u}, \mathbf{v}$, and $\mathbf{w}$ be any three vectors in R^2. Show that there are numbers a, b, and c, at least one of which is nonzero, such that $a\mathbf{u} + b\mathbf{v} + c\mathbf{w} = \mathbf{0}$.
14. Let $\mathbf{u}_1, \mathbf{u}_2, \dots, \mathbf{u}_{n+1}$ be any $n + 1$ vectors in R^n. Show that there are numbers $a_1, a_2, \dots, a_{n+1}$, at least one of which is nonzero, such that $a_1\mathbf{u}_1 + a_2\mathbf{u}_2 + \cdots + a_{n+1}\mathbf{u}_{n+1} = \mathbf{0}$.

15. Prove part 1 of Theorem 1.
16. Prove part 2 of Theorem 1.
17. Prove part 3 of Theorem 1.
18. Prove part 4 of Theorem 1.
19. Prove part 5 of Theorem 1.
20. Prove part 6 of Theorem 1.
21. Prove part 7 of Theorem 1.
22. Prove part 8 of Theorem 1.
23. Prove part 9 of Theorem 1.
24. Prove part 10 of Theorem 1.

4.2 LINEAR SPACES

In Theorem 1 of Section 4.1 we found that R^n possesses a number of arithmetic properties that do not depend on the positive integer n. The operation of addition is both commutative and associative, while the operation of scalar multiplication satisfies several distributive properties. In addition, R^n contains a zero element and every vector in R^n has an additive inverse.

We now define an abstract mathematical system that embodies the arithmetic properties of R^n.

Definition 2.   A **linear space** V over the real numbers is a nonempty set with two laws of combination, called addition and scalar multiplication, satisfying the following conditions for all **u**, **v**, and **w** in V and all real numbers a and b.

1. To every pair **u**, **v** in V there is associated a unique element in V, called the sum of **u** and **v**, which is denoted by **u** + **v**.
2. **u** + **v** = **v** + **u**.
3. (**u** + **v**) + **w** = **u** + (**v** + **w**).
4. There exists a zero element **0** in V such that **u** + **0** = **u** for all **u** in V.
5. To every element **u** in V there corresponds an additive inverse element −**u** in V such that **u** + (−**u**) = **0**.
6. To every real number a and every element **u** in V there is associated a unique element in V, called the scalar product of **u** and a, which is denoted by a**u**.
7. (ab)**u** = $a(b$**u**$)$.
8. $(a + b)$**u** = a**u** + b**u**.
9. $a($**u** + **v**$)$ = a**u** + a**v**.
10. 1**u** = **u**.

By a **scalar**, we mean a real number. If in the preceding definition we changed "real number" to "complex number" we would obtain a **linear space over the complex numbers**. For some applications it is useful to consider such linear spaces. Even though the theorems of this chapter remain true for linear spaces over the complex numbers, we shall restrict the discussion to linear spaces over the real numbers with the exception of a few examples (in later chapters) and exercises.

A word of warning needs to be given concerning the above axioms of a linear space. The plus sign used in denoting the sum of two elements of the linear space is not the same as the plus sign used in denoting the sum of two real (or complex) numbers. In particular, in property 8 of Definition 2 the plus signs are different: The plus sign on the left denotes the sum of two scalars, whereas the plus sign on the right denotes the sum of two elements (not scalars) of the linear space. Also the zero element in property 4 of Definition 2 and the number zero are different. One is an element of the linear space and the other is a scalar. When the plus sign or zero is used in context there is rarely any cause for confusion. These distinctions should be noted, however, to avoid future confusion.

Even though this generalization from concrete examples to arbitrary linear spaces may appear unnecessary, that is not the case. Once we have established a property of an arbitrary linear space, then we know that every linear space has that property. This advance work prevents us from having to establish this same property each time we encounter a new linear space. Amazing as it may seem, the proofs of theorems in an abstract setting are usually no more complicated than the analogous theorems for any specific linear space.

A linear space is also called a vector space. However, the elements of a linear space may not be n-vectors. We have chosen the name "linear space" so that the reader will not be tempted to think of linear spaces just as collections of n-vectors.

Example 1.   By Theorem 1 of Section 4.1, R^n is a linear space for every positive integer n.

Example 2.   Let V be the set of all 2-vectors whose components total zero. That is,

$$V = \left\{ \begin{bmatrix} x \\ y \end{bmatrix} : x + y = 0 \right\}$$

The set V is nonempty since $\begin{bmatrix} 0 \\ 0 \end{bmatrix}$ is an element of V. The set V also satisfies properties 2, 3, 7, 8, 9, and 10 of Definition 2, because all 2-vectors satisfy them and V is a set of 2-vectors. We now verify that the remaining properties are satisfied.

Let

$$\mathbf{x} = \begin{bmatrix} x_1 \\ x_2 \end{bmatrix} \qquad \mathbf{y} = \begin{bmatrix} y_1 \\ y_2 \end{bmatrix}$$

be any two elements of V so that

$$x_1 + x_2 = 0$$
$$y_1 + y_2 = 0$$

Then

$$\mathbf{x} + \mathbf{y} = \begin{bmatrix} x_1 \\ x_2 \end{bmatrix} + \begin{bmatrix} y_1 \\ y_2 \end{bmatrix} = \begin{bmatrix} x_1 + y_1 \\ x_2 + y_2 \end{bmatrix}$$

and

$$(x_1 + y_1) + (x_2 + y_2) = (x_1 + x_2) + (y_1 + y_2)$$
$$= 0 + 0 = 0$$

Thus $\mathbf{x} + \mathbf{y}$ is an element of V, and property 1 is satisfied.

Since $\mathbf{0} = \begin{bmatrix} 0 \\ 0 \end{bmatrix}$ is an element of V and $\mathbf{x} + \mathbf{0} = \mathbf{x}$ for every vector $\mathbf{x}$ in V, property 4 is satisfied.

It is clear that

$$-\mathbf{x} = \begin{bmatrix} -x_1 \\ -x_2 \end{bmatrix}$$

is an additive inverse for $\mathbf{x}$ since

$$\mathbf{x} + (-\mathbf{x}) = \begin{bmatrix} x_1 \\ x_2 \end{bmatrix} + \begin{bmatrix} -x_1 \\ -x_2 \end{bmatrix} = \begin{bmatrix} 0 \\ 0 \end{bmatrix} = \mathbf{0}$$

Moreover, $(-x_1) + (-x_2) = (-1)(x_1 + x_2) = (-1)(0) = 0$ so that $-x$ is an element of V. Therefore property 5 is satisfied.

If a is any number, then

$$a\mathbf{x} = a\begin{bmatrix} x_1 \\ x_2 \end{bmatrix} = \begin{bmatrix} ax_1 \\ ax_2 \end{bmatrix}$$

and

$$ax_1 + ax_2 = a(x_1 + x_2) = a(0) = 0$$

Thus $a\mathbf{x}$ is an element of V, and property 6 is satisfied.

We have shown that V satisfies all of the properties of Definition 2. Therefore V is a linear space.

Example 3.   Let V be the set of all real and valued functions defined on an interval I. If f and g are any functions in V and a is any real number, we define $f + g$ and af by

$$(f + g)(x) = f(x) + g(x), \qquad (af)(x) = af(x)$$

for all x in I. For example, the functions $f(x) = x + 4$ and $g(x) = 1/(1 - x^2)$ are functions defined on the interval $(-1, 1)$. Using the above definitions for addition and scalar multiplication, we have

$$(f + g)(x) = f(x) + g(x) = x + 4 + \frac{1}{1 - x^2}$$

and

$$\left(\frac{1}{2}f\right)(x) = \frac{1}{2}f(x) = \frac{1}{2}(x + 4) = \frac{1}{2}x + 2$$

for all x in $(-1, 1)$. We shall now show that V is a linear space.

The function $p(x) = x$ is defined on any interval I. Therefore V is not the empty set.

Now let f, g, and h be any functions in V and let a and b be any real numbers. Then

1. Since $(f + g)(x) = f(x) + g(x)$ for all x in I and f and g are functions defined on I, the function $f + g$ is also defined on I.

2. Since

$$(f + g)(x) = f(x) + g(x)$$
$$= g(x) + f(x) \quad [f(x) \text{ and } g(x) \text{ are numbers.}]$$
$$= (g + f)(x)$$

for all x in I, we have $f + g = g + f$.

3. Since

$$[(f + g) + h](x) = (f + g)(x) + h(x)$$
$$= [f(x) + g(x)] + h(x)$$
$$= f(x) + [g(x) + h(x)] \quad [f(x), g(x), \text{ and } h(x) \text{ are numbers.}]$$
$$= [f + (g + h)](x)$$

for all x in I, we have $(f + g) + h = f + (g + h)$.

4. The function z defined by $z(x) = 0$ for all x in I is a zero element for V since

$$(f + z)(x) = f(x) + z(x) = f(x) + 0 = f(x)$$

for all x in I.

5. The function p defined by $p(x) = -f(x)$ for all x in I is an additive inverse for f since

$$(f + p)(x) = f(x) + p(x) = f(x) + [-f(x)] = 0$$

for all x in I.

6. Since $(af)(x) = af(x)$ for all x in I and f is defined on I, the function af is defined on I.

7. Since

$$[(ab)f](x) = (ab)f(x)$$
$$= a[bf(x)] \quad [f(x) \text{ is a number.}]$$
$$= [a(bf)](x)$$

for all x in I, we have $(ab)f = a(bf)$.

8. Since

$$[(a + b)f](x) = (a + b)f(x)$$
$$= af(x) + bf(x) \quad [f(x) \text{ is a number.}]$$
$$= [a(bf)](x)$$

for all x in I, we have $(a + b)f = af + bf$.

9. Since

$$[a(f + g)](x) = a(f + g)(x)$$
$$= a[f(x) + g(x)]$$
$$= af(x) + ag(x) \quad [f(x) \text{ is a number.}]$$
$$= (af + ag)(x)$$

for all x in I, we have $a(f + g) = af + ag$.

10. Since

$$(1f)(x) = 1f(x) = f(x) \quad [f(x) \text{ is a number.}]$$

for all x in I, we have $1f = f$.

Thus V satisfies all of the properties of a linear space and, therefore, V is a linear space.

Example 4.   Let P_2 denote the set of all polynomial functions of degree less than or equal to 2. Since polynomial functions are defined on the interval $(-\infty, \infty)$ we can use the previous example to conclude that P_2 satisfies properties 2, 3, 7, 8, 9, and 10 of Definition 2. We shall now verify that the remaining properties are also satisfied.

Let p and q be any two polynomial functions in P_2. Then there are numbers a, b, c, d, e, and f such that

$$p(x) = ax^2 + bx + c$$
$$q(x) = dx^2 + ex + f$$

for all x. Since

$$\begin{aligned}(p + q)(x) &= p(x) + q(x) \\ &= (ax^2 + bx + c) + (dx^2 + ex + f) \\ &= (a + d)x^2 + (b + e)x + (c + f)\end{aligned}$$

the function $p + q$ is a polynomial function of degree less than or equal to 2. The degree is less than 2 if $a + d = 0$. Therefore, property 1 is satisfied.

Since the function z defined by $z(x) = 0$ for all x is the zero element of P_2, property 4 is satisfied. It is also clear that the additive inverse of any polynomial function p is the polynomial r defined by $r(x) = -p(x)$ for all x. Therefore, property 5 is satisfied.

For any number A and any function p in P_2 we have

$$(Ap)(x) = Ap(x) = A(ax^2 + bx + c) = (Aa)x^2 + (Ab)x + (Ac)$$

for all x so that the function Ap is also in P_2, and property 6 is satisfied.

Since P_2 satisfies all of the properties of a linear space, it is a linear space. Example 7 will show that the set of all polynomial functions of degree exactly equal to 2 is not a linear space.

Example 5.   An argument similar to that in the previous example shows that the set P_n of all polynomials of degree less than or equal to n is a linear space. The reader is asked in Exercise 24 to supply this argument.

Example 6.   Let V be the set of all 2-vectors whose components total one. That is,

$$V = \left\{ \begin{bmatrix} x \\ y \end{bmatrix} : x + y = 1 \right\}$$

V is not a linear space because properties 1, 4, 5, and 6 are not satisfied. In particular:

1. $\begin{bmatrix} 0 \\ 1 \end{bmatrix}$ and $\begin{bmatrix} 1 \\ 0 \end{bmatrix}$ are elements of V, but $\begin{bmatrix} 0 \\ 1 \end{bmatrix} + \begin{bmatrix} 1 \\ 0 \end{bmatrix} = \begin{bmatrix} 1 \\ 1 \end{bmatrix}$ is not. Therefore property 1 is not satisfied.

2. If $\begin{bmatrix} a \\ b \end{bmatrix}$ is a zero element of V, then

$$\begin{bmatrix} 1 \\ 0 \end{bmatrix} = \begin{bmatrix} 1 \\ 0 \end{bmatrix} + \begin{bmatrix} a \\ b \end{bmatrix} = \begin{bmatrix} 1 + a \\ b \end{bmatrix}$$

so that $1 = 1 + a$ and $0 = b$. It follows that if V has a zero element, then it is $\begin{bmatrix} 0 \\ 0 \end{bmatrix}$. But $\begin{bmatrix} 0 \\ 0 \end{bmatrix}$ is not an element of V, because $0 + 0 \neq 1$. Therefore property 4 is not satisfied.

3. Since V does not have a zero element, the concept of additive inverse does not make sense and property 5 is not satisfied.

4. Since $\begin{bmatrix} 1 \\ 0 \end{bmatrix}$ is an element of V, but $2\begin{bmatrix} 1 \\ 0 \end{bmatrix} = \begin{bmatrix} 2 \\ 0 \end{bmatrix}$ is not, property 6 is not satisfied.

Example 7.   Let V denote the set of all polynomial functions with real coefficients of exactly degree 2. The operations of addition and scalar multiplication are defined as in Example 4. Then V is not a linear space, because the sum of two second-degree polynomials need not be a second-degree polynomial. For example, the sum of $f(x) = x^2$ and $g(x) = -x^2 + x$ is a first-degree polynomial. Hence, property 1 of Definition 2 is not satisfied. The polynomial that is zero for all x is not a second-degree polynomial, so property 4 is not satisfied either.

We conclude this section with a theorem that lists several useful properties of linear spaces.

Theorem 2.   *Let V be a linear space, $\mathbf{u}$ an element of V, and c a scalar. Then*

1. *There is precisely one zero element of V.*
2. *If $\mathbf{v}$ is an element of V such that $\mathbf{u} + \mathbf{v} = \mathbf{u}$, then $\mathbf{v} = \mathbf{0}$.*
3. *There is exactly one element $\mathbf{w}$ of V such that $\mathbf{u} + \mathbf{w} = \mathbf{0}$.*
4. *$0\mathbf{x} = \mathbf{0}$.*
5. *$c\mathbf{0} = \mathbf{0}$.*
6. *If $c\mathbf{u} = \mathbf{0}$, then $c = 0$ or $\mathbf{u} = \mathbf{0}$.*
7. *$(-1)\mathbf{v} = -\mathbf{v}$.*

Proof: We will prove the first two parts of this theorem and leave the proofs of the remaining parts as exercises.

Assume that there are two elements $\mathbf{0}$ and $\mathbf{0}'$ such that $\mathbf{u} + \mathbf{0} = \mathbf{u}$ and $\mathbf{u} + \mathbf{0}' = \mathbf{u}$ for every $\mathbf{u}$ in V. Setting $\mathbf{u} = \mathbf{0}'$ in the first identity and $\mathbf{u} = \mathbf{0}$ in the second, we obtain $\mathbf{0}' + \mathbf{0} = \mathbf{0}'$ and $\mathbf{0} + \mathbf{0}' = \mathbf{0}$. Using these identities and property 2 of Definition 2, we have $\mathbf{0}' = \mathbf{0}' + \mathbf{0} = \mathbf{0} + \mathbf{0}' = \mathbf{0}$. Thus $\mathbf{0} = \mathbf{0}'$, so there is precisely one zero element. This proves part 1.

Suppose that $\mathbf{u} + \mathbf{v} = \mathbf{u}$. By property 5 of Definition 2, there is an inverse element $-\mathbf{u}$ for $\mathbf{u}$. Adding $-\mathbf{u}$ to each side of $\mathbf{u} + \mathbf{v} = \mathbf{u}$ we have

$$(-\mathbf{u}) + (\mathbf{u} + \mathbf{v}) = (-\mathbf{u}) + \mathbf{u}$$

or

$$[(-\mathbf{u}) + \mathbf{u}] + \mathbf{v} = (-\mathbf{u}) + \mathbf{u} \quad \text{(property 3)}$$

or

$$[\mathbf{u} + (-\mathbf{u})] + \mathbf{v} = \mathbf{u} + (-\mathbf{u}) \quad \text{(property 2)}$$

or

$$\mathbf{0} + \mathbf{v} = \mathbf{0} \quad \text{(property 5)}$$

or

$$\mathbf{v} = \mathbf{0} \quad \text{(properties 2 and 4)}$$

This completes the proof of part 2. ▌

───────────────────────── **Exercises** ─────────────────────────

In Exercises 1–18 determine whether the given set is a linear space. If the set is a subset of R^n or a set of functions defined on an interval, then the operations of addition and scalar multiplication are as defined in Examples 1 and 4, respectively.

1. $\left\{ \begin{bmatrix} x \\ y \end{bmatrix} : x + y = 0 \right\}$

2. $\left\{ \begin{bmatrix} x \\ y \end{bmatrix} : 2x + 3y = 0 \right\}$

3. $\left\{ \begin{bmatrix} x \\ y \end{bmatrix} : 2x - 3y = 0 \right\}$

4. $\left\{ \begin{bmatrix} x \\ y \end{bmatrix} : x - y = 1 \right\}$

5. $\left\{ \begin{bmatrix} x \\ y \end{bmatrix} : 2x - 3y = 1 \right\}$

6. $\left\{ \begin{bmatrix} x \\ y \end{bmatrix} : x^2 = y^2 \right\}$

7. $\left\{ \begin{bmatrix} x \\ y \\ z \end{bmatrix} : y = 0 \text{ and } x = z \right\}$

8. $\left\{ \begin{bmatrix} x \\ y \\ z \end{bmatrix} : x = y = z \right\}$

9. The set of all functions f defined on the interval $(-1, 1)$ such that $f(0) = 0$.

10. The set of all functions f defined on the interval $(-2, 2)$ such that $f(1) = f(-1)$.

11. (*For readers who have studied calculus.*) The set of all functions f defined on the interval $(0, \infty)$ such that $\lim_{x \to \infty} f(x) = 0$.

12. (*For readers who have studied calculus.*) The set of all continuous functions f defined on the interval $(0, \infty)$.

13. (*For readers who have studied calculus.*) The set of all twice differentiable functions f on the interval $(-\infty, \infty)$ such that $d^2 f(x)/dx^2 + f(x) = 0$.

14. (*For readers who have studied calculus.*) The set of all differentiable functions f defined on the interval $(0, 1)$ such that $df(x)/dx = 1$.

15. The set of all 2×2 matrices

$$\begin{bmatrix} a & b \\ c & d \end{bmatrix}$$

such that $a + d = b + c + 1$.

16. The set of all 2×2 matrices

$$\begin{bmatrix} a & b \\ c & d \end{bmatrix}$$

such that $a + d = 0$ and $c + b = 0$.

17. The set of all ordered pairs of numbers (a, b) where addition and scalar multiplication are defined by $(a, b) + (c, d) = (a + c, 0)$ and $c(a, b) = (ca, cb)$, respectively.

18. The set of all ordered pairs of numbers (a, b) where addition and scalar multiplication are defined by $(a, b) + (c, d) = (b + d, a + c)$ and $c(a, b) = (cb, ca)$, respectively.

19. Prove part 3 of Theorem 2.

20. Prove part 4 of Theorem 2.

21. Prove part 5 of Theorem 2.

22. Prove part 6 of Theorem 2.

23. Prove part 7 of Theorem 2.

24. Show that P_n is a linear space.

25. Let C^n denote the set of all n-vectors with complex components. The operations of addition and scalar multiplication are defined analogously to those for R^n. Show that C^n is a linear space over the complex numbers.

26. Let D^n denote the set of all n-vectors with complex components. The operations of addition and scalar multiplication are defined analogously to those for R^n. Show that D^n is a linear space over the real numbers.

27. Show that the set of all 2×2 matrices with complex components is a linear space over the complex numbers.

28. Show that the set of all $m \times n$ matrices with complex components is a linear space over the complex numbers.

4.3 SUBSPACES

Many of the linear spaces we will consider are subsets of other linear spaces. For example, the set

$$V = \left\{ \begin{bmatrix} x \\ y \end{bmatrix} : x + y = 0 \right\}$$

which was shown to be a linear space in Example 2 of Section 4.2, is a subset of R^2. In order to simplify future discussions, we shall assign a name to such linear spaces.

Definition 3. Let V be a linear space and U be a nonempty subset of V. If U is itself a linear space under the same operations of addition and scalar multiplication as V, then U is called a **subspace** of V.

Before presenting examples of subspaces we will prove a theorem that makes it relatively easy to determine whether a subset of a linear space V is a subspace of V.

Theorem 3. *A nonempty subset U of a linear space V is a subspace of V if and only if the following two conditions hold:*

1. *If* **u** *and* **v** *are any elements of U, then* **u** + **v** *is an element of U.*

2. *If c is any scalar and* **u** *is any element of U, then c***u** *is an element of U.*

Proof: If U is a subspace of V, then all of the properties (1 through 10) in Definition 2 are satisfied. Conditions 1 and 2 of Theorem 3 are properties 1 and 6, respectively.

Now suppose that U is a subset of V and conditions 1 and 2 are satisfied. Properties $2, 3, 7, 8, 9$, and 10 are satisfied because every element of U is also an element of V. Properties 1 and 6 coincide with conditions 1 and 2. If we choose $c = 0$ and then $c = -1$ in condition 2, we obtain the zero element and the inverse of each element in U (Theorem 2, parts 4 and 7). Therefore properties 5 and 6 are satisfied. The subset U is a linear space under the same operations of addition and scalar multiplication as V. Therefore U is a subspace of V. ∎

The conditions described in Theorem 3 can be restated with the following terminology:

1. U is closed under addition.
2. U is closed under scalar multiplication.

Thus a nonempty subset U of a linear space V is a subspace of V if and only if it is closed under both addition and scalar multiplication.

Every linear space contains at least one subspace: the set consisting of just the zero element of the linear space. The following examples contain less trivial examples of subspaces.

Example 1. Let U denote the set of all polynomial functions with real coefficients. In Example 3 of Section 4.2 we found that the set V of all real valued functions defined on an interval I is a linear space. Clearly U is a nonempty subset of V. It is also clear that the sum of two polynomials is a polynomial and that a constant multiple of a polynomial is a polynomial. From Theorem 3 we conclude that U is a subspace of V.

Example 2. Consider the subset U of R^2 defined by

$$U = \left\{ \begin{bmatrix} x \\ y \end{bmatrix} : 2x + 3y = 0 \right\}$$

The set U is not empty since $\begin{bmatrix} 0 \\ 0 \end{bmatrix}$ is an element of U.

Let

$$\begin{bmatrix} x \\ y \end{bmatrix} \quad \text{and} \quad \begin{bmatrix} u \\ v \end{bmatrix}$$

be any two elements of U. Then $2x + 3y = 0$ and $2u + 3v = 0$, so $2(x + u) + 3(y + v) = (2x + 3y) + (2u + 3v) = 0$. It follows that

$$\begin{bmatrix} x \\ y \end{bmatrix} + \begin{bmatrix} u \\ v \end{bmatrix} = \begin{bmatrix} x + u \\ y + v \end{bmatrix}$$

is an element of V. If c is any scalar, then $2cx + 3cy = c(2x + 3y) = 0$. It follows that

$$c\begin{bmatrix} x \\ y \end{bmatrix} = \begin{bmatrix} cx \\ cy \end{bmatrix}$$

is also an element of U. By Theorem 3 the set U is a subspace of R^2.

Example 3.   Let A be an $m \times n$ matrix with real components and let U be the set of all n-vectors $\mathbf{x}$ such that $A\mathbf{x} = \mathbf{0}$. Since $A\mathbf{0} = \mathbf{0}$ the vector $\mathbf{0}$ is in U, and U is not the empty set.

Suppose that $\mathbf{x}$ and $\mathbf{y}$ are two elements of U so that $A\mathbf{x} = \mathbf{0}$ and $A\mathbf{y} = \mathbf{0}$. Then $A(\mathbf{x} + \mathbf{y}) = A\mathbf{x} + A\mathbf{y} = \mathbf{0} + \mathbf{0} = \mathbf{0}$ and therefore $\mathbf{x} + \mathbf{y}$ is an element of U. If c is any real number, then $A(c\mathbf{x}) = c(A\mathbf{x}) = c\mathbf{0} = \mathbf{0}$ and $c\mathbf{x}$ is therefore an element of U. By Theorem 3, U is a subspace of R^n. This subspace is called the **solution space** of the system of equations $A\mathbf{x} = \mathbf{0}$.

This is an exceptionally important result because it allows us to link the concepts of Chapter 1 to the concepts in this chapter. It is left for the reader to show that the set of all solutions of $A\mathbf{x} = \mathbf{b}$ is not a subspace of R^n if $\mathbf{b} \neq \mathbf{0}$.

Example 4.   (*For readers who have studied calculus.*) Let U be the set of all twice differentiable functions f defined on an open interval I such that

$$\frac{d^2f(x)}{dx^2} + f(x) = 0$$

U is a subset of the set V of all functions defined on I, which by Example 3 of Section 4.2 is a linear space. Suppose that f and g are two functions in U so that

$$\frac{d^2f(x)}{dx^2} + f(x) = 0 \qquad \frac{d^2g(x)}{dx^2} + g(x) = 0$$

Since the derivative of a sum is the sum of the derivatives, we have

$$\frac{d^2}{dx^2}[f(x) + g(x)] + [f(x) + g(x)] = \frac{d^2f(x)}{dx^2} + \frac{d^2g(x)}{dx^2} + f(x) + g(x)$$

$$= \left[\frac{d^2f(x)}{dx^2} + f(x)\right] + \left[\frac{d^2g(x)}{dx^2} + g(x)\right]$$

$$= 0 + 0$$

$$= 0$$

Hence, $f + g$ is an element of U. Also

$$\frac{d^2}{dx^2}[cf(x)] + cf(x) = \frac{cd^2f(x)}{dx^2} + cf(x)$$

$$= c\left[\frac{d^2f(x)}{dx^2} + f(x)\right]$$

$$= 0$$

so that cf is also an element of V. By Theorem 3, U is a subspace of V.

Example 5.   Consider the subset U of R^2 defined by

$$U = \left\{ \begin{bmatrix} x \\ y \end{bmatrix} : x + y = 1 \right\}$$

Let

$$\begin{bmatrix} x \\ y \end{bmatrix} \quad \text{and} \quad \begin{bmatrix} w \\ z \end{bmatrix}$$

be any two elements of U. Then $x + y = 1$ and $w + z = 1$. Since

$$(x + w) + (y + z) = (x + y) + (w + z)$$
$$= 1 + 1 \neq 1$$

it follows that the vector

$$\begin{bmatrix} x \\ y \end{bmatrix} + \begin{bmatrix} w \\ z \end{bmatrix} = \begin{bmatrix} x + w \\ y + z \end{bmatrix}$$

is not an element of U. In particular, $\begin{bmatrix} 0 \\ 1 \end{bmatrix}$ and $\begin{bmatrix} 3 \\ -2 \end{bmatrix}$ are elements of U, but $\begin{bmatrix} 0 \\ 1 \end{bmatrix} +$ $\begin{bmatrix} 3 \\ -2 \end{bmatrix} = \begin{bmatrix} 3 \\ -1 \end{bmatrix}$ is not an element of U. Therefore U is not closed under addition. Hence U is not a subspace of R^2. In fact, U is not closed under scalar multiplication either, because $\begin{bmatrix} 0 \\ 1 \end{bmatrix}$ is an element of U but $2 \begin{bmatrix} 0 \\ 1 \end{bmatrix} = \begin{bmatrix} 0 \\ 2 \end{bmatrix}$ is not.

Example 6.   Consider the subset of P_2 defined by

$$U = \{a_2 x^2 + a_1 x + a_0 : a_0 \text{ is an integer}\}$$

If $a_2 x^2 + a_1 x + a_0$ and $b_2 x^2 + b_1 x + b_0$ are any two elements of U, then a_0 and b_0 are integers. Therefore $a_0 + b_0$ is an integer, and the polynomial $(a_2 x^2 + a_1 x + a_0) + (b_2 x^2 + b_1 x + b_0) = (a_2 + b_2)x^2 + (a_1 + b_1)x + (a_0 + b_0)$ is an element of U. Hence U is closed under addition.

If c is any real number, then $c(a_2 x^2 + a_1 x + a_0) = (ca_2)x^2 + (ca_1)x + (ca_0)$. If a_0 is an integer it is not necessarily true that ca_0 is an integer. For example, if $c = 1/2$ and $a_0 = 1$, then ca_0 is not an integer. Therefore U is not closed under scalar multiplication. The set U is not a subspace of P_2.

Example 7.   Consider the subset of R^2 defined by

$$U = \left\{ \begin{bmatrix} a \\ b \end{bmatrix} : a = b = 0 \text{ or both } a \text{ and } b \text{ are nonzero} \right\}$$

If $\begin{bmatrix} a \\ b \end{bmatrix}$ is any element of U and c is any real number, then

$$c \begin{bmatrix} a \\ b \end{bmatrix} = \begin{bmatrix} ca \\ cb \end{bmatrix}$$

If $a = b = 0$ or if $c = 0$, then $ca = cb = 0$. If a, b, and c are nonzero, then both ca and cb are nonzero. It follows that U is closed under scalar multiplication. Notice that $\begin{bmatrix} 1 \\ 1 \end{bmatrix}$ and $\begin{bmatrix} -1 \\ 1 \end{bmatrix}$ are elements of U, but

$$\begin{bmatrix} 1 \\ 1 \end{bmatrix} + \begin{bmatrix} -1 \\ 1 \end{bmatrix} = \begin{bmatrix} 0 \\ 2 \end{bmatrix}$$

is not an element of U. Therefore U is not closed under addition. The set U is not a subspace of R^2.

The previous examples show that a subset U of a linear space may be closed under either, neither, or both of the operations of addition and scalar multiplication.

At this point it is worthwhile to give a preview of what lies ahead.

1. Most of the theorems will be true for any linear space V, but we will focus our attention primarily on subspaces of R^n and subspaces of the linear space $F(I)$ of all functions defined on an open interval I.
2. Most questions concerning subspaces of R^n will be answered by solving systems of linear equations and using the techniques of Chapter 1.
3. When the linear space is a subspace of $F(I)$ there are no general techniques to answer questions. Instead we refer to the definitions and work from them. Of special importance is the fact that many of the geometric ideas from R^n can be carried over to subspaces of $F(I)$.
4. In Example 2 of Section 1.9 we found that every solution of

$$\begin{bmatrix} 1 & 2 & 3 \\ 4 & 5 & 6 \\ 7 & 8 & 9 \end{bmatrix} \mathbf{x} = \mathbf{0}$$

is a scalar multiple of

$$\begin{bmatrix} 1 \\ -2 \\ 1 \end{bmatrix}$$

Thus, in a sense, there is only one solution. We will develop this idea in the sections that follow.

Exercises

1. Show that the set V of all scalar multiples of $\begin{bmatrix} 1 \\ 2 \end{bmatrix}$ is a subspace of R^2. Show that the only subspaces of V are $\begin{bmatrix} 0 \\ 0 \end{bmatrix}$ and V itself.
2. Show that the set P_n of all polynomials of degree less than or equal to n is a subspace of the linear space of all functions defined on the interval $(-\infty, \infty)$.
3. For what value of m is R^m a subspace of R^n?
4. Is the set $\{a_2 x^2 + a_1 x + a_0 : a_2 + a_1 + a_0 = 0\}$ a subspace of P_2?
5. Is the set $\{a_2 x^2 + a_1 x + a_0 : a_2 = a_0, a_1 = 1\}$ a subspace of P_2?

6. Is the set $\left\{ \begin{bmatrix} a \\ b \end{bmatrix} : a + b = 1 \right\}$ a subspace of R^2?

7. Is the set $\left\{ \begin{bmatrix} a \\ b \end{bmatrix} : a + b = 0 \right\}$ a subspace of R^2?

8. Is the set $\left\{ \begin{bmatrix} a \\ b \\ c \end{bmatrix} : a + b = c = 0 \right\}$ a subspace of R^3?

9. Is the set $\left\{ \begin{bmatrix} a \\ b \\ c \end{bmatrix} : a - b - c = 0 \right\}$ a subspace of R^3?

10. Is the set $\left\{ \begin{bmatrix} a \\ b \\ c \end{bmatrix} : a + b = 1, c = 0 \right\}$ a subspace of R^3?

11. Is the set $\left\{ \begin{bmatrix} a \\ b \\ c \end{bmatrix} : ab = 0 \right\}$ a subspace of R^3?

12. Is the set $\left\{ \begin{bmatrix} a & b \\ c & d \end{bmatrix} : a + d = 0, \ b + c = 0 \right\}$ a subspace of the linear space of all 2×2 matrices?

13. Is the set $\left\{ \begin{bmatrix} a & b & c \\ d & e & f \end{bmatrix} : a = 3 = c, b = d = f \right\}$ a subspace of the linear space of all 2×3 matrices?

14. (*For readers who have had calculus.*) Is the set of all differentiable functions defined on $(0, 1)$ a subspace of the linear space of all functions defined on $(0, 1)$?

15. (*For readers who have had calculus.*) Is the set of all integrable functions defined on $(0, 1)$ a subspace of the linear space of all functions defined on $(0, 1)$?

16. (*For readers who have had calculus.*) Is the set of all functions defined on $(0, 2)$ such that $\lim_{x \to 1} f(x) = 1$ a subspace of the linear space of all functions defined on $(0, 2)$?

17. Is the set of all functions f defined on $[0, 1]$ such that $f(0) = f(1)$ a subspace of the linear space of all functions defined on $[0, 1]$?

18. Let A be an $m \times n$ matrix. Show that the set V of all $3 \times m$ matrices B such that $BA = 0$ is a subspace of the linear space of all $3 \times m$ matrices.

19. Let A be an $m \times n$ matrix. Show that the set $\{ \mathbf{x} : A\mathbf{x} \neq 0 \}$ is not a subspace of R^n:

20. Let $\mathbf{v}_1 = \begin{bmatrix} 1 \\ 2 \end{bmatrix}$ and $\mathbf{v}_2 = \begin{bmatrix} 2 \\ 1 \end{bmatrix}$. Is the set

$$S = \left\{ \begin{bmatrix} a \\ b \end{bmatrix} : c_1 \mathbf{v}_1 + c_2 \mathbf{v}_2 = \begin{bmatrix} a \\ b \end{bmatrix} \text{ for some numbers } c_1 \text{ and } c_2 \right\}$$

a subspace of R^2. Does $S = R^2$?

21. Let $\mathbf{v}_1$ and $\mathbf{v}_2$ be any two elements of a linear space V. Is the set

$$U = \{ u : c_1 \mathbf{v}_1 + c_2 \mathbf{v}_2 = u \text{ for some numbers } c_1 \text{ and } c_2 \}$$

a subspace of V?

22. Show that R^2 has infinitely many subspaces. (Hint: consider the set of all elements of the form $\begin{bmatrix} a \\ b \end{bmatrix}$ where $a = cb$ for some constant c.)

23. Let V be a linear space containing more than the zero element. Show that V has either exactly two subspaces ($\{0\}$ and $\{V\}$) or infinitely many subspaces. (Hint: suppose that there are nonzero elements v_1 and v_2 in V such that v_1 is not a scalar multiple of v_2. Show that the set $V_k = \{cv_1 + ckv_2 : c \text{ is a number}\}$ is a subspace of V for every scalar k.)

24. Let U be a subspace of a linear space V. Show that the zero element of V is also the zero element of U.

25. Let U and W be subspaces of a linear space V. Show that $U \cap W$ is also a subspace of V.

26. Let C^n and D^n be the linear spaces of n-vectors with complex components defined in Exercises 25 and 26 of Section 4.2. Is D^n a subspace of C^n?

27. In Exercise 25 of Section 4.2 the reader was asked to show that the set C^n of all n-vectors with complex components is a linear space over the complex numbers. Is R^n a subspace of C^n?

28. In Exercise 26 of Section 4.2 the reader was asked to show that the set D^n of all n-vectors with complex components is a linear space over the real numbers. Is R^n a subspace of D^n?

4.4 LINEAR COMBINATIONS

In Example 3 of Section 1.10 we found that

$$\mathbf{x} = \begin{bmatrix} x \\ y \\ z \end{bmatrix}$$

is a solution of

$$\begin{bmatrix} 1 & 2 & 3 \\ 2 & 4 & 6 \end{bmatrix} \mathbf{x} = 0 \tag{1}$$

if and only if $x + 2y + 3z = 0$. Thus the linear space of all solutions of equation (1) can be identified with a plane in R^3. On this plane we can place a "generalized" coordinate system that enables us to describe points of this plane by ordered pairs of numbers (see Figure 4.1).

This section and the following three sections make these ideas precise and discuss them in terms of general linear spaces, not just in terms of R^n. To begin, we will be interested in finding the "smallest" subspace of a linear space V containing a given finite set $\{v_1, v_2, \ldots, v_n\}$ of elements of V. According to Theorem 3 of Section 4.3 this

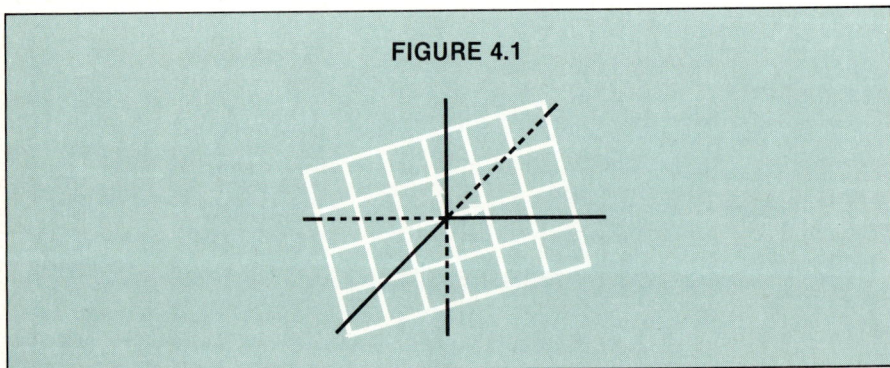

FIGURE 4.1

subspace will contain all scalar products of these elements and all possible sums of these scalar products. Therefore, we shall consider sums of the form

$$c_1 v_1 + c_2 v_2 + \cdots + c_n v_n$$

where $c_1, c_2, \ldots, c_n$ are scalars.

Definition 4. Let V be a linear space and $v_1, v_2, \ldots, v_n$ be elements of V. An element w of V is called a **linear combination** of $v_1, v_2, \ldots, v_n$ if there are scalars c_1, $c_2, \ldots, c_n$ such that

$$w = c_1 v_1 + c_2 v_2 + \cdots + c_n v_n$$

Example 1.   The 3-vector

$$w = \begin{bmatrix} -3 \\ 0 \\ -3 \end{bmatrix}$$

is a linear combination of the 3-vectors

$$u = \begin{bmatrix} 1 \\ 1 \\ 2 \end{bmatrix} \quad \text{and} \quad v = \begin{bmatrix} 3 \\ 2 \\ 5 \end{bmatrix}$$

since

$$\begin{bmatrix} -3 \\ 0 \\ -3 \end{bmatrix} = 6 \begin{bmatrix} 1 \\ 1 \\ 2 \end{bmatrix} + (-3) \begin{bmatrix} 3 \\ 2 \\ 5 \end{bmatrix}$$

That is, $w = 6u - 3v$. However, the 3-vector

$$z = \begin{bmatrix} 1 \\ 1 \\ 1 \end{bmatrix}$$

is not a linear combination of u and v. This may be proved as follows. Suppose that z is a linear combination of u and v. Then there are scalars c_1 and c_2 such that

$$z = c_1 u + c_2 v$$

That is

$$\begin{bmatrix} 1 \\ 1 \\ 1 \end{bmatrix} = c_1 \begin{bmatrix} 1 \\ 1 \\ 2 \end{bmatrix} + c_2 \begin{bmatrix} 3 \\ 2 \\ 5 \end{bmatrix}$$

$$= \begin{bmatrix} c_1 + 3c_2 \\ c_1 + 2c_2 \\ 2c_1 + 5c_2 \end{bmatrix} = \begin{bmatrix} 1 & 3 \\ 1 & 2 \\ 2 & 5 \end{bmatrix} \begin{bmatrix} c_1 \\ c_2 \end{bmatrix}$$

We now use the method of Chapter 1 to show that this system of equations has no solution. The associated augmented matrix is

$$\left[\begin{array}{cc|c} 1 & 3 & 1 \\ 1 & 2 & 1 \\ 2 & 5 & 1 \end{array}\right]$$

Then

$$\left[\begin{array}{cc|c} 1 & 3 & 1 \\ 1 & 2 & 1 \\ 2 & 5 & 1 \end{array}\right] \sim \left[\begin{array}{cc|c} 1 & 3 & 1 \\ 0 & -1 & 0 \\ 0 & -1 & -1 \end{array}\right] \quad \begin{array}{l} (-1)R_1 + R_2 \rightarrow R_2 \\ (-2)R_1 + R_3 \rightarrow R_3 \end{array}$$

$$\sim \left[\begin{array}{cc|c} 1 & 3 & 1 \\ 0 & -1 & 0 \\ 0 & 0 & -1 \end{array}\right] \quad (-1)R_2 + R_3 \rightarrow R_3$$

The last row of this augmented matrix corresponds to the equation $0 \cdot c_1 + 0 \cdot c_2 = -1$, which clearly has no solution. Therefore, there are no scalars c_1 and c_2 such that $\mathbf{z} = c_1\mathbf{u} + c_2\mathbf{v}$. The 3-vector $\mathbf{z}$ is not a linear combination of $\mathbf{u}$ and $\mathbf{v}$.

Example 2. Consider the functions

$$f(x) = x^2 + x \quad g(x) = x + 2 \quad h(x) = 3x^2 + 2x - 2$$

in the linear space of all functions defined on the interval $(-\infty, \infty)$. The function h is a linear combination of f and g since $3x^2 + 2x - 2 = 3(x^2 + x) + (-1)(x + 2)$. That is, $h(x) = 3f(x) - g(x)$.

However, the function $p(x) = x^3$ is not a linear combination of f and g. This may be proved as follows. Suppose that p is a linear combination of f and g. Then there exist scalars c_1 and c_2 such that $p(x) = c_1 f(x) + c_2 g(x)$ for all x. That is, $x^3 = c_1(x^2 + x) + c_2(x + 2)$ for all x, or equivalently, $x^3 - c_1 x^2 - (c_1 + c_2)x - 2c_2 = 0$ for all x. A consequence of the Fundamental Theorem of Algebra is that any third-degree polynomial has at most three distinct zeros. But we are considering a third-degree polynomial that has infinitely many zeros. This is impossible. Therefore there do not exist scalars c_1 and c_2 such that $p(x) = c_1 f(x) + c_2 g(x)$ for all x. The function p is not a linear combination of f and g.

Example 3. Using the trigonometric identity $\sin(a + b) = \cos a \sin b + \sin a \cos b$, we conclude that

$$\sin(1 + x) = \cos 1 \sin x + \sin 1 \cos x$$

for all x. Hence $\sin(1 + x)$ is a linear combination of $\sin x$ and $\cos x$. However, $\sin^2 x$ is not a linear combination of $\sin x$ and $\cos x$. This may be shown as follows. Suppose that $\sin^2 x$ is a linear combination of $\sin x$ and $\cos x$. Then there are numbers c_1 and c_2 such that $\sin^2 x = c_1 \sin x + c_2 \cos x$ for all x. When $x = 0$ we have $0 = c_1 \cdot 0 + c_2 \cdot 1$, so that $c_2 = 0$. Then $\sin^2 x = c_1 \sin x$ for all x. If $\sin x \neq 0$, then $\sin x = c_1$. This is impossible because c_1 is a fixed number. Therefore, there are no numbers c_1 and c_2 such that $\sin^2 x = c_1 \sin x + c_2 \cos x$. The function $\sin^2 x$ is not a linear combination of $\sin x$ and $\cos x$.

Example 4. Consider the system of equations

$$x_1 - 3x_2 + 2x_3 + x_4 = 0$$

$$3x_1 - 9x_2 + 10x_3 + 2x_4 = 0$$

(2)

$$2x_1 - 6x_2 + 8x_3 + x_4 = 0$$

$$2x_1 - 6x_2 + 4x_3 + 2x_4 = 0$$

The augmented matrix

$$A = \begin{bmatrix} 1 & -3 & 2 & 1 & 0 \\ 3 & -9 & 10 & 2 & 0 \\ 2 & -6 & 8 & 1 & 0 \\ 2 & -6 & 4 & 2 & 0 \end{bmatrix}$$

corresponds to the system of equations in (2). Elementary row operations allow us to obtain the augmented matrix

$$B = \begin{bmatrix} 1 & -3 & 0 & \frac{3}{2} & 0 \\ 0 & 0 & 1 & -\frac{1}{4} & 0 \\ 0 & 0 & 0 & 0 & 0 \\ 0 & 0 & 0 & 0 & 0 \end{bmatrix}$$

which is in reduced row-echelon form. The augmented matrix B corresponds to the system of equations

$$x_1 - 3x_2 + \frac{3}{2}x_4 = 0$$

$$x_3 - \frac{1}{4}x_4 = 0$$

which can be rearranged to

$$x_1 = 3x_2 - \frac{3}{2}x_4$$

$$x_3 = \frac{1}{4}x_4$$

Thus any solution of the system of equations in (2) has the form

$$\begin{bmatrix} x_1 \\ x_2 \\ x_3 \\ x_4 \end{bmatrix} = \begin{bmatrix} 3x_2 - \frac{3}{2}x_4 \\ x_2 \\ \frac{1}{4}x_4 \\ x_4 \end{bmatrix}$$

$$= x_2 \begin{bmatrix} 3 \\ 1 \\ 0 \\ 0 \end{bmatrix} + \frac{1}{4}x_4 \begin{bmatrix} -6 \\ 0 \\ 1 \\ 4 \end{bmatrix}$$

so that the solution space of the system of equations in (2) is the set of all linear combinations of the vectors

$$\begin{bmatrix} 3 \\ 1 \\ 0 \\ 0 \end{bmatrix} \quad \text{and} \quad \begin{bmatrix} -6 \\ 0 \\ 1 \\ 4 \end{bmatrix}$$

It is important to remember the definition of linear combination when the space is a set of functions. This definition must be applied directly as we have done in Examples 2 and 3. In the special case that the linear space is a subspace of of R^n there is a mechanical procedure for discovering whether an m-vector $\mathbf{b}$ is a linear combination of m-vectors $\mathbf{A}_1, \mathbf{A}_2, \ldots, \mathbf{A}_n$. In fact, we will show that $\mathbf{b}$ is a linear combination of $\mathbf{A}_1, \mathbf{A}_2, \ldots, \mathbf{A}_n$ if and only if the equation $A\mathbf{x} = \mathbf{b}$ is consistent, where A is the matrix having $\mathbf{A}_1, \mathbf{A}_2, \ldots, \mathbf{A}_n$ as its columns. In order to prove this we first prove a theorem that relates the concept of linear combinations to matrix multiplications. This theorem will be a useful tool later in proving important theorems in this book.

Theorem 4. *Let* $A = (a_{ij})$ *be an* $m \times n$ *matrix and let* $\mathbf{x}$ *be an* n-*vector with components* $x_1,$ $x_2, \ldots, x_n$. *If* $\mathbf{A}_1, \mathbf{A}_2, \ldots, \mathbf{A}_n$ *are the columns of* A, *then*

$$A\mathbf{x} = x_1\mathbf{A}_1 + x_2\mathbf{A}_2 + \cdots + x_n\mathbf{A}_n$$

Proof:

$$A\mathbf{x} = \begin{bmatrix} a_{11} & a_{12} & \cdots & a_{1n} \\ a_{21} & a_{22} & \cdots & a_{2n} \\ \vdots & \vdots & & \vdots \\ a_{m1} & a_{m2} & \cdots & a_{mn} \end{bmatrix} \begin{bmatrix} x_1 \\ x_2 \\ \vdots \\ x_n \end{bmatrix}$$

$$= \begin{bmatrix} a_{11}x_1 + a_{12}x_2 + \cdots + a_{1n}x_n \\ a_{21}x_1 + a_{22}x_2 + \cdots + a_{2n}x_n \\ \vdots & \vdots & \vdots \\ a_{m1}x_1 + a_{m2}x_2 + \cdots + a_{mn}x_n \end{bmatrix}$$

$$= x_1 \begin{bmatrix} a_{11} \\ a_{21} \\ \vdots \\ a_{m1} \end{bmatrix} + x_2 \begin{bmatrix} a_{12} \\ a_{22} \\ \vdots \\ a_{m2} \end{bmatrix} + \cdots + x_n \begin{bmatrix} a_{1n} \\ a_{2n} \\ \vdots \\ a_{mn} \end{bmatrix}$$

$$= x_1\mathbf{A}_1 + x_2\mathbf{A}_2 + \cdots + x_n\mathbf{A}_n \quad \blacksquare$$

Example 5.   Consider the vectors

$$\mathbf{v}_1 = \begin{bmatrix} 1 \\ 2 \\ 3 \end{bmatrix} \qquad \mathbf{v}_2 = \begin{bmatrix} 1 \\ 0 \\ 4 \end{bmatrix} \qquad \mathbf{v}_3 = \begin{bmatrix} 6 \\ 5 \\ 4 \end{bmatrix}$$

By Theorem 4 any linear combination $c_1\mathbf{v}_1 + c_2\mathbf{v}_2 + c_3\mathbf{v}_3$ of $\mathbf{v}_1, \mathbf{v}_2,$ and $\mathbf{v}_3$ can be written as

$$c_1\mathbf{v}_1 + c_2\mathbf{v}_2 + c_3\mathbf{v}_3 = A\mathbf{c}$$

where

$$A = \begin{bmatrix} 1 & 1 & 6 \\ 2 & 0 & 5 \\ 3 & 4 & 4 \end{bmatrix} \quad \text{and} \quad \mathbf{c} = \begin{bmatrix} c_1 \\ c_2 \\ c_3 \end{bmatrix}$$

Thus a 3-vector $\mathbf{b}$ is a linear combination of $\mathbf{v}_1, \mathbf{v}_2$, and $\mathbf{v}_3$ if and only if the equation $A\mathbf{c} = \mathbf{b}$ has a solution. A short calculation shows that det $A = 35 \neq 0$. By Theorem 8 of Section 2.5, the matrix A is nonsingular and the equation $A\mathbf{c} = \mathbf{b}$ has a solution for every 3-vector $\mathbf{b}$. Therefore every 3-vector is a linear combination of $\mathbf{v}_1, \mathbf{v}_2$, and $\mathbf{v}_3$.

The previous example shows that there is a relationship between the solutions of a matrix equation and the set of all linear combinations of the columns of that matrix. The following theorem describes this relationship.

Theorem 5. *Let A be an* m × n *matrix. Then the matrix equation* $A\mathbf{x} = \mathbf{b}$ *is consistent if and only if* $\mathbf{b}$ *is a linear combination of the columns of A.*

Proof: Let $\mathbf{A}_1, \mathbf{A}_2, \ldots, \mathbf{A}_n$ be the columns of A and let $x_1, x_2, \ldots, x_n$ be the components of an n-vector $\mathbf{x}$. Using Theorem 4 we find that $A\mathbf{x} = \mathbf{b}$ if and only if $x_1\mathbf{A}_1 + x_2\mathbf{A}_2 + \cdots + x_n\mathbf{A}_n = \mathbf{b}$. Therefore $A\mathbf{x} = \mathbf{b}$ is consistent if and only if $\mathbf{b}$ is a linear combination of the columns of A. ∎

Example 6. Consider the 3-vectors

$$\mathbf{v}_1 = \begin{bmatrix} 1 \\ 2 \\ 3 \end{bmatrix} \quad \mathbf{v}_2 = \begin{bmatrix} 4 \\ 5 \\ 6 \end{bmatrix} \quad \mathbf{v}_3 = \begin{bmatrix} 7 \\ 8 \\ 9 \end{bmatrix}$$

By Theorem 4 a 3-vector $\mathbf{b}$ is a linear combination of $\mathbf{v}_1, \mathbf{v}_2$, and $\mathbf{v}_3$ if and only if the matrix equation $A\mathbf{x} = \mathbf{b}$ has a solution where

$$A = \begin{bmatrix} 1 & 4 & 7 \\ 2 & 5 & 8 \\ 3 & 6 & 9 \end{bmatrix} \quad \mathbf{b} = \begin{bmatrix} b_1 \\ b_2 \\ b_3 \end{bmatrix}$$

A short calculation shows that det $A = 0$. By Theorem 9 of Section 2.5, the matrix A is singular and as such the equation $A\mathbf{x} = \mathbf{b}$ has no solution for at least one choice of the 3-vector $\mathbf{b}$. We will use the method described in Section 1.2 to determine the vectors $\mathbf{b}$ for which $A\mathbf{x} = \mathbf{b}$ has a solution.

$$\begin{bmatrix} 1 & 4 & 7 & | & b_1 \\ 2 & 5 & 8 & | & b_2 \\ 3 & 6 & 9 & | & b_3 \end{bmatrix} \sim \begin{bmatrix} 1 & 4 & 7 & | & b_1 \\ 0 & -3 & -6 & | & b_2 - 2b_1 \\ 0 & -6 & -12 & | & b_3 - 3b_1 \end{bmatrix} \quad \begin{matrix} (-2)R_1 + R_2 \to R_2 \\ (-3)R_1 + R_3 \to R_3 \end{matrix}$$

$$\sim \begin{bmatrix} 1 & 4 & 7 & | & b_1 \\ 0 & -3 & -6 & | & b_2 - 2b_1 \\ 0 & 0 & 0 & | & b_3 - 2b_2 + b_1 \end{bmatrix} \quad (-2)R_2 + R_3 \to R_3$$

Therefore $A\mathbf{x} = \mathbf{b}$ has a solution if and only if $b_3 - 2b_2 + b_1 = 0$. Consequently $A\mathbf{x} = \mathbf{b}$ has a solution if and only if the vector $\mathbf{b}$ has the form

$$\mathbf{b} = \begin{bmatrix} b_1 \\ b_2 \\ 2b_2 - b_1 \end{bmatrix} \tag{3}$$

where b_1 and b_2 are arbitrary numbers. By Theorem 5 a 3-vector $\mathbf{b}$ is a linear combination of $\mathbf{v}_1$, $\mathbf{v}_2$, and $\mathbf{v}_3$ if and only if $\mathbf{b}$ has the form in (3).

Exercises

In Exercises 1–8 write the first element as a linear combination of the remaining elements.

1. $\begin{bmatrix} 1 \\ 2 \\ 3 \end{bmatrix}$ $\begin{bmatrix} 2 \\ 4 \\ 6 \end{bmatrix}$

2. $\begin{bmatrix} 0 \\ 3 \\ 6 \end{bmatrix}$ $\begin{bmatrix} 0 \\ 1 \\ 2 \end{bmatrix}$

3. $\begin{bmatrix} 1 \\ 2 \end{bmatrix}$ $\begin{bmatrix} 1 \\ 0 \end{bmatrix}$ $\begin{bmatrix} 0 \\ 1 \end{bmatrix}$

4. $\begin{bmatrix} 1 \\ 1 \end{bmatrix}$ $\begin{bmatrix} 1 \\ 2 \end{bmatrix}$ $\begin{bmatrix} 2 \\ 1 \end{bmatrix}$

5. $x^2 + x$, $2x$, $3x^2$.

6. 1, x, $x^2 - x$, $x^2 + 3x + 1$.

7. $\begin{bmatrix} 0 & 1 \\ 2 & 0 \end{bmatrix}$ $\begin{bmatrix} 2 & 1 \\ 0 & 2 \end{bmatrix}$ $\begin{bmatrix} -1 & 1 \\ 3 & -1 \end{bmatrix}$

8. $\begin{bmatrix} 1 & 0 \\ 0 & 1 \end{bmatrix}$ $\begin{bmatrix} 0 & 1 \\ 1 & 0 \end{bmatrix}$ $\begin{bmatrix} 2 & 1 \\ 1 & 2 \end{bmatrix}$

In Exercises 9–14 find all vectors that can be written as linear combinations of the given vectors.

9. $\begin{bmatrix} 1 \\ 0 \\ 1 \end{bmatrix}$ $\begin{bmatrix} 1 \\ 1 \\ 0 \end{bmatrix}$

10. $\begin{bmatrix} 2 \\ 4 \\ 1 \end{bmatrix}$ $\begin{bmatrix} 1 \\ 3 \\ 0 \end{bmatrix}$ $\begin{bmatrix} 1 \\ 1 \\ 1 \end{bmatrix}$

11. $\begin{bmatrix} 1 \\ 4 \\ 0 \end{bmatrix}$ $\begin{bmatrix} 6 \\ 2 \\ 0 \end{bmatrix}$ $\begin{bmatrix} 1 \\ 1 \\ 0 \end{bmatrix}$

12. $\begin{bmatrix} 1 \\ 0 \\ 1 \end{bmatrix}$ $\begin{bmatrix} 0 \\ 1 \\ 1 \end{bmatrix}$ $\begin{bmatrix} 0 \\ 0 \\ 1 \end{bmatrix}$

13. $\begin{bmatrix} 1 \\ 2 \\ 3 \\ 4 \end{bmatrix}$ $\begin{bmatrix} 4 \\ 3 \\ 2 \\ 1 \end{bmatrix}$

14. $\begin{bmatrix} 2 \\ 0 \\ 1 \\ 2 \end{bmatrix}$ $\begin{bmatrix} 2 \\ 2 \\ 4 \\ 4 \end{bmatrix}$ $\begin{bmatrix} 3 \\ 1 \\ 3 \\ 4 \end{bmatrix}$

In Exercises 15–20 find all polynomials that are linear combinations of the given polynomials.

15. x, $3x^2 + 6$.

16. $x + 1$, $x - 1$.

17. $x - 1$, $x^2 + x$, x.

18. x, $x^2 + x$, $x^3 - x$.

19. $x - 1$, $x^2 + x + 1$, x^2.

20. $x^3 + x^2$, $x^2 + x$, $x^2 + x^3$, $x + 1$.

21. Let v_1 and v_2 be elements of a linear space V. Show that the set of all linear combinations of v_1 and v_2 is a subspace of V.

22. Let $v_1, v_2, \ldots, v_n$ be elements in a linear space V. Show that the set of all linear combinations of $v_1, v_2, \ldots, v_n$ is a subspace of V.

23. Find two vectors v_1 and v_2 such that every solution of

$$\begin{bmatrix} 1 & 2 & 3 \\ 2 & 4 & 6 \\ 3 & 6 & 9 \end{bmatrix} x = 0$$

is a linear combination of v_1 and v_2.

24. Find two vectors v_1 and v_2 such that every solution of

$$\begin{bmatrix} 1 & 1 & 2 & 2 \\ 3 & 0 & 1 & -1 \\ 2 & -1 & -1 & -3 \end{bmatrix} x = 0$$

is a linear combination of v_1 and v_2.

25. Show that if A and B are $m \times n$ matrices such that $Ax = Bx$ for every n-vector x, then $A = B$. (Hint: Show that the corresponding columns of A and B are equal by using Theorem 4 with x equal to the vectors shown below.)

$$\begin{bmatrix} 1 \\ 0 \\ 0 \\ \vdots \\ 0 \end{bmatrix}, \quad \begin{bmatrix} 0 \\ 1 \\ 0 \\ \vdots \\ 0 \end{bmatrix}, \quad \ldots, \quad \begin{bmatrix} 0 \\ 0 \\ 0 \\ \vdots \\ 1 \end{bmatrix}$$

26. Use Exercise 25 to show that if A is an $m \times n$ matrix such that $Ax = 0$ for every n-vector x, then A is the $m \times n$ zero matrix.

4.5 SPANNING SETS

Two concepts form the basis for most of linear algebra: One is the concept of a set of elements of a linear space spanning that linear space, and the other is the concept of linear independence. These are presented in Sections 4.5 and 4.6, respectively, and will be used throughout the remainder of this book. A little extra effort by the reader to master these concepts now will lead to greater ease in understanding the remaining material in the book.

Definition 5. Let $S = \{v_1, v_2, \ldots, v_n\}$ be a set of elements of a linear space V. If every element of V can be expressed as a linear combination of $v_1, v_2, \ldots, v_n$, then the set S is said to **span** V.

Example 1. The set of 3-vectors $\{e_1, e_2, e_3\}$ where

$$e_1 = \begin{bmatrix} 1 \\ 0 \\ 0 \end{bmatrix} \quad e_2 = \begin{bmatrix} 0 \\ 1 \\ 0 \end{bmatrix} \quad e_3 = \begin{bmatrix} 0 \\ 0 \\ 1 \end{bmatrix}$$

spans R^3 since any 3-vector

$$\mathbf{v} = \begin{bmatrix} a \\ b \\ c \end{bmatrix}$$

can be written as

$$\begin{bmatrix} a \\ b \\ c \end{bmatrix} = a\begin{bmatrix} 1 \\ 0 \\ 0 \end{bmatrix} + b\begin{bmatrix} 0 \\ 1 \\ 0 \end{bmatrix} + c\begin{bmatrix} 0 \\ 0 \\ 1 \end{bmatrix}$$

$$= a\mathbf{e}_1 + b\mathbf{e}_2 + c\mathbf{e}_3$$

which is a linear combination of $\mathbf{e}_1$, $\mathbf{e}_2$, and $\mathbf{e}_3$.

Example 2.   Clearly any polynomial $p(x) = a_2x^2 + a_1x + a_0$ is a linear combination of the polynomials 1, x, and x^2. Therefore $\{1, x, x^2\}$ spans the linear space P_2 of all polynomials of degree less than or equal to 2. However, it is not so obvious that the set $S = \{1 + x, x + x^2, 1 + x^2\}$ also spans P_2. This may be proved as follows. The set S spans P_2 if and only if any element $a + bx + cx^2$ of P_2 can be written as a linear combination of the elements of S. Thus S spans P_2 if and only if we can find scalars $c_1, c_2,$ and c_3 such that

$$c_1(1 + x) + c_2(x + x^2) + c_3(1 + x^2) = a + bx + cx^2$$

for any choice of the numbers $a, b,$ and c. This equation can be rewritten as

$$(c_1 + c_3) + (c_1 + c_2)x + (c_2 + c_3)x^2 = a + bx + cx^2$$

If we now equate the coefficients of like powers of x we obtain the system of equations

$$\begin{aligned} c_1 \quad\quad + c_3 &= a \\ c_1 + c_2 \quad\quad &= b \\ c_2 + c_3 &= c \end{aligned} \tag{1}$$

This system of equations corresponds to the matrix equation

$$\begin{bmatrix} 1 & 0 & 1 \\ 1 & 1 & 0 \\ 0 & 1 & 1 \end{bmatrix} \begin{bmatrix} c_1 \\ c_2 \\ c_3 \end{bmatrix} = \begin{bmatrix} a \\ b \\ c \end{bmatrix} \tag{2}$$

Since

$$\det \begin{bmatrix} 1 & 0 & 1 \\ 1 & 1 & 0 \\ 0 & 1 & 1 \end{bmatrix} = 2 \neq 0$$

we conclude from Theorem 8 of Section 2.5 that the matrix

$$\begin{bmatrix} 1 & 0 & 1 \\ 1 & 1 & 0 \\ 0 & 1 & 1 \end{bmatrix}$$

is nonsingular and that the equation in (2) has a solution for every choice of a, b, and c. Consequently, the system of equations in (1) has a solution for every choice of a, b, and c. Therefore, any polynomial of degree less than or equal to 2 is a linear combination of the elements in S. The set S spans P_2.

Example 3.   In Example 1 of Section 4.4 we saw that the 3-vector

$$\begin{bmatrix} 1 \\ 1 \\ 1 \end{bmatrix}$$

is not a linear combination of

$$\mathbf{u} = \begin{bmatrix} 1 \\ 1 \\ 2 \end{bmatrix} \quad \text{and} \quad \mathbf{v} = \begin{bmatrix} 3 \\ 2 \\ 5 \end{bmatrix}$$

Therefore the set $\{\mathbf{u}, \mathbf{v}\}$ does not span R^3.

Example 4.   In Example 4 of Section 4.4 we found that every solution of

$$\begin{aligned} x_1 - 3x_2 + 2x_3 + x_4 &= 0 \\ 3x_1 - 9x_2 + 10x_3 + 2x_4 &= 0 \\ 2x_1 - 6x_2 + 8x_3 + x_4 &= 0 \\ 2x_1 - 6x_2 + 4x_3 + 2x_4 &= 0 \end{aligned}$$

(3)

is a linear combination of the vectors

$$\mathbf{u} = \begin{bmatrix} 3 \\ 1 \\ 0 \\ 0 \end{bmatrix} \quad \text{and} \quad \mathbf{v} = \begin{bmatrix} -6 \\ 0 \\ 1 \\ 4 \end{bmatrix}$$

Therefore the solution space of the system of equations in (3) is spanned by $\{\mathbf{u}, \mathbf{v}\}$.

Examples 1, 2, and 3 show that a set S of elements of a linear space V may or may not span the space. If S spans V, then every element of V can be written as a linear combination of the elements of S. If S does not span V, then there is some element of V that cannot be written as a linear combination of the elements of S. The next theorem shows that even if S does not span V, the set of all linear combinations of elements of S is a subspace of V. This subspace is called the **linear space spanned by S** and is denoted by **span S**.

Theorem 6.   *Let $\mathbf{v}_1, \mathbf{v}_2, \ldots, \mathbf{v}_n$ be elements of a linear space V. Then*

1. *The set U of all linear combinations of $\mathbf{v}_1, \mathbf{v}_2, \ldots, \mathbf{v}_n$ is a subspace of V.*
2. *U is the smallest subspace of V containing $\mathbf{v}_1, \mathbf{v}_2, \ldots, \mathbf{v}_n$ in the sense that every other subspace of V containing $\mathbf{v}_1, \mathbf{v}_2, \ldots, \mathbf{v}_n$ also contains U.*

Proof: We will use Theorem 3 of Section 4.3 to prove part 1. If $\mathbf{u}$ and $\mathbf{w}$ are any two elements in the set U, then there are scalars $a_1, a_2, \ldots, a_n$ and $b_1, b_2, \ldots, b_n$ such that $\mathbf{u} = a_1\mathbf{v}_1 + a_2\mathbf{v}_2 + \cdots + a_n\mathbf{v}_n$ and $\mathbf{w} = b_1\mathbf{v}_1 + b_2\mathbf{v}_2 + \cdots + b_n\mathbf{v}_n$. Then $\mathbf{u} + \mathbf{w} = (a_1 + b_1)\mathbf{v}_1 + (a_2 + b_2)\mathbf{v}_2 + \cdots + (a_n + b_n)\mathbf{v}_n$, so that $\mathbf{u} + \mathbf{w}$ is an element of U. Hence U is closed under addition. Moreover, $c\mathbf{u} = (ca_1)\mathbf{v}_1 + (ca_2)\mathbf{v}_2 + \cdots + (ca_n)\mathbf{v}_n$ for any scalar c. Thus $c\mathbf{u}$ is also an element of U. Consequently, U is closed under scalar multiplication. By Theorem 3 of Section 4.3 the set U is a subspace of V.

Now let W be any subspace of V containing $\mathbf{v}_1, \mathbf{v}_2, \ldots, \mathbf{v}_n$. Since W is closed under addition and scalar multiplication, it must contain all linear combinations of $\mathbf{v}_1, \mathbf{v}_2, \ldots, \mathbf{v}_n$. But U is the set of all linear combinations of $\mathbf{v}_1, \mathbf{v}_2, \ldots, \mathbf{v}_n$. Therefore W contains U. ∎

Example 5. Consider the 3-vectors

$$\mathbf{v}_1 = \begin{bmatrix} 1 \\ 2 \\ 0 \end{bmatrix} \quad \text{and} \quad \mathbf{v}_2 = \begin{bmatrix} 2 \\ -3 \\ -4 \end{bmatrix}$$

By Theorem 6 the set U of all linear combinations of $\mathbf{v}_1$ and $\mathbf{v}_2$ is a subspace of R^3. Notice that any linear combination of $\mathbf{v}_1$ and $\mathbf{v}_2$ can be written as

$$a\mathbf{v}_1 + b\mathbf{v}_2 = a\begin{bmatrix} 1 \\ 2 \\ 0 \end{bmatrix} + b\begin{bmatrix} 2 \\ -3 \\ -4 \end{bmatrix}$$

$$= \begin{bmatrix} a + 2b \\ 2a - 3b \\ -4b \end{bmatrix}$$

for some choice of the scalars a and b. Therefore, the subspace U can be written as

$$U = \left\{ \begin{bmatrix} a + 2b \\ 2a - 3b \\ -4b \end{bmatrix} : a \text{ and } b \text{ real numbers} \right\}$$

Notice that

$$(-8)(a + 2b) + 4(2a - 3b) - 7(-4b) = 0$$

for every choice of the numbers a and b. Since $-8x + 4y - 7z = 0$ is the point normal form of an equation for the plane passing through $(0, 0, 0)$ and having

$$\begin{bmatrix} -8 \\ 4 \\ -7 \end{bmatrix}$$

as a normal vector, we see that the subspace U coincides with this plane. The subspace U is depicted in Figure 4.2.

FIGURE 4.2

The following theorem describes a relationship between the set of all solutions of a matrix equation $A\mathbf{x} = \mathbf{b}$ and the spanning sets for R^n.

Theorem 7. *Let $S = \{\mathbf{v}_1, \mathbf{v}_2, \ldots, \mathbf{v}_p\}$ be a set of n-vectors and let A be the $n \times p$ matrix whose columns are $\mathbf{v}_1, \mathbf{v}_2, \ldots, \mathbf{v}_p$. Then S spans R^n if and only if the equation $A\mathbf{x} = \mathbf{b}$ is consistent for every $\mathbf{b}$ in R^n.*

Proof: Let $x_1, x_2, \ldots, x_p$ be the components of an n-vector $\mathbf{x}$ and let $\mathbf{b}$ be an n-vector. By Theorem 5 of Section 4.4 we see that $x_1\mathbf{v}_1 + x_2\mathbf{v}_2 + \cdots + x_p\mathbf{v}_p = \mathbf{b}$ if and only if $A\mathbf{x} = \mathbf{b}$. Therefore, every n-vector is a linear combination of $\mathbf{v}_1, \mathbf{v}_2, \ldots, \mathbf{v}_p$ if and only if $A\mathbf{x} = \mathbf{b}$ is consistent for every n-vector $\mathbf{b}$. ∎

Example 6. Consider the set of 3-vectors

$$S = \left\{ \begin{bmatrix} 1 \\ 2 \\ 1 \end{bmatrix}, \begin{bmatrix} 3 \\ 2 \\ 1 \end{bmatrix}, \begin{bmatrix} 0 \\ 4 \\ 2 \end{bmatrix} \right\}$$

Let

$$A = \begin{bmatrix} 1 & 3 & 0 \\ 2 & 2 & 4 \\ 1 & 1 & 2 \end{bmatrix}$$

By Theorem 7 the set S spans R^3 if and only if the equation $A\mathbf{x} = \mathbf{b}$ is consistent for every 3-vector $\mathbf{b}$. A short calculation shows that $\det A = 0$. By Theorem 9 of Section 2.5, the matrix A is singular, and as such the equation $A\mathbf{x} = \mathbf{b}$ has no solution for at least one choice of the 3-vector $\mathbf{b}$. Therefore, the set S does not span R^3.

The proof of the following theorem is a direct consequence of Theorem 7 and is left as an exercise.

Theorem 8. *Let $S = \{\mathbf{v}_1, \mathbf{v}_2, \ldots, \mathbf{v}_n\}$ be a set of n-vectors and let A be the $n \times n$ matrix whose columns are $\mathbf{v}_1, \mathbf{v}_2, \ldots, \mathbf{v}_n$. Then S spans R^n if and only if A is nonsingular.*

--- **Exercises** ---

In Exercises 1–9 give computations or reasons for your answers.

1. Does $\left\{\begin{bmatrix}1\\1\end{bmatrix}, \begin{bmatrix}2\\2\end{bmatrix}\right\}$ span R^2?

2. Does $\left\{\begin{bmatrix}1\\1\end{bmatrix}, \begin{bmatrix}0\\1\end{bmatrix}\right\}$ span R^2?

3. Does $\left\{\begin{bmatrix}1\\0\end{bmatrix}, \begin{bmatrix}1\\1\end{bmatrix}, \begin{bmatrix}0\\1\end{bmatrix}\right\}$ span R^2?

4. Does $\left\{\begin{bmatrix}1\\2\end{bmatrix}, \begin{bmatrix}2\\4\end{bmatrix}, \begin{bmatrix}0\\0\end{bmatrix}\right\}$ span R^2?

5. Does $\{1, x, x + 1, x^2\}$ span P_2?
6. Does $\{1 + x, 1 - x, x^2 + 1\}$ span P_2?
7. Does $\{x^2 + x - 1, x^2 + 2x + 1, x + 2\}$ span P_2?
8. Does $\{x^2 + 1, x^2 + x + 1, x^2 - 1\}$ span P_2?

9. Does $\left\{\begin{bmatrix}1\\2\\3\end{bmatrix}, \begin{bmatrix}4\\5\\6\end{bmatrix}, \begin{bmatrix}1\\1\\1\end{bmatrix}\right\}$ span R^3?

10. Show that

$$\left\{\begin{bmatrix}1 & 0\\0 & 0\end{bmatrix}, \begin{bmatrix}0 & 1\\0 & 0\end{bmatrix}, \begin{bmatrix}0 & 0\\1 & 0\end{bmatrix}, \begin{bmatrix}0 & 0\\0 & 1\end{bmatrix}\right\}$$

spans the linear space of all 2×2 matrices.
11. Suppose that $\{v_1, v_2, \ldots, v_n\}$ spans a linear space V. Show that if v is any element of V, then $\{v_1, v_2, \ldots, v_n, v\}$ also spans V.

In Exercises 12–15 sketch the subspace of R^3 spanned by the given set of 3-vectors.

12. $\left\{\begin{bmatrix}1\\0\\1\end{bmatrix}, \begin{bmatrix}1\\1\\0\end{bmatrix}\right\}$

13. $\left\{\begin{bmatrix}1\\1\\1\end{bmatrix}, \begin{bmatrix}1\\1\\0\end{bmatrix}\right\}$

14. $\left\{\begin{bmatrix}1\\0\\1\end{bmatrix}, \begin{bmatrix}2\\0\\2\end{bmatrix}\right\}$

15. $\left\{\begin{bmatrix}2\\2\\0\end{bmatrix}, \begin{bmatrix}1\\1\\0\end{bmatrix}\right\}$

In Exercises 16–19 find two vectors that span the solution space of the given system of equations.

16. $\begin{bmatrix}1 & 2 & 3 & 4\\3 & 4 & 9 & 2\end{bmatrix}x = 0$

17. $\begin{bmatrix}2 & 4 & 2 & 3\\1 & 2 & 1 & 1\end{bmatrix}x = 0$

18. $2x_1 + 3x_2 + x_3 - x_4 = 0$
$x_1 - 2x_2 + 2x_3 + 2x_4 = 0$
$7x_1 + 8x_3 + 4x_4 = 0$

19. $x_1 + x_2 - x_3 + x_4 = 0$
$x_1 - 2x_2 + 2x_3 - x_4 = 0$
$2x_1 - x_2 + x_3 = 0$

20. If $\{v_1, v_2, \ldots, v_n\}$ spans a linear space V, is it possible for $\{v_2, v_3, \ldots, v_n\}$ to span V? Explain your answer.

21. Prove Theorem 8.

22. Suppose that $\{\mathbf{v}_1, \mathbf{v}_2, \ldots, \mathbf{v}_n\}$ spans a linear space V. Show that if $\mathbf{u}_1, \mathbf{u}_2, \ldots, \mathbf{u}_m$ are elements of V, then $\{\mathbf{v}_1, \mathbf{v}_2, \ldots, \mathbf{v}_n, \mathbf{u}_1, \mathbf{u}_2, \ldots, \mathbf{u}_m\}$ also spans V.

23. Let C^n denote the linear space over the complex numbers consisting of all n-vectors with complex components (see Exercise 25 of Section 4.2). Does the set

$$\left\{ \begin{bmatrix} 1 \\ 0 \\ 0 \\ \vdots \\ 0 \end{bmatrix}, \begin{bmatrix} 0 \\ 1 \\ 0 \\ \vdots \\ 0 \end{bmatrix}, \ldots, \begin{bmatrix} 0 \\ 0 \\ 0 \\ \vdots \\ 1 \end{bmatrix} \right\}$$

span C^n?

24. Let D^n denote the linear space over the real numbers consisting of all n-vectors with complex components (see Exercise 26 of Section 4.2). Does the set

$$\left\{ \begin{bmatrix} 1 \\ 0 \\ 0 \\ \vdots \\ 0 \end{bmatrix}, \begin{bmatrix} 0 \\ 1 \\ 0 \\ \vdots \\ 0 \end{bmatrix}, \ldots, \begin{bmatrix} 0 \\ 0 \\ 0 \\ \vdots \\ 1 \end{bmatrix} \right\}$$

span D^n?

4.6 LINEAR INDEPENDENCE

Let $\mathbf{v}_1, \mathbf{v}_2, \ldots, \mathbf{v}_n$ be elements of a linear space V. In Theorem 6 of Section 4.5 we found that span $\{\mathbf{v}_1, \mathbf{v}_2, \ldots, \mathbf{v}_n\}$ is a subspace of V. A natural question is: "What is the smallest number of elements from $\{\mathbf{v}_1, \mathbf{v}_2, \ldots, \mathbf{v}_n\}$ that span V?" To answer this question we introduce the concept of linear independence.

Consider the vectors

$$\mathbf{v}_1 = \begin{bmatrix} 1 \\ 2 \\ 3 \end{bmatrix} \qquad \mathbf{v}_2 = \begin{bmatrix} 1 \\ 1 \\ 1 \end{bmatrix} \qquad \mathbf{v}_3 = \begin{bmatrix} 1 \\ 4 \\ 7 \end{bmatrix}$$

We see that

$$\mathbf{v}_3 = \begin{bmatrix} 1 \\ 4 \\ 7 \end{bmatrix} = 3 \begin{bmatrix} 1 \\ 2 \\ 3 \end{bmatrix} + (-2) \begin{bmatrix} 1 \\ 1 \\ 1 \end{bmatrix} = 3\mathbf{v}_1 + (-2)\mathbf{v}_2$$

Since $\mathbf{v}_3$ is a linear combination of $\mathbf{v}_1$ and $\mathbf{v}_2$, any linear combination of $\mathbf{v}_1, \mathbf{v}_2,$ and $\mathbf{v}_3$ is also a linear combination of $\mathbf{v}_1$ and $\mathbf{v}_2$. This relationship is easily seen in the following equation:

$$c_1\mathbf{v}_1 + c_2\mathbf{v}_2 + c_3\mathbf{v}_3 = c_1\mathbf{v}_1 + c_2\mathbf{v}_2 + c_3(3\mathbf{v}_1 - 2\mathbf{v}_2)$$
$$= (c_1 + 3c_3)\mathbf{v}_1 + (c_2 - 2c_3)\mathbf{v}_2$$

Therefore, span $\{\mathbf{v}_1, \mathbf{v}_2, \mathbf{v}_3\}$ = span $\{\mathbf{v}_1, \mathbf{v}_2\}$, and $\mathbf{v}_3$ can be eliminated.

Sets $\{v_1, v_2, \ldots, v_n\}$ such that none of the v_i are linear combinations of the remaining elements of the set play a central role in the theory of linear spaces. This condition on the set can be phrased in a symmetrical form with which it is easier to work.

Definition 6.   A set of elements $\{v_1, v_2, \ldots, v_n\}$ of a linear space V is said to be **linearly dependent** if there are scalars $c_1, c_2, \ldots, c_n$, not all of which are zero, such that

$$c_1 v_1 + c_2 v_2 + \cdots + c_n v_n = 0 \tag{1}$$

If the set of elements is not linearly dependent, it is called **linearly independent**.

It is important to notice that a set of elements $\{v_1, v_2, \ldots, v_n\}$ of a linear space is linearly independent if and only if $c_1 = 0, c_2 = 0, \ldots, c_n = 0$ is the only choice of $c_1, c_2, \ldots, c_n$ such that $c_1 v_1 + c_2 v_2 + \cdots + c_n v_n = 0$. The coefficients $c_1 = 0$, $c_2 = 0, \ldots, c_n = 0$ always yield an identity in (1). The crucial factor for linear independence is that these are the *only* coefficients that yield the identity.

Example 1.   The set of elements $\{v_1, v_2, v_3\}$ of R^2 where

$$v_1 = \begin{bmatrix} 1 \\ 2 \end{bmatrix} \qquad v_2 = \begin{bmatrix} 3 \\ 4 \end{bmatrix} \qquad v_3 = \begin{bmatrix} 2 \\ 3 \end{bmatrix}$$

is linearly dependent since

$$v_1 + v_2 - 2v_3 = 0$$

Example 2.   Consider the set of elements $\{v_1, v_2\}$ of R^2 where

$$v_1 = \begin{bmatrix} 1 \\ 2 \end{bmatrix} \qquad v_2 = \begin{bmatrix} 3 \\ 4 \end{bmatrix}$$

We will show that this set is linearly independent. Let c_1 and c_2 be scalars such that $c_1 v_1 + c_2 v_2 = 0$. The scalars $c_1 = 0$ and $c_2 = 0$ satisfy this equation. If this is the only possible choice for c_1 and c_2 then $\{v_1, v_2\}$ is linearly independent. Otherwise $\{v_1, v_2\}$ is linearly dependent. Since

$$c_1 v_1 + c_2 v_2 = c_1 \begin{bmatrix} 1 \\ 2 \end{bmatrix} + c_2 \begin{bmatrix} 3 \\ 4 \end{bmatrix}$$

$$= \begin{bmatrix} c_1 + 3c_2 \\ 2c_1 + 4c_2 \end{bmatrix}$$

we have $c_1 v_1 + c_2 v_2 = 0$ if and only if

$$c_1 + 3c_2 = 0$$

$$2c_1 + 4c_2 = 0$$

It is easily shown that $c_1 = 0$ and $c_2 = 0$ is the only solution of this system of equations. Therefore $\{v_1, v_2\}$ is linearly independent.

Example 3.   Let V be the linear space of all functions defined on the interval $(-\infty, \infty)$. We will show that the set $\{1, x, x^2, \ldots, x^n\}$ is linearly independent. Let $c_0, c_1, \ldots, c_n$ be scalars such that

$$c_0(1) + c_1 x + \cdots + c_n x^n = 0$$

for all x in $(-\infty, \infty)$. If one of the c's is not zero, then $c_0 + c_1 x + \cdots + c_n x^n$ is a nonzero polynomial having infinitely many zeros. A consequence of the Fundamental Theorem of Algebra is that a kth degree polynomial has at most k different zeros. Therefore all of the c's must be zero. The set $\{1, x, x^2, \ldots, x^n\}$ is linearly independent.

Example 4.   Consider the subset $S = \{1 + x, x + x^2, 1 + x^2\}$ of P_2. We will show that this set is linearly independent. Let c_1, c_2, c_3 be scalars such that $c_1(1 + x) + c_2(x + x^2) + c_3(1 + x^2) = 0$ for all x. Elementary algebra allows us to rewrite this equation as $(c_1 + c_3)(1) + (c_1 + c_2)x + (c_2 + c_3)x^2 = 0$ for all x. By the previous example the set $\{1, x, x^2\}$ is linearly independent. Hence we have

$$c_1 +   c_3 = 0$$

$$c_1 + c_2 = 0$$

$$c_2 + c_3 = 0$$

It is left for the reader to show that $c_1 = 0, c_2 = 0, c_3 = 0$ is the only solution of this system of equations. Therefore the set S is linearly independent.

Linear dependence and linear independence are defined in terms of linear combinations. The following theorem gives precise relationships between these two concepts and linear combinations of elements of a linear space.

Theorem 9.   *Let $S = \{v_1, v_2, \ldots, v_n\}$ be a set of elements of a linear space V. Then*

    1. *S is linearly independent if and only if none of the v_i can be written as a linear combination of the remaining elements of S.*
    2. *S is linearly dependent if and only if at least one of v_i can be written as a linear combination of the remaining elements of S.*

Proof: Since the two statements are logically equivalent, it suffices to prove just one of them. We shall prove the second.

Suppose that one of the elements v_i is a linear combination of the remaining elements of S. Then there are scalars $c_1, \ldots, c_{i-1}, c_{i+1}, \ldots, c_n$ such that

$$v_i = c_1 v_1 + \cdots + c_{i-1} v_{i-1} + c_{i+1} v_{i+1} + \cdots + c_n v_n$$

or

$$c_1 v_1 + \cdots + c_{i-1} v_{i-1} + (-1)v_i + c_{i+1} v_{i+1} + \cdots + c_n v_n = \mathbf{0}$$

Setting $c_i = -1$ we have scalars $c_1, c_2, \ldots, c_n$, not all of which are zero, such that $c_1 v_1 + c_2 v_2 + \cdots + c_n v_n = \mathbf{0}$. Therefore S is linearly dependent.

Now suppose that S is linearly dependent. Then there exist numbers $c_1, c_2, \ldots, c_n$ not all of which are zero such that $c_1 \mathbf{v}_1 + c_2 \mathbf{v}_2 + \cdots + c_n \mathbf{v}_n = \mathbf{0}$. If $c_i \neq 0$, then

$$-c_i \mathbf{v}_i = c_1 \mathbf{v}_1 + \cdots + c_{i-1} \mathbf{v}_{i-1} + c_{i+1} \mathbf{v}_{i+1} + \cdots + c_n \mathbf{v}_n$$

or

$$\mathbf{v}_i = -\left(\frac{c_1}{c_i}\right)\mathbf{v}_i - \cdots - \left(\frac{c_{i-1}}{c_i}\right)\mathbf{v}_{i-1} - \left(\frac{c_{i+1}}{c_i}\right)\mathbf{v}_{i+1} - \cdots - \left(\frac{c_n}{c_i}\right)\mathbf{v}_n$$

Thus $\mathbf{v}_i$ is a linear combination of the remaining elements of S. This completes the proof. ∎

The following theorem relates the concept of a set of m-vectors being linearly independent to that of a matrix equation having only one solution.

Theorem 10. *Let $S = \{\mathbf{v}_1, \mathbf{v}_2, \ldots, \mathbf{v}_n\}$ be a set of m-vectors and let A be the $m \times n$ matrix whose columns are $\mathbf{v}_1, \mathbf{v}_2, \ldots, \mathbf{v}_n$. Then S is linearly independent if and only if $A\mathbf{x} = \mathbf{0}$ has $\mathbf{x} = \mathbf{0}$ as its only solution. If $m = n$, then S is linearly independent if and only if A is nonsingular.*

Proof: Let $x_1, x_2, \ldots, x_n$ be the components of an n-vector $\mathbf{x}$. By Theorem 4 of Section 4.4 we have

$$A\mathbf{x} = x_1 \mathbf{v}_1 + x_2 \mathbf{v}_2 + \cdots + x_n \mathbf{v}_n$$

By Definition 6 the set S is linearly independent if and only if $x_1 = 0, x_2 = 0, \ldots$, $x_n = 0$ are the only scalars for which $x_1 \mathbf{v}_1 + x_2 \mathbf{v}_2 + \cdots + x_n \mathbf{v}_n = \mathbf{0}$. Therefore, $x_1 \mathbf{v}_1 + x_2 \mathbf{v}_2 + \cdots + x_n \mathbf{v}_n = \mathbf{0}$ if and only if $A\mathbf{x} = \mathbf{0}$ has $\mathbf{x} = \mathbf{0}$ as its only solution. The remainder of the proof is an immediate consequence of Theorem 16 of Section 1.9. ∎

Example 5. Consider the set of 4-vectors

$$S = \left\{ \begin{bmatrix} 1 \\ 4 \\ 7 \\ 2 \end{bmatrix}, \begin{bmatrix} 2 \\ 5 \\ 8 \\ 1 \end{bmatrix}, \begin{bmatrix} 3 \\ 6 \\ 0 \\ -9 \end{bmatrix} \right\}$$

Let

$$A = \begin{bmatrix} 1 & 2 & 3 \\ 4 & 5 & 6 \\ 7 & 8 & 0 \\ 2 & 1 & -9 \end{bmatrix}$$

The augmented matrix

$$\begin{bmatrix} 1 & 2 & 3 & | & 0 \\ 4 & 5 & 6 & | & 0 \\ 7 & 8 & 0 & | & 0 \\ 2 & 1 & -9 & | & 0 \end{bmatrix}$$

corresponding to the equation $A\mathbf{x} = \mathbf{0}$, has reduced row-echelon form

$$\begin{bmatrix} 1 & 0 & 0 & 0 \\ 0 & 1 & 0 & 0 \\ 0 & 0 & 1 & 0 \\ 0 & 0 & 0 & 0 \end{bmatrix}$$

(The reader should verify this.) It follows that $\mathbf{x} = \mathbf{0}$ is the only solution of $A\mathbf{x} = \mathbf{0}$. Therefore the set S is linearly independent.

Example 6.   Consider the set $S = \{\mathbf{e}_1, \mathbf{e}_2, \dots, \mathbf{e}_n\}$ of n-vectors, where

$$\mathbf{e}_1 = \begin{bmatrix} 1 \\ 0 \\ 0 \\ \vdots \\ 0 \end{bmatrix}, \ \mathbf{e}_2 = \begin{bmatrix} 0 \\ 1 \\ 0 \\ \vdots \\ 0 \end{bmatrix}, \ \dots, \ \mathbf{e}_n = \begin{bmatrix} 0 \\ 0 \\ 0 \\ \vdots \\ 1 \end{bmatrix}$$

I_n is the matrix whose columns are $\mathbf{e}_1, \mathbf{e}_2, \dots, \mathbf{e}_n$. Since $I_n\mathbf{x} = \mathbf{x}$ for every n-vector $\mathbf{x}$, the equation $I_n\mathbf{x} = \mathbf{0}$ has $\mathbf{x} = \mathbf{0}$ as its only solution. By Theorem 10 the set S is linearly independent.

It is sometimes convenient to have Theorem 10 stated in a logically equivalent form, as shown in Theorem 11.

Theorem 11.   *Let $S = \{\mathbf{v}_1, \mathbf{v}_2, \dots, \mathbf{v}_n\}$ be a set of m-vectors and let A be the $m \times n$ matrix whose columns are $\mathbf{v}_1, \mathbf{v}_2, \dots, \mathbf{v}_n$. Then S is linearly dependent if and only if $A\mathbf{x} = \mathbf{0}$ has a solution other than $\mathbf{x} = \mathbf{0}$. If $m = n$, then S is linearly dependent if and only if A is singular.*

Example 7.   Consider the set of 3-vectors

$$S = \left\{ \begin{bmatrix} 1 \\ 4 \\ 7 \end{bmatrix}, \begin{bmatrix} 2 \\ 5 \\ 8 \end{bmatrix}, \begin{bmatrix} 3 \\ 6 \\ 9 \end{bmatrix} \right\}$$

Let

$$A = \begin{bmatrix} 1 & 2 & 3 \\ 4 & 5 & 6 \\ 7 & 8 & 9 \end{bmatrix}$$

In Example 2 of Section 1.9 we found that the matrix A is singular. Therefore the set S is linearly dependent.

Let A be an $m \times n$ matrix with $m < n$. Theorem 20 of Section 1.10 guarantees the existence of a nonzero solution of $A\mathbf{x} = \mathbf{0}$. This result coupled with Theorem 11 gives the following theorem.

Theorem 12.   Let $S = \{v_1, v_2, \ldots, v_n\}$ be a set of m-vectors. If $m < n$, then S is linearly dependent.

In the special case $m = 2$, Theorem 12 tells us that any set consisting of more than two 2-vectors is linearly dependent.

Example 8.   Theorem 12 assures us that the set

$$S = \left\{ \begin{bmatrix} 7 \\ -9 \\ 6 \\ 0 \\ 34 \end{bmatrix}, \begin{bmatrix} 5 \\ -4 \\ 18 \\ 67 \\ 9 \end{bmatrix}, \begin{bmatrix} 89 \\ 34 \\ 7 \\ -5 \\ 54 \end{bmatrix}, \begin{bmatrix} 38 \\ 72 \\ -2 \\ 19 \\ 4 \end{bmatrix}, \begin{bmatrix} 2 \\ 43 \\ 11 \\ 1 \\ -1 \end{bmatrix}, \begin{bmatrix} 72 \\ 14 \\ 34 \\ 90 \\ 76 \end{bmatrix} \right\}$$

is linearly dependent since S consists of six 5-vectors.

In summary, let $S = \{v_1, v_2, \ldots, v_n\}$ be a subset of a linear space V. Then

1. S is linearly independent if and only if none of the v_i is a linear combination of the remaining elements of S.
2. If V is a subspace of R^n, then there is a mechanical procedure for determining whether S is linearly independent (Theorem 10).
3. If V is not a subspace of R^n, then in order to determine whether S is linearly independent we must revert to Definition 6.

─────────────────────────── **Exercises** ───────────────────────────

In Exercises 1–12 give computations or reasons for your answers.

1. Is $\left\{ \begin{bmatrix} 1 \\ 2 \end{bmatrix}, \begin{bmatrix} 2 \\ 1 \end{bmatrix} \right\}$ linearly dependent?

2. Is $\left\{ \begin{bmatrix} 1 \\ 1 \end{bmatrix}, \begin{bmatrix} 3 \\ 3 \end{bmatrix} \right\}$ linearly dependent?

3. Is $\left\{ \begin{bmatrix} 1 \\ 2 \\ 3 \end{bmatrix}, \begin{bmatrix} 1 \\ 0 \\ 1 \end{bmatrix}, \begin{bmatrix} 0 \\ 1 \\ 1 \end{bmatrix} \right\}$ linearly independent?

4. Is $\left\{ \begin{bmatrix} 2 \\ 0 \\ 1 \end{bmatrix}, \begin{bmatrix} 0 \\ 0 \\ 1 \end{bmatrix}, \begin{bmatrix} 1 \\ 0 \\ 1 \end{bmatrix} \right\}$ linearly independent?

5. Is $\left\{ \begin{bmatrix} 1 \\ 0 \\ 1 \\ 0 \end{bmatrix}, \begin{bmatrix} 0 \\ 1 \\ 0 \\ 1 \end{bmatrix}, \begin{bmatrix} 1 \\ 0 \\ 0 \\ 1 \end{bmatrix} \right\}$ linearly dependent?

6. Is $\left\{ \begin{bmatrix} 1 \\ 2 \\ 0 \\ 1 \end{bmatrix}, \begin{bmatrix} 0 \\ 1 \\ 2 \\ 1 \end{bmatrix}, \begin{bmatrix} 1 \\ 0 \\ 0 \\ 1 \end{bmatrix} \right\}$ linearly dependent?

7. Is $\{x, x^2, 1 - x\}$ linearly independent?

8. Is $\{x, 1, x^2 - x, x + 1\}$ linearly independent?

9. Is $\{\cos x, \sin x, 1\}$ linearly dependent?

10. Is $\{\cos^2 x, \sin^2 x, 1, e^x\}$ linearly dependent?

11. Is

$$\left\{ \begin{bmatrix} 1 & 0 \\ 0 & 1 \end{bmatrix}, \begin{bmatrix} 0 & 1 \\ 1 & 0 \end{bmatrix}, \begin{bmatrix} 1 & 1 \\ 0 & 0 \end{bmatrix} \right\}$$

linearly independent?

12. Is

$$\left\{ \begin{bmatrix} 0 & 1 \\ 1 & 0 \end{bmatrix}, \begin{bmatrix} 2 & 3 \\ 1 & 2 \end{bmatrix}, \begin{bmatrix} 1 & 0 \\ 0 & 1 \end{bmatrix}, \begin{bmatrix} 2 & 0 \\ 0 & 0 \end{bmatrix} \right\}$$

linearly independent?

13. For which values of c is the set

$$\left\{ \begin{bmatrix} 1 \\ 0 \\ 1 \end{bmatrix}, \begin{bmatrix} c \\ 0 \\ 0 \end{bmatrix}, \begin{bmatrix} 0 \\ 1 \\ 1 \end{bmatrix} \right\}$$

linearly independent?

14. For which values of c is the set

$$\left\{ \begin{bmatrix} 1 \\ c \\ 1 \end{bmatrix}, \begin{bmatrix} c \\ 0 \\ 0 \end{bmatrix}, \begin{bmatrix} 0 \\ 0 \\ c \end{bmatrix} \right\}$$

linearly dependent?

15. Let $v_1, v_2, \ldots, v_k$ be elements of R^n and let A be an $n \times n$ nonsingular matrix. Show that $\{Av_1, Av_2, \ldots, Av_n\}$ is linearly independent if and only if $\{v_1, v_2, \ldots, v_n\}$ is linearly independent.

16. Let $S = \{v_1, v_2, \ldots, v_n\}$ be a set of elements of a linear space V. Show that if one of the elements of S is the zero element of V, then S is linearly dependent.

17. Let $\{v_1, v_2, \ldots, v_n\}$ be a linearly independent subset of a linear space V. Show that if $m < n$ then the set $\{v_1, v_2, \ldots, v_m\}$ is also linearly independent.

18. Let $\{v_1, v_2, \ldots, v_n\}$ be a linearly dependent subset of a linear space V. Show that if $u_1, u_2, \ldots, u_m$ are any elements of V then the set $\{v_1, v_2, \ldots, v_n, u_1, u_2, \ldots, u_m\}$ is also linearly dependent.

19. Show that any set of four elements of P_2 is linearly dependent.

20. Show that any set of $n + 2$ elements of P_n is linearly dependent.

21. Let v_1 and v_2 be elements of a linear space V. Show that $S = \{v_1, v_2\}$ is linearly dependent if and only if one of the elements is a scalar multiple of the other element.

22. Let C^3 be the linear space over the complex numbers consisting of all 3-vectors with complex components. Is the set of elements

$$\left\{ \begin{bmatrix} i \\ 1 \\ 0 \end{bmatrix}, \begin{bmatrix} 1 \\ -i \\ 0 \end{bmatrix} \right\}$$

of C^3 linearly dependent? Why?

23. Let D^3 be the linear space over the real numbers consisting of all 3-vectors with complex components. Is the set of elements.

$$\left\{ \begin{bmatrix} i \\ 1 \\ 0 \end{bmatrix}, \begin{bmatrix} 1 \\ -i \\ 0 \end{bmatrix} \right\}$$

of D^3 linearly dependent? Why?

4.7 BASIS

Sets with a finite number of elements that are linearly independent and span a given linear space play a fundamental role in the theory of linear spaces.

Definition 7.   A finite set $S = \{v_1, v_2, \ldots, v_n\}$ of elements of a linear space V is called a **basis** for V if

1. S is linearly independent, and
2. S spans V.

The concept of basis is exceptionally important because it lets us focus our attention on a finite number of elements rather than on the infinite number that make up the linear space.

Example 1.   Consider the set $S = \{1, x, x^2\}$. In Example 2 of Section 4.5 we showed that S spans the linear space P_2 of all polynomials of degree less than or equal to 2. Setting $n = 2$ in Example 3 of Section 4.6, we see that S is linearly independent. Therefore S is a basis for P_2. Combining Example 2 of Section 4.5 and Example 4 of Section 4.6 we find that $S' = \{1, + x, x + x^2, 1 + x^2\}$ is also a basis for P_2. Thus a linear space can have more than one basis. Indeed, every linear space consisting of more than the zero element has infinitely many bases. However, as illustrated above, any two bases of the same linear space have the same number of elements. We will prove this in Section 4.9.

Example 2.   Example 1 of Section 4.5 showed that the set $S = \{e_1, e_2, e_3\}$ of 3-vectors, where

$$e_1 = \begin{bmatrix} 1 \\ 0 \\ 0 \end{bmatrix} \quad e_2 = \begin{bmatrix} 0 \\ 1 \\ 0 \end{bmatrix} \quad e_3 = \begin{bmatrix} 0 \\ 0 \\ 1 \end{bmatrix}$$

spans R^3. Setting $n = 3$ in Example 6 of Section 4.6, we find that S is linearly independent. Therefore S is a basis for R^3.

Example 3.   One of the most commonly encountered problems involving bases of linear spaces is that of finding a basis for the solution space of a system of linear equations $Ax = 0$.

Consider the system of equations $A\mathbf{x} = \mathbf{0}$ where

$$A = \begin{bmatrix} 1 & 1 & 3 & 3 & 4 \\ 1 & 1 & -1 & -1 & 0 \\ 2 & 2 & 1 & 1 & 3 \\ 1 & 1 & 1 & 1 & 2 \\ 2 & 2 & 3 & 3 & 5 \end{bmatrix}$$

The augmented matrix corresponding to this system of equations is row equivalent to the augmented matrix

$$B = \left[\begin{array}{ccccc|c} 1 & 1 & 0 & 0 & 1 & 0 \\ 0 & 0 & 1 & 1 & 1 & 0 \\ 0 & 0 & 0 & 0 & 0 & 0 \\ 0 & 0 & 0 & 0 & 0 & 0 \\ 0 & 0 & 0 & 0 & 0 & 0 \end{array}\right]$$

(The reader should verify this.) The augmented matrix B corresponds to the system of equations

$$x_1 + x_2 + \qquad\qquad x_5 = 0$$
$$x_3 + x_4 + x_5 = 0$$

which can be rearranged as

$$x_1 = -x_2 \qquad - x_5$$
$$x_3 = \qquad -x_4 - x_5$$

Therefore any solution of $A\mathbf{x} = \mathbf{0}$ can be written as

$$\begin{bmatrix} x_1 \\ x_2 \\ x_3 \\ x_4 \\ x_5 \end{bmatrix} = \begin{bmatrix} -x_2 - x_5 \\ x_2 \\ -x_4 - x_5 \\ x_4 \\ x_5 \end{bmatrix}$$

$$= x_2 \begin{bmatrix} -1 \\ 1 \\ 0 \\ 0 \\ 0 \end{bmatrix} + x_4 \begin{bmatrix} 0 \\ 0 \\ -1 \\ 1 \\ 0 \end{bmatrix} + x_5 \begin{bmatrix} -1 \\ 0 \\ -1 \\ 0 \\ 1 \end{bmatrix}$$

Hence, every solution of $A\mathbf{x} = \mathbf{0}$ is a linear combination of

$$\mathbf{u} = \begin{bmatrix} -1 \\ 1 \\ 0 \\ 0 \\ 0 \end{bmatrix} \quad \mathbf{v} = \begin{bmatrix} 0 \\ 0 \\ -1 \\ 1 \\ 0 \end{bmatrix} \quad \mathbf{w} = \begin{bmatrix} 1 \\ 0 \\ -1 \\ 0 \\ 1 \end{bmatrix}$$

By Theorem 10 of Section 4.6 the set $\{\mathbf{u}, \mathbf{v}, \mathbf{w}\}$ is linearly independent if

$$\begin{bmatrix} -1 & 0 & -1 \\ 1 & 0 & 0 \\ 0 & -1 & -1 \\ 0 & 1 & 0 \\ 0 & 0 & 1 \end{bmatrix} \mathbf{y} = \mathbf{0}$$

has $\mathbf{y} = \mathbf{0}$ as its only solution. It is left for the reader to show that $\mathbf{y} = \mathbf{0}$ is in fact the only solution of this equation. Thus $\{\mathbf{u}, \mathbf{v}, \mathbf{w}\}$ is linearly independent and spans the solution space of $A\mathbf{x} = \mathbf{0}$. Therefore the set $\{\mathbf{u}, \mathbf{v}, \mathbf{w}\}$ is a basis for the solution space of $A\mathbf{x} = \mathbf{0}$.

If we recall some theorems that we have already proved, we realize that the problem of determining whether a set of n-vectors is a basis for R^n is equivalent to several problems we have already solved.

Theorem 13. *Let $S = \{\mathbf{v}_1, \mathbf{v}_2, \ldots, \mathbf{v}_n\}$ be a set of n-vectors and let A be the n × n matrix whose columns are $\mathbf{v}_1, \mathbf{v}_2, \ldots, \mathbf{v}_n$. Then each of the following statements implies the others:*

    1. *S is a basis for R^n.*
    2. *S is linearly independent.*
    3. *S spans R^n.*
    4. *A is nonsingular.*
    5. *$A\mathbf{x} = \mathbf{0}$ has $x = \mathbf{0}$ as its only solution.*
    6. *$A\mathbf{x} = \mathbf{b}$ is consistent for every n-vector $\mathbf{b}$.*
    7. *det A ≠ 0.*

Proof: By Theorem 16 of Section 1.9, we have $(4) \Leftrightarrow (5) \Leftrightarrow (6)$. By Theorem 8 of Section 4.5 and Theorem 10 of Section 4.6 we have $(3) \Leftrightarrow (4)$ and $(2) \Leftrightarrow (4)$. Combining all of the above we have $(2) \Leftrightarrow (3) \Leftrightarrow (4) \Leftrightarrow (5) \Leftrightarrow (6)$. Since $(2) \Leftrightarrow (3)$ we have $(2) \Rightarrow (1)$. By Definition 7 we have $(1) \Rightarrow (2)$ so that $(1) \Leftrightarrow (2)$. By Theorem 8 of Section 2.5, we have $(7) \Leftrightarrow (4)$. This completes the proof. ∎

Example 4. Let $S = \{\mathbf{v}_1, \mathbf{v}_2, \mathbf{v}_3\}$ where

$$\mathbf{v}_1 = \begin{bmatrix} 1 \\ 0 \\ 1 \end{bmatrix} \qquad \mathbf{v}_2 = \begin{bmatrix} 1 \\ 2 \\ 3 \end{bmatrix} \qquad \mathbf{v}_3 = \begin{bmatrix} 0 \\ 3 \\ 1 \end{bmatrix}$$

The matrix whose columns are $\mathbf{v}_1, \mathbf{v}_2$, and $\mathbf{v}_3$ is

$$A = \begin{bmatrix} 1 & 1 & 0 \\ 0 & 2 & 3 \\ 1 & 3 & 1 \end{bmatrix}$$

A short calculation shows that $A\mathbf{x} = \mathbf{0}$ has $\mathbf{x} = \mathbf{0}$ as its only solution. By Theorem 13 the set S is a basis for R^3.

Let $S = \{v_1, v_2, \ldots, v_n\}$ be a set of nonzero elements of a linear space V and let U be the subspace of V spanned by S. From S we can obtain a basis for U as follows:

1. Let $S_1 = \{v_1\}$. Since $v_1 \neq 0$ the set S_1 is linearly independent.
2. Consider the set $S'_2 = S_1 \cup \{v_2\}$. If S'_2 is linearly independent, let $S_2 = S'_2$. Otherwise, let $S_2 = S_1$. In either case S_2 is linearly independent and spans the same subspace as $\{v_1, v_2\}$.
3. Consider the set $S'_3 = S_2 \cup \{v_3\}$. If S'_3 is linearly independent, let $S_3 = S'_3$. Otherwise, let $S_3 = S_2$. In either case S_3 is linearly independent and spans the same subspace as $\{v_1, v_2, v_3\}$.
4. Continuing in this manner we obtain a set that is linearly independent and spans U. Therefore this set is a basis for U.

In Section 4.11 we will find that if S consists of n-vectors, then there is a more efficient method to obtain from S a linearly independent set that also spans V.

Example 5.   We will use the process described above to find a basis for the subspace of R^4 spanned by

$$S = \left\{ \begin{bmatrix} 1 \\ 0 \\ 1 \\ 0 \end{bmatrix}, \begin{bmatrix} 0 \\ 1 \\ 1 \\ 0 \end{bmatrix}, \begin{bmatrix} 3 \\ -2 \\ 1 \\ 0 \end{bmatrix}, \begin{bmatrix} 1 \\ 1 \\ 0 \\ 0 \end{bmatrix} \right\}$$

of elements of R^4. Let

$$S_1 = \left\{ \begin{bmatrix} 1 \\ 0 \\ 1 \\ 0 \end{bmatrix} \right\}, \quad S'_2 = \left\{ \begin{bmatrix} 1 \\ 0 \\ 1 \\ 0 \end{bmatrix}, \begin{bmatrix} 0 \\ 1 \\ 1 \\ 0 \end{bmatrix} \right\}$$

Evidently S'_2 is linearly independent. Hence, we let $S_2 = S'_2$ and consider the set

$$S'_3 = \left\{ \begin{bmatrix} 1 \\ 0 \\ 1 \\ 0 \end{bmatrix}, \begin{bmatrix} 0 \\ 1 \\ 1 \\ 0 \end{bmatrix}, \begin{bmatrix} 3 \\ -2 \\ 1 \\ 0 \end{bmatrix} \right\}$$

The set S'_3 is linearly dependent since

$$3 \begin{bmatrix} 1 \\ 0 \\ 1 \\ 0 \end{bmatrix} - 2 \begin{bmatrix} 0 \\ 1 \\ 1 \\ 0 \end{bmatrix} - \begin{bmatrix} 3 \\ -2 \\ 1 \\ 0 \end{bmatrix} = \begin{bmatrix} 0 \\ 0 \\ 0 \\ 0 \end{bmatrix}$$

Hence, $S_3 = S_2$. Next we consider the set

$$S'_4 = \left\{ \begin{bmatrix} 1 \\ 0 \\ 1 \\ 0 \end{bmatrix}, \begin{bmatrix} 0 \\ 1 \\ 1 \\ 0 \end{bmatrix}, \begin{bmatrix} 1 \\ 1 \\ 0 \\ 0 \end{bmatrix} \right\}$$

which can be shown to be linearly independent using Theorem 10 of the previous section. Then $S_4 = S'_4$. The set S_4 is linearly independent and spans the same subspace of R^4 as does the set S. Hence S_4 is a basis for the subspace of R^4 that is spanned by S.

──────────────── Exercises ────────────────

In Exercises 1–2 let V be the subspace of R^4 spanned by the given set. Find a linearly independent set that spans V.

1. $\left\{ \begin{bmatrix} 2 \\ 1 \\ 3 \\ 0 \end{bmatrix}, \begin{bmatrix} 4 \\ 2 \\ 6 \\ 0 \end{bmatrix}, \begin{bmatrix} 1 \\ 0 \\ 1 \\ 1 \end{bmatrix}, \begin{bmatrix} 1 \\ 1 \\ 2 \\ -1 \end{bmatrix} \right\}$

2. $\left\{ \begin{bmatrix} 1 \\ 0 \\ 1 \\ 0 \end{bmatrix}, \begin{bmatrix} 0 \\ 1 \\ 1 \\ 1 \end{bmatrix}, \begin{bmatrix} 2 \\ 3 \\ 5 \\ 3 \end{bmatrix}, \begin{bmatrix} 1 \\ -1 \\ 0 \\ -1 \end{bmatrix} \right\}$

In Exercises 3–4 let V be the subspace of P_3 spanned by the given set. Find a linearly independent set that spans V.

3. $\{x + 1, x - 1, x, x^3 + x\}$

4. $\{x^3 - x, x^3 + 1, x + 1, x - 1, x\}$

In Exercises 5–13 give computations or reasons for your answers.

5. Is $\left\{ \begin{bmatrix} 1 \\ 2 \end{bmatrix}, \begin{bmatrix} 2 \\ 1 \end{bmatrix} \right\}$ a basis for R^2?

6. Is $\left\{ \begin{bmatrix} 1 \\ 2 \end{bmatrix}, \begin{bmatrix} 2 \\ 4 \end{bmatrix} \right\}$ a basis for R^2?

7. Is $\left\{ \begin{bmatrix} 1 \\ 0 \\ 1 \end{bmatrix}, \begin{bmatrix} 1 \\ 1 \\ 0 \end{bmatrix}, \begin{bmatrix} 0 \\ 1 \\ 1 \end{bmatrix} \right\}$ a basis for R^3?

8. Is $\left\{ \begin{bmatrix} 1 \\ 2 \\ 3 \end{bmatrix}, \begin{bmatrix} 3 \\ 2 \\ 1 \end{bmatrix}, \begin{bmatrix} 0 \\ 1 \\ 0 \end{bmatrix} \right\}$ a basis for R^3?

9. Is $\{1, x + 1, x^2 - x\}$ a basis for P_2?
10. Is $\{1 + x, 1 - x, x^2\}$ a basis for P_2?
11. Is $\{x^3 + 1, x^3 - 1, x^2 + 1, x^2 - 1\}$ a basis for P_3?
12. Is $\{x^3 + 1, x^3 + x^2, x + 1, x - 1\}$ a basis for P_3?

13. Is $\left\{ \begin{bmatrix} 1 & 1 \\ 0 & 1 \end{bmatrix}, \begin{bmatrix} 2 & 1 \\ 1 & 1 \end{bmatrix}, \begin{bmatrix} 0 & 1 \\ 1 & 0 \end{bmatrix}, \begin{bmatrix} 1 & 1 \\ 1 & 0 \end{bmatrix} \right\}$ a basis for the linear space of all 2×2 matrices?

14. Find a basis for the subspace

$$U = \left\{ \begin{bmatrix} a \\ b \\ c \end{bmatrix} : a + b + c = 0 \right\}$$

of R^3.

15. Find a base for the subspace

$$U = \left\{ \begin{bmatrix} a \\ b \\ c \end{bmatrix} : a + b = 0, b + c = 0 \right\}$$

of R^3.

16. Find a basis for the subspace

$$U = \left\{ \begin{bmatrix} a \\ b \\ c \\ d \end{bmatrix} : a + 2b + 3c + d = 0 \right\}$$

of R^4.

17. Find a basis for the subspace

$$U = \left\{ \begin{bmatrix} a \\ b \\ c \\ d \end{bmatrix} : a + b = 0, b + c + d = 0 \right\}$$

of R^4.

18. Find a basis for the subspace

$$U = \{ax^2 + bx + c : a + b + c = 0\}$$

of P_2.

19. Find a basis for the subspace

$$U = \{ax^2 + bx + c : a + b = 0\}$$

of P_2.

20. Find a basis for the subspace

$$U = \left\{ \begin{bmatrix} a & c \\ b & d \end{bmatrix} : 2a + d = 0, b + c = 0 \right\}$$

of the linear space of all 2×2 matrices.

21. Find a basis for the subspace

$$U = \left\{ \begin{bmatrix} a & c \\ b & d \end{bmatrix} : a + b + c + d = 0 \right\}$$

of the linear space of all 2×2 matrices.

In Exercises 22–27 find a basis for the solution space of $A\mathbf{x} = \mathbf{0}$ where A is the given matrix.

22. $\begin{bmatrix} 1 & 2 & 3 \\ 0 & 1 & 2 \\ 2 & 4 & 6 \end{bmatrix}$ **23.** $\begin{bmatrix} 1 & 2 & 3 \\ 4 & 5 & 6 \\ 7 & 8 & 9 \end{bmatrix}$

24. $\begin{bmatrix} 1 & 0 & 1 & 2 \\ 0 & 3 & 1 & 1 \\ 1 & 0 & 0 & 1 \end{bmatrix}$
　　　　　　　　　　25. $\begin{bmatrix} 1 & 1 & 2 & 3 \\ 2 & 0 & 1 & 1 \\ 3 & 1 & 3 & 4 \end{bmatrix}$

26. $\begin{bmatrix} 1 & 1 & 1 & 1 & 1 \\ 2 & 1 & 2 & 1 & 2 \\ 1 & 3 & 1 & 3 & 1 \\ 2 & 3 & 2 & 3 & 2 \\ 4 & 4 & 4 & 4 & 4 \end{bmatrix}$
　　　　27. $\begin{bmatrix} 2 & 2 & 1 & 1 & 2 \\ 1 & 1 & 1 & 1 & 1 \\ 1 & 1 & 0 & 0 & 1 \\ 3 & 3 & 2 & 2 & 3 \\ 5 & 5 & 3 & 3 & 5 \end{bmatrix}$

4.8 COORDINATE VECTORS

Let S be a basis for a linear space V. Since S spans V, any element $\mathbf{v}$ of V can be written as a linear combination of the elements of S. The following theorem is the foundation for the development of any further theory of linear spaces.

Theorem 14. *Let $S = \{\mathbf{v}_1, \mathbf{v}_2, \dots, \mathbf{v}_n\}$ be a basis for a linear space V and let $\mathbf{v}$ be any element of V. Then there are unique scalars $c_1, c_2, \dots, c_n$ such that*

$$\mathbf{v} = c_1\mathbf{v}_1 + c_2\mathbf{v}_2 + \cdots + c_n\mathbf{v}_n$$

Proof: Suppose that an element $\mathbf{v}$ of V can be written as $\mathbf{v} = c_1\mathbf{v}_1 + c_2\mathbf{v}_2 + \cdots + c_n\mathbf{v}_n$ and as $\mathbf{v} = d_1\mathbf{v}_1 + d_2\mathbf{v}_2 + \cdots + d_n\mathbf{v}_n$. Then

$$\mathbf{0} = \mathbf{v} - \mathbf{v}$$
$$= (c_1\mathbf{v}_1 + c_2\mathbf{v}_2 + \cdots + c_n\mathbf{v}_n) - (d_1\mathbf{v}_1 + d_2\mathbf{v}_2 + \cdots + d_n\mathbf{v}_n)$$
$$= (c_1 - d_1)\mathbf{v}_1 + (c_2 - d_2)\mathbf{v}_2 + \dots + (c_n - d_n)\mathbf{v}_n$$

Since S is a basis for V, S is linearly independent. Therefore, we must have $c_1 - d_1 = 0$, $c_2 - d_2 = 0, \dots, c_n - d_n = 0$ so that $c_1 = d_1, c_2 = d_2, \dots, c_n = d_n$. This completes the proof. ∎

Let $B = \{\mathbf{v}_1, \mathbf{v}_2, \dots, \mathbf{v}_n\}$ be an ordered basis for a linear space V. By **ordered basis** we mean that B is a basis for V and the order in which the elements of B are written is fixed. If $\mathbf{v}$ is any element of V, then by Theorem 14 there are unique numbers $c_1, c_2, \dots, c_n$ such that $\mathbf{v} = c_1\mathbf{v}_1 + c_2\mathbf{v}_2 + \cdots + c_n\mathbf{v}_n$. Since the basis B is ordered we can represent the linear combination $c_1\mathbf{v}_1 + c_2\mathbf{v}_2 + \cdots + c_n\mathbf{v}_n$ by the n-vector

$$\begin{bmatrix} c_1 \\ c_2 \\ \vdots \\ c_n \end{bmatrix}_B$$

where it is understood that the first component of this vector is the coefficient of $\mathbf{v}_1$ in the linear combination, the second component is the coefficient of $\mathbf{v}_2$ in the linear combination, ..., the nth component is the coefficient of $\mathbf{v}_n$ in the linear combination. The subscript "B" on the vector tells us which basis of V we are using in our calculations. When $V = R^n$ and the ordered basis B is the natural basis

$$\left\{\begin{bmatrix} 1 \\ 0 \\ 0 \\ 0 \\ \vdots \\ 0 \end{bmatrix}, \begin{bmatrix} 0 \\ 1 \\ 0 \\ 0 \\ \vdots \\ 0 \end{bmatrix}, \begin{bmatrix} 0 \\ 0 \\ 1 \\ 0 \\ \vdots \\ 0 \end{bmatrix}, \cdots, \begin{bmatrix} 0 \\ 0 \\ 0 \\ 0 \\ \vdots \\ 1 \end{bmatrix}\right\}$$

we will not place a subscript on the vector.

In terms of this notation we have

$$\mathbf{v} = \begin{bmatrix} c_1 \\ c_2 \\ \vdots \\ c_n \end{bmatrix}_B$$

The vector

$$\begin{bmatrix} c_1 \\ c_2 \\ \vdots \\ c_n \end{bmatrix}_B$$

is called the **coordinate vector** of $\mathbf{v}$ relative to the ordered basis B. Thus to each element $\mathbf{v}$ of V we associate a unique n-vector. Conversely, given an n-vector

$$\begin{bmatrix} b_1 \\ b_2 \\ \vdots \\ b_n \end{bmatrix}$$

we can associate a unique element $b_1\mathbf{v}_1 + b_2\mathbf{v}_2 + \cdots + b_n\mathbf{v}_n$ of V. Hence every linear space (with the real numbers as scalars) having a basis with n elements can be identified with R^n.

Example 1.   In Example 1 of Section 4.7 we found that $B = \{1, x, x^2\}$ and $B' = \{1 + x, x + x^2, 1 + x^2\}$ are bases for P_2. Let $p(x) = a + bx + cx^2$ be any element of P_2. Clearly

$$p(x) = a(1) + b(x) + c(x^2)$$

so that

$$p(x) = \begin{bmatrix} a \\ b \\ c \end{bmatrix}_B$$

In order to find the coordinate vector for $p(x)$ with respect to B' we need to find scalars c_1, c_2, and c_3 such that

$$a + bx + cx^2 = c_1(1 + x) + c_2(x + x^2) + c_3(1 + x^2)$$
$$= (c_1 + c_3) + (c_1 + c_2)x + (c_2 + c_3)x^2$$

Equating the coefficients of like powers of x gives us the system of equations

$$c_1 \qquad + c_3 = a$$
$$c_1 + c_2 \qquad = b$$
$$c_2 + c_3 = c$$

A short calculation shows that

$$c_1 = \frac{1}{2}(a + b - c) \qquad c_2 = \frac{1}{2}(b + c - a) \qquad c_3 = \frac{1}{2}(a - b + c)$$

Therefore

$$p(x) = \begin{bmatrix} \frac{1}{2}(a + b - c) \\ \frac{1}{2}(b + c - a) \\ \frac{1}{2}(a - b + c) \end{bmatrix}_{B'}$$

In particular, if $p(x) = 2 - 3x + 7x^2$, then

$$p(x) = \begin{bmatrix} 2 \\ -3 \\ 7 \end{bmatrix}_B$$

$$p(x) = \begin{bmatrix} \frac{1}{2}[2 + (-3) - 7] \\ \frac{1}{2}[(-3) + 7 - 2] \\ \frac{1}{2}[2 - (-3) + 7] \end{bmatrix}_{B'}$$

$$= \begin{bmatrix} -4 \\ 1 \\ 6 \end{bmatrix}_{B'}$$

Example 2.   In Example 4 of Section 4.7 we found that

$$B = \left\{ \begin{bmatrix} 1 \\ 0 \\ 1 \end{bmatrix}, \begin{bmatrix} 1 \\ 2 \\ 3 \end{bmatrix}, \begin{bmatrix} 0 \\ 3 \\ 1 \end{bmatrix} \right\}$$

is a basis for R^3. Let

$$\mathbf{v} = \begin{bmatrix} a \\ b \\ c \end{bmatrix}$$

be any element of R^3. In order to find the coordinate vector of **v** relative to B we need to find scalars c_1, c_2, and c_3 such that

$$\begin{bmatrix} a \\ b \\ c \end{bmatrix} = c_1 \begin{bmatrix} 1 \\ 0 \\ 1 \end{bmatrix} + c_2 \begin{bmatrix} 1 \\ 2 \\ 3 \end{bmatrix} + c_3 \begin{bmatrix} 0 \\ 3 \\ 1 \end{bmatrix}$$

$$= \begin{bmatrix} 1 & 1 & 0 \\ 0 & 2 & 3 \\ 1 & 3 & 1 \end{bmatrix} \begin{bmatrix} c_1 \\ c_2 \\ c_3 \end{bmatrix}$$

It is left for the reader to show that

$$c_1 = \frac{1}{4}(7a + b - 3c) \qquad c_2 = \frac{1}{4}(3c - 3a - b) \qquad c_3 = \frac{1}{2}(a + b - c)$$

is a solution of this sytem of equations. Hence

$$\begin{bmatrix} a \\ b \\ c \end{bmatrix} = \begin{bmatrix} \frac{1}{4}(7a + b - 3c) \\ \frac{1}{4}(3c - 3a - b) \\ \frac{1}{2}(a + b - c) \end{bmatrix}_B$$

In particular

$$\begin{bmatrix} -2 \\ 4 \\ 5 \end{bmatrix} = \begin{bmatrix} \frac{1}{4}[7(-2) + 4 - 3(5)] \\ \frac{1}{4}[3(5) - 3(-2) - 4] \\ \frac{1}{2}(-2 + 4 - 5) \end{bmatrix}_B$$

$$= \begin{bmatrix} -\frac{25}{4} \\ \frac{17}{4} \\ -\frac{3}{2} \end{bmatrix}_D$$

Exercises

In Exercises 1–4 find the coordinate vector for $\begin{bmatrix} 1 \\ 2 \\ 3 \end{bmatrix}$ relative to the given ordered basis for R^3.

1. $\left\{ \begin{bmatrix} 1 \\ 0 \\ 1 \end{bmatrix}, \begin{bmatrix} 1 \\ 1 \\ 0 \end{bmatrix}, \begin{bmatrix} 0 \\ 1 \\ 1 \end{bmatrix} \right\}$

2. $\left\{ \begin{bmatrix} 2 \\ 0 \\ 0 \end{bmatrix}, \begin{bmatrix} 0 \\ 2 \\ 2 \end{bmatrix}, \begin{bmatrix} 3 \\ 3 \\ 4 \end{bmatrix} \right\}$

3. $\left\{ \begin{bmatrix} 1 \\ 1 \\ 3 \end{bmatrix}, \begin{bmatrix} 1 \\ 2 \\ 3 \end{bmatrix}, \begin{bmatrix} 1 \\ 0 \\ 1 \end{bmatrix} \right\}$

4. $\left\{ \begin{bmatrix} 1 \\ 0 \\ 2 \end{bmatrix}, \begin{bmatrix} 1 \\ 1 \\ 0 \end{bmatrix}, \begin{bmatrix} 0 \\ 1 \\ 0 \end{bmatrix} \right\}$

In Exercises 5–7 find the coordinate vector for $5 + 2x + 3x^2$ relative to the given ordered basis for P_2.

5. $\{1, x, x^2\}$
6. $\{1 + x, 1 + x^2, x + x^2\}$
7. $\{1 + x + x^2, 1 - x, 2 - x^2\}$

In Exercises 8–10 find the coordinate vector for $3 - 7x + 2x^2$ relative to the given ordered basis for P_2.

8. $\{x, 1, x^2\}$
9. $\{1 + x, 1 - x^2, x - x^2\}$
10. $\{1 + x + x^2, 1 - x - x^2, x^2\}$
11. Let V be a linear space with basis $B = \{\mathbf{u}_1, \mathbf{u}_2, \ldots, \mathbf{u}_n\}$. Show that

$$[\mathbf{u} + \mathbf{v}]_B = [\mathbf{u}]_B + [\mathbf{v}]_B$$

$$[c\mathbf{u}]_B = c[\mathbf{u}]_B$$

for every scalar c and every $\mathbf{u}, \mathbf{v}$ in V.

4.9 DIMENSION

In Section 4.7 we found that a linear space can have more than one basis. In fact, every linear space consisting of more than the zero element has infinitely many bases. In this section we prove one of the unifying theorems of linear algebra theory: Any two bases of V have the same number of elements. The number of elements n in a basis for V is a "magic" number, because (as we will prove) if S is a subset of V with n elements, then S is linearly independent if and only if S spans V. Hence, knowing that a set has the right number of elements enables us to determine whether the set is a basis by verifying only one of the two properties required for the set to be a basis.

Theorem 15. *Let $S = \{\mathbf{v}_1, \mathbf{v}_2, \ldots, \mathbf{v}_n\}$ be a basis for a linear space V. Then every basis for V has n elements.*

Proof: Let $T = \{\mathbf{u}_1, \mathbf{u}_2, \ldots, \mathbf{u}_m\}$ be another basis for V. To show that $m = n$ we will show that $m \leq n$ and that $n \leq m$.

Suppose $n < m$. Since S is a basis for V and each element of T is an element of V, we are able to write the elements of T as linear combinations of $\mathbf{v}_1, \mathbf{v}_2, \ldots, \mathbf{v}_n$:

$$
\begin{aligned}
\mathbf{u}_1 &= a_{11}\mathbf{v}_1 + a_{12}\mathbf{v}_2 + \cdots + a_{1n}\mathbf{v}_n \\
\mathbf{u}_2 &= a_{21}\mathbf{v}_1 + a_{22}\mathbf{v}_2 + \cdots + a_{2n}\mathbf{v}_n \\
&\ \ \vdots \qquad \vdots \qquad \vdots \qquad\qquad \vdots \\
\mathbf{u}_m &= a_{m1}\mathbf{v}_1 + a_{m2}\mathbf{v}_2 + \cdots + a_{mn}\mathbf{v}_n
\end{aligned}
\tag{1}
$$

We will show that the assumption $n < m$ allows us to find scalars $c_1, c_2, \ldots, c_m$, not all of which are zero, such that

$$c_1\mathbf{u}_1 + c_2\mathbf{u}_2 + \cdots + c_m\mathbf{u}_m = \mathbf{0} \tag{2}$$

If we insert the representations for $\mathbf{u}_1, \mathbf{u}_2, \ldots, \mathbf{u}_m$ found in (1) into (2) we obtain

$$
c_1(a_{11}\mathbf{v}_1 + a_{12}\mathbf{v}_2 + \cdots + a_{1n}\mathbf{v}_n) + c_2(a_{21}\mathbf{v}_1 + a_{22}\mathbf{v}_2 + \cdots + a_{2n}\mathbf{v}_n)
$$
$$
+ \cdots + c_m(a_{m1}\mathbf{v}_1 + a_{m2}\mathbf{v}_2 + \cdots + a_{mn}\mathbf{v}_n) = \mathbf{0}
$$

which can be rewritten as

$$(a_{11}c_1 + a_{21}c_2 + \cdots + a_{m1}c_m)\mathbf{v}_1 + (a_{12}c_1 + a_{22}c_2 + \cdots + a_{m2}c_m)\mathbf{v}_2$$
$$+ \cdots + (a_{1n}c_1 + a_{2n}c_2 + \cdots + a_{mn}c_m)\mathbf{v}_n = \mathbf{0}$$

Since $\mathbf{v}_1, \mathbf{v}_2, \ldots, \mathbf{v}_n$ are linearly independent, we must have

$$a_{11}c_1 + a_{21}c_2 + \cdots + a_{m1}c_m = 0$$
$$a_{12}c_1 + a_{22}c_2 + \cdots + a_{m2}c_m = 0$$
$$\vdots \qquad \vdots \qquad \qquad \vdots \qquad \vdots$$
$$a_{1n}c_1 + a_{2n}c_2 + \cdots + a_{mn}c_m = 0$$

which is a system of n equations in the m unknowns $c_1, c_2, \ldots, c_m$. Since we are assuming $n < m$, there is a nontrivial solution of this system (Theorem 20 of Section 1.10). Therefore there are scalars $c_1, c_2, \ldots, c_m$, not all of which are zero, such that the equation in (2) is satisfied. Hence T is linearly dependent. This is impossible because T is a basis for V. This means that $m \le n$.

If we interchange the roles of S and T in the above argument, we conclude that $n \le m$. Since $m \le n$ and $n \le m$, then $m = n$. This completes the proof. ∎

Example 1.   In Example 6 of Section 4.6, we showed that the set $\{\mathbf{e}_1, \mathbf{e}_2, \ldots, \mathbf{e}_n\}$ of n-vectors, where

$$\mathbf{e}_1 = \begin{bmatrix} 1 \\ 0 \\ 0 \\ \vdots \\ 0 \end{bmatrix}, \quad \mathbf{e}_2 = \begin{bmatrix} 0 \\ 1 \\ 0 \\ \vdots \\ 0 \end{bmatrix}, \quad \ldots, \quad \mathbf{e}_n = \begin{bmatrix} 0 \\ 0 \\ 0 \\ \vdots \\ 1 \end{bmatrix}$$

are linearly independent. By Theorem 13 of Section 4.7 this set is a basis for R^n. From Theorem 15 we conclude that every basis for R^n has exactly n elements.

Definition 8.   Let $\{\mathbf{v}_1, \mathbf{v}_2, \ldots, \mathbf{v}_n\}$ be a basis for a linear space V. Then V is called a **finite dimensional linear space** and is said to have **dimension** n, denoted by dim $V = n$. In other words, the number of elements in a basis is the dimension. The dimension of the zero linear space $\{0\}$ is defined to be zero. A linear space that does not have finite dimension is called an **infinite dimensional linear space**.

Example 2.   Clearly every polynomial of degree less than or equal to n is a linear combination of the elements of $S = \{1, x, x^2, \ldots, x^n\}$. By Example 3 of Section 4.6 the set S is linearly independent and, therefore, is a basis for the linear space P_n of all polynomials of degree less than or equal to n. Hence, dim $P_n = n + 1$.

Theorem 16.   *Let U be a subspace of a finite dimensional linear space V. Then U is finite dimensional and dim $U \le$ dim V.*

Proof: See Exercise 14. ∎

Example 3.   For each positive integer n the linear space P_n is a subspace of the linear space V of all polynomial functions. Using Example 2, we have $\lim_{n \to \infty} \dim P_n = \lim_{n \to \infty} (n + 1) = \infty$. From Theorem 16 we conclude that V is not finite dimensional.

In general, to show that a set $S = \{v_1, v_2, \ldots, v_n\}$ is a basis for a linear space V we must show that S is both linearly independent and spans V. In Theorem 13 of Section 4.7 we found that if $V = R^n$ then we need only verify that S has one of these properties. The same is true for any n-dimensional linear space, as we shall see from the following theorem.

Theorem 17. *Let $S = \{v_1, v_2, \ldots, v_n\}$ be a set of elements of a linear space V with dim $V = n$. Then each of the following statements implies the others.*

> **1.** *S is a basis for V.*
> **2.** *S is linearly independent.*
> **3.** *S spans V.*

Proof: Clearly part 1 implies both part 2 and part 3. Suppose part 2 holds. Then the subspace of V spanned by S has dimension equal to n. In Exercise 10 the reader is asked to show that the only subspace V with dimension n is V itself. Therefore S spans V and is a basis for V. Thus part 2 implies both part 1 and part 3.

Now suppose part 3 holds. In Section 4.7 we found that every set that spans V contains a basis for V. If S is not a basis, then V has a basis with fewer than n elements. This is impossible because dim $V = n$. Therefore S must be a basis for V and as such must be linearly independent. Thus part 3 implies both part 1 and part 2. ∎

Example 4. Consider the set $S = \{1 + x, x + x^2, 1 + x^2\}$ of polynomials in P_2. We will illustrate Theorem 17 by showing that S spans P_2 if and only if S is linearly independent. This will be done by showing that S has these properties if and only if certain systems of linear equations have solutions.

The set S spans P_2 if and only if for any polynomial $a + bx + cx^2$ in P_2 there are numbers $c_1, c_2,$ and c_3 such that

$$a + bx + cx^2 = c_1(1 + x) + c_2(x + x^2) + c_3(1 + x^2)$$
$$= (c_1 + c_3) + (c_1 + c_2)x + (c_2 + c_3)x^2$$

Since $\{1, x, x^2\}$ is a basis for P_2, Theorem 14 of Section 4.8 allows us to conclude that

$$c_1 + \qquad c_3 = a$$
$$c_1 + c_2 \qquad = b$$
$$c_2 + c_3 = c$$

Thus the set S spans P_2 if and only if the matrix equation

$$\begin{bmatrix} 1 & 0 & 1 \\ 1 & 1 & 0 \\ 0 & 1 & 1 \end{bmatrix} x = \begin{bmatrix} a \\ b \\ c \end{bmatrix} \tag{3}$$

has a solution.

The set S is linearly independent if and only if $c_1 = 0, c_2 = 0, c_3 = 0$ are the only numbers such that

$$0 = c_1(1 + x) + c_2(x + x^2) + c_3(1 + x^2)$$
$$= (c_1 + c_3) + (c_1 + c_2)x + (c_2 + c_3)x^2$$

for all x. Since $\{1, x, x^2\}$ is linearly independent, we must have

$$
\begin{aligned}
c_1 + \quad\;\; c_3 &= 0 \\
c_1 + c_2 \quad\;\; &= 0 \\
c_2 + c_3 &= 0
\end{aligned}
$$

Therefore S is linearly independent if and only if the zero vector is the only solution of

(4)
$$
\begin{bmatrix} 1 & 0 & 1 \\ 1 & 1 & 0 \\ 0 & 1 & 1 \end{bmatrix} \mathbf{x} = \mathbf{0}
$$

By Theorem 16 of Section 1.9, the equation in (3) has a solution for every choice of a, b, and c if and only if the equation in (4) has $\mathbf{x} = \mathbf{0}$ as its only solution. Therefore S spans P_2 if and only if S is linearly independent. This is precisely what is guaranteed by Theorem 17. Since

$$
\det \begin{bmatrix} 1 & 0 & 1 \\ 1 & 1 & 0 \\ 0 & 1 & 1 \end{bmatrix} = 2 \neq 0
$$

the equation in (4) has $\mathbf{x} = \mathbf{0}$ as its only solution and set S is linearly independent. Hence the set S is a basis for P_2.

The next theorem shows that any linearly independent subset of a finite dimensional linear space V, that is not a basis for V, can be expanded to a basis for V.

Theorem 18. *Let $\{v_1, v_2, \ldots, v_k\}$ be a linearly independent subset of an n-dimensional linear space V with $k < n$. Then there are elements $v_{k+1}, v_{k+2}, \ldots, v_n$ of V such that $\{v_1, v_2, \ldots, v_n\}$ is a basis for V.*

Proof: Let $B = \{u_1, u_2, \ldots, u_n\}$ be any basis for V. Let

$$
B_0 = \{v_1, v_2, \ldots, v_k, u_1, u_2, \ldots, u_n\}
$$

Evidently B_0 spans V since B spans V. We now form a subset B_1 of B_0 that also spans V. If u_n is a linear combination of the remaining elements of B_0, let B_1 denote the set B_0 with u_n removed, i.e., $B_1 = B_0 - \{u_n\}$. Otherwise, $B_1 = B_0$. Notice that B_1 spans V. Next we form a subset B_2 of B_1 that spans V. If u_{n-1} is a linear combination of the remaining elements of B_1, let B_2 denote the set B_1 with u_{n-1} removed, i.e., $B_2 = B_1 - \{u_{n-1}\}$. Otherwise, $B_2 = B_1$. Notice that B_2 spans V. Continuing in this manner we obtain a set B_n such that:

1. $v_1, v_2, \ldots, v_k$ are elements of B_n.
2. B_n spans V.
3. If u_j is an element of B_n, then u_j is not a linear combination of the remaining elements of B_n.

For ease of exposition we rename the elements of B_n that are currently denoted by u with a subscript. We denote these elements by $v_{k+1}, v_{k+2}, \ldots, v_p$ where p is some positive integer. Thus $B_n = \{v_1, v_2, \ldots, v_p\}$. We now show that B_n is linearly independent. Let $c_1, c_2, \ldots, c_p$ be any scalars such that

$$c_1 \mathbf{v}_1 + c_2 \mathbf{v}_2 + \cdots + c_p \mathbf{v}_p = \mathbf{0} \tag{5}$$

If $c_j \neq 0$ for some $j = k + 1, k + 2, \ldots, p$, then $\mathbf{v}_j$ is a linear combination of the remaining elements of B_n:

$$\mathbf{v}_j = c_j^{-1}(c_1 \mathbf{v}_1 + \cdots + c_{j-1} \mathbf{v}_{j-1} + c_{j+1} \mathbf{v}_{j+1} + \cdots + c_p \mathbf{v}_p)$$

This contradicts the third property of B_n listed above. Therefore $c_j = 0$ for $j = k + 1$, $k + 2, \ldots, p$ so that equation (5) becomes $c_1 \mathbf{v}_1 + c_2 \mathbf{v}_2 + \cdots + c_k \mathbf{v}_k = \mathbf{0}$. Since $\{\mathbf{v}_1, \mathbf{v}_2, \ldots, \mathbf{v}_k\}$ is linearly independent we must have $c_1 = 0, c_2 = 0, \ldots, c_k = 0$. Thus $c_1 = 0, c_2 = 0, \ldots, c_p = 0$ are the only scalars satisfying equation (5). Therefore B_n is linearly independent. Since B_n spans V by its construction, B_n is a basis for V. By Theorem 15 we must have $p = n$. This completes the proof. ∎

Example 5.   Consider the linearly independent subset

$$S = \left\{ \begin{bmatrix} 1 \\ 1 \\ 1 \\ 0 \end{bmatrix}, \begin{bmatrix} 1 \\ 0 \\ 1 \\ 0 \end{bmatrix} \right\}$$

of R^4. We will find a basis for R^4 containing S. To begin we set

$$B_0 = \left\{ \begin{bmatrix} 1 \\ 1 \\ 1 \\ 0 \end{bmatrix}, \begin{bmatrix} 1 \\ 0 \\ 1 \\ 0 \end{bmatrix}, \begin{bmatrix} 1 \\ 0 \\ 0 \\ 0 \end{bmatrix}, \begin{bmatrix} 0 \\ 1 \\ 0 \\ 0 \end{bmatrix}, \begin{bmatrix} 0 \\ 0 \\ 1 \\ 0 \end{bmatrix}, \begin{bmatrix} 0 \\ 0 \\ 0 \\ 1 \end{bmatrix} \right\}$$

Evidently B_0 spans R^4. Since

$$\begin{bmatrix} 0 \\ 0 \\ 0 \\ 1 \end{bmatrix}$$

is not a linear combination of the remaining elements of B_0 we set $B_1 = B_0$. Since

$$\begin{bmatrix} 0 \\ 0 \\ 1 \\ 0 \end{bmatrix} = \begin{bmatrix} 1 \\ 1 \\ 1 \\ 0 \end{bmatrix} - \begin{bmatrix} 1 \\ 0 \\ 0 \\ 0 \end{bmatrix} - \begin{bmatrix} 0 \\ 1 \\ 0 \\ 0 \end{bmatrix}$$

we delete this element from B_1 to obtain the set

$$B_2 = \left\{ \begin{bmatrix} 1 \\ 1 \\ 1 \\ 0 \end{bmatrix}, \begin{bmatrix} 1 \\ 0 \\ 1 \\ 0 \end{bmatrix}, \begin{bmatrix} 1 \\ 0 \\ 0 \\ 0 \end{bmatrix}, \begin{bmatrix} 0 \\ 1 \\ 0 \\ 0 \end{bmatrix}, \begin{bmatrix} 0 \\ 0 \\ 0 \\ 1 \end{bmatrix} \right\}$$

The set B_2 spans V since B_1 does so. Since

$$
\begin{bmatrix} 0 \\ 1 \\ 0 \\ 0 \end{bmatrix} = \begin{bmatrix} 1 \\ 1 \\ 1 \\ 0 \end{bmatrix} - \begin{bmatrix} 1 \\ 0 \\ 1 \\ 0 \end{bmatrix}
$$

we delete this element from B_2 to obtain

$$
B_3 = \left\{ \begin{bmatrix} 1 \\ 1 \\ 1 \\ 0 \end{bmatrix}, \begin{bmatrix} 1 \\ 0 \\ 1 \\ 0 \end{bmatrix}, \begin{bmatrix} 1 \\ 0 \\ 0 \\ 0 \end{bmatrix}, \begin{bmatrix} 0 \\ 0 \\ 0 \\ 1 \end{bmatrix} \right\}
$$

The set B_3 spans R^4 since B_2 does so. Since B_3 spans R^4 and contains exactly four elements it is a basis for R^4 (Theorem 17). Moreover S is a subset of B_3.

If the linear space V in Theorem 18 is R^n, then there is a simpler method for expanding $\{v_1, v_2, \ldots, v_k\}$ to a basis for V. We will describe this method in Section 4.11.

We close this section by describing properties of subsets of a linear space V that do not have the same number of elements as the dimension of V.

Theorem 19.   *Let $S = \{v_1, v_2, \ldots, v_k\}$ be a subset of an n-dimensional linear space V.*
   1. *If $k < n$, then S does not span V.*
   2. *If $n < k$, then S is linearly dependent.*

Proof: Suppose that $k < n$. In Section 4.7 we found that every set that spans V contains a basis for V. If S spans V, then V has a basis with fewer than n elements. This is impossible because dim $V = n$. Therefore S does not span V. This proves the first statement. The proof of the second statement is left as an exercise. ∎

———————————————— **Exercises** ————————————————

1. Find the dimension of the subspace

$$
U = \left\{ \begin{bmatrix} a \\ b \\ c \end{bmatrix} : a + b + c = 0 \right\}
$$

of R^3.

2. Find the dimension of the subspace

$$
U = \left\{ \begin{bmatrix} a \\ b \\ c \end{bmatrix} : a - b + 2c = 0 \right\}
$$

of R^3.

3. Find the dimension of the subspace

$$U = \left\{ \begin{bmatrix} a \\ b \\ c \\ d \end{bmatrix} : a + b - c - d = 0 \right\}$$

of R^4.

4. Find the dimension of the subspace

$$U = \left\{ \begin{bmatrix} a \\ b \\ c \\ d \end{bmatrix} : a + 2d = 0, b + 2c = 0 \right\}$$

of R^4.

5. Find the dimension of the solution space of the matrix

$$\begin{bmatrix} 1 & 2 & 3 \\ 4 & 5 & 6 \\ 7 & 8 & 9 \end{bmatrix} \mathbf{x} = \mathbf{0}$$

6. Find the dimension of the solution space of the matrix

$$\begin{bmatrix} 1 & 0 & 1 & 2 \\ 0 & 3 & 1 & 1 \\ 1 & 0 & 0 & 1 \end{bmatrix} \mathbf{x} = \mathbf{0}$$

7. Find the dimension of the subspace

$$U = \{ax^2 + bx + c : a - b + 2c = 0\}$$

of P_2.

8. Find the dimension of the subspace

$$U = \{ax^3 + bx^2 + cx + d : a + d = 0, b - c = 0\}$$

of P_3.

9. Let U and W be subspaces of a finite dimensional linear space V. Exercise 25 of Section 4.3 asks the reader to show that $U \cap W$ is a subspace of V. Show that $\dim (U \cap W) \leq \dim U$, and that $\dim (U \cap W) \leq \dim W$.

10. Let U be a subspace of an n-dimensional linear space V. Show that if $\dim U = n$, then $U = V$.

11. What is the dimension of the linear space of all $m \times n$ matrices?

12. Let $S = \{\mathbf{v}_1, \mathbf{v}_2, \ldots, \mathbf{v}_n\}$ be a basis for a linear space V and let U be a subspace of V. Is it necessarily true that a basis for U is a subset of S? Why?

13. Show that if $\{\mathbf{v}_1, \mathbf{v}_2, \ldots, \mathbf{v}_n\}$ spans a linear space V, then $\dim V \leq n$.

14. Prove Theorem 16.

15. Prove part 2 of Theorem 19.

16. Let C^3 denote the linear space over the complex numbers consisting of all 3-vectors with complex components. Find a basis for C^3. What is the dimension of C^3?

17. Let D^3 denote the linear space over the real numbers consisting of all 3-vectors with complex components. Find a basis for D^3. What is the dimension of D^3?

18. Let C^n denote the linear space over the complex numbers consisting of all n-vectors with complex components. Find a basis for C^n. What is the dimension of C^n?

19. Let D^n denote the linear space over the real numbers consisting of all n-vectors with complex components. Find a basis for D^n? What is the dimension of D^n?

4.10 CHANGE OF BASIS

In some applications involving linear spaces it is advantageous to use a basis other than the one given. For example, some bases make the calculations easier than do others. Such bases are discussed in Section 5.4.

Suppose that B is a basis for a linear space V and that we wish to use another basis B' for V. That is, we wish to write elements of V as linear combinations of the elements of B' rather than as linear combinations of the elements of B. The following theorem tells how to switch from one basis to another.

Theorem 20. *Let* $B = \{\mathbf{v}_1, \mathbf{v}_2, \ldots, \mathbf{v}_n\}$ *and* $B' = \{\mathbf{u}_1, \mathbf{u}_2, \ldots, \mathbf{u}_n\}$ *be ordered bases for a finite dimensional linear space* V. *Let* $_{B'}P_B$ *be the* $n \times n$ *matrix having* $[\mathbf{v}_1]_{B'}, [\mathbf{v}_2]_{B'}, \ldots, [\mathbf{v}_n]_{B'}$ *as its columns. Then* $_{B'}P_B$ *is nonsingular and*

$$[\mathbf{v}]_{B'} = {}_{B'}P_B[\mathbf{v}]_B$$

for every $\mathbf{v}$ *in* V.

Proof: Let $\mathbf{v}$ be any element of V. Since B is a basis for V, the element $\mathbf{v}$ can be written as a linear combination of the elements of B:

$$\mathbf{v} = x_1\mathbf{v}_1 + x_2\mathbf{v}_2 + \cdots + x_n\mathbf{v}_n$$

so that

$$[\mathbf{v}]_R = \begin{bmatrix} x_1 \\ x_2 \\ \vdots \\ x_n \end{bmatrix}$$

Since B' is a basis for V each element of B can be written as a linear combination of the elements of B':

$$\mathbf{v}_1 = a_{11}\mathbf{u}_1 + a_{21}\mathbf{u}_2 + \cdots + a_{n1}\mathbf{u}_n$$
$$\mathbf{v}_2 = a_{12}\mathbf{u}_1 + a_{22}\mathbf{u}_2 + \cdots + a_{n2}\mathbf{u}_n$$
$$\vdots \qquad \vdots \qquad \vdots \qquad \vdots$$
$$\mathbf{v}_n = a_{1n}\mathbf{u}_1 + a_{2n}\mathbf{u}_2 + \cdots + a_{nn}\mathbf{u}_n$$

so that

$$[\mathbf{v}_1]_{B'} = \begin{bmatrix} a_{11} \\ a_{21} \\ \vdots \\ a_{n1} \end{bmatrix}, \quad [\mathbf{v}_2]_{B'} = \begin{bmatrix} a_{12} \\ a_{22} \\ \vdots \\ a_{n2} \end{bmatrix}, \quad \cdots, \quad [\mathbf{v}_n]_{B'} = \begin{bmatrix} a_{1n} \\ a_{2n} \\ \vdots \\ a_{nn} \end{bmatrix}$$

and

$$_{B'}P_B = \begin{bmatrix} a_{11} & a_{12} & \cdots & a_{1n} \\ a_{21} & a_{22} & \cdots & a_{2n} \\ \vdots & \vdots & & \vdots \\ a_{n1} & a_{n2} & \cdots & a_{nn} \end{bmatrix}$$

Since

$$\begin{aligned} \mathbf{v} &= x_1\mathbf{v}_1 + x_2\mathbf{v}_2 + \cdots + x_n\mathbf{v}_n \\ &= x_1(a_{11}\mathbf{u}_1 + a_{21}\mathbf{u}_2 + \cdots + a_{n1}\mathbf{u}_n) + x_2(a_{12}\mathbf{u}_1 + a_{22}\mathbf{u}_2 + \cdots + a_{n2}\mathbf{u}_n) \\ &\quad + \cdots + x_n(a_{1n}\mathbf{u}_1 + a_{2n}\mathbf{u}_2 + \cdots + a_{nn}\mathbf{u}_n) \\ &= (a_{11}x_1 + a_{12}x_2 + \cdots + a_{1n}x_n)\mathbf{u}_1 + (a_{21}x_1 + a_{22}x_2 + \cdots + a_{2n}x_n)\mathbf{u}_2 \\ &\quad + \cdots + (a_{n1}x_1 + a_{n2}x_2 + \cdots + a_{nn}x_n)\mathbf{u}_n \end{aligned}$$

we have

$$\begin{aligned} [\mathbf{v}]_{B'} &= \begin{bmatrix} a_{11}x_1 + a_{12}x_2 + \cdots + a_{1n}x_n \\ a_{21}x_1 + a_{22}x_2 + \cdots + a_{2n}x_n \\ \vdots & \vdots & \vdots \\ a_{n1}x_1 + a_{n2}x_2 + \cdots + a_{nn}x_n \end{bmatrix} \\ &= \begin{bmatrix} a_{11} & a_{12} & \cdots & a_{1n} \\ a_{21} & a_{22} & \cdots & a_{2n} \\ \vdots & \vdots & \vdots \\ a_{n1} & a_{n2} & \cdots & a_{nn} \end{bmatrix} \begin{bmatrix} x_1 \\ x_2 \\ \vdots \\ x_n \end{bmatrix} \\ &= _{B'}P_B[\mathbf{v}]_B \end{aligned}$$

We prove that $_{B'}P_B$ is nonsingular by showing that the equation $_{B'}P_B\mathbf{x} = \mathbf{b}$ has a solution for every n-vector $\mathbf{b}$. Let

$$\mathbf{b} = \begin{bmatrix} b_1 \\ b_2 \\ \vdots \\ b_n \end{bmatrix}$$

be an n-vector and let $\mathbf{v} = b_1\mathbf{u}_1 + b_2\mathbf{u}_2 + \cdots + b_n\mathbf{u}_n$. Then

$$b = [\mathbf{v}]_{B'} = _{B'}P_B[\mathbf{v}]_B$$

so that $_{B'}P_B\mathbf{x} = \mathbf{b}$ has a solution for every n-vector $\mathbf{b}$. Therefore $_{B'}P_B$ is nonsingular. ∎

The basic idea in Theorem 20 is quite simple. The matrix $_{B'}P_B$ is obtained by computing each of its columns. The ith column of $_{B'}P_B$ is the coordinate vector $[\mathbf{v}_i]_{B'}$ with respect to the "new" basis B' of the ith element $\mathbf{v}_i$ of the "old" basis B.

Definition 9. Let B and B' be ordered bases of a finite dimensional linear space V. The matrix $_{B'}P_B$ such that $[\mathbf{v}]_{B'} = _{B'}P_B[\mathbf{v}]_B$ for all $\mathbf{v}$ in V is called the **transition matrix** from B to B'.

Throughout this section P will always be used to denote a transition matrix from one basis B of a linear space V to another basis B'. In order to keep track of the bases we subscript P on the right by the "old" basis B and on the left by the "new" basis B'. Hence ${}_{B'}P_B$ denotes the transition matrix from B to B'. Figure 4.3 demonstrates this transition pictorially.

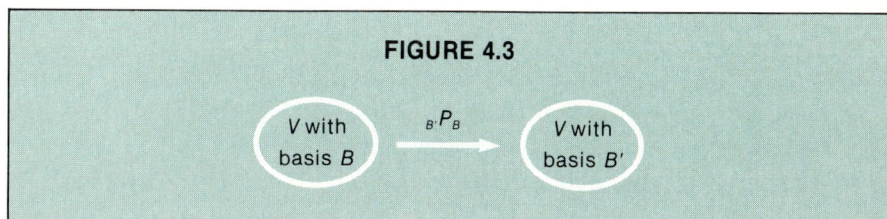

FIGURE 4.3

V with basis B $\quad \xrightarrow{\ {}_{B'}P_B\ }\quad$ V with basis B'

Example 1. Consider the ordered bases $B = \{v_1, v_2\}$ and $B' = \{u_1, u_2\}$ for R^2 where

$$\mathbf{v}_1 = \begin{bmatrix} 1 \\ 0 \end{bmatrix} \quad \mathbf{v}_2 = \begin{bmatrix} 0 \\ 1 \end{bmatrix} \quad \mathbf{u}_1 = \begin{bmatrix} 1 \\ 2 \end{bmatrix} \quad \mathbf{u}_2 = \begin{bmatrix} 2 \\ 1 \end{bmatrix}$$

To compute $[\mathbf{v}_1]_{B'}$ and $[\mathbf{v}_2]_{B'}$ we need to find scalars c_1, c_2, d_1, and d_2 such that

$$\begin{bmatrix} 1 \\ 0 \end{bmatrix} = c_1 \begin{bmatrix} 1 \\ 2 \end{bmatrix} + c_2 \begin{bmatrix} 2 \\ 1 \end{bmatrix}$$

$$\begin{bmatrix} 0 \\ 1 \end{bmatrix} = d_1 \begin{bmatrix} 1 \\ 2 \end{bmatrix} + d_2 \begin{bmatrix} 2 \\ 1 \end{bmatrix}$$

Short calculations show that $c_1 = -1/3$, $c_2 = 2/3$, $d_1 = 2/3$, and $d_2 = -1/3$. Therefore

$$\begin{bmatrix} 1 \\ 0 \end{bmatrix} = \begin{bmatrix} -\frac{1}{3} \\ \frac{2}{3} \end{bmatrix}_{B'}$$

$$\begin{bmatrix} 0 \\ 1 \end{bmatrix} = \begin{bmatrix} \frac{2}{3} \\ -\frac{1}{3} \end{bmatrix}_{B'}$$

so that

$${}_{B'}P_B = \begin{bmatrix} -\frac{1}{3} & \frac{2}{3} \\ \frac{2}{3} & -\frac{1}{3} \end{bmatrix}$$

is the transition matrix from B to B'.

Next we compute the transition matrix from B' to B. Notice that

$$\begin{bmatrix} 1 \\ 2 \end{bmatrix} = \begin{bmatrix} 1 \\ 0 \end{bmatrix} + 2 \begin{bmatrix} 0 \\ 1 \end{bmatrix}$$

$$\begin{bmatrix} 2 \\ 1 \end{bmatrix} = 2 \begin{bmatrix} 1 \\ 0 \end{bmatrix} + \begin{bmatrix} 0 \\ 1 \end{bmatrix}$$

so that

$$\begin{bmatrix} 1 \\ 2 \end{bmatrix} = \begin{bmatrix} 1 \\ 2 \end{bmatrix}_{B'}, \qquad \begin{bmatrix} 2 \\ 1 \end{bmatrix} = \begin{bmatrix} 2 \\ 1 \end{bmatrix}_{B'}$$

Therefore

$$_B P_{B'} = \begin{bmatrix} 1 & 2 \\ 2 & 1 \end{bmatrix}$$

is the transition matrix from B' to B. Further notice that $(_{B'}P_B)(_B P_{B'}) = I_2$ so that $(_B P_B)^{-1} = {}_B P_{B'}$. The following theorem shows that this is a specific example of a general property of transition matrices.

Theorem 21. *Let B and B′ be ordered bases for a finite dimensional linear space V. If $_{B'}P_B$ is the transition matrix from B to B′ and $_B P_{B'}$ is the transition matrix from B′ to B, then $_B P_{B'} = (_{B'}P_B)^{-1}$.*

Proof: For every **v** in V we have

$$(_B P_{B'})(_{B'}P_B)[\mathbf{v}]_B = {}_B P_{B'}(_{B'}P_B[\mathbf{v}]_B)$$
$$= {}_B P_{B'}[\mathbf{v}]_{B'}$$
$$= [\mathbf{v}]_B$$

It follows that $(_B P_{B'})(_{B'}P_B) = I$ so that $_B P_{B'} = (_{B'}P_B)^{-1}$. ∎

Example 2. Consider the xy-plane. Suppose that we wish to consider a new coordinate system that is obtained by rotating counterclockwise the y-axis through an angle θ, $0 \le \theta \le \pi/2$, and the x-axis through an angle ϕ, $0 \le \phi \le \pi/2$ (see Figure 4.4) where ϕ is not necessarily equal to θ. The rotated x and y axes will be denoted by x' and y', respectively. Let $\mathbf{v}_1$ and $\mathbf{v}_2$ be unit vectors directed along the x'-

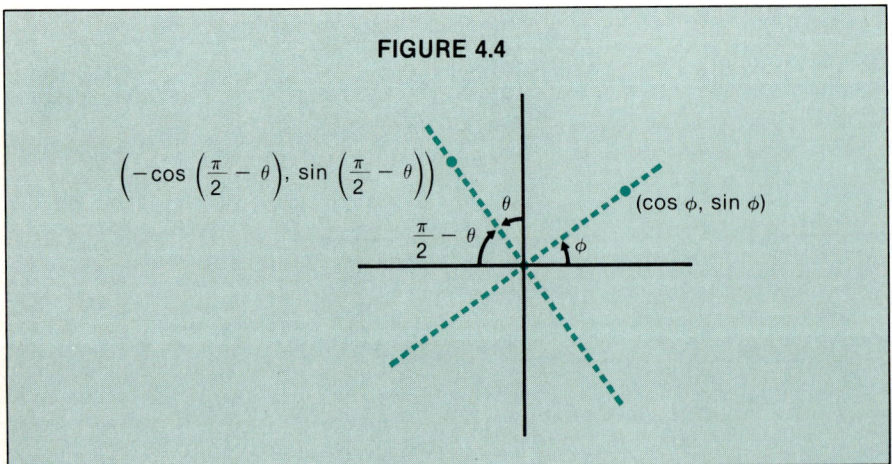

FIGURE 4.4

and y'-axes. That is, let

$$\mathbf{v}_1 = \begin{bmatrix} \cos \phi \\ \sin \phi \end{bmatrix}$$

$$= \cos \phi \begin{bmatrix} 1 \\ 0 \end{bmatrix} + \sin \phi \begin{bmatrix} 0 \\ 1 \end{bmatrix}$$

$$\mathbf{v}_2 = \begin{bmatrix} -\cos(\frac{\pi}{2} - \theta) \\ \sin(\frac{\pi}{2} - \theta) \end{bmatrix}$$

$$= \begin{bmatrix} -\sin \theta \\ \cos \theta \end{bmatrix} = -\sin \theta \begin{bmatrix} 1 \\ 0 \end{bmatrix} + \cos \theta \begin{bmatrix} 0 \\ 1 \end{bmatrix}$$

Then

$$_{B'}P_B = \begin{bmatrix} \cos \phi & -\sin \theta \\ \sin \phi & \cos \theta \end{bmatrix}$$

is the transition matrix from $B = \{\mathbf{v}_1, \mathbf{v}_2\}$ to $B' = \left\{ \begin{bmatrix} 1 \\ 0 \end{bmatrix}, \begin{bmatrix} 0 \\ 1 \end{bmatrix} \right\}$ and (see Theorem 12 of Section 1.8)

$$(_{B'}P_B)^{-1} = \frac{1}{\cos \theta \cos \phi + \sin \theta \sin \phi} \begin{bmatrix} \cos \theta & \sin \theta \\ -\sin \phi & \cos \phi \end{bmatrix}$$

$$= \frac{1}{\cos(\theta - \phi)} \begin{bmatrix} \cos \theta & \sin \theta \\ -\sin \phi & \cos \phi \end{bmatrix}$$

is the transition matrix from B' to B.

Exercise 9 asks the reader to compute directly the transition matrix from B' to B. This will illustrate that in this case it is easier to compute the transition matrix $_B P_B$ from B to B' and then form the transition matrix from B' to B by computing $(_B P_B)^{-1}$ than it is to compute directly the transition matrix from B' to B.

If the finite dimensional linear space V is R^n, the ordered bases B and B' are usually written in terms of the natural basis

$$N = \left\{ \begin{bmatrix} 1 \\ 0 \\ 0 \\ \vdots \\ 0 \end{bmatrix}, \begin{bmatrix} 0 \\ 1 \\ 0 \\ \vdots \\ 0 \end{bmatrix}, \cdots, \begin{bmatrix} 0 \\ 0 \\ 0 \\ \vdots \\ 1 \end{bmatrix} \right\}$$

Notice that the transition matrices from B to N and from B' to N are merely the matrices having the vectors of B and the vectors of B' as their columns. Thus the matrices $_N P_B$ and $_N P_{B'}$ are easily obtained. Since $_{B'} P_N$ is the transition matrix from N to B' and $_N P_B$ is the transition matrix from B to N we have that $(_{B'} P_N)(_N P_B)$ is the transition matrix from B to B'. That is,

$$_{B'}P_B = (_{B'}P_N)(_N P_B) = (_N P_{B'})^{-1}(_N P_B)$$

This identity can be depicted pictorially as in Figure 4.5. Following the arrows we get to the same place by either using $_N P_B$ followed by $(_N P_{B'})^{-1}$ or by using $_{B'}P_B$.

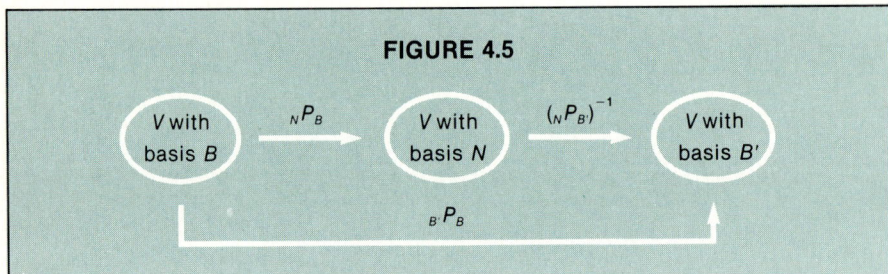

FIGURE 4.5

V with basis B $\xrightarrow{\;_N P_B\;}$ V with basis N $\xrightarrow{\;(_N P_{B'})^{-1}\;}$ V with basis B'

$_{B'}P_B$

Example 3.   Consider the ordered bases

$$B = \left\{ \begin{bmatrix} 1 \\ 1 \end{bmatrix}, \begin{bmatrix} 1 \\ -1 \end{bmatrix} \right\}, \qquad B' = \left\{ \begin{bmatrix} 1 \\ 2 \end{bmatrix}, \begin{bmatrix} 2 \\ 1 \end{bmatrix} \right\}, \qquad N = \left\{ \begin{bmatrix} 1 \\ 0 \end{bmatrix}, \begin{bmatrix} 0 \\ 1 \end{bmatrix} \right\}$$

for R^2. Then

$$_N P_B = \begin{bmatrix} 1 & 1 \\ 1 & -1 \end{bmatrix}$$

and

$$_{B'}P_N = (_N P_{B'})^{-1}$$

$$= \left(\begin{bmatrix} 1 & 2 \\ 2 & 1 \end{bmatrix} \right)^{-1}$$

$$= \begin{bmatrix} -\frac{1}{3} & \frac{2}{3} \\ \frac{2}{3} & -\frac{1}{3} \end{bmatrix}$$

Hence

$$_{B'}P_B = (_{B'}P_N)(_N P_B)$$

$$= \begin{bmatrix} -\frac{1}{3} & \frac{2}{3} \\ \frac{2}{3} & -\frac{1}{3} \end{bmatrix} \begin{bmatrix} 1 & 1 \\ 1 & -1 \end{bmatrix}$$

$$= \begin{bmatrix} \frac{1}{3} & -1 \\ \frac{1}{3} & 1 \end{bmatrix}$$

This can be verified to be the correct matrix by noting that

$$\begin{bmatrix} 1 \\ 1 \end{bmatrix} = \frac{1}{3} \begin{bmatrix} 1 \\ 2 \end{bmatrix} + \frac{1}{3} \begin{bmatrix} 2 \\ 1 \end{bmatrix} \qquad \begin{bmatrix} 1 \\ -1 \end{bmatrix} = (-1) \begin{bmatrix} 1 \\ 2 \end{bmatrix} + 1 \begin{bmatrix} 2 \\ 1 \end{bmatrix}$$

and applying Theorem 20.

─────────────────────────── **Exercises** ───────────────────────────

In Exercises 1–8 find the transition matrix $_{B'}P_B$ from the basis B to the basis B' for the given linear space. Also find the transition matrix $_BP_{B'}$ from B' to B and verify that $(_BP_{B'})^{-1} = {_{B'}P_B}$.

1. R^2; $B = \left\{ \begin{bmatrix} 1 \\ 1 \end{bmatrix}, \begin{bmatrix} -1 \\ 1 \end{bmatrix} \right\}$; $B' = \left\{ \begin{bmatrix} 1 \\ 2 \end{bmatrix}, \begin{bmatrix} 2 \\ 1 \end{bmatrix} \right\}$

2. R^2; $B = \left\{ \begin{bmatrix} 1 \\ 1 \end{bmatrix}, \begin{bmatrix} 0 \\ 1 \end{bmatrix} \right\}$; $B' = \left\{ \begin{bmatrix} 1 \\ -1 \end{bmatrix}, \begin{bmatrix} 1 \\ 0 \end{bmatrix} \right\}$

3. P_1; $B = \{1 + x, 1 - x\}$; $B' = \{2 + x, 1 + 2x\}$.
4. P_1; $B = \{1 + x, x\}$; $B' = \{1 - x, 1\}$.

5. R^3; $B = \left\{ \begin{bmatrix} 1 \\ 1 \\ 0 \end{bmatrix}, \begin{bmatrix} 1 \\ 0 \\ 1 \end{bmatrix}, \begin{bmatrix} 0 \\ 1 \\ 0 \end{bmatrix} \right\}$; $B' = \left\{ \begin{bmatrix} 1 \\ 1 \\ 1 \end{bmatrix}, \begin{bmatrix} 1 \\ 0 \\ 1 \end{bmatrix}, \begin{bmatrix} 1 \\ 0 \\ 0 \end{bmatrix} \right\}$

6. R^3; $B = \left\{ \begin{bmatrix} 1 \\ 1 \\ 0 \end{bmatrix}, \begin{bmatrix} 0 \\ 1 \\ 1 \end{bmatrix}, \begin{bmatrix} 1 \\ 0 \\ 1 \end{bmatrix} \right\}$; $B' = \left\{ \begin{bmatrix} -1 \\ 1 \\ 0 \end{bmatrix}, \begin{bmatrix} 1 \\ 1 \\ 1 \end{bmatrix}, \begin{bmatrix} 1 \\ 1 \\ 0 \end{bmatrix} \right\}$

7. P_2; $B = \{1 + x, x, 1 - x^2\}$; $B' = \{x, 1 - x, 1 + x^2\}$.
8. P_2; $B = \{1 + x^2, x^2, x + x^2\}$; $B' = \{1, x^2, x^2 + x + 1\}$.
9. Compute directly the transition matrix from B' to B in Example 2.

In Exercises 10–11 the sets B, C, and D are bases of a linear space V.

10. Determine $_BP_C$ in terms of $_DP_C$ and $_DP_B$.
11. Determine $_CP_D$ in terms of $_CP_B$ and $_DP_B$.

4.11 ROW SPACE, COLUMN SPACE, AND RANK OF A MATRIX (OPTIONAL)

To a given $m \times n$ matrix A we can associate two sets of vectors: one of which is composed of the rows of A and the other of the columns of A. In this section we investigate the basic properties of the linear spaces spanned by these two sets.

Let

$$A = \begin{bmatrix} a_{11} & a_{12} & \cdots & a_{1n} \\ a_{21} & a_{22} & \cdots & a_{2n} \\ \vdots & \vdots & & \vdots \\ a_{m1} & a_{m2} & \cdots & a_{mn} \end{bmatrix}$$

be an $m \times n$ matrix. Recall that the $1 \times n$ matrices

$$[a_{11} \ a_{12} \ \cdots \ a_{1n}]$$
$$[a_{21} \ a_{22} \ \cdots \ a_{2n}]$$
$$\vdots \quad \vdots \quad \quad \vdots$$
$$[a_{m1} \ a_{m2} \ \cdots \ a_{mn}]$$

are called the rows of A and that the $m \times 1$ matrices

$$\begin{bmatrix} a_{11} \\ a_{21} \\ \vdots \\ a_{m1} \end{bmatrix} \quad \begin{bmatrix} a_{12} \\ a_{22} \\ \vdots \\ a_{m2} \end{bmatrix} \quad \cdots \quad \begin{bmatrix} a_{1n} \\ a_{2n} \\ \vdots \\ a_{mn} \end{bmatrix}$$

are called the columns of A.

Definition 10.   Let A be an $m \times n$ matrix. The linear space spanned by the rows of A is called the **row space** of A. The linear space spanned by the columns of A is called the **column space** of A.

Theorem 5 of Section 4.4 can be reworded in terms of the column space of a matrix so that we can see the relationship between the solutions of the matrix equation $A\mathbf{x} = \mathbf{b}$ and the column space of A.

Theorem 22.   *Let A be an $m \times n$ matrix. The matrix equation $A\mathbf{x} = \mathbf{b}$ is consistent if and only if $\mathbf{b}$ is an element of the column space of A.*

This theorem clearly indicates the importance of being able to determine the column space of a matrix A. We now describe how elementary row operations can be used to obtain a basis for the column space. The method to be described is very useful because it also allows us to:

1. Find a basis for the row space of A. (This feature is unexpected because there is no apparent connection between the rows and columns of A.)
2. Prove that the row space of A and column space of A have the same dimension.
3. Find a basis for the linear space spanned by a given set of vectors that is more efficient than the one presented at the beginning of Section 4.9.

To simplify the description of this method we define two matrices A and B to be **row equivalent** if A can be changed into B by using elementary row operations. Recalling that elementary row operations can be "reversed" we conclude that A is row equivalent to B if and only if B is row equivalent to A.

The method begins by using elementary row operations to obtain a matrix B in row-echelon form that is row equivalent to A. Once B is found, bases for the row and column space for A are easily obtained as described in the following theorem.

Theorem 23.   *Let A be an $m \times n$ matrix that is row equivalent to a matrix B that is in row-echelon form.*

1. *The nonzero rows of B form a basis for the row space of A.*
2. *Let $A_1, A_2, \ldots, A_n$ be the columns of A, and let $B_1, B_2, \ldots, B_n$ be the columns of B. If $B_{i_1}, B_{i_2}, \ldots, B_{i_k}$ are the columns of B containing a nonzero leading entry of some row of B, then the corresponding columns $A_{i_1}, A_{i_2}, \ldots, A_{i_k}$ form a basis for the column space of A.*

Proof: Evidently if an elementary row operation is applied to A to obtain a matrix A', then the rows of A' are linear combinations of the rows of A and vice versa. Hence, the row space of A' is the same as the row space of A. Since B is obtained from

A by repeated application of elementary row operations, we conclude that the row spaces of A and B are identical. The very nature of row-echelon form assures that the nonzero rows of B are linearly independent. Therefore the nonzero rows of B form a basis for the row space of B and, hence, of the row space of A also. This completes the proof of part 1.

The proof of part 2 is more complicated. The interested reader is referred to Appendix 1 at the end of the book.   ∎

Example 1.   Consider the matrices

$$A = \begin{bmatrix} 1 & 1 & 1 & 2 & 2 & 1 \\ 3 & 3 & 1 & 4 & 4 & 3 \\ 1 & 1 & 1 & 1 & 2 & -1 \\ 3 & 3 & 0 & 3 & 3 & 3 \end{bmatrix} \quad \text{and} \quad B = \begin{bmatrix} 1 & 1 & 1 & 2 & 2 & 1 \\ 0 & 0 & 1 & 1 & 1 & 0 \\ 0 & 0 & 0 & -1 & 0 & -2 \\ 0 & 0 & 0 & 0 & 0 & 0 \end{bmatrix}$$

It can be easily shown by using elementary row operations that A is row equivalent to B. Since the first, third, and fourth columns of B are the columns containing nonzero leading entries, we conclude from Theorem 23 that the first, third, and fourth columns of A form a basis for the column space of A. That is

$$\left\{ \begin{bmatrix} 1 \\ 3 \\ 1 \\ 3 \end{bmatrix}, \begin{bmatrix} 1 \\ 1 \\ 1 \\ 0 \end{bmatrix}, \begin{bmatrix} 2 \\ 4 \\ 1 \\ 3 \end{bmatrix} \right\}$$

is a basis for the column space of A. Since the first three rows are the nonzero rows of B, the set $\{[1 \quad 1 \quad 1 \quad 2 \quad 2 \quad 1], [0 \quad 0 \quad 1 \quad 1 \quad 1 \quad 0], [0 \quad 0 \quad 0 \quad -1 \quad 0 \quad -2]\}$ is a basis for the row space of A.

In practice we frequently need to do calculations with the basis of a linear space. Hence, we would like to choose a basis that makes these calculations relatively easy. Many calculations are simplified if the vectors in a basis have several components equal to zero. Such a basis for the row space of the matrix A can be obtained from a matrix C in reduced row-echelon form that is row equivalent to A. It is left for the reader to show that A is row equivalent to

$$C = \begin{bmatrix} 1 & 1 & 0 & 0 & 1 & -1 \\ 0 & 0 & 1 & 0 & 1 & -2 \\ 0 & 0 & 0 & 1 & 0 & 2 \\ 0 & 0 & 0 & 0 & 0 & 0 \end{bmatrix}$$

Hence $\{[1 \quad 1 \quad 0 \quad 0 \quad 1 \quad -1], [0 \quad 0 \quad 1 \quad 0 \quad 1 \quad -2], [0 \quad 0 \quad 0 \quad 1 \quad 0 \quad 2]\}$ is a basis for the row space of A.

Notice that in the previous example the row space and the column space of A have the same dimension. This is always the case.

Theorem 24.   *The row space and column space of any matrix have the same dimension.*

Proof: Let B be a matrix in row-echelon form that is row equivalent to A. By Theorem 23 a basis for the column space of A contains as many elements as there are nonzero leading entries in the rows of B. Also by Theorem 23 a basis for the row space of A contains as many elements as there are nonzero leading entries in the rows of B. Consequently the row space and column space of A have the same dimension.   ∎

Definition 11.   The common dimension of the row space and column space of a matrix A is called the **rank** of A.

Example 2.   In Example 1 we found that the row space and column space of

$$A = \begin{bmatrix} 1 & 1 & 1 & 2 & 2 & 1 \\ 3 & 3 & 1 & 4 & 4 & 3 \\ 1 & 1 & 1 & 1 & 2 & -1 \\ 3 & 3 & 0 & 3 & 3 & 3 \end{bmatrix}$$

have dimension 3. Therefore the rank of A is 3.

The following theorem, which is almost an immediate consequence of Theorem 22, shows that it is possible to determine when a matrix equation $A\mathbf{x} = \mathbf{b}$ is consistent in terms of the ranks of matrices. Its proof is left as an exercise.

Theorem 25.   *Let A be any matrix. Then the equation $A\mathbf{x} = \mathbf{b}$ is consistent if and only if A and the augmented matrix $[A\,|\,\mathbf{b}]$ have the same rank.*

We have seen that the set of all linear combinations of the elements of a finite set $S = \{\mathbf{v}_1, \mathbf{v}_2, \dots, \mathbf{v}_k\}$ of n-vectors is a linear space V. Notice that V is the column space of the matrix A whose columns are $\mathbf{v}_1, \mathbf{v}_2, \dots, \mathbf{v}_k$. Theorem 23 describes how a basis for V may be found. It is important to notice that the basis for V found in this manner is a subset of S.

Example 3.   Consider the linear space spanned by

$$S = \left\{ \begin{bmatrix} 1 \\ 0 \\ 1 \\ 0 \end{bmatrix}, \begin{bmatrix} 0 \\ 1 \\ 1 \\ 0 \end{bmatrix}, \begin{bmatrix} 3 \\ -2 \\ 1 \\ 0 \end{bmatrix}, \begin{bmatrix} 1 \\ 1 \\ 0 \\ 0 \end{bmatrix} \right\}$$

The matrix

$$B = \begin{bmatrix} 1 & 0 & 3 & 1 \\ 0 & 1 & -2 & 1 \\ 0 & 0 & 0 & 2 \\ 0 & 0 & 0 & 0 \end{bmatrix}$$

can be shown to be row equivalent to the matrix

$$A = \begin{bmatrix} 1 & 0 & 3 & 1 \\ 0 & 1 & -2 & 1 \\ 1 & 1 & 1 & 0 \\ 0 & 0 & 0 & 0 \end{bmatrix}$$

whose columns are the elements of S.

Since the first, second, and fourth columns of B are the columns containing nonzero leading entries of the rows, the first, second, and fourth columns of A form a basis for the column space of A. That is

$$\left\{ \begin{bmatrix} 1 \\ 0 \\ 1 \\ 0 \end{bmatrix}, \begin{bmatrix} 0 \\ 1 \\ 1 \\ 0 \end{bmatrix}, \begin{bmatrix} 1 \\ 1 \\ 0 \\ 0 \end{bmatrix} \right\}$$

is a basis for the linear space spanned by S. This corresponds with our findings in Example 5 of Section 4.7.

Theorem 23 also gives us a means to expand a given linearly independent set $\{v_1, v_2, \ldots, v_k\}$ of vectors in R^n to a basis for R^n. This is done by finding a basis for the column space of the matrix A having $\{v_1, v_2, \ldots, v_k, e_1, e_2, \ldots, e_n\}$ as its first, second, $\ldots$, $(n + k)$-th columns, respectively. Since $\{e_1, e_2, \ldots, e_n\}$ is a basis for R^n, the column space of A must equal R^n. Moreover, since $\{v_1, v_2, \ldots, v_k\}$ is linearly independent, the first k columns of any matrix in row-echelon form that is row equivalent to A contain nonzero leading entries. Therefore the basis we obtain has $\{v_1, v_2, \ldots, v_k\}$ as a subset.

Example 4.　　Consider the linearly independent subset

$$S = \left\{ \begin{bmatrix} 1 \\ 1 \\ 1 \\ 0 \end{bmatrix}, \begin{bmatrix} 1 \\ 0 \\ 1 \\ 0 \end{bmatrix} \right\}$$

of R^4. We will find a basis for R^4 containing S. To begin, we let

$$A = \begin{bmatrix} 1 & 1 & 1 & 0 & 0 & 0 \\ 1 & 0 & 0 & 1 & 0 & 0 \\ 1 & 1 & 0 & 0 & 1 & 0 \\ 0 & 0 & 0 & 0 & 0 & 1 \end{bmatrix}$$

This matrix can be shown to be row equivalent to

$$B = \begin{bmatrix} 1 & 1 & 1 & 0 & 0 & 0 \\ 0 & 1 & 1 & -1 & 0 & 0 \\ 0 & 0 & 1 & 0 & -1 & 0 \\ 0 & 0 & 0 & 0 & 0 & 1 \end{bmatrix}$$

Since the first, second, third, and sixth columns of B contain the nonzero leading entries of the rows of B, the first, second, third, and sixth columns of A form a basis for the column space of A. The column space of A equals R^4 because

$$\left\{ \begin{bmatrix} 1 \\ 0 \\ 0 \\ 0 \end{bmatrix}, \begin{bmatrix} 0 \\ 1 \\ 0 \\ 0 \end{bmatrix}, \begin{bmatrix} 0 \\ 0 \\ 1 \\ 0 \end{bmatrix}, \begin{bmatrix} 0 \\ 0 \\ 0 \\ 1 \end{bmatrix} \right\}$$

is a basis for R^4. Therefore

$$\left\{ \begin{bmatrix} 1 \\ 1 \\ 1 \\ 0 \end{bmatrix}, \begin{bmatrix} 1 \\ 0 \\ 1 \\ 0 \end{bmatrix}, \begin{bmatrix} 1 \\ 0 \\ 0 \\ 0 \end{bmatrix}, \begin{bmatrix} 0 \\ 0 \\ 0 \\ 1 \end{bmatrix} \right\}$$

is a basis for R^4 that has S as a subset. This coincides with our findings in Example 5 at the end of Section 4.9.

Exercises

1. List the rows and columns of

$$\begin{bmatrix} 1 & 2 & 3 & 4 \\ 0 & -1 & 0 & 2 \\ 2 & 1 & 2 & -3 \end{bmatrix}$$

2. List the rows and columns of

$$\begin{bmatrix} 1 & 3 & 2 \\ -2 & 1 & 4 \\ 3 & 1 & -1 \\ 2 & 0 & -1 \\ 5 & 1 & 6 \end{bmatrix}$$

In Exercises 3–8 find: (a) a basis for the row space; (b) a basis for the column space; and (c) the rank of the given matrix.

3. $\begin{bmatrix} 1 & 2 & 3 \\ 4 & 5 & 6 \\ 2 & 1 & 0 \end{bmatrix}$

4. $\begin{bmatrix} 1 & 3 & 2 \\ 2 & 6 & 4 \\ 3 & 9 & 6 \end{bmatrix}$

5. $\begin{bmatrix} 1 & -1 & 2 & -3 \\ 3 & -3 & 1 & -4 \\ 2 & -2 & 0 & -2 \\ 0 & 0 & 4 & -4 \\ -1 & 1 & 2 & -1 \end{bmatrix}$

6. $\begin{bmatrix} 2 & 4 & 8 & 1 \\ -1 & -1 & -3 & 1 \\ 0 & 1 & 1 & 0 \\ 3 & 1 & 7 & 1 \\ 3 & 1 & 7 & 1 \end{bmatrix}$

7. $\begin{bmatrix} 1 & 2 & 2 & 4 & 4 \\ 2 & 4 & 1 & 3 & 5 \\ 1 & 2 & -1 & 0 & 1 \end{bmatrix}$
   8. $\begin{bmatrix} 2 & 4 & 6 & 1 & 0 \\ 2 & 4 & 6 & 1 & 2 \\ 3 & 6 & 9 & 2 & -4 \end{bmatrix}$

In Exercises 9–12 find a basis for the subspace of R^4 spanned by the given set of 4-vectors.

9. $\left\{ \begin{bmatrix} 1 \\ 1 \\ 0 \\ 1 \end{bmatrix}, \begin{bmatrix} 4 \\ 5 \\ 1 \\ 5 \end{bmatrix}, \begin{bmatrix} -1 \\ 0 \\ 1 \\ 0 \end{bmatrix}, \begin{bmatrix} 1 \\ 2 \\ 1 \\ 1 \end{bmatrix} \right\}$
   10. $\left\{ \begin{bmatrix} 1 \\ 2 \\ -1 \\ 2 \end{bmatrix}, \begin{bmatrix} 2 \\ 4 \\ -2 \\ 4 \end{bmatrix}, \begin{bmatrix} 3 \\ 6 \\ -3 \\ 6 \end{bmatrix}, \begin{bmatrix} 1 \\ 4 \\ -2 \\ 4 \end{bmatrix} \right\}$

11. $\left\{ \begin{bmatrix} 1 \\ 2 \\ 1 \\ 1 \end{bmatrix}, \begin{bmatrix} 0 \\ 1 \\ 0 \\ -1 \end{bmatrix}, \begin{bmatrix} 0 \\ 0 \\ 3 \\ 1 \end{bmatrix}, \begin{bmatrix} 3 \\ 7 \\ -3 \\ 0 \end{bmatrix} \right\}$
   12. $\left\{ \begin{bmatrix} 1 \\ 1 \\ 3 \\ 1 \end{bmatrix}, \begin{bmatrix} 3 \\ 6 \\ 8 \\ -3 \end{bmatrix}, \begin{bmatrix} 2 \\ 4 \\ 0 \\ -2 \end{bmatrix}, \begin{bmatrix} 2 \\ -1 \\ -3 \\ -1 \end{bmatrix} \right\}$

In Exercises 13–16 find a basis for R^5 that has the given set of vectors as a subset.

13. $\left\{ \begin{bmatrix} 1 \\ 1 \\ 1 \\ 1 \\ 1 \end{bmatrix}, \begin{bmatrix} 0 \\ 1 \\ 1 \\ 1 \\ 1 \end{bmatrix} \right\}$
   14. $\left\{ \begin{bmatrix} 1 \\ 1 \\ 0 \\ 1 \\ 1 \end{bmatrix}, \begin{bmatrix} 2 \\ 1 \\ 1 \\ 0 \\ 0 \end{bmatrix}, \begin{bmatrix} 0 \\ 0 \\ 1 \\ -1 \\ -1 \end{bmatrix} \right\}$

15. $\left\{ \begin{bmatrix} 0 \\ 1 \\ 1 \\ 1 \\ 0 \end{bmatrix}, \begin{bmatrix} 2 \\ 0 \\ 0 \\ 0 \\ 1 \end{bmatrix}, \begin{bmatrix} 1 \\ 1 \\ 1 \\ 1 \\ 1 \end{bmatrix} \right\}$
   16. $\left\{ \begin{bmatrix} 1 \\ 1 \\ 1 \\ 1 \\ 1 \end{bmatrix}, \begin{bmatrix} 0 \\ 1 \\ 1 \\ 1 \\ 1 \end{bmatrix}, \begin{bmatrix} 0 \\ 0 \\ 1 \\ 1 \\ 1 \end{bmatrix}, \begin{bmatrix} 1 \\ 0 \\ 1 \\ 0 \\ 0 \end{bmatrix} \right\}$

17. If A is a 3×4 matrix, show that the columns of A are linearly dependent. What is the largest possible value for rank A?

18. Let A be an $m \times n$ matrix. Show that the columns of A are linearly dependent if $m < n$ and that the rows of A are linearly dependent if $n < m$.

19. Let A be an $m \times n$ matrix. What is the largest possible value of rank A?

20. Let A be an $m \times n$ matrix. Prove that the rank of A equals the rank of A^t.

21. Prove Theorem 25.

4.12 DIMENSION THEOREM (OPTIONAL)

This section describes an intimate relationship between systems of linear equations and finite dimensional linear spaces. As we have already seen, there are two linear spaces that arise naturally when we consider systems of equations involving an $m \times n$ matrix A. The first is the solution space at $A\mathbf{x} = \mathbf{0}$

$$N_A = \{\mathbf{x} : A\mathbf{x} = \mathbf{0}\}$$

In the situation we now describe, we follow tradition and call N_A the **null space** of A.

Recall from Theorem 21 of Section 1.10 that if z and z' are any two solutions of $A\mathbf{x} = \mathbf{b}$, then there is a solution y of $A\mathbf{x} = \mathbf{0}$ such that $z' = z + y$. Hence the dimension of N_A, called the **nullity** of A, is a measure of "how nearly unique" are solutions of $A\mathbf{x} = \mathbf{b}$. The second is the linear space

$$R_A = \{\mathbf{b} : \mathbf{b} = A\mathbf{x} \text{ for some } \mathbf{x}\}$$

which, by Theorem 22 of the previous section, is the column space of A. The dimension of R_A, which is the rank of A, is a measure of "how often" the equation $A\mathbf{x} = \mathbf{b}$ can be solved. The following theorem gives the relationship between the rank and nullity of A.

Theorem 26.   *Let A be an $m \times n$ matrix. Then (rank of A) + (nullity of A) = n.*

Proof: Let $\{\mathbf{v}_1, \mathbf{v}_2, \ldots, \mathbf{v}_k\}$ be a basis for the null space of A. Using the technique illustrated in Example 4 of the previous section, we can extend this set to a basis $\{\mathbf{v}_1, \ldots, \mathbf{v}_k, \mathbf{v}_{k+1}, \ldots, \mathbf{v}_n\}$ for R^n. If the null space consists only of the zero vector, then the first set is empty and the second set can be chosen to be the natural basis for R^n. We will show that $S = \{A\mathbf{v}_{k+1}, \ldots, A\mathbf{v}_n\}$ is a basis for R_A.

Let $\mathbf{b}$ be any element of R_A. Then there is an element $\mathbf{x}$ of R^n such that $A\mathbf{x} = \mathbf{b}$. Since $\{\mathbf{v}_1, \mathbf{v}_2, \ldots, \mathbf{v}_n\}$ is a basis for R^n, there are numbers $c_1, c_2, \ldots, c_n$ such that

$$\mathbf{x} = c_1\mathbf{v}_1 + \cdots + c_k\mathbf{v}_k + c_{k+1}\mathbf{v}_{k+1} + \cdots + c_n\mathbf{v}_n$$

Then

$$\mathbf{b} = A\mathbf{x} = c_1 A\mathbf{v}_1 + \cdots + c_k A\mathbf{v}_k + c_{k+1}A\mathbf{v}_{k+1} + \cdots + c_n A\mathbf{v}_n$$
$$= c_{k+1}A\mathbf{v}_{k+1} + \cdots + c_n A\mathbf{v}_n$$

since $\mathbf{v}_1, \mathbf{v}_2, \ldots, \mathbf{v}_k$ are elements of the null space of A. Hence S spans R_A.

Now suppose that $b_{k+1}, b_{k+2}, \ldots, b_n$ are scalars such that

$$b_{k+1}A\mathbf{v}_{k+1} + b_{k+2}A\mathbf{v}_{k+2} + \cdots + b_n A\mathbf{v}_n = \mathbf{0}$$

Then

$$A(b_{k+1}\mathbf{v}_{k+1} + b_{k+2}\mathbf{v}_{k+2} + \cdots + b_n\mathbf{v}_n) = \mathbf{0}$$

so that $b_{k+1}\mathbf{v}_{k+1} + \cdots + b_n\mathbf{v}_n$ is an element of the null space of A. Therefore there are scalars $b_1, b_2, \ldots, b_k$ such that

$$b_1\mathbf{v}_1 + \cdots + b_k\mathbf{v}_k = b_{k+1}\mathbf{v}_{k+1} + \cdots + b_n\mathbf{v}_n$$

or equivalently

$$b_1\mathbf{v}_1 + \cdots + b_k\mathbf{v}_k + (-b_{k+1})\mathbf{v}_{k+1} + \cdots + (-b_n)\mathbf{v}_n = \mathbf{0}$$

Since $\{\mathbf{v}_1, \mathbf{v}_2, \cdots, \mathbf{v}_n\}$ is linearly independent, we must have $b_1 = 0, b_2 = 0, \ldots, b_n = 0$. In particular $b_{k+1} = 0, b_{k+2} = 0, \ldots, b_n = 0$ so that S is linearly independent. Hence S is a basis for R_A. Notice that the dimension of R_A is $n - k$.

We now have

$$(\text{rank of } A) + (\text{nullity of } A) = (\text{dimension of } R_A) + (\text{dimension of } N_A)$$
$$= (n - k) + k = n$$

This completes the proof. ∎

Example 1.   Consider the matrix

$$A = \begin{bmatrix} 1 & 1 & 1 & 2 & 2 & 1 \\ 3 & 3 & 1 & 4 & 4 & 3 \\ 1 & 1 & 1 & 1 & 2 & -1 \\ 3 & 3 & 0 & 3 & 3 & 3 \end{bmatrix}$$

The null space N_A for A is the set of all 6-vectors $\mathbf{x}$ such that $A\mathbf{x} = \mathbf{0}$. Using elementary row operations, we can show that the augmented matrix for $A\mathbf{x} = \mathbf{0}$ is row equivalent to

$$\left[\begin{array}{cccccc|c} 1 & 1 & 0 & 0 & 1 & -1 & 0 \\ 0 & 0 & 1 & 0 & 1 & -2 & 0 \\ 0 & 0 & 0 & 1 & 0 & 2 & 0 \\ 0 & 0 & 0 & 0 & 0 & 0 & 0 \end{array}\right]$$

The system of equations corresponding to this augmented matrix is

$$x_1 + x_2 + \qquad\qquad x_5 - \ x_6 = 0$$
$$x_3 + \qquad x_5 - 2x_6 = 0$$
$$x_4 + \qquad 2x_6 = 0$$

which can be rewritten as

$$x_1 = -x_2 - x_5 + x_6$$
$$x_3 = -x_5 + 2x_6$$
$$x_4 = -2x_6$$

Thus any vector $\mathbf{x}$ in N_A can be written as

$$\mathbf{x} = \begin{bmatrix} x_1 \\ x_2 \\ x_3 \\ x_4 \\ x_5 \\ x_6 \end{bmatrix} = \begin{bmatrix} -x_2 - x_5 + x_6 \\ x_2 \\ -x_5 + 2x_6 \\ -2x_6 \\ x_5 \\ x_6 \end{bmatrix}$$

$$= x_2 \begin{bmatrix} -1 \\ 1 \\ 0 \\ 0 \\ 0 \\ 0 \end{bmatrix} + x_5 \begin{bmatrix} -1 \\ 0 \\ -1 \\ 0 \\ 1 \\ 0 \end{bmatrix} + x_6 \begin{bmatrix} 1 \\ 0 \\ 2 \\ -2 \\ 0 \\ 1 \end{bmatrix}$$

Thus N_A is the subspace of R^6 having

$$B = \left\{ \begin{bmatrix} -1 \\ 1 \\ 0 \\ 0 \\ 0 \\ 0 \end{bmatrix}, \begin{bmatrix} -1 \\ 0 \\ -1 \\ 0 \\ 1 \\ 0 \end{bmatrix}, \begin{bmatrix} 1 \\ 0 \\ 2 \\ -2 \\ 0 \\ 1 \end{bmatrix} \right\}$$

as a basis. Therefore the nullity of A is 3.

In Example 1 of Section 4.11 we found that the column space of A has the set

$$B' = \left\{ \begin{bmatrix} 1 \\ 3 \\ 1 \\ 3 \end{bmatrix}, \begin{bmatrix} 1 \\ 1 \\ 1 \\ 0 \end{bmatrix}, \begin{bmatrix} 2 \\ 4 \\ 1 \\ 3 \end{bmatrix} \right\}$$

as a basis. Therefore the rank of A is 3. Thus (rank of A) + (nullity of A) = 3 + 3 = 6, as is guaranteed by Theorem 26.

Theorem 26 incorporates some of the important results that we have proved by other means. For example, this theorem allows us to prove:

1. A system of homogeneous equations with more unknowns (n) than equations (m) has infinitely many solutions.

2. If $m = n$, then $A\mathbf{x} = \mathbf{b}$ is consistent for every n-vector $\mathbf{b}$ if and only if $A\mathbf{x} = \mathbf{0}$ has $\mathbf{x} = \mathbf{0}$ as its only solution.

For any $m \times n$ matrix the rank of A is at most m. In the special case that $m < n$ Theorem 26 assures us that nullity of $A = n -$ rank of $A \geq n - m > 0$. Hence $A\mathbf{x} = \mathbf{0}$ has a solution other than $\mathbf{x} = \mathbf{0}$. Therefore $A\mathbf{x} = \mathbf{b}$ has infinitely many solutions whenever it is consistent (Theorem 22 of Section 1.10, which does not depend on Theorem 20 of the same section). Since $A\mathbf{x} = \mathbf{0}$ is always consistent, the equation $A\mathbf{x} = \mathbf{0}$ has infinitely many solutions.

In the special case that $m = n$ Theorem 26 tells us that the rank of A equals n if and only if the nullity of A equals zero. This is another way of saying that $A\mathbf{x} = \mathbf{b}$ has a solution for every n-vector $\mathbf{b}$ if and only if $A\mathbf{x} = \mathbf{0}$ has $\mathbf{x} = \mathbf{0}$ as its only solution. This result was first encountered in Theorem 2 of Section 1.5 and has been used and illustrated several times since then.

─────────────────── **Exercises** ───────────────────

In Exercises 1–8 verify Theorem 26 for the given matrix by finding the dimensions of the null space and column space.

1. $\begin{bmatrix} 1 & 2 \\ 2 & 4 \end{bmatrix}$

2. $\begin{bmatrix} 1 & 2 \\ 2 & 1 \end{bmatrix}$

3. $\begin{bmatrix} 1 & 2 & 1 \\ 2 & 0 & 3 \end{bmatrix}$

4. $\begin{bmatrix} 1 & 0 & 1 & 0 \\ 2 & 3 & 0 & 1 \end{bmatrix}$

5. $\begin{bmatrix} 1 & 0 \\ 0 & 1 \\ 1 & 1 \end{bmatrix}$ **6.** $\begin{bmatrix} 1 & 0 & 1 \\ 0 & 1 & 0 \\ 1 & 1 & 2 \end{bmatrix}$

7. $\begin{bmatrix} 1 & 0 & 1 & 0 & 0 \\ 0 & 2 & 0 & 1 & 0 \\ 1 & 3 & 0 & 0 & 1 \\ 1 & 0 & 0 & 1 & 0 \end{bmatrix}$ **8.** $\begin{bmatrix} 1 & 0 & 1 \\ 0 & 2 & 0 \\ 3 & 0 & 2 \\ 6 & 4 & 4 \end{bmatrix}$

9. Let A be an 8×6 matrix such that $A\mathbf{x} = \mathbf{0}$ has only the trivial solution. What is the rank of A? Is $A\mathbf{x} = \mathbf{b}$ consistent for all $\mathbf{b}$ in R^8?

10. Let A be a 6×8 matrix. What are the largest and smallest possible values for the nullity of A? If $A\mathbf{x} = \mathbf{b}$ is consistent for all $\mathbf{b}$ in R^6 what is the nullity of A?

——————Supplementary Exercises for Chapter 4——————

Consider the following statements. If a statement is necessarily true, explain why it is true citing appropriate theorems and definitions whenever possible. Otherwise give an example to show that the statement is not necessarily true.

In statements 1–22, $S = \{\mathbf{v}_1, \mathbf{v}_2, ..., \mathbf{v}_k\}$ is a subset of a linear space V, and $\mathbf{v}_{k+1}$ is an element of V.

1. The set of all linear combinations of the elements of S is a linear subspace of V.

2. If S spans V, then $\{\mathbf{v}_1, \mathbf{v}_2, ..., \mathbf{v}_k, \mathbf{v}_{k+1}\}$ also spans V.

3. If S is linearly independent, then $\{\mathbf{v}_1, \mathbf{v}_2, ..., \mathbf{v}_k, \mathbf{v}_{k+1}\}$ is also linearly independent.

4. If S spans V, then $\{\mathbf{v}_1, \mathbf{v}_2, ..., \mathbf{v}_{k-1}\}$ also spans V.

5. If S is linearly independent, then $\{\mathbf{v}_1, \mathbf{v}_2, ..., \mathbf{v}_{k-1}\}$ is also linearly independent.

6. If S is linearly dependent, then $\{\mathbf{v}_1, \mathbf{v}_2, ..., \mathbf{v}_k, \mathbf{v}_{k+1}\}$ is also linearly dependent.

7. If S is linearly dependent, then $\{\mathbf{v}_1, \mathbf{v}_2, ..., \mathbf{v}_{k-1}\}$ is also linearly dependent.

8. If S is linearly independent, then S is a basis for V.

9. If S spans V, then S is a basis for V.

10. If $\dim V = k$, then S is a basis for V.

11. If $\dim V = k$ and S is linearly independent, then S is a basis for V.

12. If $\dim V = k$ and S spans V, then S is a basis for V.

13. If $\dim V \neq k$, then S is not a basis for V.

14. If $\dim V > k$, then S cannot be linearly independent.

15. If $\dim V > k$, then S cannot span V.

16. If $\dim V > k$, then S cannot be linearly dependent.

17. If $\dim V < k$, then S cannot be linearly independent.

18. If $\dim V < k$, then S cannot span V.

19. If $\dim V < k$, then S cannot be linearly dependent.

20. If $\dim V < k$, then at least one element of S is a linear combination of the remaining elements of S.

21. If S is a basis for V, then $k = \dim V$.

22. If U is a linear subspace of V and V is finite dimensional, then $\dim U \leq \dim V$.

In statements 22–28, B is a matrix in row-echelon form that can be obtained from the matrix A by using elementary row operations.

23. The row rank of A equals the row rank of B.

24. The row space of A equals the row space of B.

25. The column space of A equals the column space of B.

26. The nonzero rows of B form a basis for the row space of A.

27. The nonzero rows of A form a basis for the row space of B.

28. The nonzero columns of B form a basis for the column space of A.

In statements 29–33, A is an $m \times n$ matrix.

29. The set of all solutions of $A\mathbf{x} = \mathbf{0}$ is a subspace of R^m.

30. The set of all solutions of $A\mathbf{x} = \mathbf{0}$ is a subspace of R^n.

31. The set of all solutions of $A\mathbf{x} = \mathbf{b}$ with $\mathbf{b} \neq \mathbf{0}$ is never a subspace of R^n.

32. Let $\mathbf{v}$ be any solution of $A\mathbf{x} = \mathbf{b}$ and $\mathbf{h}$ be any solution of $A\mathbf{x} = \mathbf{0}$, then $\mathbf{v} + \mathbf{h}$ is a solution of $A\mathbf{x} = \mathbf{b}$.

33. Let $\mathbf{u}$ and $\mathbf{v}$ be solutions of $A\mathbf{x} = \mathbf{b}$. Then there is a solution $\mathbf{h}$ of $A\mathbf{x} = \mathbf{0}$ such that $\mathbf{u} = \mathbf{v} + \mathbf{h}$.

In statements 34–39, A is an $n \times n$ matrix.

34. A is nonsingular if and only if the rank of A is n.

35. It is possible to have the rank of A greater than n.

36. A is nonsingular if and only if the columns of A span R^n.

37. A is nonsingular if and only if the row space of A has dimension n.

38. A is nonsingular if and only if the columns of A are linearly dependent.

39. A is nonsingular if and only if the rows of A are linearly independent.

5

Inner Product Spaces

5.1 INNER PRODUCT

In Sections 3.3 and 3.5 we introduced the concept of inner product in the linear spaces R^2 and R^3. Inspired by the usefulness of the inner product in these spaces we now extend this concept to other linear spaces. This will enable us to describe geometric properties such as length, angle, and orthogonality in these linear spaces. We will consider only linear spaces having the real numbers as scalars. Linear spaces having the complex numbers as scalars are considered in the exercises. We begin with a general definition of inner product.

Definition 1. An **inner product** on a linear space V having the real numbers as scalars is a function that associates with each pair of elements $\mathbf{u}$ and $\mathbf{v}$ of V a real number $\langle \mathbf{u}, \mathbf{v} \rangle$ in such a way that the following axioms are satisfied for all elements $\mathbf{u}$, $\mathbf{v}$, and $\mathbf{w}$ of V and all scalars c:

1. $\langle \mathbf{u}, \mathbf{v} \rangle = \langle \mathbf{v}, \mathbf{u} \rangle$
2. $\langle \mathbf{u} + \mathbf{v}, \mathbf{w} \rangle = \langle \mathbf{u}, \mathbf{w} \rangle + \langle \mathbf{v}, \mathbf{w} \rangle$
3. $\langle c\mathbf{u}, \mathbf{v} \rangle = c\langle \mathbf{u}, \mathbf{v} \rangle$
4. $\langle \mathbf{u}, \mathbf{u} \rangle \geq 0$
5. $\langle \mathbf{u}, \mathbf{u} \rangle = 0$ if and only if $\mathbf{u} = \mathbf{0}$

Example 1. Let V be the linear space R^2. In Section 3.3 we found that the real valued function on pairs

$$\mathbf{u} = \begin{bmatrix} u_1 \\ u_2 \end{bmatrix} \qquad \mathbf{v} = \begin{bmatrix} v_1 \\ v_2 \end{bmatrix}$$

of vectors in R^2 defined by

$$\mathbf{u} \cdot \mathbf{v} = u_1 v_1 + u_2 v_2$$

satisfies the following properties:

1. $\mathbf{u} \cdot \mathbf{v} = \mathbf{v} \cdot \mathbf{u}$
2. $(\mathbf{u} + \mathbf{v}) \cdot \mathbf{w} = \mathbf{u} \cdot \mathbf{w} + \mathbf{v} \cdot \mathbf{w}$
3. $(c\mathbf{u}) \cdot \mathbf{v} = c(\mathbf{u} \cdot \mathbf{v})$
4. $\mathbf{u} \cdot \mathbf{u} \geq 0$
5. $\mathbf{u} \cdot \mathbf{u} = 0$ if and only if $\mathbf{u} = \mathbf{0}$

Therefore $\mathbf{u} \cdot \mathbf{v}$ is an inner product on R^2.

Similarly, in Section 3.5 we showed that the real valued function on pairs of vectors

$$\mathbf{u} = \begin{bmatrix} u_1 \\ u_2 \\ u_3 \end{bmatrix} \quad \text{and} \quad \mathbf{v} = \begin{bmatrix} v_1 \\ v_2 \\ v_3 \end{bmatrix}$$

in R^3 defined by

$$\mathbf{u} \cdot \mathbf{v} = u_1 v_1 + u_2 v_2 + u_3 v_3$$

satisfies the axioms in Definition 1. Therefore $\mathbf{u} \cdot \mathbf{v}$ is an inner product on R^3.

In fact the real valued function on pairs of vectors

$$\mathbf{u} = \begin{bmatrix} u_1 \\ u_2 \\ \vdots \\ u_n \end{bmatrix} \quad \text{and} \quad \mathbf{v} = \begin{bmatrix} v_1 \\ v_2 \\ \vdots \\ v_n \end{bmatrix}$$

in R^n defined by

$$\langle \mathbf{u}, \mathbf{v} \rangle = u_1 v_1 + u_2 v_2 + \cdots + u_n v_n \tag{1}$$

can be shown to be an inner product on R^n (see Exercise 17). This inner product is called the **Euclidean inner product** on R^n. Notice that the Euclidean inner product coincides with the inner products given earlier on R^2 and R^3. Exercise 18 asks the reader to show that the Euclidean inner product of two n-vectors $\mathbf{u}$ and $\mathbf{v}$ can also be written as

$$\langle \mathbf{u}, \mathbf{v} \rangle = \mathbf{u}^t \mathbf{v}$$

or

$$\langle \mathbf{u}, \mathbf{v} \rangle = \mathbf{v}^t \mathbf{u}$$

where $\mathbf{u}^t$ and $\mathbf{v}^t$ denote the transposes of $\mathbf{u}$ and $\mathbf{v}$, respectively, and $\mathbf{u}$, $\mathbf{v}$, $\mathbf{u}^t$, and $\mathbf{v}^t$ are viewed as matrices.

Example 2. Let

$$\mathbf{u} = \begin{bmatrix} u_1 \\ u_2 \end{bmatrix} \quad \text{and} \quad \mathbf{v} = \begin{bmatrix} v_1 \\ v_2 \end{bmatrix}$$

be any two vectors in R^2, and consider the function

(2) $$\langle \mathbf{u}, \mathbf{v} \rangle = 5u_1 v_1 + 4u_2 v_2$$

We will show that this function is an inner product on R^2. Evidently

$$\langle \mathbf{u}, \mathbf{v} \rangle = 5u_1 v_1 + 4u_2 v_2$$
$$= 5v_1 u_1 + 4v_2 u_2$$
$$= \langle \mathbf{v}, \mathbf{u} \rangle$$

and

$$\langle c\mathbf{u}, \mathbf{v} \rangle = 5(cu_1)v_1 + 4(cu_2)v_2$$
$$= c(5u_1 v_1 + 4u_2 v_2)$$
$$= c\langle \mathbf{u}, \mathbf{v} \rangle$$

so that axioms 1 and 3 in Definition 1 are satisfied. Since

$$\langle \mathbf{u}, \mathbf{u} \rangle = 5u_1^2 + 4u_2^2$$

we see that $\langle \mathbf{u}, \mathbf{u} \rangle > 0$ if either $u_1 \neq 0$ or $u_2 \neq 0$. It easily follows that $\langle \mathbf{u}, \mathbf{u} \rangle > 0$ if $u \neq 0$; and $\langle \mathbf{u}, \mathbf{u} \rangle = 0$ if $\mathbf{u} = \mathbf{0}$, so that axioms 4 and 5 are satisfied. If

$$\mathbf{w} = \begin{bmatrix} w_1 \\ w_2 \end{bmatrix}$$

is any vector in R^2, then

$$\langle \mathbf{u} + \mathbf{v}, \mathbf{w} \rangle = 5(u_1 + v_1)w_1 + 4(u_2 + v_2)w_2$$
$$= (5u_1 w_1 + 4u_2 w_2) + (5v_1 w_1 + 4v_2 w_2)$$
$$= \langle \mathbf{u}, \mathbf{w} \rangle + \langle \mathbf{v}, \mathbf{w} \rangle$$

Thus, axiom 2 is satisfied. The function defined in (2) is an inner product on R^2. Therefore it is possible to define an inner product, other than the Euclidean inner product, on R^2.

A linear space on which we have chosen a specific inner product is called an **inner product space**. For an arbitrary inner product space we will denote the designated inner product by $\langle \cdot, \cdot \rangle$. Unless stated to the contrary, we always assume that R^n is given the Euclidean inner product.

Example 3.   Consider the linear space P_2 of all polynomials of degree less than or equal to 2 and the function

(3) $$\langle f, g \rangle = a_0 b_0 + a_1 b_1 + a_2 b_2$$

where

$$f(x) = a_0 + a_1 x + a_2 x^2, \; g(x) = b_0 + b_1 x + b_2 x^2$$

are any two elements of P_2. Evidently

$$\langle f, g \rangle = a_0 b_0 + a_1 b_1 + a_2 b_2$$
$$= b_0 a_0 + b_1 a_1 + a_2 b_2$$
$$= \langle g, f \rangle$$

and

$$\langle cf, g \rangle = (ca_0)b_0 + (ca_1)b_1 + (ca_2)b_2$$
$$= c(a_0 b_0 + a_1 b_1 + a_2 b_2)$$
$$= c\langle f, g \rangle$$

so that axioms 1 and 3 of Definition 1 are satisfied. Since

$$\langle f, f \rangle = a_0^2 + a_1^2 + a_2^2$$

we see that $\langle f, f \rangle > 0$ if $a_0 \neq 0$, $a_1 \neq 0$, or $a_2 \neq 0$; and $\langle f, f \rangle = 0$ if $a_0 = a_1 = a_2 = 0$. It easily follows that $\langle f, f \rangle > 0$ if $f(x) \neq 0$ for some x and $\langle f, f \rangle = 0$ if $f(x) = 0$ for all x. Therefore axioms 4 and 5 are satisfied. If

$$h(x) = c_0 + c_1 x + c_2 x^2$$

is any element of P_2, then

$$\langle f + g, h \rangle = (a_0 + b_0)c_0 + (a_1 + b_1)c_1 + (a_2 + b_2)c_2$$
$$= (a_0 c_0 + a_1 c_1 + a_2 c_2) + (b_0 c_0 + b_1 c_1 + b_2 c_2)$$
$$= \langle f, h \rangle + \langle g, h \rangle$$

so that axiom 2 is satisfied. The function defined in (3) is an inner product of P_2.

Example 4.   (*For readers who have studied calculus.*) Let f and g be continuous functions on the interval $[a, b]$ and consider the function

$$\langle f, g \rangle = \int_a^b f(x)g(x)\, dx \tag{4}$$

We will show that this function is an inner product on the linear space $C[a, b]$ of all continuous functions on the interval $[a, b]$. Evidently

$$\langle f, g \rangle = \int_a^b f(x)g(x)\, dx$$
$$= \int_a^b g(x)f(x)\, dx$$
$$= \langle g, f \rangle$$

and

$$\langle cf, g \rangle = \int_a^b cf(x)g(x)\, dx$$
$$= c \int_a^b f(x)g(x)\, dx$$
$$= c\langle f, g \rangle$$

so that axioms 1 and 3 of Definition 1 are satisfied.
    Since f is continuous and $[f(x)]^2 \geq 0$ for all x in $[a, b]$ we conclude that

$$\int_a^b [f(x)]^2\, dx > 0$$

if $f(x) \neq 0$, so that $\langle f, f \rangle > 0$ if $f(x) \neq 0$. Clearly $\langle f, f \rangle = 0$ if $f(x) \equiv 0$. Therefore axioms 4 and 5 are satisfied. If h is any element of $C[a, b]$, then

$$
\begin{aligned}
\langle f + g, h \rangle &= \int_a^b [f(x) + g(x)]h(x)\, dx \\
&= \int_a^b f(x)h(x)\, dx + \int_a^b g(x)h(x)\, dx \\
&= \langle f, h \rangle + \langle g, h \rangle
\end{aligned}
$$

so that axiom 2 is satisfied. The function defined in (4) is an inner product on $C[a, b]$.

The following theorem lists some of the basic properties of an inner product.

Theorem 1.   *Let V be an inner product space. Then*

1. $\langle \mathbf{0}, \mathbf{v} \rangle = 0$
2. $\langle \mathbf{w}, \mathbf{u} + \mathbf{v} \rangle = \langle \mathbf{w}, \mathbf{u} \rangle + \langle \mathbf{w}, \mathbf{v} \rangle$
3. $\langle \mathbf{u}, k\mathbf{v} \rangle = k\langle \mathbf{u}, \mathbf{v} \rangle$

for all elements $\mathbf{u}$, $\mathbf{v}$, and $\mathbf{w}$ of V and scalars k.

Proof: We will prove the first identity and leave the others as exercises. Using the axioms in Definition 1, we have

$$
\begin{aligned}
\langle \mathbf{0}, \mathbf{v} \rangle &= \langle \mathbf{v} + (-1)\mathbf{v}, \mathbf{v} \rangle \\
&= \langle \mathbf{v}, \mathbf{v} \rangle + \langle (-1)\mathbf{v}, \mathbf{v} \rangle & \text{axiom 2} \\
&= \langle \mathbf{v}, \mathbf{v} \rangle + (-1)\langle \mathbf{v}, \mathbf{v} \rangle & \text{axiom 3} \\
&= 0 \quad \blacksquare
\end{aligned}
$$

Exercises

In Exercises 1–6 compute the Euclidean inner product of each of the following pairs of vectors.

1. $\begin{bmatrix} 1 \\ 2 \\ 3 \end{bmatrix}$ $\begin{bmatrix} 7 \\ -4 \\ 3 \end{bmatrix}$

2. $\begin{bmatrix} 1 \\ -6 \\ 4 \end{bmatrix}$ $\begin{bmatrix} 7 \\ 0 \\ 1 \end{bmatrix}$

3. $\begin{bmatrix} 1 \\ 3 \\ 7 \\ -2 \end{bmatrix}$ $\begin{bmatrix} 2 \\ 5 \\ 7 \\ 9 \end{bmatrix}$

4. $\begin{bmatrix} 3 \\ -8 \\ 2 \\ -1 \end{bmatrix}$ $\begin{bmatrix} 7 \\ -6 \\ 6 \\ -8 \end{bmatrix}$

5. $\begin{bmatrix} 1 \\ 0 \\ 3 \\ 0 \\ 2 \end{bmatrix}$ $\begin{bmatrix} 5 \\ 4 \\ 0 \\ 2 \\ 5 \end{bmatrix}$

6. $\begin{bmatrix} -2 \\ 6 \\ 1 \\ 4 \\ 3 \end{bmatrix}$ $\begin{bmatrix} 2 \\ -8 \\ 5 \\ 7 \\ 4 \end{bmatrix}$

In Exercises 7–10 determine which of the given functions are inner products on R^3 where

$$\mathbf{u} = \begin{bmatrix} u_1 \\ u_2 \\ u_3 \end{bmatrix} \quad \text{and} \quad \mathbf{v} = \begin{bmatrix} v_1 \\ v_2 \\ v_3 \end{bmatrix}$$

7. $\langle \mathbf{u}, \mathbf{v} \rangle = 2u_1v_1 + 3u_2v_2 + 4u_3v_3$
8. $\langle \mathbf{u}, \mathbf{v} \rangle = u_1v_1 + 5u_2v_2 + 7u_3v_3$
9. $\langle \mathbf{u}, \mathbf{v} \rangle = u_1v_2 + u_2v_1$
10. $\langle \mathbf{u}, \mathbf{v} \rangle = u_1v_3 + u_2v_2 + u_3v_1$
11. Let p and q be polynomials in P_2. Show that $\langle p(x), q(x) \rangle = p(-1)q(-1) + p(0)q(0) + p(1)q(1)$ defines an inner product on P_2.
12. Let M_{22} denote the linear space of all 2×2 matrices with real components and let

$$A = \begin{bmatrix} a_1 & a_2 \\ a_3 & a_4 \end{bmatrix} \quad \text{and} \quad B = \begin{bmatrix} b_1 & b_2 \\ b_3 & b_4 \end{bmatrix}$$

be any two elements of M_{22}. Show that $\langle A, B \rangle = a_1b_1 + a_2b_2 + a_3b_3 + a_4b_4$ defines an inner product on M_{22}.
13. Let $p(x) = a_0 + a_1x + a_2x^2$ and $q(x) = b_0 + b_1x + b_2x^2$ be any two elements of P_2. Does $\langle p, q \rangle = a_0b_0 - a_1b_1 + a_2b_2$ define an inner product on P_2? Explain.
14. Show that $\langle \mathbf{w}, \mathbf{u} + \mathbf{v} \rangle = \langle \mathbf{w}, \mathbf{u} \rangle + \langle \mathbf{w}, \mathbf{v} \rangle$ for any inner product on a linear space having the real numbers as scalars.
15. Let $\mathbf{u}$ and $\mathbf{v}$ be elements of an inner product space having the real numbers as scalars. Show that $\langle \mathbf{u}, c\mathbf{v} \rangle = c\langle \mathbf{u}, \mathbf{v} \rangle$ for any scalar c.
16. (*For readers who have studied calculus.*) Compute the inner product of $f(x) = x^2$ and $g(x) = x^3 + 3$ on the interval $[0, 1]$ using the inner product in Example 4.
17. Show that the function defined in (1) is an inner product on R^n.
18. Show that the Euclidean inner product defined in (1) can be written as $\mathbf{v}^t\mathbf{u}$ or as $\mathbf{u}^t\mathbf{v}$.

Let V be a linear space having the complex numbers as scalars. An inner product on V is defined as in Definition 1 with the exception that axiom 1 is replaced by $\langle \mathbf{v}, \mathbf{u} \rangle = \overline{\langle \mathbf{u}, \mathbf{v} \rangle}$, where $\overline{\langle \mathbf{u}, \mathbf{v} \rangle}$ denotes the complex conjugate of $\langle \mathbf{u}, \mathbf{v} \rangle$, In Exercises 19 and 20 show that the given function is an inner product on the given linear space.

19. Let C^n denote the linear space of all n-vectors with complex components and $\langle \mathbf{u}, \mathbf{v} \rangle = u_1\bar{v}_1 + u_2\bar{v}_2 + \cdots + u_n\bar{v}_n$ where

$$\mathbf{u} = \begin{bmatrix} u_1 \\ u_2 \\ \vdots \\ u_n \end{bmatrix} \quad \text{and} \quad \mathbf{v} = \begin{bmatrix} v_1 \\ v_2 \\ \vdots \\ v_n \end{bmatrix}$$

20. Let V denote the set of all polynomials with complex coefficients and $\langle p, q \rangle = a_0\bar{b}_0 + a_1\bar{b}_1 + a_2\bar{b}_2$ where $p(x) = a_0 + a_1x + a_2x^2$ and $q(x) = b_0 + b_1x + b_2x^2$.

5.2 CAUCHY–SCHWARZ INEQUALITY

An inner product on a linear space V enables us to define geometric concepts in V that are analogous to those in R^2. We begin by defining the norm of an element of an inner product space. As in R^2 and R^3 the norm of an element of an inner product space may be interpreted as the element's length.

Definition 2. Let $\mathbf{v}$ be an element of an inner product space V. The norm $\|\mathbf{v}\|$ of $\mathbf{v}$ is defined by

$$\|\mathbf{v}\| = \langle \mathbf{v}, \mathbf{v}\rangle^{1/2}$$

Example 1. The Euclidean norm of any element

$$\mathbf{u} = \begin{bmatrix} u_1 \\ u_2 \\ \vdots \\ u_n \end{bmatrix}$$

of R^n is given by

$$\|\mathbf{u}\| = \sqrt{u_1^2 + u_2^2 + \cdots + u_n^2}$$

For example

$$\left\| \begin{bmatrix} 1 \\ 2 \\ 3 \\ 4 \end{bmatrix} \right\| = \sqrt{1^2 + 2^2 + 3^2 + 4^2} = \sqrt{30}$$

Example 2. (*For readers who have studied calculus.*) Consider the element $\cos x$ of the linear space $C[0, \pi]$ on which is defined the inner product (see Example 4 of Section 5.1)

$$\langle f, g \rangle = \int_0^\pi f(x)g(x)\, dx$$

Then

$$\|\cos x\| = \left(\int_0^\pi \cos^2 x\, dx \right)^{1/2}$$

$$= \left(\int_0^\pi \frac{1 + \cos 2x}{2}\, dx \right)^{1/2}$$

$$= \left(\frac{\pi}{2} \right)^{1/2}$$

In Theorems 4 and 10 of Chapter 3 we showed that $\mathbf{u} \cdot \mathbf{v} = \|\mathbf{u}\|\,\|\mathbf{v}\| \cos\theta$ for $\mathbf{u}$ and $\mathbf{v}$ that are either 2-vectors or 3-vectors. Since $|\cos\theta| \leq 1$ we conclude that $|\mathbf{u} \cdot \mathbf{v}| \leq \|\mathbf{u}\|\,\|\mathbf{v}\|$. The following theorem shows that this inequality holds for any inner product.

Theorem 2 (*Cauchy–Schwarz inequality*). *If $\mathbf{u}$ and $\mathbf{v}$ are any elements of an inner product space, then*

$$|\langle \mathbf{u}, \mathbf{v}\rangle| \leq \|\mathbf{u}\|\,\|\mathbf{v}\|$$

Proof: If $\mathbf{u} = \mathbf{0}$, then $\langle \mathbf{u}, \mathbf{v}\rangle = \langle \mathbf{0}, \mathbf{v}\rangle = 0$ (part 1 of Theorem 1) so that in this case the inequality holds. Now suppose that $\mathbf{u} \neq \mathbf{0}$ and consider the element $t\mathbf{u} + \mathbf{v}$ where

t is a scalar. Repeated use of the axioms for an inner product (Definition 1) allows us to show that

$$0 \leq \langle t\mathbf{u} + \mathbf{v}, \ t\mathbf{u} + \mathbf{v} \rangle = \langle \mathbf{u}, \mathbf{u} \rangle t^2 + 2\langle \mathbf{u}, \mathbf{v} \rangle t + \langle \mathbf{v}, \mathbf{v} \rangle \tag{1}$$

for any real numbers t (see Exercise 5). Setting $a = \langle \mathbf{u}, \mathbf{u} \rangle$, $b = 2\langle \mathbf{u}, \mathbf{v} \rangle$, and $c = \langle \mathbf{v}, \mathbf{v} \rangle$ we see from the preceding inequality that

$$0 \leq at^2 + bt + c$$

for all real numbers t. Since $\mathbf{u} \neq 0$, the coefficient a of t^2 is nonzero. It follows that the polynomial $at^2 + bt + c$ is of second degree and has either no real roots or a single repeated root. This can happen only if the discriminant $b^2 - 4ac$ is nonpositive; i.e.,

$$b^2 - 4ac \leq 0$$

Replacing a, b, and c by their values in terms of $\mathbf{u}$ and $\mathbf{v}$, we obtain

$$(2\langle \mathbf{u}, \mathbf{v} \rangle)^2 - 4\langle \mathbf{u}, \mathbf{u} \rangle\langle \mathbf{v}, \mathbf{v} \rangle \leq 0$$

Elementary algebra now yields

$$\langle \mathbf{u}, \mathbf{v} \rangle^2 \leq \langle \mathbf{u}, \mathbf{u} \rangle\langle \mathbf{v}, \mathbf{v} \rangle$$

so that

$$|\langle \mathbf{u}, \mathbf{v} \rangle| \leq \langle \mathbf{u}, \mathbf{u} \rangle^{1/2}\langle \mathbf{v}, \mathbf{v} \rangle^{1/2} = \|\mathbf{u}\| \, \|\mathbf{v}\| \quad \blacksquare$$

Example 3. Applying the Cauchy–Schwarz inequality to R^n we obtain the inequality

$$(u_1v_1 + u_2v_2 + \cdots + u_nv_n)^2 \leq (u_1^2 + u_2^2 + \cdots + u_n^2)(v_1^2 + v_2^2 + \cdots + v_n^2)$$

for any real numbers $u_1, u_2, \ldots, u_n, v_1, v_2, \ldots, v_n$. This inequality is known as **Cauchy's inequality**.

Example 4. (*For readers who have studied calculus.*) Let $\langle f, g \rangle$ be the inner product defined on the linear space $C[a, b]$ defined in Example 4 of Section 5.1. The Cauchy–Schwarz inequality yields

$$\left(\int_a^b f(x)g(x) \, dx \right)^2 \leq \int_a^b [f(x)]^2 \, dx \int_a^b [g(x)]^2 \, dx$$

for any continuous functions f and g on $[a, b]$.

The following theorem lists some of the basic properties of any norm.

Theorem 3. *If V is an inner product space, then for every $\mathbf{u}$ and $\mathbf{v}$ in V and every scalar c*

1. $0 \leq \|\mathbf{u}\|$
2. $\|\mathbf{u}\| = 0$ if and only if $\mathbf{u} = 0$
3. $\|c\mathbf{u}\| = |c| \, \|\mathbf{u}\|$
4. $\|\mathbf{u} + \mathbf{v}\| \leq \|\mathbf{u}\| + \|\mathbf{v}\|$ (*Triangle inequality*)

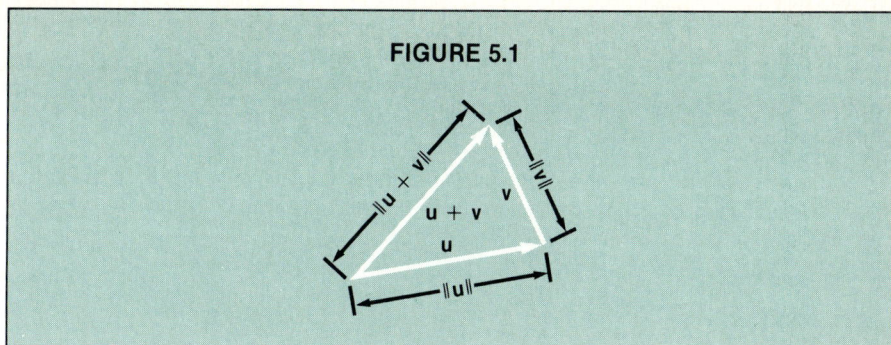

FIGURE 5.1

Proof: We will prove the last property and leave the proofs of the others as exercises.

$$\|u + v\|^2 = \langle u + v, u + v \rangle$$
$$= \langle u, u \rangle + 2\langle u, v \rangle + \langle v, v \rangle$$
$$\le \|u\|^2 + 2|\langle u, v \rangle| + \|v\|^2$$
$$\le \|u\|^2 + 2\|u\| \|v\| + \|v\|^2$$
$$= (\|u\| + \|v\|)^2$$

Since $0 \le \|u + v\|$ we can obtain the desired inequality by taking the square root of each side of the last inequality. ∎

Recall that in R^2 the norm of a vector represents the vector's length. In R^2 the triangle inequality has a simple geometric interpretation: the sum of the lengths of two sides of a triangle is greater than or equal to the length of the remaining side (see Figure 5.1).

Exercises

In Exercises 1–4 compute the Euclidean norm of the given vector.

1. $\begin{bmatrix} 1 \\ 3 \\ -2 \end{bmatrix}$

2. $\begin{bmatrix} 2 \\ 5 \\ 7 \end{bmatrix}$

3. $\begin{bmatrix} 2 \\ -4 \\ 3 \\ 1 \end{bmatrix}$

4. $\begin{bmatrix} -6 \\ -4 \\ -3 \\ 0 \end{bmatrix}$

5. Establish the identity in (1).

6. (*For readers who have studied calculus.*) Let f and g be continuous functions on $[-1, 1]$. Show that

$$\int_0^1 [f(x) + g(x)]^2 \, dx \le \left[\left(\int_0^1 [f(x)]^2 \, dx \right)^{1/2} + \left(\int_0^1 [g(x)]^2 \, dx \right)^{1/2} \right]^2$$

7. Prove part 1 of Theorem 3.
8. Prove part 2 of Theorem 3.
9. Prove part 3 of Theorem 3.
10. Show that $\|\mathbf{u} + \mathbf{v}\|^2 - \|\mathbf{u} - \mathbf{v}\|^2 = 4\langle\mathbf{u}, \mathbf{v}\rangle$.
11. Show that $\|\mathbf{u} + \mathbf{v}\|^2 + \|\mathbf{u} - \mathbf{v}\|^2 = 2\|\mathbf{u}\|^2 + 2\|\mathbf{v}\|^2$.

A norm can be defined without reference to an inner product. When this is done the four properties listed in Theorem 3 are taken to be the axioms of a norm. In Exercises 12–13 show that the given function is a norm on the given linear space.

12. Let

$$\mathbf{u} = \begin{bmatrix} u_1 \\ u_2 \\ \vdots \\ u_n \end{bmatrix}$$

be any element of R^n and $\|\mathbf{u}\| = \max\{|u_i| : i = 1, 2, \ldots n\}$.
13. Let V denote the linear space of all real valued continuous functions defined on $[0, 1]$ and $\|f\| = \max\{|f(x)| : 0 \le x \le 1\}$ for any f in V.

5.3 ORTHOGONALITY

In Chapter 3 we found that it is possible to determine the angle between two nonzero vectors in R^2 and R^3 in terms of the inner product and norms of the two vectors. The Cauchy–Schwarz inequality (Theorem 2 of the previous section) allows us to do the same in any inner product space. The Cauchy–Schwarz inequality assures us that for any nonzero elements $\mathbf{u}$ and $\mathbf{v}$ of an inner product space V we have

$$\frac{|\langle\mathbf{u}, \mathbf{v}\rangle|}{\|\mathbf{u}\| \, \|\mathbf{v}\|} \le 1$$

or equivalently

$$-1 \le \frac{\langle\mathbf{u}, \mathbf{v}\rangle}{\|\mathbf{u}\| \, \|\mathbf{v}\|} \le 1$$

Consequently there is a unique angle θ such that

$$\cos\theta = \frac{\langle\mathbf{u}, \mathbf{v}\rangle}{\|\mathbf{u}\| \, \|\mathbf{v}\|} \quad \text{and} \quad 0 \le \theta \le \pi \tag{1}$$

This angle θ is called the *angle between* $\mathbf{u}$ *and* $\mathbf{v}$. Observe that in R^2 and R^3 the angle between vectors defined in (1) coincides with the angle between vectors as defined in Chapter 3 (see Theorems 4 and 10 of Chapter 3). Also notice that if $\mathbf{u}$ and $\mathbf{v}$ are nonzero elements of an inner product space then $\langle\mathbf{u}, \mathbf{v}\rangle = 0$ if and only if the angle θ between $\mathbf{u}$ and $\mathbf{v}$ is $\pi/2$.

Definition 3. Two elements $\mathbf{u}$ and $\mathbf{v}$ of an inner product space are said to be **orthogonal** if $\langle\mathbf{u}, \mathbf{v}\rangle = 0$.

Example 1. Consider the linear space P_2 of all polynomials of degree less than or equal to 2 on which is defined the inner product

$$\langle a_0 + a_1 x + a_2 x^2, b_0 + b_1 x + b_2 x^2 \rangle = a_0 b_0 + a_1 b_1 + a_2 b_2$$

(see Example 3 of Section 5.1). Then the angle θ between $1 + 2x + 5x^2$ and $4 - 3x + x^2$ satisfies the equation

$$\cos \theta = \frac{(1)(4) + (2)(-3) + (5)(1)}{\sqrt{1^2 + 2^2 + 5^2} \sqrt{4^2 + (-3)^2 + 1^2}}$$

$$= \frac{3}{2\sqrt{195}}$$

so that $\theta \approx 1.4632$ radians.

Example 2. The angle θ between the 4-vectors

$$\mathbf{u} = \begin{bmatrix} 1 \\ 2 \\ -1 \\ 4 \end{bmatrix} \quad \text{and} \quad \mathbf{v} = \begin{bmatrix} -3 \\ 5 \\ 7 \\ 2 \end{bmatrix}$$

satisfies

$$\cos \theta = \frac{(1)(-3) + (2)(5) + (-1)(7) + (4)(2)}{\sqrt{1^2 + 2^2 + (-1)^2 + 4^2} \sqrt{(-3)^2 + 5^2 + 7^2 + 2^2}}$$

$$= \frac{8}{\sqrt{1914}}$$

so that $\theta \approx 1.3869$ radians.

Example 3. (*For readers who have studied calculus.*) Consider the elements $\sin x$ and $\cos x$ of the linear space $C[0, \pi]$ of all continuous functions on the interval $[0, \pi]$. These elements are orthogonal with respect to the inner product

$$\langle f, g \rangle = \int_0^\pi f(x) g(x) \, dx$$

(see Example 4 of Section 5.1) since

$$\langle \sin x, \cos x \rangle = \int_0^\pi \sin x \cos x \, dx$$

$$= \frac{1}{2} \sin^2 x \Big|_0^\pi = 0$$

Example 4. Care must be taken when stating that two given elements of a linear space V are orthogonal. As we saw in Section 5.1, it may be possible to define more than one inner product on V. In such a case a pair of elements may be orthogonal with respect to one inner product and not orthogonal with respect to another. For example, the vectors

$$\begin{bmatrix} 1 \\ 1 \end{bmatrix} \quad \text{and} \quad \begin{bmatrix} -1 \\ 1 \end{bmatrix}$$

in R^2 are orthogonal with respect to the Euclidean inner product since $(1)(-1) + (1)(1) = 0$, but they are not orthogonal with respect to the inner product

$$\left\langle \begin{bmatrix} u_1 \\ u_2 \end{bmatrix}, \begin{bmatrix} v_1 \\ v_2 \end{bmatrix} \right\rangle = 5u_1v_1 + 4u_2v_2$$

(see Example 2 of Section 5.1) since

$$5(1)(-1) + 4(1)(1) = -1 \neq 0$$

It is left as an exercise for the reader to show that the vectors

$$\begin{bmatrix} 1 \\ 1 \end{bmatrix} \quad \text{and} \quad \begin{bmatrix} -4 \\ 5 \end{bmatrix}$$

are orthogonal with respect to the latter inner product but are not orthogonal with respect to the Euclidean inner product.

Thus, when we state that elements of a linear space are orthogonal, we must also state the inner product with respect to which they are orthogonal. Remember that unless stated to the contrary, the Euclidean inner product is the inner product we use on R^n.

We close this section with a generalization of the Pythagorean theorem from geometry. Before stating this generalization we will rephrase the Pythagorean theorem in terms of vector notation. Let the triangle ABC (see Figure 5.2) be a right triangle and let $\mathbf{u}$, $\mathbf{v}$, and $\mathbf{w}$ be vectors having as representations the directed line segments $\overrightarrow{AB}$, $\overrightarrow{BC}$, and $\overrightarrow{AC}$, respectively. Since $\mathbf{w} = \mathbf{u} + \mathbf{v}$ and the norm of a vector is its length, the Pythagorean Theorem yields

$$\|\mathbf{u} + \mathbf{v}\|^2 = \|\mathbf{u}\|^2 + \|\mathbf{v}\|^2$$

We now show that this identity is necessary and sufficient for two elements $\mathbf{u}$ and $\mathbf{v}$ of an inner product space to be orthogonal.

FIGURE 5.2

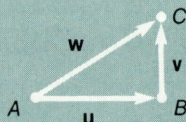

Theorem 4. *Let* **u** *and* **v** *be any elements of an inner product space V. Then* **u** *and* **v** *are orthogonal if and only if*

(2)
$$\|\mathbf{u} + \mathbf{v}\|^2 = \|\mathbf{u}\|^2 + \|\mathbf{v}\|^2$$

Proof:

$$\|\mathbf{u} + \mathbf{v}\|^2 = \langle \mathbf{u} + \mathbf{v}, \mathbf{u} + \mathbf{v} \rangle$$
$$= \langle \mathbf{u}, \mathbf{u} \rangle + 2\langle \mathbf{u}, \mathbf{v} \rangle + \langle \mathbf{v}, \mathbf{v} \rangle$$
$$= \|\mathbf{u}\|^2 + 2\langle \mathbf{u}, \mathbf{v} \rangle + \|\mathbf{v}\|^2$$

Thus the identity in (2) holds if and only if $\langle \mathbf{u}, \mathbf{v} \rangle = 0$, i.e., if and only if **u** and **v** are orthogonal. ∎

Exercises

In Exercises 1–4 determine the angle between the given vectors with respect to the Euclidean inner product.

1. $\begin{bmatrix} 1 \\ 2 \\ 0 \\ 1 \end{bmatrix}$ $\begin{bmatrix} -2 \\ 3 \\ 1 \\ 2 \end{bmatrix}$

2. $\begin{bmatrix} -2 \\ 3 \\ 1 \\ 4 \end{bmatrix}$ $\begin{bmatrix} -3 \\ 1 \\ 4 \\ 6 \end{bmatrix}$

3. $\begin{bmatrix} 1 \\ 2 \\ 5 \\ 4 \\ -3 \end{bmatrix}$ $\begin{bmatrix} 2 \\ 8 \\ -6 \\ -7 \\ 2 \end{bmatrix}$

4. $\begin{bmatrix} 5 \\ 2 \\ 1 \\ 0 \\ 4 \end{bmatrix}$ $\begin{bmatrix} -7 \\ -8 \\ 0 \\ 0 \\ 3 \end{bmatrix}$

In Exercises 5–6 determine the values of k so that the given vectors are orthogonal with respect to the Euclidean inner product.

5. $\begin{bmatrix} 2 \\ 3 \\ k \\ 4 \end{bmatrix}$ $\begin{bmatrix} 1 \\ k \\ 3 \\ -5 \end{bmatrix}$

6. $\begin{bmatrix} 2 \\ 8 \\ 4 \\ k \end{bmatrix}$ $\begin{bmatrix} 3 \\ -6 \\ 2 \\ k \end{bmatrix}$

In Exercises 7–8 determine the angle between the given elements of P_2 with respect to the inner product $\langle p, q \rangle = p(-1)q(-1) + p(0)q(0) + p(1)q(1)$ (see Exercise 11 of Section 5.1).

7. $x^2 + 2x + 3, x^2 + 1$.

8. $2x^2 + x - 3, 4x^2 - x$.

In Exercises 9–10 determine the angle between the given 2×2 matrices with respect to the inner product given in Exercise 12 of Section 5.1.

9. $\begin{bmatrix} 1 & 2 \\ 3 & 4 \end{bmatrix}$ $\begin{bmatrix} 5 & -3 \\ 1 & 2 \end{bmatrix}$

10. $\begin{bmatrix} 5 & -4 \\ 2 & 6 \end{bmatrix}$ $\begin{bmatrix} -2 & -1 \\ 1 & -5 \end{bmatrix}$

11. Let **u**, **v**, and **w** be elements of an inner product space. Show that if **u** is orthogonal to both **v** and **w**, then **u** is orthogonal to every linear combination $a\mathbf{v} + b\mathbf{w}$ of **v** and **w**.

12. Show that $\|\mathbf{u} + \mathbf{v}\|^2 = \|\mathbf{u}\|^2 + \|\mathbf{v}\|^2 + 2\|\mathbf{u}\|\,\|\mathbf{v}\|\cos\theta$ where θ is the angle between $\mathbf{u}$ and $\mathbf{v}$.

13. (*For readers who have studied calculus.*) For $n = 1, 2, \ldots$, let $f_n(x) = \sin nx$, and let $g_n(x) = \cos nx$. Show that with respect to the inner product in Example 3,
    (a) f_n and g_m are orthogonal.
    (b) f_n and f_m are orthogonal if $n \neq m$.
    (c) g_n and g_m are orthogonal if $n \neq m$.

14. Let $\mathbf{v}$ be a nonzero element of an inner product space V. Show that the set W of all elements of V orthogonal to $\mathbf{v}$ is a subspace of V.

15. Let $\mathbf{v}_1, \mathbf{v}_2, \ldots, \mathbf{v}_k$ be elements of an inner product space V. Show that the set W of all elements of V orthogonal to $\mathbf{v}_1, \mathbf{v}_2, \ldots,$ and $\mathbf{v}_k$ is a subspace of V. The subspace W is called the **orthogonal complement** of the linear space U spanned by $\mathbf{v}_1, \mathbf{v}_2, \ldots, \mathbf{v}_k$.

5.4 ORTHONORMAL BASES

In many applications involving linear spaces it is necessary to write selected elements of a linear space as linear combinations of the elements in a basis. This frequently requires finding the solution of a linear system of equations. For example, suppose that we wish to write a given n-vector $\mathbf{b}$ as a linear combination of the elements of a basis $\{\mathbf{v}_1, \mathbf{v}_2, \ldots, \mathbf{v}_n\}$ for R^n. In Theorem 5 of Section 4.4 we found that this is equivalent to solving the matrix equation

$$A\mathbf{x} = \mathbf{b}$$

where A is the matrix having $\mathbf{v}_1, \mathbf{v}_2, \ldots, \mathbf{v}_n$ as its columns. If n is large it may be difficult, or at least time consuming, to solve this equation. However, if

$$\mathbf{v}_1 = \begin{bmatrix} 1 \\ 0 \\ 0 \\ \vdots \\ 0 \end{bmatrix}, \quad \mathbf{v}_2 = \begin{bmatrix} 0 \\ 1 \\ 0 \\ \vdots \\ 0 \end{bmatrix}, \quad \ldots, \quad \mathbf{v}_n = \begin{bmatrix} 0 \\ 0 \\ 0 \\ \vdots \\ 1 \end{bmatrix} \tag{1}$$

it is easy to write

$$\mathbf{b} = \begin{bmatrix} b_1 \\ b_2 \\ \vdots \\ b_n \end{bmatrix}$$

as a linear combination of the $\mathbf{v}$'s: $\mathbf{b} = b_1\mathbf{v}_1 + b_2\mathbf{v}_2 + \cdots + b_n\mathbf{v}_n$. Thus some bases are easier to use in computations than are others. This section demonstrates how such bases can be constructed. We begin by noting that the set of vectors $\{\mathbf{v}_1, \mathbf{v}_2, \ldots, \mathbf{v}_n\}$, where $\mathbf{v}_1, \mathbf{v}_2, \ldots, \mathbf{v}_n$ are as in (1), has the property

$$\langle \mathbf{v}_i, \mathbf{v}_j \rangle = \begin{cases} 1 & \text{if } i = j \\ 0 & \text{if } i \neq j \end{cases}$$

This observation leads to the following definition.

Definition 4. A set S of elements of an inner product space is called **orthogonal** if any two distinct elements are orthogonal. If, in addition, the norm of each element of S is 1, then S is called **orthonormal**.

Example 1. Consider the set $S = \{v_1, v_2, v_3\}$ of 3-vectors where

$$v_1 = \frac{1}{\sqrt{2}}\begin{bmatrix} 1 \\ 0 \\ -1 \end{bmatrix} \quad v_2 = \frac{1}{\sqrt{3}}\begin{bmatrix} 1 \\ 1 \\ 1 \end{bmatrix} \quad v_3 = \frac{1}{\sqrt{6}}\begin{bmatrix} 1 \\ -2 \\ 1 \end{bmatrix}$$

Straightforward calculations show that $\langle v_1, v_1 \rangle = \langle v_2, v_2 \rangle = \langle v_3, v_3 \rangle = 1$, and $\langle v_1, v_2 \rangle = \langle v_1, v_3 \rangle = \langle v_2, v_3 \rangle = 0$. Therefore the set S is orthonormal.

Example 2. If $\{u_1, u_2, \ldots, u_n\}$ is an orthogonal set of nonzero elements of an inner product space, then $\{v_1, v_2, \ldots, v_n\}$ is an orthonormal set where

$$v_1 = \frac{1}{\|u_1\|}u_1, \quad v_2 = \frac{1}{\|u_2\|}u_2, \ldots, v_n = \frac{1}{\|u_n\|}u_n$$

This is easily verified by noting that

$$\langle v_i, v_j \rangle = \langle \frac{1}{\|u_i\|}u_i, \frac{1}{\|u_j\|}u_j \rangle$$

$$= \frac{1}{\|u_i\|\,\|u_j\|}\langle u_i, u_j \rangle$$

$$= \begin{cases} 0 & \text{if } i \neq j \\ 1 & \text{if } i = j \end{cases}$$

since $\langle u_i, u_i \rangle = \|u_i\|^2$.

The following theorem gives the relationship between orthogonality and linear independence.

Theorem 5. *Let $S = \{v_1, v_2, \ldots, v_n\}$ be an orthogonal set of nonzero elements of an inner product space. Then S is linearly independent.*

Proof: We use the standard approach to show that S is linearly independent. That is, we let $c_1, c_2, \ldots, c_n$ be any scalars such that

$$(2) \qquad\qquad c_1 v_1 + c_2 v_2 + \cdots + c_n v_n = 0$$

and show that each $c_i = 0$. For any i, $i = 1, 2, \ldots, n$,

$$0 = \langle 0, v_i \rangle$$
$$= \langle c_1 v_1 + c_2 v_2 + \cdots + c_n v_n, v_i \rangle$$
$$(3) \qquad = c_1 \langle v_1, v_i \rangle + c_2 \langle v_2, v_i \rangle + \cdots + c_n \langle v_n, v_i \rangle$$

Since S is orthogonal we have $\langle v_j, v_i \rangle = 0$ if $i \neq j$, so that the equation in (3) can be rewritten as $0 = c_i \langle v_i, v_i \rangle$. We must have $\langle v_i, v_i \rangle \neq 0$ because $v_i \neq 0$. Therefore $c_i = 0$ for $i = 1, 2, \ldots, n$. Since $c_1 = 0, c_2 = 0, \ldots, c_n = 0$ are the only scalars satisfying the equation in (2), the set S is linearly independent. ∎

Example 3.   In Example 1 we found that the set

$$S = \left\{ \frac{1}{\sqrt{2}} \begin{bmatrix} 1 \\ 0 \\ -1 \end{bmatrix}, \ \frac{1}{\sqrt{3}} \begin{bmatrix} 1 \\ 1 \\ 1 \end{bmatrix}, \ \frac{1}{\sqrt{6}} \begin{bmatrix} 1 \\ -2 \\ 1 \end{bmatrix} \right\}$$

is orthonormal. By Theorem 5 the set S is linearly independent. Recall that any linearly independent set of n elements from an n-dimensional linear space is a basis for that space (Theorem 17 of Section 4.9). Thus S is a basis for R^3.

The interest in orthonormal sets that are also bases for inner product spaces is based, at least in part, on the exceptional computational properties possessed by these sets. Suppose that we wish to write a given element of a linear space V as a linear combination of the elements of a basis for V. To find the coefficients of the linear combination it is usually necessary to solve a system of equations. If the basis is orthonormal, then it is easy to find these coefficients.

Theorem 6.   *If* $\{v_1, v_2, \ldots, v_n\}$ *is an orthonormal basis for an inner product space* V, *and* u *is any element of* V, *then*

$$u = \langle u, v_1 \rangle v_1 + \langle u, v_2 \rangle v_2 + \cdots + \langle u, v_n \rangle v_n$$

Proof: Since $\{v_1, v_2, \ldots, v_n\}$ is a basis for V, there are scalars $c_1, c_2, \ldots, c_n$ such that

$$u = c_1 v_1 + c_2 v_2 + \cdots + c_n v_n$$

Then for any $i, i = 1, 2, \cdots, n$,

$$\langle u, v_i \rangle = \langle c_1 v_1 + c_2 v_2 + \cdots + c_n v_n, v_i \rangle \tag{4}$$
$$= c_1 \langle v_1, v_i \rangle + c_2 \langle v_2, v_i \rangle + \cdots + c_n \langle v_n, v_i \rangle$$

Since $\{v_1, v_2, \ldots, v_n\}$ is orthonormal, $\{v_j, v_i\} = 0$ if $i \neq j$, and $\langle v_i, v_i \rangle = 1$. Hence the equation in (4) may be rewritten as

$$\langle u, v_i \rangle = c_i$$

for $i = 1, 2, \ldots, n$. Therefore $u = \langle u, v_1 \rangle v_1 + \langle u_1, v_2 \rangle v_2 + \cdots + \langle u, v_n \rangle v_n$. ∎

Example 4.   Consider the orthonormal basis $\{v_1, v_2, v_3\}$ for R^3, where

$$v_1 = \frac{1}{\sqrt{2}} \begin{bmatrix} 1 \\ 0 \\ -1 \end{bmatrix} \quad v_2 = \frac{1}{\sqrt{3}} \begin{bmatrix} 1 \\ 1 \\ 1 \end{bmatrix} \quad v_3 = \frac{1}{\sqrt{6}} \begin{bmatrix} 1 \\ -2 \\ 1 \end{bmatrix}$$

(see Example 3). By Theorem 6 the vector

$$u = \begin{bmatrix} 4 \\ 8 \\ 11 \end{bmatrix}$$

can be written as

$$\begin{bmatrix} 4 \\ 8 \\ 11 \end{bmatrix} = \langle \mathbf{u}, \mathbf{v}_1 \rangle \mathbf{v}_1 + \langle \mathbf{u}, \mathbf{v}_2 \rangle \mathbf{v}_2 + \langle \mathbf{u}, \mathbf{v}_3 \rangle \mathbf{v}_3$$

$$= -\frac{7}{\sqrt{2}} \mathbf{v}_1 + \frac{23}{\sqrt{3}} \mathbf{v}_2 - \frac{1}{\sqrt{6}} \mathbf{v}_3$$

In light of the relative ease of writing linear combinations by using an orthonormal basis, it is desirable to choose such a basis for any inner product space under consideration. Fortunately, there is an algorithm for computing an orthonormal basis from any given basis for an inner product space. We will first give the algorithm and then discuss its various steps.

Gram–Schmidt Process. Let $\{\mathbf{u}_1, \mathbf{u}_2, \dots, \mathbf{u}_n\}$ be any basis for an inner product space V.

Step 1. Let $\mathbf{v}_1 = \dfrac{1}{\|\mathbf{u}_1\|} \mathbf{u}_1$

Step 2. Let $\mathbf{w}_2 = \mathbf{u}_2 - \langle \mathbf{u}_2, \mathbf{v}_1 \rangle \mathbf{v}_1$

Step 3. Let $\mathbf{v}_2 = \dfrac{1}{\|\mathbf{w}_2\|} \mathbf{w}_2$

Step 4. Let $\mathbf{w}_3 = \mathbf{u}_3 - \langle \mathbf{u}_3, \mathbf{v}_1 \rangle \mathbf{v}_1 - \langle \mathbf{u}_3, \mathbf{v}_2 \rangle \mathbf{v}_2$

Step 5. Let $\mathbf{v}_3 = \dfrac{1}{\|\mathbf{w}_3\|} \mathbf{w}_3$

Step 6. Continue in this manner for $k = 4, 5, \dots, n$ by letting

$$\mathbf{w}_k = \mathbf{u}_k - \langle \mathbf{u}_k, \mathbf{v}_1 \rangle \mathbf{v}_1 - \langle \mathbf{u}_k, \mathbf{v}_2 \rangle \mathbf{v}_2 - \cdots - \langle \mathbf{u}_k, \mathbf{v}_{k-1} \rangle \mathbf{v}_{k-1}$$

and then letting

$$\mathbf{v}_k = \frac{1}{\|\mathbf{w}_k\|} \mathbf{w}_k$$

The set $\{\mathbf{v}_1, \mathbf{v}_2, \dots, \mathbf{v}_k\}$ is an orthonormal basis for V.

The first step of the Gram–Schmidt process gives us a vector $\mathbf{v}_1$ having norm 1 and spanning the same subspace of V as $\mathbf{u}_1$. Since $\{\mathbf{u}_1, \mathbf{u}_2, \dots, \mathbf{u}_n\}$ is a basis, none of its elements is the zero vector. In particular, $\mathbf{u}_1 \neq 0$ so that $\|\mathbf{u}_1\| \neq 0$.

In the second step we find an element $\mathbf{w}_2$ that is orthogonal to every element in the subspace V_1 of V spanned by $\{\mathbf{v}_1\}$. Hence $\mathbf{w}_2$ and $\mathbf{v}_1$ are orthogonal. The element $\mathbf{w}_2$ cannot be the zero element, because if it were then

$$0 = \mathbf{w}_2 = \mathbf{u}_2 - \langle \mathbf{u}_2, \mathbf{v}_1 \rangle \mathbf{v}_1$$

$$= \mathbf{u}_2 - \frac{\langle \mathbf{u}_2, \mathbf{v}_1 \rangle}{\|\mathbf{u}_1\|} \mathbf{u}_1$$

which shows that $\{\mathbf{u}_1, \mathbf{u}_2\}$ is linearly dependent. This is impossible because $\{\mathbf{u}_1, \mathbf{u}_2, \ldots, \mathbf{u}_n\}$ is linearly independent. Therefore $\mathbf{w}_2$ is not the zero element.

The third step gives us an element of norm 1 that is orthogonal to $\mathbf{v}_1$. Notice that both $\mathbf{v}_1$ and $\mathbf{v}_2$ are linear combinations of $\mathbf{u}_1$ and $\mathbf{u}_2$. Since $\mathbf{v}_1$ and $\mathbf{v}_2$ are nonzero and orthogonal, the set $\{\mathbf{v}_1, \mathbf{v}_2\}$ is linearly independent (Theorem 5). Therefore $\{\mathbf{v}_1, \mathbf{v}_2\}$ must span the same subspace of V as $\{\mathbf{u}_1, \mathbf{u}_2\}$.

Suppose that we have obtained elements $\mathbf{v}_1, \mathbf{v}_2, \ldots, \mathbf{v}_{k-1}$ by using this process. We then construct a nonzero element $\mathbf{w}_k$ that is orthogonal to every element in the subspace V_{k-1} of V spanned by $\{\mathbf{v}_1, \mathbf{v}_2, \ldots, \mathbf{v}_{k-1}\}$. (This is proved in the next section.) From $\mathbf{w}_k$ we obtain an element $\mathbf{v}_k$ having norm 1 that is orthogonal to $\mathbf{v}_1, \mathbf{v}_2, \ldots, \mathbf{v}_{k-1}$ and is a linear combination of $\mathbf{u}_1, \mathbf{u}_2, \ldots, \mathbf{u}_k$. Thus we have constructed an orthonormal basis $\{\mathbf{v}_1, \mathbf{v}_2, \ldots, \mathbf{v}_k\}$ for the subspace of V spanned by $\{\mathbf{u}_1, \mathbf{u}_2, \ldots, \mathbf{u}_k\}$. Continuing in this manner until $k = n$, we obtain an orthonormal basis for V.

The Gram–Schmidt process gives us a proof of the following theorem.

Theorem 7. *Every finite dimensional inner product space V consisting of more than the zero element has an orthonormal basis.*

Example 5. We will use the Gram–Schmidt process to obtain an orthonormal basis for the subspace V of R^4 spanned by $S = \{\mathbf{u}_1, \mathbf{u}_2, \mathbf{u}_3\}$ where

$$\mathbf{u}_1 = \begin{bmatrix} 1 \\ 1 \\ 0 \\ 1 \end{bmatrix} \qquad \mathbf{u}_2 = \begin{bmatrix} 1 \\ 0 \\ 0 \\ 1 \end{bmatrix} \qquad \mathbf{u}_3 = \begin{bmatrix} 0 \\ 1 \\ 1 \\ 1 \end{bmatrix}$$

A short calculation shows that the set S is linearly independent. We begin by setting

$$\mathbf{v}_1 = \frac{1}{\|\mathbf{u}_1\|}\mathbf{u}_1 = \frac{1}{\sqrt{3}}\begin{bmatrix} 1 \\ 1 \\ 0 \\ 1 \end{bmatrix}$$

Then

$$\mathbf{w}_2 = \mathbf{u}_2 - \langle \mathbf{u}_2, \mathbf{v}_1 \rangle \mathbf{v}_1$$

$$= \begin{bmatrix} 1 \\ 0 \\ 0 \\ 1 \end{bmatrix} - \frac{2}{\sqrt{3}}\left(\frac{1}{\sqrt{3}}\begin{bmatrix} 1 \\ 1 \\ 0 \\ 1 \end{bmatrix} \right)$$

$$= \frac{1}{3}\begin{bmatrix} 1 \\ -2 \\ 0 \\ 1 \end{bmatrix}$$

so that

$$\mathbf{v}_2 = \frac{1}{\|\mathbf{w}_2\|}\mathbf{w}_2 = \frac{1}{\sqrt{6}}\begin{bmatrix} 1 \\ -2 \\ 0 \\ 1 \end{bmatrix}$$

Finally,

$$\mathbf{w}_3 = \mathbf{u}_3 - \langle \mathbf{u}_3, \mathbf{v}_1 \rangle \mathbf{v}_1 - \langle \mathbf{u}_3, \mathbf{v}_2 \rangle \mathbf{v}_2$$

$$= \begin{bmatrix} 0 \\ 1 \\ 1 \\ 1 \end{bmatrix} - \frac{2}{\sqrt{3}}\left(\frac{1}{\sqrt{3}}\begin{bmatrix} 1 \\ 1 \\ 0 \\ 1 \end{bmatrix} \right) - \left(-\frac{1}{\sqrt{6}} \right)\left(\frac{1}{\sqrt{6}}\begin{bmatrix} 1 \\ -2 \\ 0 \\ 1 \end{bmatrix} \right)$$

$$= \frac{1}{2}\begin{bmatrix} -1 \\ 0 \\ 2 \\ 1 \end{bmatrix}$$

so that

$$\mathbf{v}_3 = \frac{1}{\|\mathbf{w}_3\|}\mathbf{w}_3 = \frac{1}{\sqrt{6}}\begin{bmatrix} -1 \\ 0 \\ 2 \\ 1 \end{bmatrix}$$

The set $\{\mathbf{v}_1, \mathbf{v}_2, \mathbf{v}_3\}$ is an orthonormal basis for V.

―――――――――――――――――― **Exercises** ――――――――――――――――――

In Exercises 1–8 determine whether the given set of vectors is orthogonal, orthonormal, or neither with respect to the Euclidean inner product.

1. $\left\{ \begin{bmatrix} 1 \\ 0 \end{bmatrix}, \begin{bmatrix} 0 \\ 3 \end{bmatrix} \right\}$

2. $\left\{ \begin{bmatrix} 1 \\ 2 \end{bmatrix}, \begin{bmatrix} 0 \\ 3 \end{bmatrix} \right\}$

3. $\left\{ \begin{bmatrix} 1 \\ 1 \end{bmatrix}, \begin{bmatrix} -1 \\ 2 \end{bmatrix} \right\}$

4. $\left\{ \begin{bmatrix} \frac{1}{\sqrt{2}} \\ \frac{1}{\sqrt{2}} \end{bmatrix}, \begin{bmatrix} \frac{-1}{\sqrt{2}} \\ \frac{1}{\sqrt{2}} \end{bmatrix} \right\}$

5. $\left\{ \begin{bmatrix} \frac{1}{\sqrt{6}} \\ \frac{-2}{\sqrt{6}} \\ \frac{1}{\sqrt{6}} \end{bmatrix}, \begin{bmatrix} \frac{1}{\sqrt{3}} \\ \frac{1}{\sqrt{3}} \\ \frac{1}{\sqrt{3}} \end{bmatrix}, \begin{bmatrix} \frac{1}{\sqrt{2}} \\ 0 \\ \frac{-1}{\sqrt{2}} \end{bmatrix} \right\}$

6. $\left\{ \begin{bmatrix} 1 \\ 0 \\ 1 \end{bmatrix}, \begin{bmatrix} 0 \\ 1 \\ 0 \end{bmatrix}, \begin{bmatrix} -1 \\ 0 \\ 1 \end{bmatrix} \right\}$

7. $\left\{ \begin{bmatrix} \frac{1}{\sqrt{2}} \\ 0 \\ \frac{1}{\sqrt{2}} \\ 0 \end{bmatrix}, \begin{bmatrix} 0 \\ \frac{1}{\sqrt{2}} \\ 0 \\ \frac{1}{\sqrt{2}} \end{bmatrix}, \begin{bmatrix} \frac{1}{\sqrt{2}} \\ \frac{-1}{\sqrt{2}} \\ \frac{-1}{\sqrt{2}} \\ \frac{1}{\sqrt{2}} \end{bmatrix} \right\}$

8. $\left\{ \begin{bmatrix} 1 \\ 0 \\ 0 \\ 0 \end{bmatrix}, \begin{bmatrix} 1 \\ 1 \\ -1 \\ 1 \end{bmatrix}, \begin{bmatrix} -1 \\ 1 \\ 1 \\ 1 \end{bmatrix} \right\}$

In Exercises 9–10 determine which of the given sets of elements of P_2 are orthogonal, orthonormal, or neither with respect to the inner product given in Exercise 11 of Section 5.1.

9. $\{x^2 + 1, x^2 + x - 1, x^2 - 2x - 1\}$

10. $\left\{ \frac{1}{\sqrt{3}}(x^2 + x + 1), \frac{1}{\sqrt{2}}(x^2 - 1), \frac{1}{\sqrt{6}}(x^2 - 2x + 1) \right\}$

11. Write the vector

$$\mathbf{u} = \begin{bmatrix} 1 \\ 2 \\ 3 \end{bmatrix}$$

as a linear combination of the elements of the set given in Exercise 5.

12. Write the vector

$$\mathbf{u} = \begin{bmatrix} 2 \\ 3 \\ 5 \end{bmatrix}$$

as a linear combination of the elements in the set given in Exercise 5.

In Exercises 13–16 use the Gram–Schmidt process to find an orthonormal basis for the linear space spanned by the given set of vectors.

13. $\left\{ \begin{bmatrix} 1 \\ 0 \\ 1 \end{bmatrix}, \begin{bmatrix} 0 \\ 1 \\ 1 \end{bmatrix} \right\}$

14. $\left\{ \begin{bmatrix} 1 \\ 1 \\ 0 \end{bmatrix}, \begin{bmatrix} 1 \\ 1 \\ 1 \end{bmatrix} \right\}$

15. $\left\{ \begin{bmatrix} 1 \\ 0 \\ 1 \\ 0 \end{bmatrix}, \begin{bmatrix} 1 \\ 1 \\ 0 \\ 0 \end{bmatrix}, \begin{bmatrix} 1 \\ 0 \\ 0 \\ 1 \end{bmatrix} \right\}$

16. $\left\{ \begin{bmatrix} 1 \\ 1 \\ 0 \\ 1 \end{bmatrix}, \begin{bmatrix} 1 \\ 0 \\ 1 \\ 1 \end{bmatrix}, \begin{bmatrix} 0 \\ 1 \\ 1 \\ 0 \end{bmatrix} \right\}$

17. Given the basis $\left\{ \begin{bmatrix} 1 \\ 0 \end{bmatrix}, \begin{bmatrix} 0 \\ 1 \end{bmatrix} \right\}$ for R^2, find an orthonormal basis for R^2 with respect to the inner product given in Example 2 of Section 5.1.

18. (*For readers who have studied calculus.*) Given the basis $\{1, x, x^2\}$ for P_2, find an orthonormal basis for P_2 with respect to the inner product given in Example 4 of Section 5.1 with $a = 0$ and $b = 1$.

19. (*For readers who have studied calculus.*) Given the basis $\{1, x, x^2\}$ for P_2, find an orthonormal basis for P_2 with respect to the inner product given in Example 4 of Section 5.1 with $a = -1$ and $b = 1$.

20. Given the basis $\{1, x, x^2\}$ for P_2, find an orthonormal basis for P_2 with respect to the inner product given in Exercise 11 of Section 5.1.

21. Let $\{v_1, v_2, v_3\}$ be an orthonormal basis for an inner product space V. Show that $\|w\|^2 = \langle w, v_1 \rangle^2 + \langle w, v_2 \rangle^2 + \langle w, v_3 \rangle^2$ for every element w in V.

5.5 PROJECTION AND APPROXIMATION (OPTIONAL)

One of the important problems in functional and numerical analysis is to approximate an element **v** of an inner product space V as closely as possible by an element from a finite dimensional subspace W of V. By this we mean that we want to find an element **u** of W such that

$$\|v - u\| < \|v - w\|$$

for every element **w** of W different from **u**. As we will see, it is easy to find this element of W that "best" approximates **v**. To begin, we generalize the concept of orthogonal projection encountered in Section 3.3 to define the projection, $\text{proj}_W v$, of a vector **v** onto a subspace W of V. After investigating properties of $\text{proj}_W v$ we show that this element is the element of W that "best" approximates **v**.

Let $\{v_1, v_2, \ldots, v_n\}$ be an orthonormal basis for W. If **v** is any element of W, then we know (by Theorem 6 of Section 5.4) that

$$v = \langle v, v_1 \rangle v_1 + \langle v, v_2 \rangle v_2 + \cdots + \langle v, v_n \rangle v_n$$

Even if **v** is not an element of W, but an element of V, it is still worthwhile to consider the linear combination

$$\langle v, v_1 \rangle v_1 + \langle v, v_2 \rangle v_2 + \cdots + \langle v, v_n \rangle v_n$$

called the **projection of v on** W and denoted by $\text{proj}_W v$. The difference between **v** and its projection on W (i.e., $v - \text{proj}_W v$) is called the **component of v orthogonal to** W. The significance of this last definition is shown by the following theorem.

Theorem 8. *Let W be a finite dimensional subspace of an inner product space V. If **v** is any element of V and if $B = \{v_1, v_2, \ldots, v_n\}$ is an orthonormal basis of W, then the component of **v** orthogonal to W,*

$$v - \text{proj}_W v = v - (\langle v, v_1 \rangle v_1 + \langle v, v_2 \rangle v_2 + \cdots + \langle v, v_n \rangle v_n)$$

is orthogonal to every element of W.

Proof: We will prove the theorem in the case $n = 2$. The general case may be proved in a similar manner. Since B is a basis for W, every element **w** of W has the form

$$w = c_1 v_1 + c_2 v_2$$

$$\begin{aligned}
\langle v - \text{proj}_W v, w \rangle &= \langle v - \langle v, v_1 \rangle v_1 - \langle v, v_2 \rangle v_2, c_1 v_1 + c_2 v_2 \rangle \\
&= \langle v, c_1 v_1 + c_2 v_2 \rangle - \langle v, v_1 \rangle \langle v_1, c_1 v_1 + c_2 v_2 \rangle \\
&\quad - \langle v, v_2 \rangle \langle v_2, c_1 v_1 + c_2 v_2 \rangle \\
&= c_1 \langle v, v_1 \rangle + c_2 \langle v, v_2 \rangle - c_1 \langle v, v_1 \rangle \langle v_1, v_1 \rangle - c_2 \langle v, v_1 \rangle \langle v_1, v_2 \rangle \\
&\quad - c_1 \langle v, v_2 \rangle \langle v_2, v_1 \rangle - c_2 \langle v, v_2 \rangle \langle v_2, v_2 \rangle
\end{aligned}$$

(1)

Since B is orthonormal

$$\langle \mathbf{v}_i, \mathbf{v}_j \rangle = \begin{cases} 1 & i = j \\ 0 & i \neq j \end{cases}$$

It now follows directly from equation (1) that $\langle \mathbf{v} - \text{proj}_W \mathbf{v}, \mathbf{w} \rangle = 0$ for every $\mathbf{w}$ in W. This completes the proof. ∎

When discussing the Gram–Schmidt process in Section 5.4, we stated that the element

$$\mathbf{w}_k = \mathbf{u}_k - \langle \mathbf{u}_k, \mathbf{v}_1 \rangle \mathbf{v}_1 - \langle \mathbf{u}_k, \mathbf{v}_2 \rangle \mathbf{v}_2 - \cdots - \langle \mathbf{u}_k, \mathbf{v}_{k-1} \rangle \mathbf{v}_{k-1}$$

is orthogonal to every element of the linear space spanned by $\{\mathbf{v}_1, \mathbf{v}_2, \ldots, \mathbf{v}_k\}$. Theorem 8 shows this to be true.

Example 1.   Let W be the subspace of R^3 having $\{\mathbf{v}_1, \mathbf{v}_2\}$ as an orthonormal basis where

$$\mathbf{v}_1 = \frac{1}{\sqrt{2}} \begin{bmatrix} 1 \\ 0 \\ 1 \end{bmatrix} \qquad \mathbf{v}_2 = \frac{1}{\sqrt{3}} \begin{bmatrix} 1 \\ 1 \\ -1 \end{bmatrix}$$

The projection of

$$\mathbf{v} = \begin{bmatrix} 3 \\ 2 \\ 3 \end{bmatrix}$$

on W is given by

$$\text{proj}_W \mathbf{v} = \langle \mathbf{v}, \mathbf{v}_1 \rangle \mathbf{v}_1 + \langle \mathbf{v}, \mathbf{v}_2 \rangle \mathbf{v}_2$$

$$= \frac{6}{\sqrt{2}} \left(\frac{1}{\sqrt{2}} \begin{bmatrix} 1 \\ 0 \\ 1 \end{bmatrix} \right) + \frac{2}{\sqrt{3}} \left(\frac{1}{\sqrt{3}} \begin{bmatrix} 1 \\ 1 \\ -1 \end{bmatrix} \right)$$

$$= \frac{1}{3} \begin{bmatrix} 11 \\ 2 \\ 7 \end{bmatrix}$$

and the component of $\mathbf{v}$ orthogonal to W is given by

$$\mathbf{v} - \text{proj}_W \mathbf{v} = \begin{bmatrix} 3 \\ 2 \\ 3 \end{bmatrix} - \frac{1}{3} \begin{bmatrix} 11 \\ 2 \\ 7 \end{bmatrix}$$

$$= \frac{1}{3} \begin{bmatrix} -2 \\ 4 \\ 2 \end{bmatrix}$$

It is left for the reader to check that $\mathbf{v} - \text{proj}_W\mathbf{v}$ is in fact orthogonal to both $\mathbf{v}_1$ and $\mathbf{v}_2$.

Theorem 9. *Let W be a finite dimensional subspace of an inner product space V. If* $\mathbf{v}$ *is any element of V, then* $\text{proj}_W\mathbf{v}$ *is the best approximation to* $\mathbf{v}$ *from W in the sense that*

$$\|\mathbf{v} - \text{proj}_W\mathbf{v}\| < \|\mathbf{v} - \mathbf{w}\|$$

for every element $\mathbf{w}$ *of W different from* $\text{proj}_W\mathbf{v}$.

Proof: Let $\mathbf{w}$ be any element of W, then $\mathbf{v} - \mathbf{w}$ can be written as

$$\mathbf{v} - \mathbf{w} = (\mathbf{v} - \text{proj}_W\mathbf{v}) + (\text{proj}_W\mathbf{v} - \mathbf{w})$$

Since $\text{proj}_W\mathbf{v}$ and $\mathbf{w}$ are elements of W, their difference is also an element of W. By Theorem 8 the elements $\mathbf{v} - \text{proj}_W\mathbf{v}$ and $\text{proj}_W\mathbf{v} - \mathbf{w}$ are orthogonal. Using the generalization of the Pythagorean theorem found in Theorem 4 of Section 5.3, we have

$$\|\mathbf{v} - \mathbf{w}\|^2 = \|\mathbf{v} - \text{proj}_W\mathbf{v}\|^2 + \|\text{proj}_W\mathbf{v} - \mathbf{w}\|^2$$

If $\mathbf{w} \neq \text{proj}_W\mathbf{v}$, then $\|\text{proj}_W\mathbf{v} - \mathbf{w}\| > 0$ so that

$$\|\mathbf{v} - \mathbf{w}\|^2 > \|\mathbf{v} - \text{proj}_W\mathbf{v}\|^2$$

Recalling that the norm of any element is nonnegative, we conclude that

$$\|\mathbf{v} - \mathbf{w}\| > \|\mathbf{v} - \text{proj}_W\mathbf{v}\|$$

whenever $\mathbf{w} \neq \text{proj}_W\mathbf{v}$, which is clearly equivalent to the desired inequality.   ∎

Example 2.   By Example 1 the vector

$$\text{proj}_W\mathbf{v} = \frac{1}{3}\begin{bmatrix} 11 \\ 2 \\ 7 \end{bmatrix}$$

is the vector in the subspace W of R^3 spanned by

$$\left\{ \frac{1}{\sqrt{2}}\begin{bmatrix} 1 \\ 0 \\ 1 \end{bmatrix}, \frac{1}{\sqrt{3}}\begin{bmatrix} 1 \\ 1 \\ -1 \end{bmatrix} \right\}$$

that best approximates the vector

$$\mathbf{v} = \begin{bmatrix} 3 \\ 2 \\ 3 \end{bmatrix}$$

Example 3.   (*For readers who have studied calculus.*) One of the most commonly encountered approximation problems is that of approximating a continuous function on the interval $[-\pi, \pi]$ by a linear combination of functions of the form

$$f_0(x) = \frac{1}{\sqrt{2\pi}}$$

$$f_1(x) = \frac{1}{\sqrt{\pi}} \cos x \qquad f_2(x) = \frac{1}{\sqrt{\pi}} \sin x$$

$$f_3(x) = \frac{1}{\sqrt{\pi}} \cos 2x \qquad f_4(x) = \frac{1}{\sqrt{\pi}} \sin 2x$$

$$f_5(x) = \frac{1}{\sqrt{\pi}} \cos 3x \qquad f_6(x) = \frac{1}{\sqrt{\pi}} \sin 3x$$

$$\vdots \qquad\qquad \vdots$$

$$f_{2k-1}(x) = \frac{1}{\sqrt{\pi}} \cos kx \quad f_{2k}(x) = \frac{1}{\sqrt{\pi}} \sin kx$$

where k is a positive integer. The inner product space for this problem is the linear space V of all continuous functions on $[-\pi, \pi]$ endowed with the inner product

$$\langle f, g \rangle = \int_{-\pi}^{\pi} f(x)g(x)\, dx \tag{2}$$

(See Example 4 of Section 5.1.) The finite dimensional subspace is the subspace W spanned by $B = \{f_0, f_1, f_2, \dots, f_{2k}\}$.

Using the trigonometric identities

$$\sin a \cos b = \frac{1}{2}[\sin(a - b) + \sin(a + b)]$$

$$\cos a \cos b = \frac{1}{2}[\cos(a - b) + \cos(a + b)]$$

$$\sin a \sin b = \frac{1}{2}[\cos(a - b) - \cos(a + b)]$$

we can show that if m and n are positive integers, then

$$\int_{-\pi}^{\pi} \cos mx \sin nx\, dx = 0$$

$$\int_{-\pi}^{\pi} \cos mx \cos nx\, dx = \begin{cases} 0 & \text{if } m \neq n \\ \pi & \text{if } m = n \end{cases}$$

$$\int_{-\pi}^{\pi} \sin mx \sin nx\, dx = \begin{cases} 0 & \text{if } m \neq n \\ \pi & \text{if } m = n \end{cases}$$

Straightforward calculations using these integral identities show that B is an orthonormal set and as such is a basis for W. Hence, if f is any continuous function on $[-\pi, \pi]$, then

$$\mathrm{proj}_W f = \langle f, f_0 \rangle f_0 + \langle f, f_1 \rangle f_1 + \cdots + \langle f, f_{2k} \rangle f_{2k}$$
$$= a_0 + a_1 \cos x + b_1 \sin x + a_2 \cos 2x + b_2 \sin 2x$$
$$+ \cdots + a_k \cos kx + b_k \sin kx$$

(3)
$$= a_0 + \sum_{j=1}^{k} (a_j \cos jx + b_j \sin jx)$$

where

$$a_0 = \frac{1}{2\pi} \int_{-\pi}^{\pi} f(x)\, dx$$

$$a_j = \frac{1}{\pi} \int_{-\pi}^{\pi} f(x) \cos jx\, dx \qquad \text{for} \quad j = 1, 2, \dots, k$$

$$b_j = \frac{1}{\pi} \int_{-\pi}^{\pi} f(x) \sin jx\, dx \qquad \text{for} \quad j = 1, 2, \dots, k$$

The right side of equation (3) is called a **finite Fourier approximation** for f.

For example, if $f(x) = x$, then

$$a_0 = \frac{1}{2\pi} \int_{-\pi}^{\pi} x\, dx = 0$$

$$a_j = \frac{1}{\pi} \int_{-\pi}^{\pi} x \cos jx\, dx$$

$$= \frac{1}{\pi} \left(\frac{\cos jx}{j^2} + \frac{x \sin jx}{j} \right) \Big|_{-\pi}^{\pi}$$

$$= 0 \qquad \text{for} \quad j = 1, 2, \dots, k$$

$$a_j = \frac{1}{\pi} \int_{-\pi}^{\pi} x \sin jx\, dx$$

$$= \frac{1}{\pi} \left(\frac{\sin jx}{j^2} - \frac{x \cos jx}{j} \right) \Big|_{-\pi}^{\pi}$$

$$= -\frac{2}{j} \cos j\pi = \frac{2}{j}(-1)^{j+1} \qquad \text{for} \quad j = 1, 2, \dots, k$$

so that

$$\mathrm{proj}_W f = 2 \sin x - \sin 2x + \frac{2}{3} \sin 3x + \cdots + \frac{2}{k}(-1)^{k+1} \sin kx$$

$$= 2 \sum_{j=1}^{k} (-1)^{j+1} \frac{1}{j} \sin jx$$

It seems natural to expect the approximations to f given in equation (3) to improve as k increases. This is precisely what happens. In fact, it can be shown that

$$\lim_{k \to \infty} \left\| f(x) - \left[a_0 + \sum_{j=1}^{k} (a_j \cos jx + b_j \sin jx) \right] \right\| = 0$$

where the norm is the one determined by the inner product in equation (2). The series

$$a_0 + \sum_{j=1}^{\infty} (a_j \cos jx + b_j \sin jx)$$

is called the **Fourier series** for f on $[-\pi, \pi]$. Fourier series are used extensively in engineering, mathematical physics, and mathematics to solve problems of various types. Unfortunately it is not necessarily true that

$$f(x) = a_0 + \sum_{j=1}^{\infty} (a_j \cos jx + b_j \sin jx) \tag{4}$$

for all x in $[-\pi, \pi]$. However, if we restrict our attention to functions f having a continuous derivative on $[-\pi, \pi]$, then the identity in (4) does hold for all x in $[-\pi, \pi]$.

───────────────────── **Exercises** ─────────────────────

In Exercises 1–8 find the element of the subspace W of R^3 that best approximates v, where W is spanned by B. Notice that in Exercises 7–8 the sets B are not orthonormal.

1. $v = \begin{bmatrix} 1 \\ 2 \\ 3 \end{bmatrix}$ $\quad B = \left\{ \dfrac{1}{\sqrt{3}} \begin{bmatrix} 1 \\ -1 \\ 1 \end{bmatrix}, \dfrac{1}{\sqrt{6}} \begin{bmatrix} 1 \\ 2 \\ 1 \end{bmatrix} \right\}$

2. $v = \begin{bmatrix} 3 \\ -2 \\ 4 \end{bmatrix}$ $\quad B = \left\{ \dfrac{1}{\sqrt{3}} \begin{bmatrix} 1 \\ -1 \\ 1 \end{bmatrix}, \dfrac{1}{\sqrt{6}} \begin{bmatrix} 1 \\ 2 \\ 1 \end{bmatrix} \right\}$

3. $v = \begin{bmatrix} 2 \\ 5 \\ 1 \end{bmatrix}$ $\quad B = \left\{ \dfrac{1}{\sqrt{3}} \begin{bmatrix} 1 \\ 1 \\ 1 \end{bmatrix}, \dfrac{1}{\sqrt{2}} \begin{bmatrix} 0 \\ 1 \\ -1 \end{bmatrix} \right\}$

4. $v = \begin{bmatrix} -1 \\ 0 \\ 2 \end{bmatrix}$ $\quad B = \left\{ \dfrac{1}{\sqrt{3}} \begin{bmatrix} 1 \\ 1 \\ 1 \end{bmatrix}, \dfrac{1}{\sqrt{2}} \begin{bmatrix} 0 \\ 1 \\ -1 \end{bmatrix} \right\}$

5. $v = \begin{bmatrix} 3 \\ 5 \\ 2 \end{bmatrix}$ $\quad B = \left\{ \dfrac{1}{3\sqrt{5}} \begin{bmatrix} 2 \\ 5 \\ -4 \end{bmatrix} \right\}$

6. $v = \begin{bmatrix} -2 \\ 1 \\ 4 \end{bmatrix}$ $\quad B = \left\{ \dfrac{1}{\sqrt{2}} \begin{bmatrix} 1 \\ 0 \\ 1 \end{bmatrix} \right\}$

7. $v = \begin{bmatrix} 2 \\ 5 \\ 4 \end{bmatrix}$ $\quad B = \left\{ \begin{bmatrix} 1 \\ 1 \\ 0 \end{bmatrix}, \begin{bmatrix} 0 \\ 1 \\ 1 \end{bmatrix} \right\}$

8. $v = \begin{bmatrix} -1 \\ 2 \\ -3 \end{bmatrix}$ $\quad B = \left\{ \begin{bmatrix} 1 \\ 1 \\ 1 \end{bmatrix}, \begin{bmatrix} 1 \\ 0 \\ 1 \end{bmatrix} \right\}$

In Exercises 9–12 find the element of the subspace W of P_2 that best approximates $p(x)$, where W is spanned by B and the inner product on P_2 is defined by

$$\langle f, g \rangle = f(-1)g(-1) + f(0)g(0) + f(1)g(1)$$

(See Exercise 11 of Section 5.1.) Notice that in Exercises 11 and 12 the sets B are not orthonormal.

9. $p(x) = x + 1;$ $\quad B = \left\{ \dfrac{1}{\sqrt{2}} x, \dfrac{1}{\sqrt{2}} x^2 \right\}.$

10. $p(x) = x^2 + x + 3;$ $\quad B = \left\{ \dfrac{1}{\sqrt{2}} x, \dfrac{1}{\sqrt{2}} x^2 \right\}.$

11. $p(x) = x^2;$ $\quad B = \{ x + 1, x^2 - 1 \}.$

12. $p(x) = 3x + 4;$ $\quad B = \{ x + 1, x^2 - 1 \}.$

In Exercises 13–16 find the element of the space W of P_2 that best approximates $p(x)$, where W is spanned by B and the inner product on P_2 is defined by

$$\langle a_2 x^2 + a_1 x + a_0, b_2 x^2 + b_1 x + b_0 \rangle = a_2 b_2 + a_1 b_1 + a_0 b_0$$

(See Example 3 of Section 5.1.) Notice that in Exercises 15 and 16 the sets B are not orthonormal.

13. $p(x) = x + 1;$ $\quad B = \{1, x^2\}.$
14. $p(x) = x^2 + x + 3;$ $\quad B = \{1, x^2\}.$
15. $p(x) = x^2;$ $\quad B = \{ x + 1, x^2 - 1 \}.$
16. $p(x) = 3x + 4;$ $\quad B = \{ x + 1, x^2 - 1 \}.$
17. Compute the Fourier series for $f(x) = 2x + 3$.
18. Compute the Fourier series for $f(x) = e^x$.

5.6 THE METHOD OF LEAST SQUARES (OPTIONAL)*

Suppose that we have collected data that consists of the ordered pairs of numbers $(0, 1), (1, 3), (2, 5.5),$ and $(3, 8)$. When these pairs are plotted in the x, y-plane it appears that they may lie on a straight line (see Figure 5.3). This leads us to conjecture that

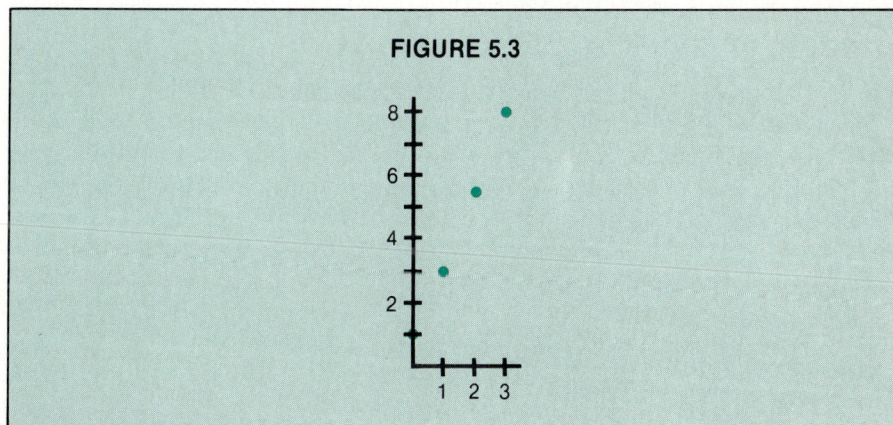

FIGURE 5.3

---

* This section requires material from Section 5.5.

there is an equation

$$y = ax + b$$

relating the components of the ordered pairs. That is, we hypothesize that there are numbers a and b such that

$$1 = a(0) + b = \quad\ b$$
$$3 = a(1) + b = \ a + b$$
$$5.5 = a(2) + b = 2a + b$$
$$8 = a(3) + b = 3a + b$$

This system of linear equations can be rewritten as

$$\begin{bmatrix} 0 & 1 \\ 1 & 1 \\ 2 & 1 \\ 3 & 1 \end{bmatrix} \begin{bmatrix} a \\ b \end{bmatrix} = \begin{bmatrix} 1 \\ 3 \\ 5.5 \\ 8 \end{bmatrix} \tag{1}$$

By Theorem 25 of Section 4.11 this equation has a solution if and only if the matrices

$$\begin{bmatrix} 0 & 1 \\ 1 & 1 \\ 2 & 1 \\ 3 & 1 \end{bmatrix} \quad \text{and} \quad \left[\begin{array}{cc|c} 0 & 1 & 1 \\ 1 & 1 & 3 \\ 2 & 1 & 5.5 \\ 3 & 1 & 8 \end{array}\right]$$

have the same rank. Using elementary row operations we find that these matrices are row equivalent to

$$\begin{bmatrix} 1 & 0 \\ 0 & 1 \\ 0 & 0 \\ 0 & 0 \end{bmatrix} \quad \text{and} \quad \left[\begin{array}{cc|c} 1 & 0 & 2 \\ 0 & 1 & 1 \\ 0 & 0 & .5 \\ 0 & 0 & 1 \end{array}\right]$$

Hence the matrices have ranks 2 and 3, respectively, so that equation (1) has no solution. This means that our data points do not lie on a straight line.

Even though the data points do not lie on a straight line, Figure 5.3 leads us to hypothesize that there are small errors in our data that prevent the points from lying on a straight line. Even if we assume that this is the case, there is no way of determining this theoretical line without knowing the errors in the data. However, there is a standard method of approximating the equation of this theoretical line.

We want to choose the numbers a and b so that

$$\begin{bmatrix} 0 & 1 \\ 1 & 1 \\ 2 & 1 \\ 3 & 1 \end{bmatrix} \begin{bmatrix} a \\ b \end{bmatrix}$$

is as close as possible to

$$\begin{bmatrix} 1 \\ 3 \\ 5.5 \\ 8 \end{bmatrix}$$

That is, we want to choose a and b so that

$$\left\| \begin{bmatrix} 0 & 1 \\ 1 & 1 \\ 2 & 1 \\ 3 & 1 \end{bmatrix} \begin{bmatrix} a \\ b \end{bmatrix} - \begin{bmatrix} 1 \\ 3 \\ 5.5 \\ 8 \end{bmatrix} \right\|$$

is as small as possible. We could proceed and compute these values of a and b. However, it is no more difficult to consider a general situation, and this is what we will do.

Let A be an $m \times n$ matrix and $\mathbf{b}$ be an m-vector. In Theorem 22 of Section 4.11 we found that the set $V = \{A\mathbf{x} : \mathbf{x}$ an n-vector$\}$ is a subspace of R^n. If $\mathbf{b}$ is an element of V, then the equation $A\mathbf{x} = \mathbf{b}$ is consistent. If $\mathbf{b}$ is not an element of V, the equation $A\mathbf{x} = \mathbf{b}$ is not consistent. If $A\mathbf{x} = \mathbf{b}$ is not consistent, it is frequently desirable to find an n-vector $\mathbf{y}$ that comes as "close as possible" to being a solution of $A\mathbf{x} = \mathbf{b}$. By "as close as possible" we mean that

(2) $$\|A\mathbf{y} - \mathbf{b}\| \le \|A\mathbf{x} - \mathbf{b}\|$$

for every n-vector $\mathbf{x}$. Since every element of V has the form $A\mathbf{x}$ we conclude from Theorem 9 of Section 5.5 that $A\mathbf{y}$ is the projection of $\mathbf{b}$ onto V. That is, $A\mathbf{y} = \text{proj}_V \mathbf{b}$. In order to determine $\mathbf{y}$ we recall that $\mathbf{b} - \text{proj}_V \mathbf{b}$ ($= \mathbf{b} - A\mathbf{y}$) is orthogonal to every element of V (Theorem 8 of Section 5.5). Then for every n-vector $\mathbf{x}$

$$\begin{aligned} 0 &= \langle A\mathbf{x}, \mathbf{b} - A\mathbf{y} \rangle \\ &= (A\mathbf{x})^t (\mathbf{b} - A\mathbf{y}) \\ &= \mathbf{x}^t A^t (\mathbf{b} - A\mathbf{y}) \\ &= \mathbf{x}^t (A^t \mathbf{b} - A^t A\mathbf{y}) \end{aligned}$$

so that

$$\begin{aligned} 0 &= [\mathbf{x}^t (A^t \mathbf{b} - A^t A\mathbf{y})]^t \\ &= (A^t \mathbf{b} - A^t A\mathbf{y})^t \mathbf{x} \end{aligned}$$

for every n-vector $\mathbf{x}$. By Exercise 26 of Section 4.4 we have

$$(A^t \mathbf{b} - A^t A\mathbf{y})^t = \mathbf{0}$$

Consequently

$$A^t \mathbf{b} - A^t A\mathbf{y} = \mathbf{0}$$

or equivalently

(3) $$A^t A\mathbf{y} = A^t \mathbf{b}$$

which is called the **normal equation** for $A\mathbf{x} = \mathbf{b}$. We have shown that any vector $\mathbf{y}$ satisfying the inequality in (2) is a solution of equation (3).

Notice that $A^t A$ is an $n \times n$ matrix. If $A^t A$ is nonsingular, then there is only one solution $\mathbf{y}$ of equation (3) and, hence, $\mathbf{y}$ is the only vector satisfying the inequality in (2). If $A^t A$ is singular, then it is more difficult to find the vector $\mathbf{y}$ that satisfies the inequality in (2). In most elementary applications $A^t A$ is nonsingular. Hence we will restrict our attention to this case.

If we set

$$A\mathbf{y} = \begin{bmatrix} y_1' \\ y_2' \\ \vdots \\ y_m' \end{bmatrix} \qquad A\mathbf{x} = \begin{bmatrix} x_1' \\ x_2' \\ \vdots \\ x_m' \end{bmatrix} \qquad \mathbf{b} = \begin{bmatrix} b_1 \\ b_2 \\ \vdots \\ b_m \end{bmatrix}$$

Then the inequality in (2) can be rewritten as

$$\sqrt{(y_1' - b_1)^2 + (y_2' - b_2)^2 + \cdots + (y_m' - b_m)^2} \\ \leq \sqrt{x_1' - b_1)^2 + (x_2' - b_2)^2 + \cdots + (x_m' - b_m)^2}$$

Thus the vector $\mathbf{y}$ is chosen so that the function

$$f(\mathbf{x}) = \sqrt{x_1' - b_1)^2 + (x_2' - b_2)^2 + \cdots + (x_n' - b_n)^2}$$

consisting of the sum of squares, assumes its least value when $\mathbf{x} = \mathbf{y}$. For this reason the solution $\mathbf{y}$ of equation (3) is called the **least squares solution** of $A\mathbf{x} = \mathbf{b}$.

Example 1.   Equation (1) can be rewritten in the form $A\mathbf{x} = \mathbf{b}$ if we choose

$$A = \begin{bmatrix} 0 & 1 \\ 1 & 1 \\ 2 & 1 \\ 3 & 1 \end{bmatrix}, \qquad b = \begin{bmatrix} 1 \\ 3 \\ 5.5 \\ 8 \end{bmatrix}$$

Since

$$A^t A = \begin{bmatrix} 14 & 6 \\ 6 & 4 \end{bmatrix} \quad \text{and} \quad A^t \mathbf{b} = \begin{bmatrix} 38 \\ 17.5 \end{bmatrix}$$

the normal equation for $A\mathbf{x} = \mathbf{b}$ is

$$\begin{bmatrix} 14 & 6 \\ 6 & 4 \end{bmatrix} \mathbf{y} = \begin{bmatrix} 38 \\ 17.5 \end{bmatrix} \tag{4}$$

which has

$$\mathbf{y} = \begin{bmatrix} 2.35 \\ .85 \end{bmatrix}$$

as its only solution. Thus

$$y = 2.35x + .85$$

is the equation for the straight line that comes closest (in the sense described above) to passing through the four data points. The following table and Figure 5.4 show how well this line approximates the data.

x	y = 2.35x + .85	Second Coordinate of the Data Point (x, y)
0	.85	1
1	3.20	3
2	5.55	5.5
3	7.9	8

FIGURE 5.4

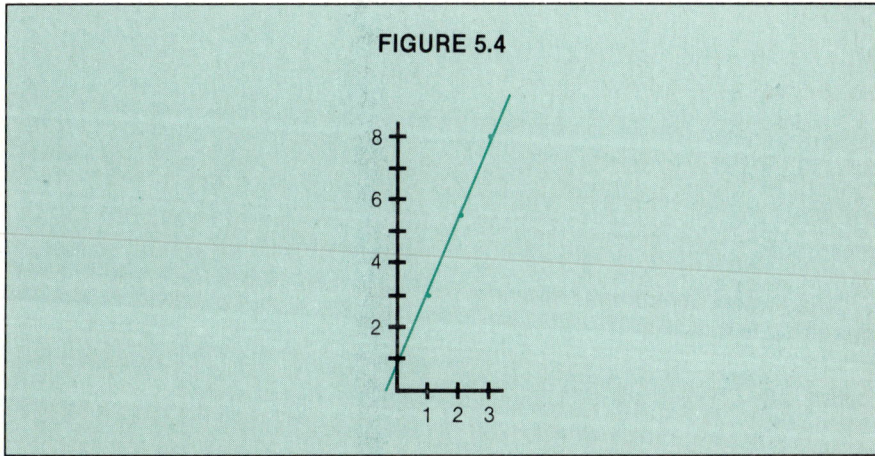

The example at the beginning of this section is typical of many problems that arise in applications. We are given m ordered pairs $(x_1, y_1), (x_2, y_2), \ldots, (x_m, y_m)$ and we wish to find a line that comes as close as possible to passing through all of the data points. That is, we want to choose numbers a and b that are as close as possible to a solution of the system of linear equations

$$ax_1 + b = y_1$$
$$ax_2 + b = y_2$$
$$\vdots \qquad \vdots \qquad \vdots$$
$$ax_m + b = y_m$$

which can be rewritten as

(5)
$$\begin{bmatrix} x_1 & 1 \\ x_2 & 1 \\ \vdots & \vdots \\ x_m & 1 \end{bmatrix} \begin{bmatrix} a \\ b \end{bmatrix} = \begin{bmatrix} y_1 \\ y_2 \\ \vdots \\ y_m \end{bmatrix}$$

The normal equation (5) is

$$
\begin{bmatrix} \sum_{i=1}^{m} x_i^2 & \sum_{i=1}^{m} x_i \\ \sum_{i=1}^{m} x_i & m \end{bmatrix} \begin{bmatrix} a \\ b \end{bmatrix} = \begin{bmatrix} \sum_{i=1}^{m} x_i y_i \\ \sum_{i=1}^{m} y_i \end{bmatrix}
\tag{6}
$$

Recalling that $A^{-1}\mathbf{b}$ is the solution of $A\mathbf{x} = \mathbf{b}$ whenever A is nonsingular, we use Theorem 11 of Section 1.8 to find that

$$
a = \frac{m \sum_{i=1}^{m} x_i y_i - \sum_{i=1}^{m} x_i \sum_{i=1}^{m} y_i}{m \sum_{i=1}^{m} x_i^2 - \left(\sum_{i=1}^{m} x_i \right)^2}
$$

$$
b = \frac{\sum_{i=1}^{m} x_i^2 \sum_{i=1}^{m} y_i - \sum_{i=1}^{m} x_i \sum_{i=1}^{m} x_i y_i}{m \sum_{i=1}^{m} x_i^2 - \left(\sum_{i=1}^{m} x_i \right)^2}
$$

whenever $m \sum_{i=1}^{m} x_i^2 - \left(\sum_{i=1}^{m} x_i \right)^2 \neq 0$. It is left as an exercise for the reader to show that the normal equation in (6) coincides with that in (4) when we consider the equation in (1).

Example 2.   Figures for the population of the United States at ten-year intervals from 1810 to 1910 are given in Table 1.

We will take the year 1810 to be $x = 0$ and measure x in years. When we plot the population data in the x, y-plane we see that the points are not close to lying on a straight line (see Figure 5.5). However, if we plot the logarithm of the population figures, then the data points appear to lie on a straight line (see Figure 5.6).

TABLE 1

Year	Population (in Millions)	Logarithm of Population (Rounded)
1810	7.2	1.97
1820	9.6	2.26
1830	12.8	2.55
1840	17.1	2.84
1850	22.7	3.12
1860	30.2	3.41
1870	40.3	3.70
1880	53.6	3.98
1890	71.4	4.27
1900	95.0	4.55
1910	126.0	4.84

FIGURE 5.5

Population
(in millions)

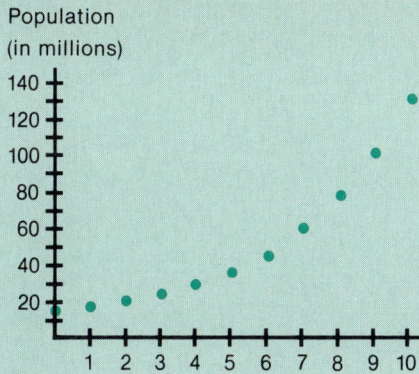

FIGURE 5.6

Logarithm
of population

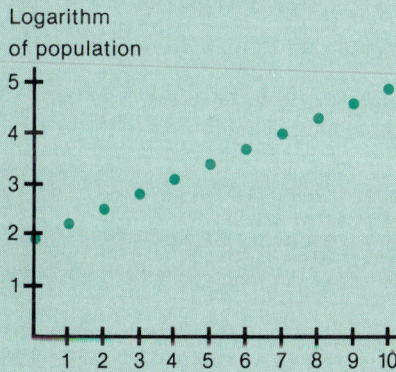

Using the logarithm of the population figures, equation (5) becomes

$$
\begin{bmatrix}
0 & 1 \\
10 & 1 \\
20 & 1 \\
30 & 1 \\
40 & 1 \\
50 & 1 \\
60 & 1 \\
70 & 1 \\
80 & 1 \\
90 & 1 \\
100 & 1
\end{bmatrix}
\begin{bmatrix} a \\ b \end{bmatrix}
=
\begin{bmatrix}
1.97 \\
2.26 \\
2.55 \\
2.84 \\
3.12 \\
3.41 \\
3.70 \\
3.98 \\
4.27 \\
4.55 \\
4.84
\end{bmatrix}
$$

TABLE 2

x	Year	P(x) (Rounded)	.02866x + 1.9750 (Rounded)
0	1810	7.2	1.97
10	1820	9.6	2.26
20	1830	12.8	2.55
30	1840	17.0	2.83
40	1850	22.7	3.12
50	1860	30.2	3.41
60	1870	40.2	3.69
70	1880	53.6	3.98
80	1890	71.4	4.27
90	1900	95.0	4.55
100	1910	126.6	4.84

and the corresponding normal equation is

$$\begin{bmatrix} 38,500 & 550 \\ 550 & 11 \end{bmatrix} \begin{bmatrix} a \\ b \end{bmatrix} = \begin{bmatrix} 2189.80 \\ 37.49 \end{bmatrix}$$

Solving this equation we find that (to five decimal places)

$$a = .02866 \qquad b = 1.9750$$

The last column of Table 2 shows how well the values of the function

$$y = .02866x + 1.9750$$

approximate the logarithm of the populations given in Table 1. This leads us to hypothesize that the population $P(x)$, x years after 1810, satisfies

$$\ln P(x) = .02866x + 1.9750$$

Therefore

$$P(x) = e^{.02866x + 1.9750}$$
$$= e^{1.9750} e^{.02866x}$$
$$= 7.2066 e^{.02866x}$$

Comparing the third column of Table 2 with the population figures in Table 1 we see that the function $P(x)$ gives excellent approximations to the population x years after 1810.

Exercises

In Exercises 1–6 find the least squares solution of the given equation.

1. $\begin{bmatrix} 1 & 2 \\ 2 & 4 \\ 3 & 5 \end{bmatrix} \mathbf{x} = \begin{bmatrix} 0 \\ 1 \\ 0 \end{bmatrix}$

2. $\begin{bmatrix} 0 & 2 \\ 3 & 1 \\ 1 & 2 \end{bmatrix} \mathbf{x} = \begin{bmatrix} 1 \\ 1 \\ 1 \end{bmatrix}$

3. $\begin{bmatrix} 1 & 1 \\ 2 & 1 \\ 3 & 1 \\ 4 & 1 \end{bmatrix} \mathbf{x} = \begin{bmatrix} 0 \\ 1 \\ 0 \\ 1 \end{bmatrix}$
4. $\begin{bmatrix} 2 & 1 \\ 4 & 1 \\ 6 & 1 \\ 8 & 1 \end{bmatrix} \mathbf{x} = \begin{bmatrix} 1 \\ 0 \\ 1 \\ 1 \end{bmatrix}$

5. $\begin{bmatrix} 1 & 2 & 3 \\ 0 & 1 & 4 \\ -2 & 1 & 3 \\ 1 & -2 & 1 \end{bmatrix} \mathbf{x} = \begin{bmatrix} 0 \\ 0 \\ 1 \\ 0 \end{bmatrix}$
6. $\begin{bmatrix} 3 & 0 & 4 \\ -2 & 2 & 3 \\ 0 & 1 & -2 \\ -1 & 2 & 0 \end{bmatrix} \mathbf{x} = \begin{bmatrix} 3 \\ 0 \\ 1 \\ 4 \end{bmatrix}$

7. U.S. population figures (in millions) in 1920, 1930, 1940, and 1950 were 106, 123.2, 132.2, and 151.3, respectively. Use the method illustrated in Example 2 to find a function of the form $P(x) = Ae^{bx}$ that approximates the population between 1920 and 1950.

8. U.S. population figures (in millions) in 1950, 1960, and 1970 were 151.3, 179.3, and 203.2, respectively. Use the method illustrated in Example 2 to find a function of the form $P(x) = Ae^{bx}$ that approximates the population between 1950 and 1970.

9. Suppose that the equation $A\mathbf{x} = \mathbf{b}$ has exactly one solution. Show that this solution is also a least squares solution for $A\mathbf{x} = \mathbf{b}$.

6

Linear Transformations

6.1 LINEAR TRANSFORMATIONS

In this chapter we study a special class of functions that map one linear space into another. This class of functions is absolutely fundamental for a wide range of problems in most branches of applied mathematics. We begin our study by recalling the definition of function.

A **function** f from a set A into a set B is a rule that associates with each element x of A a unique element $f(x)$ of B. The element $f(x)$ of B is called the **image** of x under f. We will denote that f is a function from A into B by $f : A \rightarrow B$ and say that f maps A into B.

In this chapter we are not interested in arbitrary functions from one set into another, but in a special class of functions from one linear space into another linear space. The functions in this class are those that are compatible with the arithmetic operations in each linear space. This is made precise in the following definition.

Definition 1.   A function $T : V \rightarrow W$ from a linear space V into a linear space W is called a **linear transformation** of V into W if

1. $T(\mathbf{u} + \mathbf{v}) = T(\mathbf{u}) + T(\mathbf{v})$ for all elements $\mathbf{u}$ and $\mathbf{v}$ of V.
2. $T(c\mathbf{u}) = cT(\mathbf{u})$ for all elements $\mathbf{u}$ of V and all scalars c.

Before considering examples of linear transformations, the reader should understand that the plus sign on the left side of the first equation denotes addition in the linear space V, while the plus sign on the right side denotes addition in the linear space W.

Example 1.   Consider the function $T: R^3 \to R^2$ defined by

$$T\left(\begin{bmatrix} a \\ b \\ c \end{bmatrix}\right) = \begin{bmatrix} 3a + c \\ b - 2c \end{bmatrix}$$

Let

$$\mathbf{u} = \begin{bmatrix} x_1 \\ x_2 \\ x_3 \end{bmatrix} \quad \text{and} \quad \mathbf{v} = \begin{bmatrix} y_1 \\ y_2 \\ y_3 \end{bmatrix}$$

be any elements of R^3. Then

$$T(\mathbf{u} + \mathbf{v}) = T\left(\begin{bmatrix} x_1 + y_1 \\ x_2 + y_2 \\ x_3 + y_3 \end{bmatrix}\right) = \begin{bmatrix} 3(x_1 + y_1) + (x_3 + y_3) \\ (x_2 + y_2) - 2(x_3 + y_3) \end{bmatrix}$$

Also

$$T(\mathbf{u}) + T(\mathbf{v}) = T\left(\begin{bmatrix} x_1 \\ x_2 \\ x_3 \end{bmatrix}\right) + T\left(\begin{bmatrix} y_1 \\ y_2 \\ y_3 \end{bmatrix}\right)$$

$$= \begin{bmatrix} 3x_1 + x_3 \\ x_2 - 2x_3 \end{bmatrix} + \begin{bmatrix} 3y_1 + y_3 \\ y_2 - 2y_3 \end{bmatrix}$$

$$= \begin{bmatrix} 3(x_1 + y_1) + (x_3 + y_3) \\ (x_2 + y_2) - 2(x_3 + y_3) \end{bmatrix}$$

so that

$$T(\mathbf{u} + \mathbf{v}) = T(\mathbf{u}) + T(\mathbf{v})$$

Furthermore

$$T(c\mathbf{u}) = T\left(\begin{bmatrix} cx_1 \\ cx_2 \\ cx_3 \end{bmatrix}\right)$$

$$= \begin{bmatrix} 3cx_1 + cx_3 \\ cx_2 - 2cx_3 \end{bmatrix}$$

$$= c\begin{bmatrix} 3x_1 + x_3 \\ x_2 - 2x_3 \end{bmatrix}$$

$$= cT(\mathbf{u})$$

Therefore T is a linear transformation of R^3 into R^2.

Example 2.   Let A be an $m \times n$ matrix and consider the function $T: R^n \to R^m$ defined by

$$T(\mathbf{x}) = A\mathbf{x}$$

for every n-vector $\mathbf{x}$. Then

$$T(\mathbf{x} + \mathbf{y}) = A(\mathbf{x} + \mathbf{y}) = A\mathbf{x} + A\mathbf{y} = T(\mathbf{x}) + T(\mathbf{y})$$

$$T(c\mathbf{x}) = A(c\mathbf{x}) = c(A\mathbf{x}) = cT(\mathbf{x})$$

Therefore T is a linear transformation.

Example 3.   Consider the function $R_\theta: R^2 \to R^2$ that rotates each point counterclockwise about the origin through an angle θ. We will show that R_θ is a linear transformation.

Let $\mathbf{u}$ and $\mathbf{v}$ be any 2-vectors, which we represent as in Figure 6.1a. When the vectors are rotated the entire triangle formed by $\mathbf{u}, \mathbf{v}$, and $\mathbf{u} + \mathbf{v}$ is rotated (see Figure 6.1b). Hence $R_\theta(\mathbf{u} + \mathbf{v}) = R_\theta(\mathbf{u}) + R_\theta(\mathbf{v})$.

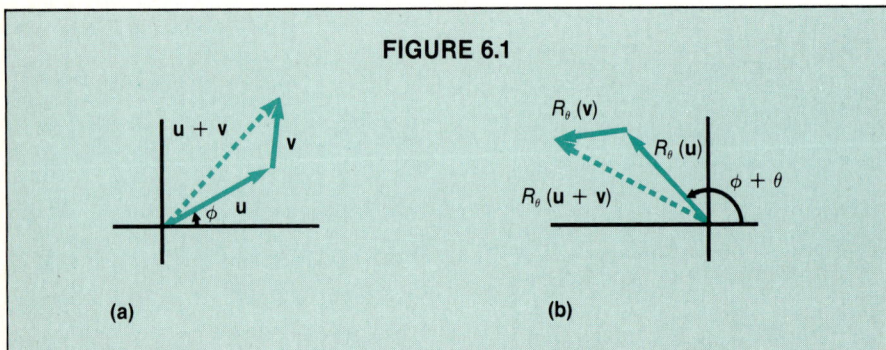

FIGURE 6.1

(a)   (b)

Let c be a positive number and $\mathbf{u}$ be a nonzero vector. Then $\mathbf{u}$ and $c\mathbf{u}$ have the same direction. Therefore the rotated vectors $R_\theta(\mathbf{u})$ and $R_\theta(c\mathbf{u})$ have the same direction. Moreover, since the rotated vector does not change length, we have

$$\| R_\theta(c\mathbf{u}) \| = \| c\mathbf{u} \| = |c| \, \| \mathbf{u} \| = |c| \, \| R_\theta(\mathbf{u}) \| = \| cR_\theta(\mathbf{u}) \|$$

Thus $R_\theta(c\mathbf{u})$ and $cR_\theta(\mathbf{u})$ have the same length and direction. Hence $R_\theta(c\mathbf{u}) = cR_\theta(\mathbf{u})$ whenever $c > 0$ and $\mathbf{u} \neq \mathbf{0}$. A similar argument shows that $R_\theta(c\mathbf{u}) = cR_\theta(\mathbf{u})$ whenever $0 > c$ and $\mathbf{u} \neq \mathbf{0}$. If $c = 0$ or $\mathbf{u} = \mathbf{0}$, then $c\mathbf{u} = \mathbf{0}$. The zero vector is not changed by a rotation about the origin. Therefore $R_\theta(c\mathbf{u}) = cR_\theta(\mathbf{u})$ whenever $c = 0$ or $\mathbf{u} = \mathbf{0}$. Thus for all c and all $\mathbf{u}$ in R^2 we have

$$R_\theta(c\mathbf{u}) = cR_\theta(\mathbf{u})$$

and R_θ is a linear transformation of R^2 into R^2.

Example 4.   Let L be a line that passes through the origin in the xy-plane. For each vector

$$\mathbf{u} = \begin{bmatrix} a \\ b \end{bmatrix}$$

let $T(\mathbf{u}) = \mathbf{u}'$ where

$$\mathbf{u}' = \begin{bmatrix} c \\ d \end{bmatrix}$$

is the vector such that L is the perpendicular bisector of the line segment connecting the points (a, b) and (c, d) (see Figure 6.2). The function T is called the **reflection through** L.

If $\mathbf{u}$ and $\mathbf{v}$ are any two 2-vectors, then the sum $\mathbf{u} + \mathbf{v}$ can be represented as a diagonal of a parallelogram (see Figure 6.3). The reflection through L reflects the entire parallelogram. Hence $T(\mathbf{u} + \mathbf{v}) = T(\mathbf{u}) + T(\mathbf{v})$. An argument similar to that used for rotations shows that $T(c\mathbf{u}) = cT(\mathbf{u})$ for all numbers c and all $\mathbf{u}$ in R^2. Hence the reflection through L is a linear transformation.

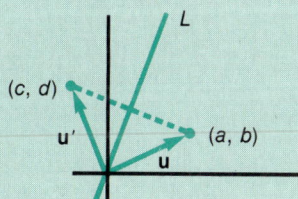

FIGURE 6.2

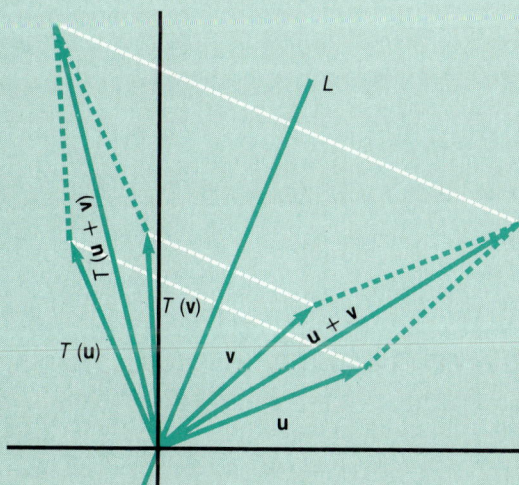

FIGURE 6.3

Example 5.   Consider the function $T: P_2 \to P_1$ defined by

$$T(ax^2 + bx + c) = 2ax + b$$

Let

$$f(x) = a_1 x^2 + b_1 x + c_1$$

$$g(x) = a_2 x^2 + b_2 x + c_2$$

be any elements of P_2. Then

$$T[f(x) + g(x)] = T[(a_1 + a_2)x^2 + (b_1 + b_2)x + (c_1 + c_2)]$$
$$= 2(a_1 + a_2)x + (b_1 + b_2)$$

$$T[f(x)] + T[g(x)] = (2a_1 x + b_1) + (2a_2 x + b_2)$$
$$= 2(a_1 + a_2)x + (b_1 + b_2)$$

so that $T[f(x) + g(x)] = T[f(x)] + T[g(x)]$. Also

$$T[cf(x)] = T(ca_1 x^2 + cb_1 x + cc_1)$$
$$= 2ca_1 x + cb_1$$
$$= c(2a_1 x + b_1)$$
$$= cT[f(x)].$$

Therefore T is a linear transformation of P_2 into P_1. The reader who has studied calculus should note that $T(f)$ is the derivative of f.

Example 6.   Consider the function $T: R^2 \to R^2$ defined by

$$T\left(\begin{bmatrix} x_1 \\ x_2 \end{bmatrix}\right) = \begin{bmatrix} x_1 + 1 \\ x_2 + 3 \end{bmatrix}$$

Then

$$T\left(c\begin{bmatrix} x_1 \\ x_2 \end{bmatrix}\right) = T\left(\begin{bmatrix} cx_1 \\ cx_2 \end{bmatrix}\right)$$
$$= \begin{bmatrix} cx_1 + 1 \\ cx_2 + 3 \end{bmatrix}$$

and

$$cT\left(\begin{bmatrix} x_1 \\ x_2 \end{bmatrix}\right) = c\begin{bmatrix} x_1 + 1 \\ x_2 + 3 \end{bmatrix}$$
$$= \begin{bmatrix} cx_1 + c \\ cx_2 + 3c \end{bmatrix}$$

Thus if $c \neq 1$, then

$$T\left(c\begin{bmatrix} x_1 \\ x_2 \end{bmatrix}\right) \neq cT\left(\begin{bmatrix} x_1 \\ x_2 \end{bmatrix}\right)$$

In particular

$$T\left(3\begin{bmatrix} 2 \\ 7 \end{bmatrix}\right) = \begin{bmatrix} 7 \\ 24 \end{bmatrix} \quad \text{and} \quad 3T\left(\begin{bmatrix} 2 \\ 7 \end{bmatrix}\right) = 3\begin{bmatrix} 3 \\ 10 \end{bmatrix} = \begin{bmatrix} 9 \\ 30 \end{bmatrix}$$

are not equal. Therefore T is not a linear transformation of R^2 into R^2.

Example 7. (*For readers who have studied calculus.*) Let V denote the linear space of all differentiable functions on an open interval (a, b) and let W denote the linear space of all functions on (a, b). Consider the function $D: V \to W$ defined by

$$D(f) = f'$$

where f' denotes the derivative of f. Then for any functions f and g in V

$$\begin{aligned} D(f + g) &= (f + g)' \\ &= f' + g' \\ &= D(f) + D(g) \end{aligned}$$

Also

$$D(cf) = (cf)' = cf' = cD(f)$$

Hence, D is a linear transformation of V into W.

Example 8. (*For readers who have studied calculus*). Let $C[a, b]$ denote the linear space of all continuous functions on an interval $[a, b]$ and let R denote the real numbers. Consider the function $I: C[a, b] \to R$ defined by

$$I(f) = \int_a^b f(x)\, dx$$

Then for any functions f and g in $C[a, b]$

$$\begin{aligned} I(f + g) &= \int_a^b (f + g)(x)\, dx \\ &= \int_a^b \lfloor f(x) + g(x)\rfloor\, dx \\ &= \int_a^b f(x)\, dx + \int_a^b g(x)\, dx \\ &= I(f) + I(g) \end{aligned}$$

Also

$$\begin{aligned} I(cf) &= \int_a^b (cf)(x)\, dx \\ &= \int_a^b cf(x)\, dx \\ &= c\int_a^b f(x)\, dx \\ &= cI(f) \end{aligned}$$

Hence, I is a linear transformation of $C[a, b]$ into R.

Example 9.   The function $f: R \to R$ defined by $f(x) = x^2$ is not a linear transformation because, in general, $(x + y)^2 \neq x^2 + y^2$. For example, $5 = 2 + 3$; but $f(5) = 5^2 = 25 \neq 13 = 2^2 + 3^2 = f(2) + f(3)$.

Example 10.   The function $T: R^2 \to R$ defined by $T(\mathbf{u}) = \|\mathbf{u}\|$ is not a linear transformation because, in general, $\|\mathbf{u} + \mathbf{v}\| \neq \|\mathbf{u}\| + \|\mathbf{v}\|$. For example, if

$$\mathbf{u} = \begin{bmatrix} 1 \\ 2 \end{bmatrix} \qquad \mathbf{v} = \begin{bmatrix} 3 \\ 4 \end{bmatrix}$$

then

$$\mathbf{u} + \mathbf{v} = \begin{bmatrix} 4 \\ 6 \end{bmatrix}$$

and we have $\|\mathbf{u} + \mathbf{v}\| = 2\sqrt{13} \neq \sqrt{5} + 5 = \|\mathbf{u}\| + \|\mathbf{v}\|$.

───────────────── **Exercises** ─────────────────

In Exercises 1–8 determine whether the given function $T: R^2 \to R^3$ is a linear transformation. If the function is not a linear transformation, show why it is not.

1. $T\left(\begin{bmatrix} x \\ y \end{bmatrix}\right) = \begin{bmatrix} 2x + y \\ x - 3y \end{bmatrix}$

2. $T\left(\begin{bmatrix} x \\ y \end{bmatrix}\right) = \begin{bmatrix} 3x + 4y \\ 6x + 7y \end{bmatrix}$

3. $T\left(\begin{bmatrix} x \\ y \end{bmatrix}\right) = \begin{bmatrix} 2x + 3y + 4 \\ 5x + 3y + 1 \end{bmatrix}$

4. $T\left(\begin{bmatrix} x \\ y \end{bmatrix}\right) = \begin{bmatrix} 5x^2 + y \\ 7x + 2y \end{bmatrix}$

5. $T\left(\begin{bmatrix} x \\ y \end{bmatrix}\right) = \begin{bmatrix} 4x - 6y \\ 9x + 8y \end{bmatrix}$

6. $T\left(\begin{bmatrix} x \\ y \end{bmatrix}\right) = \begin{bmatrix} x + 2y \\ 2x + y \end{bmatrix}$

7. $T\left(\begin{bmatrix} x \\ y \end{bmatrix}\right) = \begin{bmatrix} \cos x \\ \sin y \end{bmatrix}$

8. $T\left(\begin{bmatrix} x \\ y \end{bmatrix}\right) = \begin{bmatrix} x^2 + y^2 \\ xy \end{bmatrix}$

In Exercises 9–12 determine whether the given function $T: R^4 \to R^2$ is a linear transformation. If the function is not a linear transformation, show why it is not.

9. $T\left(\begin{bmatrix} a \\ b \\ c \\ d \end{bmatrix}\right) = \begin{bmatrix} a + b \\ c + d \end{bmatrix}$

10. $T\left(\begin{bmatrix} a \\ b \\ c \\ d \end{bmatrix}\right) = \begin{bmatrix} ad \\ bc \end{bmatrix}$

11. $T\left(\begin{bmatrix} a \\ b \\ c \\ d \end{bmatrix}\right) = \begin{bmatrix} ad - bc \\ 0 \end{bmatrix}$

12. $T\left(\begin{bmatrix} a \\ b \\ c \\ d \end{bmatrix}\right) = \begin{bmatrix} a + b + c + d \\ a - b + c - d \end{bmatrix}$

In Exercises 13–16 determine whether the given function $T: P_2 \to P_3$ is a linear transformation. If the function is not a linear transformation, show why it is not.

13. $T(ax^2 + bx + c) = 2ax + b$

14. $T(ax^2 + bx + c) = (1/3)ax^3 + (1/2)bx^2 + cx$

15. $T(ax^2 + bx + c) = a(x - 1)^2 + b(x - 1) + c$
16. $T(ax^2 + bx + c) = a + bx + cx^2$
17. Let $T: R^2 \to R^2$ be the function defined by

$$T\left(\begin{bmatrix} x \\ y \end{bmatrix}\right) = \begin{bmatrix} x \\ -y \end{bmatrix}$$

Show that T is a linear transformation. Interpret the effect of T geometrically.
18. Let $T: R^2 \to R^2$ be the function defined by

$$T\left(\begin{bmatrix} x \\ y \end{bmatrix}\right) = \begin{bmatrix} -x \\ y \end{bmatrix}$$

Show that T is a linear transformation. Interpret the effect of T geometrically.

6.2 AN EXAMPLE CONCERNING THE SPECIAL THEORY OF RELATIVITY (OPTIONAL)

We begin this section by considering an example illustrating a basic premise of Newtonian mechanics. Namely, velocities are relative to the observer. If a particle is fired from a moving vehicle in the direction of motion of that vehicle, then the velocity of the projectile as measured by a stationary observer is the sum of the velocity, v_1, of the vehicle relative to the observer and the velocity, v_2, of the particle relative to the vehicle. However, the velocity of the same particle as measured by an observer on the vehicle is v_2. Thus the two observers measure different velocities for the projectile. We now examine in detail how observers in motion with respect to each other view an event.

Suppose that an event is witnessed by two observers. One of the observers is in a coordinate system S and finds that the event took place at a point (x, y, z) at time t. The other is in a different coordinate system S' that is moving with constant velocity v in the direction of the positive x-axis. (Figure 6.4 depicts two such coordinate systems.) This second observer finds that the event took place at point (x', y', z') at

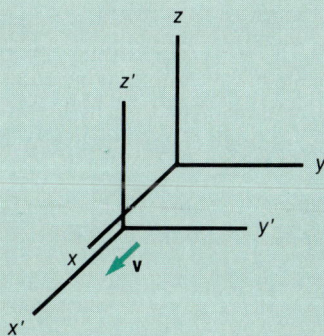

FIGURE 6.4

time t'. Further suppose that time is measured in both systems in the same manner beginning when the origins of the two coordinate systems coincided.

In accordance with traditional reasoning, Newtonian mechanics, the x measurement made in S at time t exceeds similar measurements in S' by an amount vt since vt is the distance traveled by S' relative to S in time t. Hence $x' = x - vt$. Since there is no motion of S' relative to S in the y and z directions, $y' = y$ and $z' = z$. Also time measured in S is the same as that in S'. The equations

$$x' = x - vt$$

$$y' = y$$

$$z' = z$$

$$t' = t$$

or equivalently

$$\begin{bmatrix} x' \\ y' \\ z' \\ t' \end{bmatrix} = \begin{bmatrix} 1 & 0 & 0 & -v \\ 0 & 1 & 0 & 0 \\ 0 & 0 & 1 & 0 \\ 0 & 0 & 0 & 1 \end{bmatrix} \begin{bmatrix} x \\ y \\ z \\ t \end{bmatrix}$$

relate the measurements in S to those in S'. This is called the **Galilean transformation** of coordinates.

In 1887 A. A. Michelson and E. W. Morley reported an experiment that challenged this view of motion and with it the foundations of physics. They showed that the velocity of light is the same for all observers regardless of their motion relative to the path of the light. This and work by several other people led Albert Einstein in 1905 to base his special theory of relativity on the following two postulates:[*]

1. The laws of physics can be expressed by the same set of equations for all frames of reference that are moving at constant velocity relative to each other.
2. The speed of light in a vacuum has the same value for all observers.

The first postulate, the principle of relativity, is also basic to Newtonian mechanics. The second postulate is seemingly contradictory to both Newtonian mechanics and the first postulate. Einstein showed that in fact these two postulates yield a consistent theory of the physical universe.

We now investigate how these two postulates affect our view of the moving coordinate systems S and S'. Since $x' = x - vt$ in Newtonian mechanics, any change based on the above postulates must yield an identity that is nearly $x' = x - vt$ whenever the relative velocity is small. The simplest case is when

$$x' = k(x - vt) \tag{1}$$

where k is a positive function of v and possibly x and t.

---

[*] A. Einstein, *The Meaning of Relativity*, 5th ed. Princeton, N.J.: Princeton University Press, 1955, pp. 25, 27.

By the first postulate, the measurements made in S' relative to S must have the same form as in (1) except for the sign of the velocity:

$$(2) \qquad\qquad x = k(x' + vt')$$

As before, $y' = y$, and $z' = z$; but we now have no reason to believe $t' = t$.

If we substitute the value of x' in equation (1) into equation (2) we find that $x = k^2(x - vt) + kvt'$. Solving this equation for t' yields

$$(3) \qquad\qquad t' = kt + \frac{1 - k^2}{kv}x$$

To evaluate the function k we use the second postulate. Since both observers determine the same speed c of light, the x-coordinate and x'-coordinate of an impulse of light emitted at $t = 0$ from the common origin of S and S' must satisfy

$$(4) \qquad\qquad x = ct$$

$$(5) \qquad\qquad x' = ct'$$

If we multiply each side of equation (3) by c and use equations (5) and (1) we obtain

$$k(x - vt) = ckt + \frac{1 - k^2}{kv}cx$$

By equation (4) we can replace x by ct so that

$$k(ct - vt) = ckt + \frac{1 - k^2}{kv}c^2t$$

or equivalently

$$-kv = \frac{1 - k^2}{kv}c^2$$

If we multiply each side of this equation by kv and rearrange terms, we obtain

$$(c^2 - v^2)k^2 = c^2$$

so that

$$k^2 = \frac{c^2}{c^2 - v^2} = \frac{1}{1 - (v/c)^2}$$

Recalling that k is positive we have

$$k = \frac{1}{\sqrt{1 - (v/c)^2}}$$

When this value for k is inserted into equations (1) and (3) we obtain the equations

$$x' = \frac{x - vt}{\sqrt{1 - (v/c)^2}}$$

$$y' = y$$

$$z' = z$$

$$t' = kt + \frac{1 - k^2}{kv} x$$

$$= kt + \frac{kx}{v}\left(\frac{1}{k^2} - 1\right)$$

$$= kt + \frac{kx}{v}\left\{\left[1 - \left(\frac{v}{c}\right)^2\right] - 1\right\}$$

$$= kt - \frac{kvx}{c^2}$$

$$= \frac{t - vx/c^2}{\sqrt{1 - (v/c)^2}}$$

or equivalently,

$$\begin{bmatrix} x' \\ y' \\ z' \\ t' \end{bmatrix} = \begin{bmatrix} \dfrac{1}{\sqrt{1 - (v/c)^2}} & 0 & 0 & \dfrac{-v}{\sqrt{1 - (v/c)^2}} \\ 0 & 1 & 0 & 0 \\ 0 & 0 & 1 & 0 \\ \dfrac{-v/c^2}{\sqrt{1 - (v/c)^2}} & 0 & 0 & \dfrac{1}{\sqrt{1 - (v/c)^2}} \end{bmatrix} \begin{bmatrix} x \\ y \\ z \\ t \end{bmatrix}$$

which relate the measurements in S to those in S'. This is called the **Lorentz transformation** in honor of the Dutch physicist H. A. Lorentz, who in 1903 used this transformation in his theory of electromagnetism.

Notice that both the Galilean transformation and the Lorentz transformation are linear transformations.

6.3 BASIC PROPERTIES OF LINEAR TRANSFORMATIONS

The following theorem contains the basic properties of linear transformations upon which the remainder of this chapter is based. In particular, the third and fourth parts will be used repeatedly in the sections to come.

Theorem 1. *Let $T: V \to W$ be a linear transformation of a linear space V into a linear space W. Then*

1. $T(\mathbf{0}) = \mathbf{0}$
2. $T(-\mathbf{v}) = -T(\mathbf{v})$ *for all $\mathbf{v}$ in V.*
3. $T(a\mathbf{u} + b\mathbf{v}) = aT(\mathbf{u}) + bT(\mathbf{v})$ *for all $\mathbf{u}$ and $\mathbf{v}$ in V and all scalars a and b.*
4. *For each positive integer k, $T(c_1\mathbf{v}_1 + c_2\mathbf{v}_2 + \cdots + c_k\mathbf{v}_k) = c_1 T(\mathbf{v}_1) + c_2 T(\mathbf{v}_2) + \cdots + c_k T(\mathbf{v}_k)$ for all $\mathbf{v}_1, \mathbf{v}_2, \dots, \mathbf{v}_k$ in V and all scalars $c_1, c_2, \dots, c_k$.*

Proof: For any scalar c we have $T(c\mathbf{v}) = cT(\mathbf{v})$ for all $\mathbf{v}$ in V. The first two properties are obtained by first setting $c = 0$ and then setting $c = -1$. Using both axioms for a linear transformation we have

$$T(a\mathbf{u} + b\mathbf{v}) = T(a\mathbf{u}) + T(b\mathbf{v})$$
$$= aT(\mathbf{u}) + bT(\mathbf{v})$$

for all **u** and **v** in V and all scalars a and b. This establishes the third property. Repeated use of this property yields the fourth property. ∎

As illustrated by the following examples, if the effect of a linear transformation $T: V \rightarrow W$ is known on the elements of a basis for V, then the effect of T on any element of V can be determined.

Example 1.   Suppose that $T: R^2 \rightarrow R^2$ is a linear transformation such that

$$T\left(\begin{bmatrix} 1 \\ 0 \end{bmatrix}\right) = \begin{bmatrix} 1 \\ 2 \end{bmatrix} \qquad T\left(\begin{bmatrix} 0 \\ 1 \end{bmatrix}\right) = \begin{bmatrix} 3 \\ 4 \end{bmatrix}$$

Then for any element $\begin{bmatrix} a \\ b \end{bmatrix}$ of R^2 we have

$$\begin{aligned} T\left(\begin{bmatrix} a \\ b \end{bmatrix}\right) &= T\left(a\begin{bmatrix} 1 \\ 0 \end{bmatrix} + b\begin{bmatrix} 0 \\ 1 \end{bmatrix}\right) \\ &= aT\left(\begin{bmatrix} 1 \\ 0 \end{bmatrix}\right) + bT\left(\begin{bmatrix} 0 \\ 1 \end{bmatrix}\right) \\ &= a\begin{bmatrix} 1 \\ 2 \end{bmatrix} + b\begin{bmatrix} 3 \\ 4 \end{bmatrix} \\ &= \begin{bmatrix} a + 3b \\ 2a + 4b \end{bmatrix} \end{aligned}$$

In particular,

$$T\left(\begin{bmatrix} 6 \\ -5 \end{bmatrix}\right) = \begin{bmatrix} 6 + 3(-5) \\ 2(6) + 4(-5) \end{bmatrix} = \begin{bmatrix} -9 \\ -8 \end{bmatrix}$$

Notice that

$$\begin{bmatrix} 1 & 3 \\ 2 & 4 \end{bmatrix}\begin{bmatrix} a \\ b \end{bmatrix} = \begin{bmatrix} a + 3b \\ 2a + 4b \end{bmatrix}$$

so that

$$T\left(\begin{bmatrix} a \\ b \end{bmatrix}\right) = \begin{bmatrix} 1 & 3 \\ 2 & 4 \end{bmatrix}\begin{bmatrix} a \\ b \end{bmatrix}$$

for every $\begin{bmatrix} a \\ b \end{bmatrix}$ in R^2.

Example 2.   Suppose that $T: R^2 \rightarrow R^3$ is a linear transformation such that

$$T\left(\begin{bmatrix} 1 \\ 2 \end{bmatrix}\right) = \begin{bmatrix} 1 \\ 2 \\ 3 \end{bmatrix} \qquad T\left(\begin{bmatrix} -1 \\ 2 \end{bmatrix}\right) = \begin{bmatrix} -3 \\ 0 \\ 5 \end{bmatrix}$$

Evidently $\left\{\begin{bmatrix} 1 \\ 2 \end{bmatrix}, \begin{bmatrix} -1 \\ 2 \end{bmatrix}\right\}$ is linearly independent and, therefore, is a basis for R^2

(Why?). If $\begin{bmatrix} a \\ b \end{bmatrix}$ is any element of R^2, then

$$T\left(\begin{bmatrix} a \\ b \end{bmatrix}\right) = T\left(\frac{b + 2a}{4}\begin{bmatrix} 1 \\ 2 \end{bmatrix} + \frac{b - 2a}{4}\begin{bmatrix} -1 \\ 2 \end{bmatrix}\right)$$

$$= \frac{b + 2a}{4} T\left(\begin{bmatrix} 1 \\ 2 \end{bmatrix}\right) + \frac{b - 2a}{4} T\left(\begin{bmatrix} -1 \\ 2 \end{bmatrix}\right)$$

$$= \frac{b + 2a}{4}\begin{bmatrix} 1 \\ 2 \\ 3 \end{bmatrix} + \frac{b - 2a}{4}\begin{bmatrix} -3 \\ 0 \\ 5 \end{bmatrix}$$

$$= \begin{bmatrix} 2a - \frac{1}{2}b \\ a + \frac{1}{2}b \\ -a + 2b \end{bmatrix}$$

In particular,

$$T\left(\begin{bmatrix} 3 \\ 4 \end{bmatrix}\right) = \begin{bmatrix} 2(3) - \frac{1}{2}(4) \\ 3 + \frac{1}{2}(4) \\ -3 + 2(4) \end{bmatrix} = \begin{bmatrix} 4 \\ 5 \\ 5 \end{bmatrix}$$

Notice that

$$\begin{bmatrix} 2 & -\frac{1}{2} \\ 1 & \frac{1}{2} \\ -1 & 2 \end{bmatrix}\begin{bmatrix} a \\ b \end{bmatrix} = \begin{bmatrix} 2a - \frac{1}{2}b \\ a + \frac{1}{2}b \\ -a + 2b \end{bmatrix}$$

so that

$$T\left(\begin{bmatrix} a \\ b \end{bmatrix}\right) = \begin{bmatrix} 2 & -\frac{1}{2} \\ 1 & \frac{1}{2} \\ -1 & 2 \end{bmatrix}\begin{bmatrix} a \\ b \end{bmatrix}$$

for every $\begin{bmatrix} a \\ b \end{bmatrix}$ in R^2.

In the previous two examples we saw that the given linear transformation can be represented by matrix multiplication. In the next section we shall see that every linear transformation from a finite dimensional linear space V into a linear space W can be represented by matrix multiplication.

─────────────── **Exercises** ───────────────

1. Let $T: R^2 \to R^2$ be a linear transformation such that

$$T\left(\begin{bmatrix} 1 \\ 0 \end{bmatrix}\right) = \begin{bmatrix} 1 \\ 3 \end{bmatrix} \quad \text{and} \quad T\left(\begin{bmatrix} 0 \\ 1 \end{bmatrix}\right) = \begin{bmatrix} -2 \\ 4 \end{bmatrix}$$

(a) Find $T\left(\begin{bmatrix} -2 \\ 4 \end{bmatrix}\right)$

(b) Find $T\left(\begin{bmatrix} x \\ y \end{bmatrix}\right)$

(c) Find a matrix A such that

$$T\left(\begin{bmatrix} x \\ y \end{bmatrix}\right) = A\begin{bmatrix} x \\ y \end{bmatrix}$$

2. Let $T: R^3 \to R^2$ be a linear transformation such that

$$T\left(\begin{bmatrix} 1 \\ 0 \\ 0 \end{bmatrix}\right) = \begin{bmatrix} 1 \\ 0 \\ 1 \end{bmatrix} \quad T\left(\begin{bmatrix} 0 \\ 1 \\ 0 \end{bmatrix}\right) = \begin{bmatrix} 2 \\ 3 \\ 4 \end{bmatrix} \quad T\left(\begin{bmatrix} 0 \\ 0 \\ 1 \end{bmatrix}\right) = \begin{bmatrix} -4 \\ 3 \\ -5 \end{bmatrix}$$

(a) Find $T\left(\begin{bmatrix} 7 \\ 9 \\ 0 \end{bmatrix}\right)$

(b) Find $T\left(\begin{bmatrix} x \\ y \\ z \end{bmatrix}\right)$

(c) Find a matrix A such that $T\left(\begin{bmatrix} x \\ y \\ z \end{bmatrix}\right) = A\begin{bmatrix} x \\ y \\ z \end{bmatrix}$

3. Let $T: R^2 \to R^3$ be a linear transformation such that

$$T\left(\begin{bmatrix} 1 \\ 1 \end{bmatrix}\right) = \begin{bmatrix} 0 \\ 4 \\ 3 \end{bmatrix} \quad T\left(\begin{bmatrix} 1 \\ -1 \end{bmatrix}\right) = \begin{bmatrix} 2 \\ 5 \\ 1 \end{bmatrix}$$

(a) Find $T\left(\begin{bmatrix} 7 \\ 6 \end{bmatrix}\right)$

(b) Find $T\left(\begin{bmatrix} x \\ y \end{bmatrix}\right)$

(c) Find a matrix A such that

$$T\left(\begin{bmatrix} x \\ y \\ z \end{bmatrix}\right) = A \begin{bmatrix} x \\ y \\ z \end{bmatrix}$$

4. Let $T: R^3 \to R^2$ be a linear transformation such that

$$T\left(\begin{bmatrix} 1 \\ 1 \\ 0 \end{bmatrix}\right) = \begin{bmatrix} 2 \\ 1 \end{bmatrix} \quad T\left(\begin{bmatrix} 0 \\ 1 \\ 1 \end{bmatrix}\right) = \begin{bmatrix} -2 \\ 3 \end{bmatrix} \quad T\left(\begin{bmatrix} 1 \\ 0 \\ 1 \end{bmatrix}\right) = \begin{bmatrix} 4 \\ 5 \end{bmatrix}$$

(a) Find $T\left(\begin{bmatrix} 1 \\ 2 \\ 3 \end{bmatrix}\right)$

(b) Find $T\left(\begin{bmatrix} x \\ y \\ z \end{bmatrix}\right)$

(c) Find a matrix A such that

$$T\left(\begin{bmatrix} x \\ y \\ z \end{bmatrix}\right) = A \begin{bmatrix} x \\ y \\ z \end{bmatrix}$$

5. Let $T: P_2 \to P_2$ be a linear transformation such that $T(x^2) = 2x + 4$; $T(x) = x^2 - 3$; and $T(1) = -3x + 5$.
  (a) Find $T(2x^2 + 3x - 7)$.
  (b) Find $T(ax^2 + bx + c)$.
6. Let $T: P_2 \to P_2$ be a linear transformation such that $T(x^2) = 3x^2 + 4$; $T(x) = x - 4$; and $T(1) = x^2 + x$.
  (a) Find $T(5x^2 - 2x + 3)$.
  (b) Find $T(ax^2 + bx + c)$.

In Exercises 7–12 let $T: V \to W$ be a linear transformation of a linear space V into a linear space W.

7. Show that the set $\{w : w = T(v) \text{ for some } v \text{ in } V\}$ is a subspace of W.
8. Show that the set $\{v : T(v) = 0\}$ is a subspace of V.
9. Let $\{v_1, v_2, \ldots, v_k\}$ be a linearly dependent subset of V. Show that $\{T(v_1), T(v_2), \ldots, T(v_k)\}$ is a linearly dependent subset of W.
10. Suppose that $v = 0$ is the only element of V such that $T(v) = 0$. Let $\{v_1, v_2, \ldots, v_k\}$ be a linearly independent subset of V. Show that $\{T(v_1), T(v_2), \ldots, T(v_k)\}$ is a linearly independent subset of W.
11. Suppose that $\{v_1, v_2, \ldots, v_k\}$ is a basis for V and that $T(v_1) = 0$, $T(v_2) = 0, \ldots, T(v_k) = 0$. Show that $T(v) = 0$ for every v in V.
12. Suppose that $\{v_1, v_2, \ldots, v_k\}$ is a basis for V and that $T(v_1) = v_1, T(v_2) = v_2, \ldots, T(v_k) = v_k$. Show that $T(v) = v$ for every v in V.

6.4 REPRESENTATIONS OF LINEAR TRANSFORMATIONS, I

The examples in the first two sections of this chapter show that linear transformations can be used to describe many useful mathematical operations. In several of these examples it is possible to represent the linear transformation in a simple fashion: namely by matrix multiplication. In this section we shall learn how to obtain this matrix representation. We begin by considering the case in which the linear transformation is from R^n to R^m.

Theorem 2. *Let $T: R^n \to R^m$ be a linear transformation of R^n into R^m. If A is the matrix whose columns are*

$$T\left(\begin{bmatrix} 1 \\ 0 \\ 0 \\ \vdots \\ 0 \end{bmatrix}\right), \quad T\left(\begin{bmatrix} 0 \\ 1 \\ 0 \\ \vdots \\ 0 \end{bmatrix}\right), \quad \dots, \quad T\left(\begin{bmatrix} 0 \\ 0 \\ 0 \\ \vdots \\ 1 \end{bmatrix}\right)$$

then $T(\mathbf{x}) = A\mathbf{x}$ for every $\mathbf{x}$ in R^n. Moreover, if A' is an $m \times n$ matrix such that $T(\mathbf{x}) = A'\mathbf{x}$ for every $\mathbf{x}$ in R^n, then $A' = A$.

Proof: Let A be the matrix described in the statement of the theorem and let

$$\mathbf{x} = \begin{bmatrix} x_1 \\ x_2 \\ \vdots \\ x_n \end{bmatrix}$$

Using part 4 of Theorem 1 and Theorem 4 of Section 4.4, we have

$$T(\mathbf{x}) = T\left(x_1 \begin{bmatrix} 1 \\ 0 \\ 0 \\ \vdots \\ 0 \end{bmatrix} + x_2 \begin{bmatrix} 0 \\ 1 \\ 0 \\ \vdots \\ 0 \end{bmatrix} + \dots + x_n \begin{bmatrix} 0 \\ 0 \\ 0 \\ \vdots \\ 1 \end{bmatrix}\right)$$

$$= x_1 T\left(\begin{bmatrix} 1 \\ 0 \\ 0 \\ \vdots \\ 0 \end{bmatrix}\right) + x_2 T\left(\begin{bmatrix} 0 \\ 1 \\ 0 \\ \vdots \\ 0 \end{bmatrix}\right) + \dots + x_n T\left(\begin{bmatrix} 0 \\ 0 \\ 0 \\ \vdots \\ 1 \end{bmatrix}\right)$$

$$= A\mathbf{x}$$

If A' is an $m \times n$ matrix such that $T(\mathbf{x}) = A'\mathbf{x}$ for every $\mathbf{x}$ in R^n, then $A\mathbf{x} = A'\mathbf{x}$ for all $\mathbf{x}$ in R^n. By Exercise 25 of Section 4.4 the matrices A and A' are identical. ∎

Definition 2.   Let $T: R^n \to R^m$ be a linear transformation. The $m \times n$ matrix A such that $T(\mathbf{x}) = A\mathbf{x}$ for every $\mathbf{x}$ in R^n is called the **standard matrix** for T.

Example 1.   In Example 1 of Section 6.1 we found that the function $T: R^3 \to R^2$ defined by

$$T\left(\begin{bmatrix} x_1 \\ x_2 \\ x_3 \end{bmatrix}\right) = \begin{bmatrix} 3x_1 + x_3 \\ x_2 - 2x_3 \end{bmatrix}$$

is a linear transformation. Since

$$T\left(\begin{bmatrix} 1 \\ 0 \\ 0 \end{bmatrix}\right) = \begin{bmatrix} 3 \\ 0 \end{bmatrix}, \quad T\left(\begin{bmatrix} 0 \\ 1 \\ 0 \end{bmatrix}\right) = \begin{bmatrix} 0 \\ 1 \end{bmatrix}, \quad T\left(\begin{bmatrix} 0 \\ 0 \\ 1 \end{bmatrix}\right) = \begin{bmatrix} 1 \\ -2 \end{bmatrix}$$

the standard matrix for T is

$$A = \begin{bmatrix} 3 & 0 & 1 \\ 0 & 1 & -2 \end{bmatrix}$$

Example 2.   In Example 3 of Section 6.1 we found that the function R_θ: $R^2 \to R^2$ that rotates each point counterclockwise about the origin through an angle θ is a linear transformation. From Figure 6.5 we see that

$$R_\theta\left(\begin{bmatrix} 1 \\ 0 \end{bmatrix}\right) = \begin{bmatrix} \cos\theta \\ \sin\theta \end{bmatrix}$$

$$R_\theta\left(\begin{bmatrix} 0 \\ 1 \end{bmatrix}\right) = \begin{bmatrix} \cos(\frac{\pi}{2} + \theta) \\ \sin(\frac{\pi}{2} + \theta) \end{bmatrix} = \begin{bmatrix} -\sin\theta \\ \cos\theta \end{bmatrix}$$

The standard matrix for R_θ is

$$\begin{bmatrix} \cos\theta & -\sin\theta \\ \sin\theta & \cos\theta \end{bmatrix}$$

FIGURE 6.5

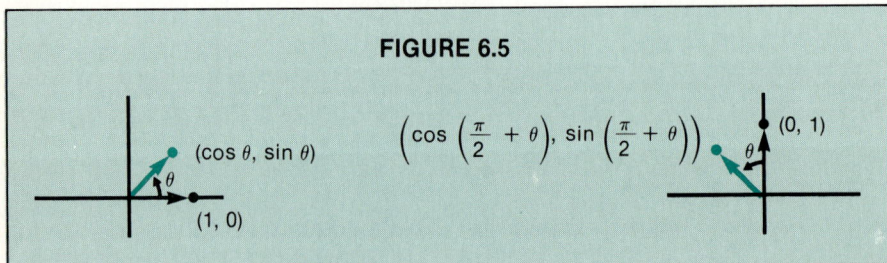

Example 3.   Let L be the graph of $y = x$. The reflection through L (see Example 4 of Section 6.1) carries a point (a, b) into the point (b, a). Hence

$$T\left(\begin{bmatrix} 1 \\ 0 \end{bmatrix}\right) = \begin{bmatrix} 0 \\ 1 \end{bmatrix} \qquad T\left(\begin{bmatrix} 0 \\ 1 \end{bmatrix}\right) = \begin{bmatrix} 1 \\ 0 \end{bmatrix}$$

so that

$$\begin{bmatrix} 0 & 1 \\ 1 & 0 \end{bmatrix}$$

is the standard matrix for this reflection.

In the previous theorem we found that any linear transformation from R^n into R^m can be represented by matrix multiplication. We will now show that the same is true for any linear transformation from one finite dimensional linear space into another linear space. To construct the matrix representing such a linear transformation we recall the concept of coordinate vectors introduced in Section 4.8. In that section we associated with each element $\mathbf{v}$ of a finite dimensional linear space with ordered basis $B = \{\mathbf{v}_1, \mathbf{v}_2, \ldots, \mathbf{v}_n\}$ the n-vector

$$[\mathbf{v}]_B = \begin{bmatrix} c_1 \\ c_2 \\ \vdots \\ c_n \end{bmatrix}_B$$

where $c_1, c_2, \ldots, c_n$ are the numbers such that $\mathbf{v} = c_1 \mathbf{v}_1 + c_2 \mathbf{v}_2 + \cdots + c_n \mathbf{v}_n$. The subscript "$B$" tells us which basis for V we are using in our calculations.

Theorem 3.   *Let $T: V \to W$ be a linear transformation from a finite dimensional linear space V into a finite dimensional linear space W. Let $B = \{\mathbf{v}_1, \mathbf{v}_2, \ldots, \mathbf{v}_n\}$ be an ordered basis of V, $B' = \{\mathbf{w}_1, \mathbf{w}_2, \ldots, \mathbf{w}_m\}$ be an ordered basis of W, and A the $m \times n$ matrix having $[T(\mathbf{v}_1)]_{B'}$, $[T(\mathbf{v}_2)]_{B'}, \ldots, [T(\mathbf{v}_n)]_{B'}$ as its columns. Then*

(1) $$[T(\mathbf{v})]_{B'} = A[\mathbf{v}]_B$$

for every $\mathbf{v}$ in V.

Proof: Let

$$[\mathbf{v}]_B = \begin{bmatrix} c_1 \\ c_2 \\ \vdots \\ c_n \end{bmatrix}_B$$

Using Theorem 4 of Section 4.4 and Exercise 11 of Section 4.8 we have

$$\begin{aligned}
A[\mathbf{v}]_B &= c_1 [T(\mathbf{v}_1)]_{B'} + c_2 [T(\mathbf{v}_2)]_{B'} + \cdots + c_n [T(\mathbf{v}_n)]_{B'} \\
&= [c_1 T(\mathbf{v}_1) + c_2 T(\mathbf{v}_2) + \cdots + c_n T(\mathbf{v}_n)]_{B'} \\
&= [T(c_1 \mathbf{v}_1 + c_2 \mathbf{v}_2 + \cdots + c_n \mathbf{v}_n)]_{B'} \\
&= [T(\mathbf{v})]_{B'} \quad \blacksquare
\end{aligned}$$

The basic idea in Theorem 3 is simple. The matrix A is obtained by computing each of its columns. To get the ith column of A we take the image $T(v_i)$ of the ith basis element v_i in basis B of V and compute its coordinate vector $[T(v_i)]_{B'}$ with respect to the basis B' of W.

Definition 3. Let $T: V \rightarrow W$ be a linear transformation of a finite dimensional linear space V into a finite linear space W. Let B be an ordered basis for V, and let B' be an ordered basis for W. The matrix $_{B'}A_B$ such that

$$[T(\mathbf{v})]_{B'} = {}_{B'}A_B[\mathbf{v}]_B$$

is called the **matrix representation** of T relative to B and B'.

The subscripts have been added to the matrix representing the linear transformation T so that we will be able to keep track of the bases we are using in the calculations. Notice that the right subscript of $_{B'}A_B$ is the same as the subscript on the vector $[\mathbf{v}]_B$ it is multiplying, while the left subscript is the same as the subscript of the vector $[T(\mathbf{v})]_{B'}$ on the left of the identity.

Example 4. In Example 5 of Section 6.1 we found that the function $T: P_2 \rightarrow P_1$ defined by

$$T(ax^2 + bx + c) = 2ax + b$$

is a linear transformation. Let $B = \{x^2, x, 1\}$ and $B' = \{x, 1\}$ be ordered bases of P_2 and P_1, respectively. Since $T(x^2) = 2x$, $T(x) = 1$, and $T(1) = 0$, we have

$$[T(x^2)]_{B'} = \begin{bmatrix} 2 \\ 0 \end{bmatrix}_{B'}, \qquad [T(x)]_{B'} = \begin{bmatrix} 0 \\ 1 \end{bmatrix}_{B'}, \qquad [T(1)]_{B'} = \begin{bmatrix} 0 \\ 0 \end{bmatrix}_{B'},$$

so that

$$_{B'}A_B = \begin{bmatrix} 2 & 0 & 0 \\ 0 & 1 & 0 \end{bmatrix}$$

is the matrix representation of T relative to B and B'. Hence if $f(x) = ax^2 + bx + c$ is any element of P_2, then

$$[T(f(x))]_{B'} = \begin{bmatrix} 2 & 0 & 0 \\ 0 & 1 & 0 \end{bmatrix} [f(x)]_B$$

This is easily checked by noting that

$$[T(f(x))]_{B'} = [2ax + b]_{B'} = \begin{bmatrix} 2a \\ b \end{bmatrix}_{B'}$$

$$[f(x)]_B = [ax^2 + bx + c]_B = \begin{bmatrix} a \\ b \\ c \end{bmatrix}_B$$

and

$$\begin{bmatrix} 2a \\ b \end{bmatrix}_{B'} = \begin{bmatrix} 2 & 0 & 0 \\ 0 & 1 & 0 \end{bmatrix} \begin{bmatrix} a \\ b \\ c \end{bmatrix}_B$$

Example 5.   In Example 1 of Section 6.1 we found that the function $T: R^3 \to R^2$ defined by

$$T\left(\begin{bmatrix} x_1 \\ x_2 \\ x_3 \end{bmatrix}\right) = \begin{bmatrix} 3x_1 + x_3 \\ x_2 - 2x_3 \end{bmatrix}$$

is a linear transformation. We will find the matrix representation of T relative to the ordered bases

$$B = \left\{ \begin{bmatrix} 1 \\ 0 \\ 0 \end{bmatrix}, \begin{bmatrix} 1 \\ 1 \\ 0 \end{bmatrix}, \begin{bmatrix} 1 \\ 1 \\ 1 \end{bmatrix} \right\} \qquad \text{and} \qquad B' = \left\{ \begin{bmatrix} 1 \\ 1 \end{bmatrix}, \begin{bmatrix} 2 \\ 1 \end{bmatrix} \right\}$$

of R^3 and R^2, respectively. To begin,

$$T\left(\begin{bmatrix} 1 \\ 0 \\ 0 \end{bmatrix}\right) = \begin{bmatrix} 3 \\ 0 \end{bmatrix} \qquad T\left(\begin{bmatrix} 1 \\ 1 \\ 0 \end{bmatrix}\right) = \begin{bmatrix} 3 \\ 1 \end{bmatrix} \qquad T\left(\begin{bmatrix} 1 \\ 1 \\ 1 \end{bmatrix}\right) = \begin{bmatrix} 4 \\ -1 \end{bmatrix}$$

We now need to compute the coordinate vector with respect to B' for each of these vectors. To do this we need to find scalars $a_1, a_2, b_1, b_2, c_1, c_2$ such that

$$\begin{bmatrix} 3 \\ 0 \end{bmatrix} = a_1 \begin{bmatrix} 1 \\ 1 \end{bmatrix} + a_2 \begin{bmatrix} 2 \\ 1 \end{bmatrix}$$

$$\begin{bmatrix} 3 \\ 1 \end{bmatrix} = b_1 \begin{bmatrix} 1 \\ 1 \end{bmatrix} + b_2 \begin{bmatrix} 2 \\ 1 \end{bmatrix}$$

$$\begin{bmatrix} 4 \\ -1 \end{bmatrix} = c_1 \begin{bmatrix} 1 \\ 1 \end{bmatrix} + c_2 \begin{bmatrix} 2 \\ 1 \end{bmatrix}$$

It is left for the reader to verify that

$$a_1 = -3, \quad a_2 = 3, \quad b_1 = -1, \quad b_2 = 2, \quad c_1 = -6, \quad c_2 = 5$$

so that

$$T\left(\begin{bmatrix} 1 \\ 0 \\ 0 \end{bmatrix}\right) = \begin{bmatrix} -3 \\ 3 \end{bmatrix}_{B'} \qquad T\left(\begin{bmatrix} 1 \\ 1 \\ 0 \end{bmatrix}\right) = \begin{bmatrix} -1 \\ 2 \end{bmatrix}_{B'} \qquad T\left(\begin{bmatrix} 1 \\ 1 \\ 1 \end{bmatrix}\right) = \begin{bmatrix} -6 \\ 5 \end{bmatrix}_{B'}$$

Therefore

$$_{B'}A_B = \begin{bmatrix} -3 & -1 & -6 \\ 3 & 2 & 5 \end{bmatrix}$$

is the matrix representation of T with respect to B and B' so that

$$[T(\mathbf{v})]_{B'} = {}_{B'}A_B[\mathbf{v}]_B$$

for every $\mathbf{v}$ in R^3. For example, consider the element

$$\mathbf{v} = \begin{bmatrix} 4 \\ 2 \\ 3 \end{bmatrix}$$

of R^3. Since

$$\begin{bmatrix} 4 \\ 2 \\ 3 \end{bmatrix} = 2\begin{bmatrix} 1 \\ 0 \\ 0 \end{bmatrix} - \begin{bmatrix} 1 \\ 1 \\ 0 \end{bmatrix} + 3\begin{bmatrix} 1 \\ 1 \\ 1 \end{bmatrix}$$

we have

$$\begin{bmatrix} 4 \\ 2 \\ 3 \end{bmatrix} = \begin{bmatrix} 2 \\ -1 \\ 3 \end{bmatrix}_B$$

and

$$[T(\mathbf{v})]_{B'} = \begin{bmatrix} -3 & -1 & -6 \\ 3 & 2 & 5 \end{bmatrix}\begin{bmatrix} 2 \\ -1 \\ 3 \end{bmatrix}_B$$

$$= \begin{bmatrix} -23 \\ 19 \end{bmatrix}_{B'}$$

This may be easily checked by noting that

$$T(\mathbf{v}) = \begin{bmatrix} 15 \\ -4 \end{bmatrix} = -23\begin{bmatrix} 1 \\ 1 \end{bmatrix} + 19\begin{bmatrix} 2 \\ 1 \end{bmatrix} = \begin{bmatrix} -23 \\ 19 \end{bmatrix}_{B'}$$

Exercises

In Exercises 1–8 find the standard matrix for each of the given linear transformations of R^m into R^n for appropriate values of m and n.

1. $T\left(\begin{bmatrix} x \\ y \end{bmatrix}\right) = \begin{bmatrix} 3x + 2y \\ -5x + 4y \end{bmatrix}$

2. $T\left(\begin{bmatrix} x \\ y \end{bmatrix}\right) = \begin{bmatrix} 8x - 4y \\ -4x + 7y \end{bmatrix}$

3. $T\left(\begin{bmatrix} x \\ y \\ z \end{bmatrix}\right) = \begin{bmatrix} 2x + 3z \\ 3y + 2z \\ 2x + 5y \end{bmatrix}$

4. $T\left(\begin{bmatrix} x \\ y \\ z \end{bmatrix}\right) = \begin{bmatrix} -7x + 5y - 8z \\ 6x + 2y + 9z \\ -8x + 7y - 6z \end{bmatrix}$

5. $T\left(\begin{bmatrix} x \\ y \end{bmatrix}\right) = \begin{bmatrix} x + 7y \\ 3x - 2y \\ 4x + 5y \end{bmatrix}$

6. $T\left(\begin{bmatrix} x \\ y \\ z \end{bmatrix}\right) = \begin{bmatrix} 3x + 5y - 7z \\ -2x - 3y + z \end{bmatrix}$

7. $T\left(\begin{bmatrix} x_1 \\ x_2 \\ x_3 \\ x_4 \end{bmatrix}\right) = \begin{bmatrix} x_1 - x_2 \\ x_3 - x_4 \\ x_2 + x_3 \end{bmatrix}$

8. $T\left(\begin{bmatrix} x_1 \\ x_2 \\ x_3 \\ x_4 \end{bmatrix}\right) = \begin{bmatrix} 2x_1 + 3x_2 + x_3 \\ -x_2 + 3x_3 - x_4 \\ 3x_1 - 5x_2 + x_4 \\ 5x_1 + x_2 + x_3 \end{bmatrix}$

9. Let $T: R^3 \to R^2$ be a linear transformation such that

$$T\left(\begin{bmatrix} 1 \\ 1 \\ 0 \end{bmatrix}\right) = \begin{bmatrix} 2 \\ 1 \end{bmatrix} \qquad T\left(\begin{bmatrix} 1 \\ 0 \\ 1 \end{bmatrix}\right) = \begin{bmatrix} 3 \\ -2 \end{bmatrix} \qquad T\left(\begin{bmatrix} 0 \\ 1 \\ 1 \end{bmatrix}\right) = \begin{bmatrix} -4 \\ 1 \end{bmatrix}$$

Find the standard matrix for T. Find $T\left(\begin{bmatrix} 1 \\ 2 \\ 3 \end{bmatrix}\right)$

10. Let $T: R^3 \to R^3$ be a linear transformation such that

$$T\left(\begin{bmatrix} 0 \\ 1 \\ 1 \end{bmatrix}\right) = \begin{bmatrix} 1 \\ 0 \\ 2 \end{bmatrix} \qquad T\left(\begin{bmatrix} 1 \\ 1 \\ 1 \end{bmatrix}\right) = \begin{bmatrix} 2 \\ 0 \\ 4 \end{bmatrix} \qquad T\left(\begin{bmatrix} 1 \\ -1 \\ 1 \end{bmatrix}\right) = \begin{bmatrix} 3 \\ 5 \\ 9 \end{bmatrix}$$

Find the standard matrix for T. Find $T\left(\begin{bmatrix} 1 \\ 2 \\ 3 \end{bmatrix}\right)$

In Exercises 11–14 find the matrix representation of the given linear transformation $T: R^3 \to P_2$ with respect to the ordered bases

$$B = \left\{ \begin{bmatrix} 1 \\ 0 \\ 0 \end{bmatrix}, \begin{bmatrix} 0 \\ 1 \\ 0 \end{bmatrix}, \begin{bmatrix} 0 \\ 0 \\ 1 \end{bmatrix} \right\} \qquad \text{for } R^3 \text{ and } B' = \{1, x, x^2\} \text{ for } P_2.$$

11. $T\left(\begin{bmatrix} a \\ b \\ c \end{bmatrix}\right) = ax^2 + bx + c$

12. $T\left(\begin{bmatrix} a \\ b \\ c \end{bmatrix}\right) = (a + b)x - c$

13. $T\left(\begin{bmatrix} a \\ b \\ c \end{bmatrix}\right) = bx^2 + ax + 2c$

14. $T\left(\begin{bmatrix} a \\ b \\ c \end{bmatrix}\right) = (a + b)x^2 + bx + (a - c)$

In Exercises 15–18 find the matrix representation of the given linear transformation $T:R^3 \to P_2$ with respect to the ordered bases

$$B = \left\{ \begin{bmatrix} 1 \\ 0 \\ 0 \end{bmatrix}, \begin{bmatrix} 1 \\ 1 \\ 0 \end{bmatrix}, \begin{bmatrix} 0 \\ 1 \\ 1 \end{bmatrix} \right\} \text{ for } R^3 \text{ and } B' = \{1 + x, 1 + x^2, x\} \text{ for } P_2.$$

15. $T\left(\begin{bmatrix} a \\ b \\ c \end{bmatrix} \right) = ax^2 + bx + c$

16. $T\left(\begin{bmatrix} a \\ b \\ c \end{bmatrix} \right) = (a + b)x - c$

17. $T\left(\begin{bmatrix} a \\ b \\ c \end{bmatrix} \right) = bx^2 + ax + 2c$

18. $T\left(\begin{bmatrix} a \\ b \\ c \end{bmatrix} \right) = (a + b)x^2 + bx + (a - c)$

In Exercises 19–20 find the matrix representation of the linear transformation $T:R^3 \to R^2$ defined by

$$T\left(\begin{bmatrix} x \\ y \\ z \end{bmatrix} \right) = \begin{bmatrix} 1 & 2 & -1 \\ 3 & 4 & -2 \end{bmatrix} \begin{bmatrix} x \\ y \\ z \end{bmatrix}$$

with respect to the given ordered bases B for R^3 and B' for R^2.

19. $B = \left\{ \begin{bmatrix} 1 \\ 0 \\ 1 \end{bmatrix}, \begin{bmatrix} 0 \\ 1 \\ 0 \end{bmatrix}, \begin{bmatrix} 1 \\ 0 \\ 0 \end{bmatrix} \right\}$ $B' = \left\{ \begin{bmatrix} 1 \\ 1 \end{bmatrix}, \begin{bmatrix} -1 \\ 1 \end{bmatrix} \right\}$

20. $B = \left\{ \begin{bmatrix} 1 \\ 0 \\ 0 \end{bmatrix}, \begin{bmatrix} 1 \\ 1 \\ 0 \end{bmatrix}, \begin{bmatrix} 1 \\ 0 \\ 1 \end{bmatrix} \right\}$ $B' = \left\{ \begin{bmatrix} 1 \\ 1 \end{bmatrix}, \begin{bmatrix} 0 \\ 1 \end{bmatrix} \right\}$

In Exercises 21–24 find the matrix representation for the given linear transformation $T:P_2 \to P_3$ with respect to the given ordered bases B for P_2 and B' for P_3.

21. $T(ax^2 + bx + c) = 2ax + b; B = \{1, x, x^2\}; B' = \{1, x, x^2, x^3\}.$
22. $T(ax^2 + bx + c) = \frac{1}{3}ax^3 + \frac{1}{2}bx^2 + cx; B = \{1, x, x^2\}; B' = \{1, x, x^2, x^3\}.$
23. $T(ax^2 + bx + c) = 2ax^2 + (a + b)x + 3c; B = \{1 + x, 1 - x, x^2\};$
    $B' = \{1, x + x^2, x^2 + x^3, x^3\}.$
24. $T(ax^2 + bx + c) = ax^2 + bx + c; B = \{1 + x, 1 - x, x^2\}; B' = \{1, x + x^2, x^2 + x^3, x^3\}.$

In Exercises 25–27 find the matrix representation for the linear transformation $T: R^2 \to R^2$ defined by

$$T\left(\begin{bmatrix} x \\ y \end{bmatrix}\right) = \begin{bmatrix} 1 & 2 \\ 3 & 4 \end{bmatrix}\begin{bmatrix} x \\ y \end{bmatrix}$$

with respect to the given ordered basis B for R^2, i.e., $_BA_B$.

25. $B = \left\{ \begin{bmatrix} 1 \\ 1 \end{bmatrix}, \begin{bmatrix} 1 \\ -1 \end{bmatrix} \right\}$

26. $B = \left\{ \begin{bmatrix} 2 \\ 1 \end{bmatrix}, \begin{bmatrix} 1 \\ 2 \end{bmatrix} \right\}$

27. $B = \left\{ \begin{bmatrix} 1 \\ 0 \end{bmatrix}, \begin{bmatrix} 1 \\ 1 \end{bmatrix} \right\}$

6.5 COMPOSITION OF LINEAR TRANSFORMATIONS

Let $T: U \to V$ and $S: V \to W$ be linear transformations of U into V and V into W where U, V, and W are linear spaces. If $\mathbf{u}$ is any element of U, then $\mathbf{v} = T(\mathbf{u})$ is an element of V and $\mathbf{w} = S(\mathbf{v}) = S[T(\mathbf{u})]$ is an element of W. Thus the function $S \circ T$, called the **composition** of S with T, defined by

$$S \circ T(\mathbf{u}) = S[T(\mathbf{u})]$$

is a function from U into W. This may be viewed pictorially as in Figure 6.6. Following the arrows we get to the same place either by using T followed by S or by using $S \circ T$.

Using the linear properties of T and S we can show that $S \circ T$ is a linear transformation as follows:

$$\begin{aligned} S \circ T(\mathbf{u} + \mathbf{v}) &= S[T(\mathbf{u} + \mathbf{v})] \\ &= S[T(\mathbf{u}) + T(\mathbf{v})] \\ &= S[T(\mathbf{u})] + S[T(\mathbf{v})] \\ &= S \circ T(\mathbf{u}) + S \circ T(\mathbf{v}) \end{aligned}$$

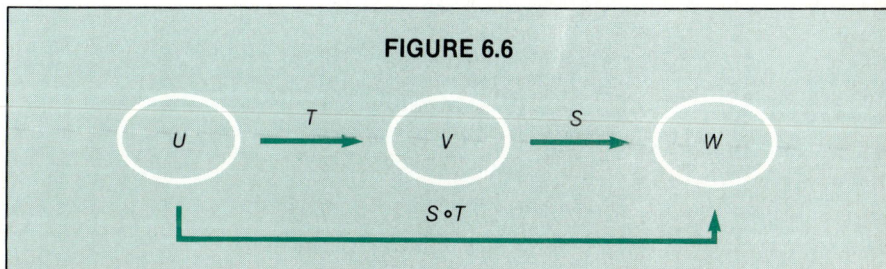

FIGURE 6.6

for every **u** and **v** in U, and

$$
\begin{aligned}
S \circ T(c\mathbf{u}) &= S[T(c\mathbf{u})] \\
&= S[cT(\mathbf{u})] \\
&= cS[T(\mathbf{u})] \\
&= c[S \circ T(\mathbf{u})]
\end{aligned}
$$

for every **u** in U and every scalar c. We have proved the following theorem.

Theorem 4.   Let $T: U \to V$ and $S: V \to W$ be linear transformations of U into V and V into W where U, V, and W are linear spaces. Then $S \circ T$ is a linear transformation of U into W.

In the special case that $U = R^n$, $V = R^m$, and $W = R^p$, Theorem 2 of the previous section assures us that there are matrices A, B, and C such that $T(\mathbf{u}) = A\mathbf{u}$ for all **u** in U, $S(\mathbf{v}) = B\mathbf{v}$ for all **v** in V, and $S \circ T(\mathbf{u}) = C\mathbf{u}$ for all **u** in U. Then

$$
\begin{aligned}
C\mathbf{u} &= S \circ T(\mathbf{u}) \\
&= S[T(\mathbf{u})] \\
&= S(A\mathbf{u}) \\
&= BA\mathbf{u}
\end{aligned}
$$

for all **u** in U. By Exercise 25 of Section 4.4 we have $C = BA$. Thus the definition of matrix multiplication given in Section 1.7 was chosen so that the standard matrix for $S \circ T$ equals the product of the standard matrices for S and T.

Example 1.   In Example 2 of the previous section we found that the linear transformation $R_\theta: R^2 \to R^2$ that rotates each point counterclockwise about the origin through an angle θ has

$$
\begin{bmatrix} \cos \theta & -\sin \theta \\ \sin \theta & \cos \theta \end{bmatrix}
$$

as its standard matrix. If we rotate first through an angle ϕ and then through an angle θ, we obtain the same results as though we rotated through an angle $\theta + \phi$. That is,

$$
R_{\theta + \phi} = R_\theta \circ R_\phi
$$

so that

$$
\begin{aligned}
\begin{bmatrix} \cos(\theta + \phi) & -\sin(\theta + \phi) \\ \sin(\theta + \phi) & \cos(\theta + \phi) \end{bmatrix} &= \begin{bmatrix} \cos \theta & -\sin \theta \\ \sin \theta & \cos \theta \end{bmatrix} \begin{bmatrix} \cos \phi & -\sin \phi \\ \sin \phi & \cos \phi \end{bmatrix} \\
&= \begin{bmatrix} \cos \theta \cos \phi - \sin \theta \sin \phi & -\cos \theta \sin \phi - \sin \theta \cos \phi \\ \sin \theta \cos \phi + \cos \theta \sin \phi & -\sin \theta \sin \phi + \cos \theta \cos \phi \end{bmatrix}
\end{aligned}
$$

We conclude that

$$
\cos(\theta + \phi) = \cos \theta \cos \phi - \sin \theta \sin \phi
$$

$$
\sin(\theta + \phi) = \sin \theta \cos \phi + \cos \theta \sin \phi
$$

which are well known trigonometric identities.

In many applications linear transformations describe how a system changes from one state to another. In such a situation the composition of linear transformations corresponds to how the system reacts to successive changes of state.

Example 2.   In 1202 A.D. Leonardo Fibonacci published a work entitled "*Liber abaci*" which contained the following problem:

> How many pairs of rabbits can be produced from a single pair in a year if every month each pair begets a new pair which from the second month on becomes productive?

Initially there is one pair of rabbits. At the end of one month there is still one pair of rabbits. At this time the pair reproduces so that at the end of the second month there are two pairs of rabbits. During the next month the first pair reproduces, but the second does not because it is not old enough. Continuing in this manner we see that at the end of n months the number of pairs present will be the number of pairs present one month earlier plus the number of pairs born during the month. The number of pairs born during the month is the number of pairs that are old enough to reproduce; that is, the number of pairs that are more than one month old. The number of pairs forms the sequence 1, 1, 2, 3, 5, 8, 13, 21, ... The numbers p_k of this sequence, called **Fibonacci numbers**, satisfy the recurrence relation

$$P_k = p_{k-1} + p_{k-2}$$

for $k \geq 3$ with $p_2 = 1$ and $p_1 = 1$. This formula can be written in matrix notation as

$$P_k = AP_{k-1}$$

where

$$P_k = \begin{bmatrix} p_{k-1} \\ p_k \end{bmatrix} \qquad A = \begin{bmatrix} 0 & 1 \\ 1 & 1 \end{bmatrix}$$

for $k \geq 3$ and

$$P_2 = \begin{bmatrix} 1 \\ 1 \end{bmatrix}$$

The second component of P_k is the number of pairs at the end of k months, while the first component is the number of pairs one month earlier. Thus the matrix A describes how the population changes from one month to the next. Notice that

$$
\begin{aligned}
P_3 &= & AP_2 & \\
P_4 &= & AP_3 &= A^2 P_2 \\
P_5 &= & AP_4 &= A^3 P_2 \\
&\vdots & \vdots & \quad \vdots \\
P_k &= & AP_{k-1} &= A^{k-2} P_2
\end{aligned}
$$

The number of pairs of rabbits at the end of k months can be determined by applying the linear transformation represented by the matrix A successively $k - 2$ times.

Hence knowing p_{k-1} is the same as knowing

$$\begin{bmatrix} 0 & 1 \\ 1 & 1 \end{bmatrix}^{k-2}$$

The Fibonacci numbers p_k arise in unexpected circumstances. In the nineteenth century, botanists noted that spiral leaf arrangements commonly occur in groups expressed by fractions describing how the leaves are spaced on the stem. Corn and many grasses have leaves that are in two rows on opposite sides of the stem. Thus each leaf is $\frac{1}{2}$ of the circumference of the stem from the next leaf above or below. In birch and some other trees and shrubs, adjacent leaves are separated by $\frac{1}{3}$ of a circumference. In some other trees, such as oak, cherry, and apple, adjacent leaves are separated by $\frac{2}{5}$ of a circumference. In rose and blackberry the fraction is $\frac{3}{8}$. Other common spiral leaf arrangements have fractions $\frac{5}{13}$ and $\frac{8}{21}$. Thus leaf arrangements fall into groups described by the fractions $\frac{1}{2}, \frac{1}{3}, \frac{2}{5}, \frac{3}{8}, \frac{5}{13}, \frac{8}{21}, \dots$ Notice that each numerator is the sum of the two previous numerators and that each denominator is the sum of the two previous denominators. Hence both the numerators and denominators are Fibonacci numbers. In fact, each of the above fractions is of the form p_k/p_{k+2}. In Section 7.3 we will find an expression for p_k.

--------------------------- **Exercises** ---------------------------

1. Let T denote the reflection about the line $y = x$ (see Example 3 of Section 6.4). Show that $T \circ R_{\pi/4} \neq R_{\pi/4} \circ T$ geometrically and by constructing the standard matrix for each transformation.
2. Let T be the reflection about the line $x = 0$. Show that $T \circ R_{\pi/4} \neq R_{\pi/4} \circ T$ geometrically and by constructing the standard matrix for each transformation.
3. Let U be a finite dimensional subspace of a linear space V. Show that the function $T: V \to U$ defined by $T(\mathbf{u}) = \text{proj}_U \mathbf{u}$ (see Section 5.5) is a linear transformation such that $T^2 = T$.
4. For what values of θ does the standard matrix for R_θ^3 equal I_2.

In Exercises 5–10 let $T: R^2 \to R^2$ be the reflection about the line $y = x$ (see Example 3 of Section 6.4). Compute the standard matrix for the given linear transformation.

5. $T \circ R_{\pi/3}$	**6.** $R_{\pi/2} \circ T$	**7.** $R_{\pi/3} \circ T \circ R_{\pi/3}$
8. $T \circ R_{\pi/2} \circ T$	**9.** $T \circ R_{\pi/3} \circ T$	**10.** $R_{\pi/2} \circ T \circ R_{\pi/2}$

6.6 REPRESENTATIONS OF LINEAR TRANSFORMATIONS, II (OPTIONAL)

Suppose that we have a linear transformation $T: V \to V$ of a finite dimensional linear space into itself. If B is any basis for V, then from Theorem 3 of Section 6.3 (with $W = V$ and $B' = B$) we know that there is a matrix $_B A_B$ such that

$$[T(\mathbf{v})]_B = {}_B A_B [\mathbf{v}]_B$$

Likewise if B' is another basis for V, then there is a matrix $_{B'} A_{B'}$ such that

$$[T(\mathbf{v})]_{B'} = {}_{B'} A_{B'} [\mathbf{v}]_{B'}$$

The following theorem shows how the matrix representing T changes when we change from one basis for V to another.

Theorem 5. Let $T : V \to V$ be a linear transformation of a finite dimensional linear space into itself. If $_B A_B$ is the matrix representation for T relative to an ordered basis B and $_{B'} A_{B'}$ is the matrix representation for T relative to an ordered basis B', then

$$_{B'} A_{B'} = (_B P_{B'})^{-1} (_B A_B)(_B P_{B'})$$

where $_B P_{B'}$ is the transition matrix from B' to B.

Proof: If $_B P_{B'}$ is the transition matrix from B' to B, then $(_B P_{B'})^{-1}$ is the transition matrix from B to B' (Theorem 21 of Section 4.10) so that $[\mathbf{u}]_{B'} = (_B P_{B'})^{-1}[\mathbf{u}]_B$ for every $\mathbf{u}$ in V. Since $T(\mathbf{v})$ is an element of V for every $\mathbf{v}$ in V we have

$$\begin{aligned} [T(\mathbf{v})]_{B'} &= (_B P_{B'})^{-1}[T(\mathbf{v})]_B \\ &= (_B P_{B'})^{-1} {}_B A_B [\mathbf{v}]_B \\ &= (_B P_{B'})^{-1}(_B A_B)(_B P_{B'})[\mathbf{v}]_{B'} \end{aligned}$$

Thus $(_B P_{B'})^{-1}(_B A_B)(_B P_{B'})$ is the matrix representation of T relative to B'. Therefore $_{B'} A_{B'} = (_B P_{B'})^{-1}(_B A_B)(_B P_{B'})$. ∎

Theorem 5 can be depicted pictorially as seen in Figure 6.7. Following the arrows, we get to the same place either by using $_{B'} A_{B'}$ or by using $_B P_{B'}$ followed by $_B A_B$ followed by $_{B'} P_B$. That is,

$$\begin{aligned} _{B'} A_{B'} &= (_{B'} P_B)(_B A_B)(_B P_{B'}) \\ &= (_B P_{B'})^{-1}(_B A_B)(_B P_{B'}) \end{aligned}$$

In Theorem 5 if the basis B is the natural basis N for R^n, if B' is any basis for R^n, and if A is the standard matrix for T, then

$$\begin{aligned} [T(\mathbf{v})]_{B'} &= {}_{B'} A_{B'} [\mathbf{v}]_{B'} \\ &= (_N P_{B'})^{-1} A(_N P_{B'})[\mathbf{v}]_{B'} \end{aligned}$$

FIGURE 6.7

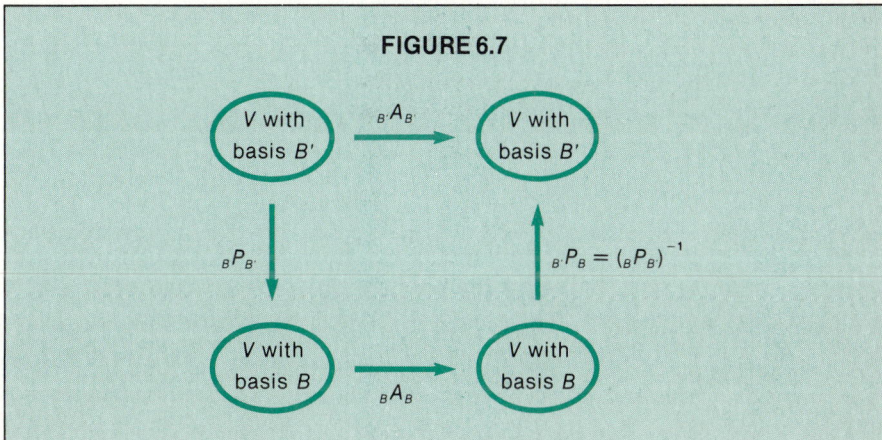

Example 1.   The function $T: R^2 \to R^2$ defined by

$$T\left(\begin{bmatrix} x_1 \\ x_2 \end{bmatrix}\right) = \begin{bmatrix} x_1 + 2x_2 \\ 2x_1 + x_2 \end{bmatrix}$$

is easily shown to be a linear transformation. We will find the matrix representations $_BA_B$ and $_{B'}A_{B'}$ for T relative to the ordered bases

$$B = \left\{ \begin{bmatrix} 1 \\ 0 \end{bmatrix}, \begin{bmatrix} 0 \\ 1 \end{bmatrix} \right\} \qquad \text{and} \qquad B' = \left\{ \begin{bmatrix} 1 \\ -1 \end{bmatrix}, \begin{bmatrix} 1 \\ 1 \end{bmatrix} \right\}$$

Then we will find the transition matrix from B' to B and verify that $_{B'}A_{B'} = (_BP_{B'})^{-1}(_BA_B)(_BP_{B'})$ as is guaranteed by Theorem 5. To begin

$$T\left(\begin{bmatrix} 1 \\ 0 \end{bmatrix}\right) = \begin{bmatrix} 1 \\ 2 \end{bmatrix} = \begin{bmatrix} 1 \\ 0 \end{bmatrix} + 2\begin{bmatrix} 0 \\ 1 \end{bmatrix}$$

$$T\left(\begin{bmatrix} 0 \\ 1 \end{bmatrix}\right) = \begin{bmatrix} 2 \\ 1 \end{bmatrix} = 2\begin{bmatrix} 1 \\ 0 \end{bmatrix} + \begin{bmatrix} 0 \\ 1 \end{bmatrix}$$

so that

$$T\left(\begin{bmatrix} 1 \\ 0 \end{bmatrix}\right)_B = \begin{bmatrix} 1 \\ 2 \end{bmatrix}_B \qquad T\left(\begin{bmatrix} 0 \\ 1 \end{bmatrix}\right)_B = \begin{bmatrix} 2 \\ 1 \end{bmatrix}_B$$

Therefore

$$_BA_B = \begin{bmatrix} 1 & 2 \\ 2 & 1 \end{bmatrix}$$

Next

$$T\left(\begin{bmatrix} 1 \\ -1 \end{bmatrix}\right) = \begin{bmatrix} -1 \\ 1 \end{bmatrix} = (-1)\begin{bmatrix} 1 \\ -1 \end{bmatrix} + (0)\begin{bmatrix} 1 \\ 1 \end{bmatrix}$$

$$T\left(\begin{bmatrix} 1 \\ 1 \end{bmatrix}\right) = \begin{bmatrix} 3 \\ 3 \end{bmatrix} = (0)\begin{bmatrix} 1 \\ -1 \end{bmatrix} + 3\begin{bmatrix} 1 \\ 1 \end{bmatrix}$$

so that

$$T\left(\begin{bmatrix} 1 \\ -1 \end{bmatrix}\right)_{B'} = \begin{bmatrix} -1 \\ 0 \end{bmatrix}_{B'} \qquad T\left(\begin{bmatrix} 1 \\ 1 \end{bmatrix}\right)_{B'} = \begin{bmatrix} 0 \\ 3 \end{bmatrix}_{B'}$$

Therefore

$$_{B'}A_{B'} = \begin{bmatrix} -1 & 0 \\ 0 & 3 \end{bmatrix}$$

Since

$$\begin{bmatrix} 1 \\ -1 \end{bmatrix} = \begin{bmatrix} 1 \\ 0 \end{bmatrix} + (-1)\begin{bmatrix} 0 \\ 1 \end{bmatrix}$$

and

$$\begin{bmatrix} 1 \\ 1 \end{bmatrix} = \begin{bmatrix} 1 \\ 0 \end{bmatrix} + \begin{bmatrix} 0 \\ 1 \end{bmatrix}$$

we have

$$_BP_{B'} = \begin{bmatrix} 1 & 1 \\ -1 & 1 \end{bmatrix}$$

is the transition matrix from B' to B. Therefore

$$_{B'}P_B = (_BP_{B'})^{-1} = \begin{bmatrix} \frac{1}{2} & -\frac{1}{2} \\ \frac{1}{2} & \frac{1}{2} \end{bmatrix}$$

is the transition matrix from B to B'. A straightforward calculation shows that

$$_{B'}A_{B'} = (_{B'}P_B)(_BA_B)(_BP_{B'})$$

$$= \begin{bmatrix} \frac{1}{2} & -\frac{1}{2} \\ \frac{1}{2} & \frac{1}{2} \end{bmatrix} \begin{bmatrix} 1 & 2 \\ 2 & 1 \end{bmatrix} \begin{bmatrix} 1 & 1 \\ -1 & 1 \end{bmatrix}$$

$$= \begin{bmatrix} -1 & 0 \\ 0 & 3 \end{bmatrix}$$

Exercises

In Exercises 1–2 the linear transformation T is the one given in Example 1. Find the matrix representation for T with respect to the given basis B' first by using Theorem 5 with B as the natural basis N and again by using Theorem 3 of Section 6.4 with $B = B'$.

1. $B' = \left\{ \begin{bmatrix} 1 \\ 1 \end{bmatrix}, \begin{bmatrix} -1 \\ 1 \end{bmatrix} \right\}$ **2.** $B' = \left\{ \begin{bmatrix} 1 \\ 1 \end{bmatrix}, \begin{bmatrix} 0 \\ 1 \end{bmatrix} \right\}$

In Exercises 3–4 let $T: P_2 \to P_2$ be the linear transformation defined by $T(ax^2 + bx + c) = 2ax + b$. Find the matrix representation for T with respect to the given basis B' first by using Theorem 5 with $B = \{1, x, x^2\}$ and again by using Theorem 3 of Section 6.4 with $B = B'$.

3. $B' = \{1 + x, x, 1 - x^2\}$ **4.** $B' = \{1, x^2, x^2 + x + 1\}$

In Exercises 5–6 find the matrix representation for the linear transformation $T: R^3 \to R^3$ with respect to the given basis B' if the matrix representation of T with respect to the given basis B is

$$\begin{bmatrix} 1 & 2 & -1 \\ 2 & 0 & 1 \\ -1 & 1 & 2 \end{bmatrix}$$

5. $B = \left\{ \begin{bmatrix} 1 \\ 0 \\ 1 \end{bmatrix}, \begin{bmatrix} 0 \\ 1 \\ 1 \end{bmatrix}, \begin{bmatrix} 1 \\ 0 \\ 0 \end{bmatrix} \right\}$ $B' = \left\{ \begin{bmatrix} 1 \\ 0 \\ 0 \end{bmatrix}, \begin{bmatrix} 0 \\ 1 \\ 1 \end{bmatrix}, \begin{bmatrix} 0 \\ -1 \\ 1 \end{bmatrix} \right\}$

6. $B = \left\{ \begin{bmatrix} 1 \\ 0 \\ 0 \end{bmatrix}, \begin{bmatrix} 1 \\ 1 \\ 0 \end{bmatrix}, \begin{bmatrix} 0 \\ 1 \\ 1 \end{bmatrix} \right\}$ $B' = \left\{ \begin{bmatrix} 0 \\ 1 \\ 0 \end{bmatrix}, \begin{bmatrix} 0 \\ 1 \\ 1 \end{bmatrix}, \begin{bmatrix} 1 \\ -1 \\ 0 \end{bmatrix} \right\}$

In Exercises 7–8 find the matrix representation for the linear transformation $T: P_2 \to P_2$ with respect to the given basis B' if the matrix representation of T with respect to the given basis B is

$$\begin{bmatrix} 1 & -1 & 2 \\ 2 & 1 & 0 \\ -1 & 0 & 3 \end{bmatrix}$$

7. $B = \{1, x - 1, x^2 - x\};$ $B' = \{x, x + 1, x^2 + x\}.$
8. $B = \{1 + x, 1 - x, x^2\};$ $B' = \{1 + x^2, 1 - x^2, x\}.$

An $n \times n$ matrix A is said to be **similar** to an $n \times n$ matrix B if there is a nonsingular matrix P such that $A = P^{-1}BP$.

9. Show that if A is similar to B, then B is similar to A.
10. Show that if A is similar to B and if B is similar to C, then A is similar to C.
11. Show that if A is similar to B, then $\det A = \det B$.

6.7 DIMENSION THEOREM REVISITED (OPTIONAL)

Consider the linear transformation $S: R^n \to R^m$ defined by

$$S(\mathbf{x}) = A\mathbf{x} \tag{1}$$

where A is an $m \times n$ matrix. In Section 4.12 we found that (rank of A) + (nullity of A) = n. This result has a generalization to linear transformations of a finite dimensional linear space into another linear space. As in the case of matrices, we will relate the dimensions of the linear spaces (which will subsequently be shown to be linear spaces)

$$\ker(T) = \{\mathbf{v}: T(\mathbf{v}) = \mathbf{0}\}$$

called the **kernel** of the linear transformation $T: V \to W$, and

$$R(T) = \{\mathbf{w}: \mathbf{w} = T(\mathbf{v}) \text{ for some } \mathbf{v} \text{ in } V\}$$

called the **range** of T.

If a linear transformation is defined by matrix multiplication as in (1), then its kernel and range are the null space and column space of the matrix A. Thus for this type of linear transformations the kernel and range are linear spaces. We will now show that this is the case for any linear transformation.

Theorem 6. *Let $T: V \to W$ be a linear transformation of a linear space V into a linear space W. Then*

      1. *The kernel of T is a subspace of V.*
      2. *The range of T is a subspace of W.*

Proof: Recall that a subset of a linear space is a subspace if it is closed under addition and scalar multiplication (Theorem 3 of Section 4.3). Notice that the kernel and range of T are not empty since $T(\mathbf{0}) = \mathbf{0}$. Let $\mathbf{u}$ and $\mathbf{v}$ be any two elements of the kernel of T and let c be any scalar. Then $T(\mathbf{u}) = \mathbf{0}$ and $T(\mathbf{v}) = \mathbf{0}$ so that

$$T(\mathbf{u} + \mathbf{v}) = T(\mathbf{u}) + T(\mathbf{v}) = \mathbf{0} + \mathbf{0} = \mathbf{0}$$

$$T(c\mathbf{u}) = cT(\mathbf{u}) = c\mathbf{0} = \mathbf{0}$$

Thus the kernel of T is closed under addition and scalar multiplication. Therefore the kernel of T is a subspace of V.

Now let $\mathbf{w}_1$ and $\mathbf{w}_2$ be any two elements of the range of T. Then there are elements $\mathbf{v}_1$ and $\mathbf{v}_2$ of V such that $\mathbf{w}_1 = T(\mathbf{v}_1)$ and $\mathbf{w}_2 = T(\mathbf{v}_2)$. Then

$$\mathbf{w}_1 + \mathbf{w}_2 = T(\mathbf{v}_1) + T(\mathbf{v}_2) = T(\mathbf{v}_1 + \mathbf{v}_2)$$

$$c\mathbf{w}_1 = cT(\mathbf{v}_1) = T(c\mathbf{v}_1)$$

for any scalar c. These identities show that $\mathbf{w}_1 + \mathbf{w}_2$ and $c\mathbf{w}_1$ are the images of $\mathbf{v}_1 + \mathbf{v}_2$ and $c\mathbf{v}_1$, respectively. Therefore $\mathbf{w}_1 + \mathbf{w}_2$ and $c\mathbf{w}_1$ are elements of the range of T. Hence the range is closed under addition and scalar multiplication, and the range of T is a subspace of W. ∎

    Definition 4. Let $T: V \to W$ be a linear transformation of a linear space V into a linear space W. The dimension of the kernel of T is called the **nullity** of T, and the dimension of the range of T is called the **rank** of T.

    The following theorem extends Theorem 26 of Section 4.12 to linear transformations of a finite dimensional linear space into another linear space.

Theorem 7. *If $T: V \to W$ is a linear transformation of a finite dimensional linear space V into a linear space W (not necessarily finite dimensional), then*

$$(rank\ of\ T) + (nullity\ of\ T) = \dim V$$

Proof: Let $B = \{\mathbf{v}_1, \mathbf{v}_2, \ldots, \mathbf{v}_n\}$ be a basis for V and let $\mathbf{u}$ be any element of the range of T. Then there is a $\mathbf{v}$ in V such that $\mathbf{u} = T(\mathbf{v})$. There are also scalars $c_1, c_2, \ldots, c_n$ such that $\mathbf{v} = c_1\mathbf{v}_1 + c_2\mathbf{v}_2 + \cdots + c_n\mathbf{v}_n$ so that

$$\mathbf{u} = T(\mathbf{v})$$

$$= T(c_1\mathbf{v}_1 + c_2\mathbf{v}_2 + \cdots + c_n\mathbf{v}_n)$$

$$= c_1T(\mathbf{v}_1) + c_2T(\mathbf{v}_2) + \cdots + c_nT(\mathbf{v}_n)$$

The last identity shows that every element of the range is a linear combination of $T(\mathbf{v}_1), T(\mathbf{v}_2), \ldots, T(\mathbf{v}_n)$. Thus the range of T must be finite dimensional. The linear transformation T can now be viewed as being from one finite dimensional linear space into another finite dimensional linear space. Let B' be a basis for the range of

T. By Theorem 3 of Section 6.4 there is a matrix A such that

$$[T(\mathbf{v})]_{B'} = A[\mathbf{v}]_B$$

The problem is now reduced to considering a linear transformation defined in terms of a matrix. Consequently the desired result follows from Theorem 26 of Section 4.12. ∎

This theorem gives us an alternate way of proving that an $n \times n$ matrix A is nonsingular if and only if A^{-1} exists. In order to establish this result we recall some elementary concepts concerning functions.

A function $f: X \to Y$ of a set X into a set Y is called **one-to-one** if $f(x) = f(y)$ implies $x = y$. The fundamental property of one-to-one functions is that their inverses exist. That is, a function $f: X \to Y$ is one-to-one if and only if there is a function $f^{-1}: Z \to X$, where $Z = \{y : y = f(x)$ for some $x\}$, such that

1. $f^{-1} \circ f(x) = x$ for every x in X.
2. $f \circ f^{-1}(z) = z$ for every z in Z.

Hence, $x = f^{-1}(z)$ if and only if $z = f(x)$. We now carry these ideas over to matrix theory.

Let A be an $n \times n$ matrix and consider the linear transformation $T: R^n \to R^n$ defined by

$$T(\mathbf{x}) = A\mathbf{x}$$

There are vectors $\mathbf{y}$ and $\mathbf{z}$ such that $T(\mathbf{y}) = T(\mathbf{z})$ if and only if $A\mathbf{y} = A\mathbf{z}$. That is, $T(\mathbf{y}) = T(\mathbf{z})$ if and only if $A(\mathbf{y} - \mathbf{z}) = \mathbf{0}$. Hence, T is one-to-one if and only if the equation $A\mathbf{x} = \mathbf{0}$ has $\mathbf{x} = \mathbf{0}$ as its only solution. By Theorem 2 of Section 1.5, the equation $A\mathbf{x} = \mathbf{0}$ has precisely one solution if and only if the equation $A\mathbf{x} = \mathbf{b}$ has a solution for every n-vector $\mathbf{b}$. Thus T is one-to-one if and only if A is nonsingular. Therefore T^{-1} exists if and only if A is nonsingular.

Our goal now is to show that T^{-1} exists if and only if A^{-1} exists. If A^{-1} exists, it is easy to show that T^{-1} is defined by $T^{-1}(\mathbf{x}) = A^{-1}\mathbf{x}$. Now suppose that T^{-1} exists. Then T is one-to-one so that $\ker(T) = \{0\}$ and the nullity of T is zero. By Theorem 7,

$$\text{rank of } T = \dim R^n = n$$

Hence $R(T) = R^n$, so T^{-1} is a function from R^n into R^n. We now show that T^{-1} is a linear transformation. Let $\mathbf{x}$ and $\mathbf{y}$ be any n-vectors and c be any scalar. Then there exist n-vectors $\mathbf{x}'$ and $\mathbf{y}'$ such that $\mathbf{x} = T(\mathbf{x}')$ and $\mathbf{y} = T(\mathbf{y}')$. Hence we also have $\mathbf{x}' = T^{-1}(\mathbf{x})$ and $\mathbf{y}' = T^{-1}(\mathbf{y})$. Now

$$T^{-1}(\mathbf{x} + \mathbf{y}) = T^{-1}[T(\mathbf{x}') + T(\mathbf{y}')] = T^{-1}[T(\mathbf{x}' + \mathbf{y}')]$$
$$= \mathbf{x}' + \mathbf{y}' = T^{-1}(\mathbf{x}) + T^{-1}(\mathbf{y})$$
$$T^{-1}(c\mathbf{x}) = T^{-1}[cT(\mathbf{x}')] = T^{-1}[T(c\mathbf{x}')] = c\mathbf{x}' = cT^{-1}(\mathbf{x})$$

so that T^{-1} is a linear transformation of R^n into R^n. By Theorem 2 of Section 6.4 there is an $n \times n$ matrix B such that $T^{-1}(\mathbf{x}) = B\mathbf{x}$ for every $\mathbf{x}$ in R^n. Then for every $\mathbf{x}$

in R^n we have

$$\mathbf{x} = T^{-1} \circ T(\mathbf{x}) = T^{-1}[T(\mathbf{x})] = T^{-1}(A\mathbf{x}) = BA\mathbf{x}$$

$$\mathbf{x} = T \circ T^{-1}(\mathbf{x}) = T[T^{-1}(\mathbf{x})] = T(B\mathbf{x}) = AB\mathbf{x}$$

so that $BA = I_n = AB$. Therefore $B = A^{-1}$. Thus T^{-1} exists if and only if A^{-1} exists.

Combining the conclusions of the previous two paragraphs we have that A is nonsingular if and only if A^{-1} exists.

─────────────────────────── **Exercises** ───────────────────────────

In Exercises 1–6 find bases for the kernel and range of the given linear transformation $T: R^4 \to R^3$. Also verify that the conclusion of Theorem 7 holds. Notice that in each exercise there is a matrix A such that $T(\mathbf{x}) = A\mathbf{x}$ for all $\mathbf{x}$ in R^4.

1. $T\left(\begin{bmatrix} x_1 \\ x_2 \\ x_3 \\ x_4 \end{bmatrix}\right) = \begin{bmatrix} x_1 + 2x_2 \\ x_2 - 3x_3 \\ x_3 + 4x_4 \end{bmatrix}$
$\qquad$
2. $T\left(\begin{bmatrix} x_1 \\ x_2 \\ x_3 \\ x_4 \end{bmatrix}\right) = \begin{bmatrix} 2x_2 - 5x_4 \\ x_1 + 3x_2 \\ x_3 \end{bmatrix}$

3. $T\left(\begin{bmatrix} x_1 \\ x_2 \\ x_3 \\ x_4 \end{bmatrix}\right) = \begin{bmatrix} 2x_1 + 3x_3 - x_4 \\ x_3 - x_4 \\ x_1 + 2x_3 - x_4 \end{bmatrix}$
$\qquad$
4. $T\left(\begin{bmatrix} x_1 \\ x_2 \\ x_3 \\ x_4 \end{bmatrix}\right) = \begin{bmatrix} x_1 - x_2 + x_3 + x_4 \\ 2x_1 + x_2 - x_3 - x_4 \\ x_1 + x_2 - x_3 - x_4 \end{bmatrix}$

5. $T\left(\begin{bmatrix} x_1 \\ x_2 \\ x_3 \\ x_4 \end{bmatrix}\right) = \begin{bmatrix} x_2 + 2x_3 + x_4 \\ 0 \\ x_1 - x_3 - x_4 \end{bmatrix}$
$\qquad$
6. $T\left(\begin{bmatrix} x_1 \\ x_2 \\ x_3 \\ x_4 \end{bmatrix}\right) = \begin{bmatrix} -2x_1 + x_2 + 2x_3 \\ x_2 + 2x_3 \\ 0 \end{bmatrix}$

In Exercises 7–10 find bases for the kernel and range of the given linear transformation $T: P_2 \to P_2$. Also verify that the conclusion of Theorem 7 holds.

7. $T(ax^2 + bx + c) = 2ax + b$
$\qquad\qquad$
8. $T(ax^2 + bx + c) = cx^2 + bx + a$

9. $T(ax^2 + bx + c) = ax^2$
$\qquad\qquad$
10. $T(ax^2 + bx + c) = (c + b)x$

7

Eigenvalues and Eigenvectors

7.1 EIGENVALUES AND EIGENVECTORS

Consider the linear transformation $T: R^2 \rightarrow R^2$ defined by

$$T(\mathbf{x}) = A\mathbf{x}$$

where

$$A = \begin{bmatrix} 1 & 2 \\ 2 & 1 \end{bmatrix}$$

We would like to describe in a simple fashion how this linear transformation affects vectors in R^2. This is not easy to do if we restrict ourselves to working with the natural basis for R^2. Suppose that we do not have this restriction and choose the basis

$$B = \left\{ \begin{bmatrix} 1 \\ 1 \end{bmatrix}, \begin{bmatrix} 1 \\ -1 \end{bmatrix} \right\}$$

(Why this particular basis is chosen will be explained later.) Notice that any 2-vector

$$\mathbf{x} = \begin{bmatrix} a \\ b \end{bmatrix}$$

can be written as

$$\mathbf{x} = \begin{bmatrix} a \\ b \end{bmatrix} = \frac{a+b}{2}\begin{bmatrix} 1 \\ 1 \end{bmatrix} + \frac{a-b}{2}\begin{bmatrix} 1 \\ -1 \end{bmatrix}$$

$$= \begin{bmatrix} \dfrac{a+b}{2} \\[2mm] \dfrac{a-b}{2} \end{bmatrix}_B$$

and that

(1) $$A\begin{bmatrix} 1 \\ 1 \end{bmatrix} = 3\begin{bmatrix} 1 \\ 1 \end{bmatrix} \qquad A\begin{bmatrix} 1 \\ -1 \end{bmatrix} = (-1)\begin{bmatrix} 1 \\ -1 \end{bmatrix}$$

Hence

$$A\mathbf{x} = A\begin{bmatrix} a \\ b \end{bmatrix} = \frac{a+b}{2}A\begin{bmatrix} 1 \\ 1 \end{bmatrix} + \frac{a-b}{2}A\begin{bmatrix} 1 \\ -1 \end{bmatrix}$$

$$= \frac{a+b}{2}3\begin{bmatrix} 1 \\ 1 \end{bmatrix} + \frac{a-b}{2}(-1)\begin{bmatrix} 1 \\ -1 \end{bmatrix}$$

$$= \begin{bmatrix} 3\left(\dfrac{a+b}{2}\right) \\[2mm] -\dfrac{a-b}{2} \end{bmatrix}_B$$

Evidently the transformation T multiplies the first component of $[\mathbf{x}]_B$ by 3 and the second component by -1. Hence the matrix representation of T relative to the basis B is

$$\begin{bmatrix} 3 & 0 \\ 0 & -1 \end{bmatrix}$$

Thus the basis B allows us to easily describe the action of T.

The key to the simple description of T is contained in the equations in (1). These equations show that T merely multiplies each basis element by a scalar. This suggests that if we seek a simple description of a linear transformation, then we should investigate the nonzero elements on which the linear transformation acts as scalar multiplication.

Definition 1.   Let $T: V \to W$ be a linear transformation of a linear space V into a linear space W having V as a subspace. A number λ is called an **eigenvalue** of T if there is a nonzero element $\mathbf{v}$ of V such that

$$T(\mathbf{v}) = \lambda\mathbf{v}$$

If V is a linear space over the real numbers, then only real numbers λ are allowed. If V is a linear space over the complex numbers, then complex numbers λ are allowed.

With the exception of the exercises we will find eigenvalues only for linear transformations defined as multiplication by an $n \times n$ matrix.

Example 1. Consider the linear transformation $R_\theta : R^2 \to R^2$ that rotates each vector counterclockwise about the origin through an angle θ. In Example 2 of Section 6.4 we found that

$$R_\theta(x) = \begin{bmatrix} \cos \theta & -\sin \theta \\ \sin \theta & \cos \theta \end{bmatrix} x$$

for every x in R^2. If a number λ is an eigenvalue of R_θ, then

$$R_\theta(\mathbf{x}) = \lambda \mathbf{x}$$

for some nonzero element of R^2. Notice that λ must be real since the domain of R_θ is R^2. Then

$$\begin{bmatrix} \cos \theta & -\sin \theta \\ \sin \theta & \cos \theta \end{bmatrix} \mathbf{x} = \lambda \mathbf{x}$$

$$= \lambda I_2 \mathbf{x}$$

for some nonzero element of R^2. Hence

$$\mathbf{0} = \begin{bmatrix} \lambda & 0 \\ 0 & \lambda \end{bmatrix} \mathbf{x} - \begin{bmatrix} \cos \theta & -\sin \theta \\ \sin \theta & \cos \theta \end{bmatrix} \mathbf{x}$$

$$= \begin{bmatrix} \lambda - \cos \theta & \sin \theta \\ -\sin \theta & \lambda - \cos \theta \end{bmatrix} \mathbf{x}$$

for some nonzero element $\mathbf{x}$ of R^2. By Theorem 17 of Section 1.9 the matrix

$$\begin{bmatrix} \lambda - \cos \theta & \sin \theta \\ -\sin \theta & \lambda - \cos \theta \end{bmatrix}$$

is singular and, hence, its determinant is zero (Theorem 9 of Section 2.5). Therefore λ is a real root of

$$0 = \det \begin{bmatrix} \lambda - \cos \theta & \sin \theta \\ -\sin \theta & \lambda - \cos \theta \end{bmatrix}$$

$$= \lambda^2 - (2 \cos \theta)\lambda + 1$$

Using the quadratic formula, we find that

$$\lambda = \cos \theta \pm \sqrt{\cos^2 \theta - 1}$$

Since $\cos^2 \theta \le 1$ for all θ, the linear transformation R_θ has eigenvalues if and only if $\cos^2 \theta = 1$. Thus R_θ has eigenvalues if and only if $\theta = 2n\pi$ or $\theta = (2n + 1)\pi$, where n is a positive integer. In the first case we have

$$R_\theta(\mathbf{x}) = \begin{bmatrix} 1 & 0 \\ 0 & 1 \end{bmatrix} \mathbf{x} = \mathbf{x}$$

for every **x** in R^2, and in the second case

$$R_\theta(\mathbf{x}) = \begin{bmatrix} -1 & 0 \\ 0 & -1 \end{bmatrix} \mathbf{x} = -\mathbf{x}$$

for every **x** in R^2. The only rotations having eigenvalues are those that leave every vector unchanged and those that reverse the direction of every vector. Hence not every linear transformation has an eigenvalue.

Given an $n \times n$ matrix A with real components, we can define a linear transformation $T: R^n \to R^n$ by

$$T(\mathbf{x}) = A\mathbf{x}$$

It is also possible to define a linear transformation $T': C^n \to C^n$, where C^n is the linear space of all n-tuples of complex numbers (see Exercise 25 of Section 4.2), by

$$T'(\mathbf{x}) = A\mathbf{x}$$

This leads to an ambiguous situation if the matrix A is given without stating the domain of the linear transformation. In the first case the eigenvalues, if any, of A will be real numbers, while in the second case the eigenvalues may be complex numbers. This distinction is important because the linear transformation T may have no eigenvalues, while the linear transformation T' always has exactly n eigenvalues, if we count them properly. For the most part we will do calculations with matrices for which the linear transformation T' has real numbers as its only eigenvalues. (The major exceptions to this occur in Example 4, designated exercises, and Section 7.4.) For such matrices the eigenvalues of the linear transformations T and T' are identical. In order for the theorems of this section and the following sections to hold for all $n \times n$ matrices and to be stated concisely, we need to consider the eigenvalues of T' rather than those of T. With this in mind we define a number to be an **eigenvalue** of the matrix A if it is also an eigenvalue of the linear transformation T'. We make no assumption in this definition that T' has only real eigenvalues. If λ is an eigenvalue of A and **v** is a nonzero vector such that

(2) $$A\mathbf{v} = \lambda\mathbf{v}$$

then **v** is called an **eigenvector** of A associated with λ.

At first sight it may appear that finding the eigenvalues and eigenvectors of an $n \times n$ matrix A would require solving one equation, equation (2), which contains two unknowns: λ and **v**. In fact, this is not the case. To show this we begin by noting that the equation $A\mathbf{v} = \lambda\mathbf{v}$ is equivalent to the following equation (where I denotes the $n \times n$ identity matrix).

(3) $$(\lambda I - A)\mathbf{v} = \mathbf{0}$$

Thus finding the eigenvalues of A is equivalent to finding numbers λ such that the matrix equation in (3) has nontrivial solutions. Recall that if B is an $n \times n$ matrix then

1. The equation $B\mathbf{x} = \mathbf{0}$ has a nontrivial solution if and only if B is singular (Theorem 17 of Section 1.9).
2. B is singular if and only if det $B = 0$ (Theorem 9 of Section 2.5)

Therefore λ is an eigenvalue of A if and only if

$$\det(\lambda I - A) = 0$$

This equation is called the **characteristic equation** for A. Using either the definition of determinant (Definition 2 of Section 2.1) or cofactor expansions, we can show that

1. $\det(\lambda I - A)$ is a polynomial of degree n in λ with 1 as the coefficient of λ^n.
2. If A has integer or real components, then $\det(\lambda I - A)$ has integer or real coefficients, respectively.

The polynomial $\det(\lambda I - A)$ is called the **characteristic polynomial** of A. A consequence of the Fundamental Theorem of Algebra is that any polynomial of degree $n, n \geq 1$, has precisely n roots (allowing for complex roots and multiplicities). Summarizing the above discussion we have the following theorem.

Theorem 1.   *Let A be an $n \times n$ matrix. Then*

1. *λ is an eigenvalue of A if and only if λ is a root of $\det(\lambda I - A) = 0$.*
2. *$\det(\lambda I - A)$ is a polynomial of degree n in λ with 1 as the coefficient of λ^n.*
3. *A has precisely n eigenvalues (counting multiplicities).*
4. *The real roots of $\det(\lambda I - A) = 0$ are the eigenvalues of the linear transformation $T:R^n \to R^n$ defined by $T(\mathbf{x}) = A\mathbf{x}$.*

Once an eigenvalue λ_1 of A is found, the associated eigenvectors are found by solving the homogeneous equation $(\lambda_1 I - A)\mathbf{v} = \mathbf{0}$. These eigenvectors are the nonzero elements of the solution space of $(\lambda_1 I - A)\mathbf{v} = \mathbf{0}$. This linear space must contain a nonzero vector because the matrix $\lambda_1 I - A$ is singular. Thus computing the eigenvalues and eigenvectors of a matrix A consists of finding roots of a polynomial and finding the solution spaces of appropriate matrix equations.

Example 2.   Consider the matrix

$$A = \begin{bmatrix} 1 & 2 \\ 2 & 1 \end{bmatrix}$$

A short calculation shows that

$$\det(\lambda I - A) = \det \begin{bmatrix} \lambda - 1 & -2 \\ -2 & \lambda - 1 \end{bmatrix}$$

$$= \lambda^2 - 2\lambda - 3$$

$$= (\lambda + 1)(\lambda - 3)$$

Therefore the eigenvalues of A are -1 and 3.

To find the eigenvectors of A associated with -1 we solve the equation $(\lambda I - A)\mathbf{v} = \mathbf{0}$ when $\lambda = -1$:

$$\mathbf{0} = [(-1)I - A]\mathbf{v}$$

$$= \begin{bmatrix} (-1) - 1 & -2 \\ -2 & (-1) - 1 \end{bmatrix} \mathbf{v}$$

$$= \begin{bmatrix} -2 & -2 \\ -2 & -2 \end{bmatrix} \mathbf{v} \tag{4}$$

The augmented matrix associated with this equation is

$$\left[\begin{array}{rr|r} -2 & -2 & 0 \\ -2 & -2 & 0 \end{array}\right]$$

Using elementary row operations we easily find that this matrix is row equivalent to

$$\left[\begin{array}{rr|r} 1 & 1 & 0 \\ 0 & 0 & 0 \end{array}\right]$$

which corresponds to the single equation $v_1 + v_2 = 0$. Since $v_2 = -v_1$ any solution of equation (4) has the form

$$\left[\begin{array}{c} v_1 \\ v_2 \end{array}\right] = v_1 \left[\begin{array}{r} 1 \\ -1 \end{array}\right]$$

Therefore the eigenvectors of A associated with -1 are all the nonzero (remember an eigenvector cannot be a zero vector) scalar multiples of

$$\mathbf{u} = \left[\begin{array}{r} 1 \\ -1 \end{array}\right]$$

In order to find the eigenvectors of A associated with 3 we solve the equation $(\lambda I - A)\mathbf{v} = \mathbf{0}$ when $\lambda = 3$.

$$\mathbf{0} = (3I - A)\mathbf{v}$$

$$= \left[\begin{array}{cc} 3-1 & -2 \\ -2 & 3-1 \end{array}\right]\mathbf{v}$$

(5)
$$= \left[\begin{array}{rr} 2 & -2 \\ -2 & 2 \end{array}\right]\mathbf{v}$$

The augmented matrix associated with this equation is

$$\left[\begin{array}{rr|r} 2 & -2 & 0 \\ -2 & 2 & 0 \end{array}\right]$$

Using elementary row operations we easily find that this matrix is row equivalent to

$$\left[\begin{array}{rr|r} 1 & -1 & 0 \\ 0 & 0 & 0 \end{array}\right]$$

which corresponds to the single equation $v_1 - v_2 = 0$. Since $v_2 = v_1$ any solution of equation (5) has the form

$$\left[\begin{array}{c} v_1 \\ v_2 \end{array}\right] = v_1 \left[\begin{array}{c} 1 \\ 1 \end{array}\right]$$

Therefore the eigenvectors of A associated with 3 are all the nonzero scalar multiples of

$$\mathbf{w} = \begin{bmatrix} 1 \\ 1 \end{bmatrix}$$

Notice that for this 2×2 matrix A we have two distinct eigenvalues and two linear independent eigenvectors. This is not always the case, as the following example shows.

Example 3.   Consider the matrix

$$A = \begin{bmatrix} 3 & -1 \\ 4 & -1 \end{bmatrix}$$

Then

$$\det(\lambda I - A) = \det \begin{bmatrix} \lambda - 3 & 1 \\ -4 & \lambda + 1 \end{bmatrix}$$

$$= \lambda^2 - 2\lambda + 1$$

$$= (\lambda - 1)^2$$

Thus 1 is the only eigenvalue of A and has multiplicity two. When $\lambda = 1$ the equation $(\lambda I - A)\mathbf{v} = \mathbf{0}$ can be rewritten as

$$\mathbf{0} = [(1)I - A]\mathbf{v}$$

$$= \begin{bmatrix} 1 - 3 & 1 \\ -4 & 1 + 1 \end{bmatrix} \mathbf{v}$$

$$= \begin{bmatrix} -2 & 1 \\ -4 & 2 \end{bmatrix} \mathbf{v} \tag{6}$$

The augmented matrix associated with this equation is

$$\begin{bmatrix} -2 & 1 & | & 0 \\ -4 & 2 & | & 0 \end{bmatrix}$$

Using elementary row operations we easily find that this equation is row equivalent to

$$\begin{bmatrix} 1 & -\frac{1}{2} & | & 0 \\ 0 & 0 & | & 0 \end{bmatrix}$$

which corresponds to the single equation $v_1 - \frac{1}{2}v_2 = 0$. Since $v_2 = 2v_1$ any solution of equation (6) has the form

$$\begin{bmatrix} v_1 \\ v_2 \end{bmatrix} = v_1 \begin{bmatrix} 1 \\ 2 \end{bmatrix}$$

Therefore the eigenvectors of A are all the nonzero scalar multiples of

$$\mathbf{u} = \begin{bmatrix} 1 \\ 2 \end{bmatrix}$$

Example 4.   Consider the matrix

$$A = \begin{bmatrix} 7 & 4 & -4 \\ 4 & -8 & -1 \\ -4 & -1 & -8 \end{bmatrix}$$

A straightforward calculation shows that

$$\det(\lambda I - A) = \lambda^3 + 9\lambda^2 - 81\lambda - 729$$
$$= (\lambda - 9)(\lambda + 9)^2$$

so that 9, -9, and -9 are the eigenvalues of A. It is left as an exercise for the reader to show that the eigenvectors of A associated with 9 are all the nonzero scalar multiples of

$$\begin{bmatrix} 4 \\ 1 \\ -1 \end{bmatrix}$$

When $\lambda = -9$ the equation $(\lambda I - A)\mathbf{v} = \mathbf{0}$ can be rewritten as

$$\mathbf{0} = [(-9)I - A]\mathbf{v}$$

(7)
$$= \begin{bmatrix} -16 & -4 & 4 \\ -4 & -1 & 1 \\ 4 & 1 & -1 \end{bmatrix} \mathbf{v}$$

The augmented matrix for this equation is

$$\begin{bmatrix} -16 & -4 & 4 & | & 0 \\ -4 & -1 & 1 & | & 0 \\ 4 & 1 & -1 & | & 0 \end{bmatrix}$$

Using elementary row operations we easily find that this matrix is row equivalent to the augmented matrix

$$\begin{bmatrix} 4 & 1 & -1 & 0 \\ 0 & 0 & 0 & 0 \\ 0 & 0 & 0 & 0 \end{bmatrix}$$

which corresponds to the single equation

$$4v_1 + v_2 - v_3 = 0$$

Since $v_3 = 4v_1 + v_2$ any solution $\mathbf{v}$ of the equation in (7) can be written as

$$\mathbf{v} = \begin{bmatrix} v_1 \\ v_2 \\ v_3 \end{bmatrix} = \begin{bmatrix} v_1 \\ v_2 \\ 4v_1 + v_2 \end{bmatrix}$$

$$= v_1 \begin{bmatrix} 1 \\ 0 \\ 4 \end{bmatrix} + v_2 \begin{bmatrix} 0 \\ 1 \\ 1 \end{bmatrix} \tag{8}$$

Therefore the eigenvectors of A associated with -9 are all the nonzero linear combinations given in (8). Thus for this matrix we have two linearly independent eigenvectors, namely

$$\begin{bmatrix} 1 \\ 0 \\ 4 \end{bmatrix} \quad \text{and} \quad \begin{bmatrix} 0 \\ 1 \\ 1 \end{bmatrix}$$

associated with the eigenvalue -9 of multiplicity two. As we saw in Example 2, this is not always the case.

We close this section with an example showing that the eigenvalues of a matrix need not be real.

Example 5.   Consider the linear transformation $T: C^n \to C^n$ defined by $T(\mathbf{x}) = A\mathbf{x}$ where

$$A = \begin{bmatrix} 0 & 1 \\ -1 & 0 \end{bmatrix}$$

Evidently

$$\det(\lambda I - A) = \det \begin{bmatrix} \lambda & -1 \\ 1 & \lambda \end{bmatrix}$$

$$= \lambda^2 + 1$$

$$= (\lambda - i)(\lambda + i)$$

so that i and $-i$ are the eigenvalues of A. When $\lambda = i$ the equation $(\lambda I - A)\mathbf{v} = 0$ becomes

$$\mathbf{0} = \begin{bmatrix} i & -1 \\ 1 & i \end{bmatrix} \mathbf{v}$$

The solutions of this system are all complex multiples of

$$\mathbf{u} = \begin{bmatrix} 1 \\ i \end{bmatrix}$$

Therefore the eigenvectors of A associated with i are all the nonzero complex multiples of $\mathbf{u}$. Similarly it can be shown that the eigenvectors of A associated with

$-i$ are all the nonzero complex multiples of

$$\begin{bmatrix} 1 \\ -i \end{bmatrix}$$

Complex eigenvalues of a matrix with real components lead us to consider vectors with complex components. While such transformations are considered in some of the exercises (see Exercises 25–28), we do not wish to emphasize their properties. Exercises involving matrices with complex eigenvalues will be clearly marked.

―――――――――――――――――――― Exercises ――――――――――――――――――――

In Exercises 1–16 find the eigenvalues and associated eigenvectors of the given matrices.

1. $\begin{bmatrix} 1 & 0 \\ 0 & 2 \end{bmatrix}$ **2.** $\begin{bmatrix} 0 & 3 \\ -3 & 0 \end{bmatrix}$

3. $\begin{bmatrix} 2 & 3 \\ 1 & 4 \end{bmatrix}$ **4.** $\begin{bmatrix} 3 & 3 \\ 1 & 5 \end{bmatrix}$

5. $\begin{bmatrix} 2 & -1 \\ 1 & 4 \end{bmatrix}$ **6.** $\begin{bmatrix} 3 & -1 \\ 1 & 5 \end{bmatrix}$

7. $\begin{bmatrix} 2 & 0 & -2 \\ 3 & 4 & 3 \\ 2 & 0 & 6 \end{bmatrix}$ **8.** $\begin{bmatrix} 1 & 1 & 1 \\ 0 & 3 & 0 \\ -2 & 2 & -2 \end{bmatrix}$

9. $\begin{bmatrix} 5 & 1 & 2 \\ -4 & 0 & -2 \\ -4 & -1 & -1 \end{bmatrix}$ **10.** $\begin{bmatrix} 5 & 2 & 2 \\ -6 & -3 & -2 \\ 3 & 2 & 0 \end{bmatrix}$

11. $\begin{bmatrix} 1 & 1 & 1 \\ 0 & 2 & 2 \\ 1 & -1 & 3 \end{bmatrix}$ **12.** $\begin{bmatrix} 3 & -1 & 3 \\ 0 & 2 & 2 \\ 3 & -3 & 5 \end{bmatrix}$

13. $\begin{bmatrix} 2 & 1 & 1 \\ 1 & 2 & 1 \\ 1 & 1 & 2 \end{bmatrix}$ **14.** $\begin{bmatrix} 3 & 2 & 1 \\ 1 & 4 & 1 \\ 1 & 2 & 3 \end{bmatrix}$

15. $\begin{bmatrix} 2 & 3 & 0 & 0 \\ 1 & 4 & 0 & 0 \\ 0 & 0 & 0 & 2 \\ 0 & 0 & 2 & 0 \end{bmatrix}$ **16.** $\begin{bmatrix} 2 & -1 & 0 & 0 \\ 1 & 4 & 0 & 0 \\ 0 & 0 & 3 & -1 \\ 0 & 0 & 1 & 5 \end{bmatrix}$

17. Show that the eigenvalues of a diagonal matrix are the numbers on the diagonal.
18. Show that the eigenvalues of a triangular matrix are the numbers on the diagonal.

In Exercises 19–22 find the eigenvalues and eigenvectors of the given linear transformation from R^2 into R^2.

19. $T\left(\begin{bmatrix} x \\ y \end{bmatrix}\right) = \begin{bmatrix} 2x + 3y \\ 2x + 7y \end{bmatrix}$

20. $T\left(\begin{bmatrix} x \\ y \end{bmatrix}\right) = \begin{bmatrix} -3x + 2y \\ -5x + 4y \end{bmatrix}$

21. $T\left(\begin{bmatrix} x \\ y \end{bmatrix}\right) = \begin{bmatrix} x + y \\ x + y \end{bmatrix}$

22. $T\left(\begin{bmatrix} x \\ y \end{bmatrix}\right) = \begin{bmatrix} 2x + y \\ x + 2y \end{bmatrix}$

23. Find the eigenvalues of the linear transformation

$$T: P_2 \to P_2$$

defined by $T(ax^2 + bx + c) = cx^2 + bx + a$.

24. (*For readers who have had calculus.*) Let V denote the linear space of all differentiable functions on $(-\infty, \infty)$. Find the eigenvalues of the linear transformation $T: V \to V$ defined by $T(f) = f'$.

In Exercises 25–28 the given matrix has complex eigenvalues. Find the eigenvalues and the eigenvectors of the given matrix.

25. $\begin{bmatrix} 2 & 2 \\ -2 & 2 \end{bmatrix}$

26. $\begin{bmatrix} 2 & 1 \\ -2 & 0 \end{bmatrix}$

27. $\begin{bmatrix} 0 & 0 & 1 \\ 0 & 2 & 0 \\ -1 & 0 & 0 \end{bmatrix}$

28. $\begin{bmatrix} 0 & 0 & 3 \\ 0 & 3 & 0 \\ -3 & 0 & 0 \end{bmatrix}$

7.2 PROPERTIES OF EIGENVALUES

In the previous section we found that the complex eigenvalues of an $n \times n$ matrix A are the n roots (counting multiplicities) $\lambda_1, \lambda_2, \ldots, \lambda_n$ of the characteristic polynomial $\det(\lambda I - A)$. Recalling that 1 is the coefficient of λ^n we can factor the characteristic polynomial so that

$$\det(\lambda I - A) = (\lambda - \lambda_1)(\lambda - \lambda_2)\ldots(\lambda - \lambda_n)$$

for all λ. In particular, when $\lambda = 0$ we have

$$\det(-A) = (-\lambda_1)(-\lambda_2)\ldots(-\lambda_n)$$

Since $\det(-A) = (-1)^n \det A$ (see Exercise 14 of Section 2.1) we conclude that

$$\det A = \lambda_1 \lambda_2 \ldots \lambda_n$$

Thus the determinant of A is the product of its eigenvalues. Hence $\det A = 0$ if and only if one of its eigenvalues is 0. Therefore A is singular if and only if one of its eigenvalues is zero. These results are important, so we state them as a theorem.

Theorem 2.   Let A be an $n \times n$ matrix.

1. If $\lambda_1, \lambda_2, \ldots, \lambda_n$ are the eigenvalues (counting multiplicities) of A, then $\det A = \lambda_1 \lambda_2 \ldots \lambda_n$.
2. A is singular if and only if 0 is an eigenvalue of A.

Example 1.   In Example 2 of the previous section we found that $\lambda_1 = -1$ and $\lambda_2 = 3$ are the eigenvalues of

$$A = \begin{bmatrix} 1 & 2 \\ 2 & 1 \end{bmatrix}$$

Then

$$\det A = (1)(1) - (2)(2)$$
$$= -3 = (-1)(3) = \lambda_1 \lambda_2$$

Example 2.   In Example 4 of the previous section we found that $\lambda_1 = -9, \lambda_2 = -9$, and $\lambda_3 = 9$ are the eigenvalues of

$$A = \begin{bmatrix} 7 & 4 & -4 \\ 4 & -8 & -1 \\ -4 & -1 & -8 \end{bmatrix}$$

Straightforward calculations show

$$\det A = 729$$
$$= (-9)(-9)(9) = \lambda_1 \lambda_2 \lambda_3$$

Example 3.   In Example 5 of the previous section we found that $\lambda_1 = i$ and $\lambda_2 = -i$ are the eigenvalues of

$$A = \begin{bmatrix} 0 & 1 \\ -1 & 0 \end{bmatrix}$$

Then

$$\det A = (0)(0) - (1)(-1) = 1 = (i)(-i) = \lambda_1 \lambda_2$$

Notice that in each of the previous examples the sum of the eigenvalues of the matrix is the same as the sum of the diagonal components of the matrix. The sum $a_{11} + a_{22} + \cdots + a_{nn}$ of the diagonal components of an $n \times n$ matrix $A = (a_{ij})$ is called the **trace** of A and is denoted by $\operatorname{tr}(A)$. The following theorem shows that our observation concerning the equality of the sum of the eigenvalues and the trace of A holds for any $n \times n$ matrix A.

Theorem 3.   *Let $\lambda_1, \lambda_2, \ldots, \lambda_n$ be the eigenvalues (counting multiplicities) of an $n \times n$ matrix A. Then*

$$\operatorname{tr}(A) = \lambda_1 + \lambda_2 + \cdots + \lambda_n$$

Proof: The proof of this theorem is not difficult, but it requires a knowledge of the relationship between the roots of a polynomial and the coefficients of that polynomial. For simplicity we will prove only the case when $n = 2$. The characteristic polynomial $p(\lambda)$ can be written as

$$p(\lambda) = (\lambda - \lambda_1)(\lambda - \lambda_2)$$
$$= \lambda^2 - (\lambda_1 + \lambda_2)\lambda + \lambda_1 \lambda_2$$

and as

$$p(\lambda) = \det \begin{bmatrix} \lambda - a_{11} & a_{12} \\ a_{21} & \lambda - a_{22} \end{bmatrix}$$

$$= \lambda^2 - (a_{11} + a_{22})\lambda + (a_{11}a_{22} - a_{12}a_{21})$$

The coefficients of λ must be the same in each representation of $p(\lambda)$. Therefore $\text{tr}(A) = a_{11} + a_{22} = \lambda_1 + \lambda_2$. ∎

We can use this theorem to guard against computational errors, such as sign mistakes, when determining the eigenvalues of a matrix. For example, suppose that in factoring the characteristic polynomial $p(\lambda) = \lambda^2 - 2\lambda - 3$ of

$$A = \begin{bmatrix} 1 & 2 \\ 2 & 1 \end{bmatrix}$$

we obtained $p(\lambda) = (\lambda - 1)(\lambda + 3)$ instead of the correct factorization $p(\lambda) = (\lambda + 1)(\lambda - 3)$. Then we would obtain the incorrect numbers $\mu_1 = 1$ and $\mu_2 = -3$ as eigenvalues. Theorem 3 allows us to easily detect this error because

$$\text{tr}(A) = 1 + 1 = 2 \neq 1 + (-3) = \mu_1 + \mu_2.$$

The final theorem of this section relates the eigenvalues of A to those of A^t and A^{-1} (if it exists).

Theorem 4.   *Let A be an $n \times n$ matrix with eigenvalues $\lambda_1, \lambda_2, \ldots, \lambda_n$ (counting multiplicities).*

   1. *A and A^t have the same eigenvalues.*
   2. *If A is nonsingular, then $\lambda_1^{-1}, \lambda_2^{-1}, \ldots, \lambda_n^{-1}$ are the eigenvalues of A^{-1}.*

Proof: We will prove the first part while leaving the second as an exercise. To begin we recall that the determinant of a matrix is the same as the determinant of the transpose of that matrix (Theorem 3 of Section 2.3). Using Exercise 19 of Section 2.3 we have

$$\det(\lambda I - A) = \det(\lambda I - A)^t$$
$$= \det(\lambda I^t - A^t)$$
$$= \det(\lambda I - A^t)$$

so that A and A^t have the same characteristic polynomial. Therefore A and A^t have the same eigenvalues. ∎

Example 4.   In Example 3 of the previous section we found that 1 is an eigenvalue of multiplicity two of

$$A = \begin{bmatrix} 3 & -1 \\ 4 & -1 \end{bmatrix}$$

The characteristic polynomial of A^t is

$$\det(\lambda I - A^t) = \det \begin{bmatrix} \lambda - 3 & -4 \\ 1 & \lambda + 1 \end{bmatrix}$$

$$= (\lambda - 1)^2$$

Thus 1 is an eigenvalue of multiplicity two of A^t, as is guaranteed by part 1 of Theorem 4.

Example 5.   In Example 2 of the previous section we found that -1 and 3 are the eigenvalues of

$$A = \begin{bmatrix} 1 & 2 \\ 2 & 1 \end{bmatrix}$$

with associated eigenvectors

$$\mathbf{u} = \begin{bmatrix} 1 \\ -1 \end{bmatrix} \qquad \mathbf{v} = \begin{bmatrix} 1 \\ 1 \end{bmatrix}$$

respectively. That is,

$$A\mathbf{u} = -\mathbf{u} \qquad A\mathbf{v} = 3\mathbf{v}$$

Then

$$\mathbf{v} = (A^{-1}A)\mathbf{v} = A^{-1}(A\mathbf{v}) = A^{-1}(3\mathbf{v}) = 3A^{-1}\mathbf{v}$$

so that

$$A^{-1}\mathbf{v} = \frac{1}{3}\mathbf{v}$$

A similar calculation shows that

$$A^{-1}\mathbf{u} = -\mathbf{u}$$

We have shown that 3^{-1} and $-1^{-1}(= -1)$ are the eigenvalues of A^{-1}. Thus the eigenvalues of A^{-1} are the reciprocals of the eigenvalues of A, as is guaranteed by part 2 of Theorem 4.

————————————————— **Exercises** —————————————————

In Exercises 1–8 verify the conclusions of Theorems 2 (part 1), 3, and 4 (part 1) for the given matrices.

1. $\begin{bmatrix} 1 & 2 \\ 0 & 4 \end{bmatrix}$ **2.** $\begin{bmatrix} 1 & 2 \\ 2 & 4 \end{bmatrix}$

3. $\begin{bmatrix} 1 & 3 \\ 3 & 0 \end{bmatrix}$ **4.** $\begin{bmatrix} 4 & -2 \\ -2 & 2 \end{bmatrix}$

5. $\begin{bmatrix} 1 & 0 & 2 \\ 0 & 2 & 3 \\ 0 & 3 & 1 \end{bmatrix}$ **6.** $\begin{bmatrix} 4 & 0 & 5 \\ 0 & 6 & 0 \\ 5 & 0 & 0 \end{bmatrix}$

7. $\begin{bmatrix} 2 & 4 & 0 & 0 \\ 4 & 3 & 0 & 0 \\ 0 & 0 & -5 & 0 \\ 0 & 0 & 0 & 9 \end{bmatrix}$ **8.** $\begin{bmatrix} 7 & 0 & 0 & 0 \\ 0 & 1 & 1 & 0 \\ 0 & 1 & 2 & 0 \\ 0 & 0 & 0 & 6 \end{bmatrix}$

9. Prove part 2 of Theorem 4. That is, show that if A is nonsingular, then the eigenvalues of A^{-1} are the reciprocals of the eigenvalues of A.

10. Show that if λ is an eigenvalue of A, then λ^2 is an eigenvalue of A^2. (Hint: Multiply each side of $A\mathbf{v} = \lambda\mathbf{v}$ by A.)

11. Show that if λ is an eigenvalue of A and k is a positive integer, then λ^k is an eigenvalue of A^k.

12. Let c be a nonzero number and A be an $n \times n$ matrix. Show that the eigenvalues of cA are c times the eigenvalues of A.

13. Let A be an $n \times n$ matrix having a number λ as an eigenvalue. By Exercise 12, the number $c\lambda$ is an eigenvalue of cA for any nonzero c. Show that if $\mathbf{v}$ is an eigenvector of A associated with λ, then $\mathbf{v}$ is also an eigenvector of cA associated with $c\lambda$.

14. If λ^2 is an eigenvalue of A^2, is it necessarily true that λ is an eigenvalue of A?

15. If λ_1 and λ_2 are eigenvalues of a matrix A, is it necessarily true that $\lambda_1 + \lambda_2$ is an eigenvalue of A?

16. Suppose that λ is an eigenvalue of an $n \times n$ matrix A and that $c\lambda$ is also an eigenvalue for every real number c. What is the value of λ?

17. Show that if A is similar to B (see explanation preceding Exercise 9 of Section 6.6), then A and B have the same eigenvalues. (Hint: Show that if $A = P^{-1}BP$, then $\lambda I - A = P^{-1}(\lambda I - B)P$.)

7.3 DIAGONALIZATION

At the beginning of Section 7.1 we found that the action of the linear transformation $T: R^2 \to R^2$ defined by

$$T(\mathbf{x}) = \begin{bmatrix} 1 & 2 \\ 2 & 1 \end{bmatrix} \mathbf{x}$$

is easily described if we use the basis

$$B = \left\{ \begin{bmatrix} 1 \\ 1 \end{bmatrix}, \begin{bmatrix} 1 \\ -1 \end{bmatrix} \right\}$$

instead of the natural basis for R^2. In Example 2 of the same section we found that the elements of B are eigenvectors of

$$A = \begin{bmatrix} 1 & 2 \\ 2 & 1 \end{bmatrix}$$

In general, given a linear transformation $T: R^n \to R^n$ it is advantageous to use a basis for R^n that is composed of eigenvectors of the standard matrix for T. Unfortunately, this is not always possible. In this section we determine when such a basis exists and the representation of T relative to such a basis. We begin with the following theorem.

Theorem 5.   *Let A be an $n \times n$ matrix. If $\lambda_1, \lambda_2, \ldots, \lambda_k$ are distinct real eigenvalues of A and $\mathbf{v}_1, \mathbf{v}_2, \ldots, \mathbf{v}_k$ are eigenvectors of A associated with $\lambda_1, \lambda_2, \ldots, \lambda_k$, respectively, then $\{\mathbf{v}_1, \mathbf{v}_2, \ldots, \mathbf{v}_k\}$ is linearly independent.*

Proof: Suppose that $\{v_1, v_2, \ldots, v_k\}$ is linearly dependent. Since $v_1 \neq 0$, the set $\{v_1\}$ is linearly independent. Let p be the largest integer such that $\{v_1, v_2, \ldots, v_p\}$ is linearly independent. Evidently $1 \leq p < k$ and the set $\{v_1, v_2, \ldots, v_{p+1}\}$ is linearly dependent. Let $c_1, c_2, \ldots, c_{p+1}$ be scalars such that

(1)
$$c_1 v_1 + c_2 v_2 + \cdots + c_p v_p + c_{p+1} v_{p+1} = 0$$

Then

$$
\begin{aligned}
0 = A0 &= A(c_1 v_1 + c_2 v_2 + \cdots + c_p v_p + c_{p+1} v_{p+1}) \\
&= c_1 A v_1 + c_2 A v_2 + \cdots + c_p A v_p + c_{p+1} A v_{p+1} \\
\text{(2)} \quad &= c_1 \lambda_1 v_1 + c_2 \lambda_2 v_2 + \cdots + c_p \lambda_p v_p + c_{p+1} \lambda_{p+1} v_{p+1}
\end{aligned}
$$

We now multiply each side of equation (1) by λ_{p+1} and solve for $c_{p+1} \lambda_{p+1} v_{p+1}$:

$$c_{p+1} \lambda_{p+1} v_{p+1} = -\lambda_{p+1}(c_1 v_1 + c_2 v_2 + \cdots + c_p v_p)$$

Using this expression for $c_{p+1} \lambda_{p+1} v_{p+1}$ in (2) yields

$$
\begin{aligned}
0 &= c_1 \lambda_1 v_1 + c_2 \lambda_2 v_2 + \cdots + c_p \lambda_p v_p - \lambda_{p+1}(c_1 v_1 + c_2 v_2 + \cdots + c_p v_p) \\
&= c_1(\lambda_1 - \lambda_{p+1})v_1 + c_2(\lambda_2 - \lambda_{p+1})v_2 + \cdots + c_p(\lambda_p - \lambda_{p+1})v_p
\end{aligned}
$$

Since $\{v_1, v_2, \ldots, v_p\}$ is linearly independent we must have $c_1(\lambda_1 - \lambda_{p+1}) = 0$, $c_2(\lambda_2 - \lambda_{p+1}) = 0, \ldots, c_p(\lambda_p - \lambda_{p+1}) = 0$. By assumption $\lambda_1, \lambda_2, \ldots, \lambda_k$ are distinct. Therefore $\lambda_i - \lambda_{p+1} \neq 0$ for $i = 1, 2, \ldots, p$ and we must have $c_1 = 0, c_2 = 0, \ldots, c_p = 0$. Equation (1) now becomes

$$c_{p+1} v_{p+1} = 0$$

Since $v_{p+1} \neq 0$ we must have $c_{p+1} = 0$. We have shown that $c_1 = 0, c_2 = 0, \ldots, c_{p+1} = 0$ are the only scalars satisfying equation (1). Consequently $\{v_1, v_2, \ldots, v_{p+1}\}$ is linearly independent. This contradicts our assumption that $\{v_1, v_2, \ldots, v_{p+1}\}$ is linearly dependent, which is based on the assumption that $\{v_1, v_2, \ldots, v_k\}$ is linearly dependent. Therefore $\{v_1, v_2, \ldots, v_k\}$ must be linearly independent. This completes the proof. ∎

Of special interest is the following immediate consequence of Theorem 5.

Theorem 6. *Let A be an $n \times n$ matrix with n distinct real eigenvalues. Then there is a basis of R^n consisting of eigenvectors of A.*

Example 1. In Example 2 of Section 7.1 we considered a matrix having distinct eigenvalues -1 and 3 with

$$v_1 = \begin{bmatrix} 1 \\ -1 \end{bmatrix} \quad \text{and} \quad v_2 = \begin{bmatrix} 1 \\ 1 \end{bmatrix}$$

as associated eigenvectors. Clearly the set $\{v_1, v_2\}$ is linearly independent and, therefore, is a basis for R^2.

Example 2.   It is important to note that an $n \times n$ matrix may have n linearly independent eigenvectors without having distinct n eigenvalues. For example, the matrix

$$A = \begin{bmatrix} 7 & 4 & -4 \\ 4 & -8 & -1 \\ -4 & -1 & -8 \end{bmatrix}$$

considered in Example 4 of Section 7.1 has

$$\mathbf{u} = \begin{bmatrix} 4 \\ 1 \\ -1 \end{bmatrix} \qquad \mathbf{v} = \begin{bmatrix} 1 \\ 0 \\ 4 \end{bmatrix} \qquad \mathbf{w} = \begin{bmatrix} 0 \\ 1 \\ 1 \end{bmatrix}$$

as eigenvectors. By Theorem 10 of Section 4.6 the set $\{\mathbf{u}, \mathbf{v}, \mathbf{w}\}$ is linearly independent if the matrix B having $\mathbf{u}$, $\mathbf{v}$, and $\mathbf{w}$ as its columns is nonsingular. Hence this set of 3-vectors is linearly independent if det $B \neq 0$. Since

$$\det \begin{bmatrix} 4 & 1 & 0 \\ 1 & 0 & 1 \\ -1 & 4 & 1 \end{bmatrix} = -\det \begin{bmatrix} 1 & 0 & 1 \\ 0 & 1 & -4 \\ 0 & 4 & 2 \end{bmatrix} \qquad \begin{array}{l} R_1 \leftrightarrow R_2 \\ (-4)R_1 + R_2 \to R_2 \\ R_1 + R_3 \to R_3 \end{array}$$

$$= -\det \begin{bmatrix} 1 & -4 \\ 4 & 2 \end{bmatrix} \qquad \begin{array}{l} \text{Cofactor expansion} \\ \text{along the first column.} \end{array}$$

$$= -18$$

we conclude that the set $\{\mathbf{u}, \mathbf{v}, \mathbf{w}\}$ of eigenvectors of A is linearly independent and, therefore, is a basis of R^3. Hence, it is possible for a basis for R^n to consist of eigenvectors of an $n \times n$ matrix without the matrix having n distinct eigenvalues.

We have already noted that the linear transformation $T': R^2 \to R^2$ defined by $T'(\mathbf{x}) = A'\mathbf{x}$, where

$$A' = \begin{bmatrix} 1 & 2 \\ 2 & 1 \end{bmatrix}$$

is represented by a diagonal matrix with respect to the basis

$$\left\{ \begin{bmatrix} 1 \\ 1 \end{bmatrix}, \begin{bmatrix} 1 \\ -1 \end{bmatrix} \right\}$$

for R^2 consisting of eigenvectors of A'. In fact, every linear transformation $T: R^n \to R^n$ defined by $T(\mathbf{x}) = A\mathbf{x}$, where A is an $n \times n$ matrix, can be represented by a diagonal matrix if there is a basis for R^n consisting of eigenvectors of A. To see this, let $B = \{\mathbf{v}_1, \mathbf{v}_2, \ldots, \mathbf{v}_n\}$ be an ordered basis for R^n consisting of eigenvectors of A that are associated with the eigenvalues $\lambda_1, \lambda_2, \ldots, \lambda_n$, respectively. We make no assumption that the eigenvalues are distinct. If $\mathbf{x}$ is any n-vector, then there are numbers $c_1, c_2, \ldots, c_n$ such that

$$\mathbf{x} = c_1\mathbf{v}_1 + c_2\mathbf{v}_2 + \cdots + c_n\mathbf{v}_n = \begin{bmatrix} c_1 \\ c_2 \\ \vdots \\ c_n \end{bmatrix}_B$$

and

$$
\begin{aligned}
A\mathbf{x} &= c_1 A\mathbf{v}_1 + c_2 A\mathbf{v}_2 + \cdots + c_n A\mathbf{v}_n \\
&= c_1\lambda_1\mathbf{v}_1 + c_2\lambda_2\mathbf{v}_2 + \cdots + c_n\lambda_n\mathbf{v}_n \\
&= \begin{bmatrix} \lambda_1 c_1 \\ \lambda_2 c_2 \\ \vdots \\ \lambda_n c_n \end{bmatrix}_B = \begin{bmatrix} \lambda_1 & 0 & 0 & \cdots & 0 \\ 0 & \lambda_2 & 0 & \cdots & 0 \\ \vdots & \vdots & \vdots & & \vdots \\ 0 & 0 & 0 & \cdots & \lambda_n \end{bmatrix}\begin{bmatrix} c_1 \\ c_2 \\ \vdots \\ c_n \end{bmatrix}_B
\end{aligned}
$$

Therefore the matrix representation of T with respect to the basis B is the diagonal matrix having $\lambda_1, \lambda_2, \ldots, \lambda_n$ on its diagonal. Thus if R^n has a basis B consisting of eigenvectors of A, then T can be represented by a diagonal matrix with respect to the basis B. The converse of this is also true. That is, if there is an ordered basis $B' = \{\mathbf{u}_1, \mathbf{u}_2, \ldots, \mathbf{u}_n\}$ of R^n such that the matrix representation ${}_{B'}A_{B'}$ for T with respect to B' is a diagonal matrix,

$$
{}_{B'}A_{B'} = \begin{bmatrix} d_1 & 0 & 0 & \cdots & 0 \\ 0 & d_2 & 0 & \cdots & 0 \\ \vdots & \vdots & \vdots & & \vdots \\ 0 & 0 & 0 & \cdots & d_n \end{bmatrix}
$$

then each $\mathbf{u}_i$ is an eigenvector of A associated with the eigenvalue d_i. In order to prove this we notice that

(3)
$$A\mathbf{u}_i = T(\mathbf{u}_i) = {}_{B'}A_{B'}\begin{bmatrix} 0 \\ \vdots \\ 1 \\ \vdots \\ 0 \end{bmatrix}_{B'} = \begin{bmatrix} 0 \\ \vdots \\ d_i \\ \vdots \\ 0 \end{bmatrix}_{B'} = d_i\mathbf{u}_i$$

Since B' is a basis none of its elements is the zero vector. Thus d_i is an eigenvalue of A with $\mathbf{u}_i$ as an associated eigenvector. This shows that if A can be represented by a diagonal matrix D with respect to a basis B' of R^n, then the basis consists of eigenvectors of A and the diagonal elements of D are eigenvalues of A.

Notice further that $A\mathbf{u}_i$ is the ith column of AP, where P is the matrix having $\mathbf{u}_1, \mathbf{u}_2, \ldots, \mathbf{u}_n$ as its columns. It is left for the reader to show that $d_i\mathbf{u}_i$ is the ith column of PD. By equation (3) we have $A\mathbf{u}_i = d_i\mathbf{u}_i$ for each i so that $AP = PD$. Since the columns of P are linearly independent, P is nonsingular. (Theorem 13 of Section 4.7) and we have $P^{-1}AP = D$. We have proved the following theorem.

Theorem 7. *Consider the linear transformation $T: R^n \to R^n$ defined by $T(\mathbf{x}) = A\mathbf{x}$ where A is an $n \times n$ matrix.*

1. *If T has n linearly independent eigenvectors $\mathbf{v}_1, \mathbf{v}_2, \ldots, \mathbf{v}_n$ associated with eigenvalues $\lambda_1, \lambda_2, \ldots, \lambda_n$, respectively, then the representation of T with respect to the ordered basis $B = \{\mathbf{v}_1, \mathbf{v}_2, \ldots, \mathbf{v}_n\}$ is*

$$\begin{bmatrix} \lambda_1 & 0 & 0 & \cdots & 0 \\ 0 & \lambda_2 & 0 & \cdots & 0 \\ \vdots & \vdots & \vdots & & \vdots \\ 0 & 0 & 0 & \cdots & \lambda_n \end{bmatrix}$$

2. *If T can be represented by a diagonal matrix D, with $d_1, d_2, \ldots, d_n$ as its diagonal elements, with respect to an ordered basis $B' = \{\mathbf{u}_1, \mathbf{u}_2, \ldots, \mathbf{u}_n\}$, then each d_i is an eigenvalue of A with $\mathbf{u}_i$ as an associated eigenvector. Moreover, if P is the matrix having $\mathbf{u}_1, \mathbf{u}_2, \ldots, \mathbf{u}_n$ as its columns, then $P^{-1}AP = D$.*

Example 3. In Example 4 of Section 7.1 and Example 2 of this section we found that R^3 has a basis

$$B = \left\{ \begin{bmatrix} 4 \\ 1 \\ -1 \end{bmatrix}, \begin{bmatrix} 1 \\ 0 \\ 4 \end{bmatrix}, \begin{bmatrix} 0 \\ 1 \\ 1 \end{bmatrix} \right\}$$

consisting of eigenvectors of

$$A = \begin{bmatrix} 7 & 4 & -4 \\ 4 & -8 & -1 \\ -4 & -1 & -8 \end{bmatrix}$$

associated with the eigenvalues 9, -9, and -9 respectively. Therefore the matrix representation with respect to the ordered basis B of the linear transformation $T: R^3 \to R^3$ defined by $T(\mathbf{x}) = A\mathbf{x}$ is

$$_B A_B = \begin{bmatrix} 9 & 0 & 0 \\ 0 & -9 & 0 \\ 0 & 0 & -9 \end{bmatrix}$$

Moreover $P^{-1}AP = {}_B A_B$, where

$$P = \begin{bmatrix} 4 & 1 & 0 \\ 1 & 0 & 1 \\ -1 & 4 & 1 \end{bmatrix}$$

Example 4. In Example 3 of Section 7.1 we found that the matrix

$$A = \begin{bmatrix} 3 & -1 \\ 4 & -1 \end{bmatrix}$$

does not have two linearly independent eigenvectors. Therefore the linear transformation $T : R^2 \to R^2$ defined by $T(\mathbf{x}) = A\mathbf{x}$ cannot be represented by a diagonal matrix with respect to any basis for R^2.

Definition 2.   An $n \times n$ matrix A is said to be **diagonalizable** if there is an $n \times n$ diagonal matrix D and a nonsingular matrix P such that $P^{-1}AP = D$.

An immediate consequence of Theorem 7 is the following theorem that tells us precisely which matrices are diagonalizable.

Theorem 8.   *An $n \times n$ matrix A is diagonalizable if and only if A has n linearly independent eigenvectors.*

Example 5.   In Example 2 of Section 6.5 we found that the Fibonacci numbers p_k satisfy

(4)
$$P_k = A^{k-2} \begin{bmatrix} 1 \\ 1 \end{bmatrix}$$

where

$$P_k = \begin{bmatrix} p_{k-1} \\ p_k \end{bmatrix} \qquad A = \begin{bmatrix} 0 & 1 \\ 1 & 1 \end{bmatrix}$$

We now find an explicit formula for p_k. To begin we diagonalize A. The eigenvalues of A are the roots of

$$0 = \det \begin{bmatrix} \lambda & -1 \\ -1 & \lambda - 1 \end{bmatrix} = \lambda^2 - \lambda - 1$$

Hence

$$\lambda_1 = \frac{1 + \sqrt{5}}{2} \qquad \lambda_2 = \frac{1 - \sqrt{5}}{1}$$

are the eigenvalues of A. When $\lambda = \lambda_1$ the equation $(\lambda I - A)\mathbf{v} = \mathbf{0}$ becomes

$$\begin{bmatrix} \dfrac{1 + \sqrt{5}}{2} & -1 \\ -1 & \dfrac{-1 + \sqrt{5}}{2} \end{bmatrix} \begin{bmatrix} v_1 \\ v_2 \end{bmatrix} = \begin{bmatrix} 0 \\ 0 \end{bmatrix}$$

so that

$$\frac{1 + \sqrt{5}}{2} v_1 - v_2 = 0$$

Hence

$$\mathbf{u} = \begin{bmatrix} 1 \\ \dfrac{1 + \sqrt{5}}{2} \end{bmatrix} = \begin{bmatrix} 1 \\ \lambda_1 \end{bmatrix}$$

is an eigenvector of A associated with λ_1. A similar argument shows that

$$\mathbf{v} = \begin{bmatrix} 1 \\ \lambda_2 \end{bmatrix}$$

is an eigenvector of A associated with λ_2. Since $\lambda_1 \neq \lambda_2$ the set $\{\mathbf{u}, \mathbf{v}\}$ is linearly independent. By Theorem 7, $Q^{-1}AQ = D$, where

$$Q = \begin{bmatrix} 1 & 1 \\ \lambda_1 & \lambda_2 \end{bmatrix}, \qquad D = \begin{bmatrix} \lambda_1 & 0 \\ 0 & \lambda_2 \end{bmatrix}$$

Thus $A = QDQ^{-1}$. Notice that

$$\begin{aligned} A^2 &= AA = (QDQ^{-1})(QDQ^{-1}) = QD^2Q^{-1} \\ A^3 &= AA^2 = (QDQ^{-1})(QD^2Q^{-1}) = QD^3Q^{-1} \\ &\vdots \\ A^{k-2} &= QD^{k-2}Q^{-1} \end{aligned}$$

Moreover

$$D^{k-2} = \begin{bmatrix} \lambda_1^{k-2} & 0 \\ 0 & \lambda_2^{k-2} \end{bmatrix}$$

and, by Theorem 12 of Section 1.8,

$$Q^{-1} = \frac{1}{\lambda_2 - \lambda_1} \begin{bmatrix} \lambda_2 & -1 \\ -\lambda_1 & 1 \end{bmatrix}$$

From equation (4) we now have

$$\begin{aligned} P_k &= A^{k-2} \begin{bmatrix} 1 \\ 1 \end{bmatrix} \\ &= QD^{k-2}Q^{-1} \begin{bmatrix} 1 \\ 1 \end{bmatrix} \\ &= \frac{1}{\lambda_2 - \lambda_1} \begin{bmatrix} 1 & 1 \\ \lambda_1 & \lambda_2 \end{bmatrix} \begin{bmatrix} \lambda_1^{k-2} & 0 \\ 0 & \lambda_2^{k-2} \end{bmatrix} \begin{bmatrix} \lambda_2 & -1 \\ -\lambda_1 & 1 \end{bmatrix} \begin{bmatrix} 1 \\ 1 \end{bmatrix} \\ &= \frac{1}{\lambda_2 - \lambda_1} \begin{bmatrix} \lambda_1^{k-2}\lambda_2 - \lambda_1\lambda_2^{k-2} - \lambda_1^{k-2} + \lambda_2^{k-2} \\ \lambda_1^{k-1}\lambda_2 - \lambda_1\lambda_2^{k-1} - \lambda_1^{k-1} + \lambda_2^{k-1} \end{bmatrix} \end{aligned}$$

so that

$$p_k = \frac{1}{\lambda_2 - \lambda_1}(\lambda_1^{k-1}\lambda_2 - \lambda_1\lambda_2^{k-1} - \lambda_1^{k-1} + \lambda_2^{k-1})$$

$$= -\frac{1}{\sqrt{5}}\left(\frac{1-\sqrt{5}}{2}\lambda_1^{k-1} - \frac{1+\sqrt{5}}{2}\lambda_2^{k-1} - \lambda_1^{k-1} + \lambda_2^{k-1}\right)$$

$$= -\frac{1}{\sqrt{5}}\left(-\frac{1+\sqrt{5}}{2}\lambda_1^{k-1} + \frac{1-\sqrt{5}}{2}\lambda_2^{k-1}\right)$$

$$= \frac{1}{\sqrt{5}}(\lambda_1^k - \lambda_2^k)$$

$$= \frac{(1+\sqrt{5})^k - (1-\sqrt{5})^k}{2^k\sqrt{5}}$$

$$= \frac{1}{\sqrt{5}}\left(\frac{1+\sqrt{5}}{2}\right)^k - \frac{1}{\sqrt{5}}\left(\frac{1-\sqrt{5}}{2}\right)^k$$

Thus we have found an explicit formula for p_k. Since $|(1-\sqrt{5})/2| < 1$, we have

$$\left|\frac{1}{\sqrt{5}}\left(\frac{1-\sqrt{5}}{2}\right)^k\right| < \frac{1}{\sqrt{5}}\left(\frac{\sqrt{5}-1}{2}\right) < .14$$

Recalling that p_k is an integer, we conclude that it must be the integer closest to

$$\frac{1}{\sqrt{5}}\left(\frac{1+\sqrt{5}}{2}\right)^k$$

───────────────────────── **Exercises** ─────────────────────────

In Exercises 1–8 determine whether the matrix A is diagonalizable. If it is diagonalizable, find a matrix P such that $P^{-1}AP$ is a diagonal matrix and determine $P^{-1}AP$.

1. $\begin{bmatrix} 0 & 1 \\ 1 & 0 \end{bmatrix}$

2. $\begin{bmatrix} 2 & 3 \\ -1 & 6 \end{bmatrix}$

3. $\begin{bmatrix} 3 & 1 \\ -1 & 1 \end{bmatrix}$

4. $\begin{bmatrix} 4 & 1 \\ -1 & 2 \end{bmatrix}$

5. $\begin{bmatrix} 2 & 0 & 3 \\ 3 & 5 & -3 \\ -1 & 0 & 6 \end{bmatrix}$

6. $\begin{bmatrix} 2 & 6 & 1 \\ 0 & 5 & 0 \\ 3 & 6 & 4 \end{bmatrix}$

7. $\begin{bmatrix} 3 & 1 & 0 & 0 \\ -1 & 1 & 0 & 0 \\ 0 & 0 & 0 & 2 \\ 0 & 0 & -2 & 0 \end{bmatrix}$

8. $\begin{bmatrix} 6 & 0 & 0 & 5 \\ 1 & 2 & 0 & 5 \\ 2 & 0 & 5 & 8 \\ 1 & 0 & 0 & 2 \end{bmatrix}$

9. Let $T: R^2 \to R^2$ be the linear transformation defined by

$$T\left(\begin{bmatrix} x \\ y \end{bmatrix}\right) = \begin{bmatrix} 3 & 4 \\ -3 & -1 \end{bmatrix} \begin{bmatrix} x \\ y \end{bmatrix}$$

Find a basis for R^2 with respect to which the matrix representation for T is diagonal.

10. Let $T: R^3 \to R^3$ be the linear transformation defined by

$$T\left(\begin{bmatrix} x \\ y \\ z \end{bmatrix}\right) = \begin{bmatrix} 2 & 0 & 3 \\ 3 & 5 & -3 \\ -1 & 0 & 6 \end{bmatrix} \begin{bmatrix} x \\ y \\ z \end{bmatrix}$$

Find a basis for R^3 with respect to which the matrix representation for T is diagonal.

11. Let $T: P_2 \to P_2$ be the linear transformation defined by

$$T(a_2 x^2 + a_1 x + a_0) = (a_2 + a_1 + a_0)x^2 + 2a_1 x + 3a_0$$

Find a basis for P_2 with respect to which the matrix representation for T is diagonal.

12. Let $T: P_2 \to P_2$ be the linear transformation defined by

$$T(a_2 x^2 + a_1 x + a_0) = (a_0 + a_1)x^2 + (a_1 + a_2)x + (a_0 + a_2)$$

Find a basis for P_2 with respect to which the matrix representation for T is diagonal.

13. Show that

$$A = \begin{bmatrix} a & b \\ c & d \end{bmatrix}$$

is diagonalizable if $(a - d)^2 + 4bc > 0$.

7.4 GERSHGORIN'S THEOREM (OPTIONAL)

In Theorem 2 of Section 7.2 we found that an $n \times n$ matrix is singular if and only if 0 is an eigenvalue of A. If we could approximate the eigenvalues of A well enough to tell that none are zero, then we would know that A is nonsingular. This would be useful information when solving a matrix equation $A\mathbf{x} = \mathbf{b}$. In other applications also it is not necessary to know the exact eigenvalues, but to know merely that they have some property. For example, certain computational procedures involving matrices require that the eigenvalues of the matrix have absolute values less than one. In applications involving differential equations it is sometimes necessary to know whether the eigenvalues of a certain matrix have negative real parts. The following theorem gives an easily obtained estimation for where the eigenvalues lie in the complex plane. Before stating the theorem we recall a geometrical interpretation of an inequality involving complex numbers.

Let z_0 be any complex number and let r be any nonnegative real number. Then the set of all complex numbers z such that $|z - z_0| \le r$ consists of

1. The single point z_0 if $r = 0$.
2. All points in the disk of radius r centered at z_0 if $r \ne 0$.

Theorem 9 (***Gershgorin's Theorem***). *Let* $A = (a_{ij})$ *be an* $n \times n$ *matrix and let* D_i *denote the disk in the complex plane with center* a_{ii} *and radius the sum of the absolute values of all nondiagonal components of the ith row of* A; *that is*

$$D_i = \left\{ z : |z - a_{ii}| \le \sum_{\substack{j=1 \\ j \ne i}}^{n} |a_{ij}| \right\}$$

Then every eigenvalue of A *is contained in at least one of these disks. Moreover, the union of any* k *disks that does not intersect the remaining* $n - k$ *disks contains precisely* k (*counting multiplicities*) *eigenvalues of* A.

Proof: Let λ be any eigenvalue of A and let $\mathbf{v}$ be an associated eigenvector. Since $\mathbf{v}$ is not the zero vector, at least one of its components is nonzero. Let v_k be a component of $\mathbf{v}$ with the largest absolute value. That is, $0 < |v_k| = \max_{1 \le j \le n} |v_j|$. Since $\mathbf{v}$ is an eigenvector of A associated with λ, we have $A\mathbf{v} = \lambda\mathbf{v}$. Hence the kth components of $A\mathbf{v}$ and $\lambda\mathbf{v}$ must be identical:

$$\sum_{j=1}^{n} a_{kj}v_j = \lambda v_k$$

An algebraic rearrangement of the terms of this identity yields

$$(\lambda - a_{kk})v_k = \sum_{\substack{j=1 \\ j \ne k}}^{n} a_{kj}v_j$$

Thus

$$|\lambda - a_{kk}||v_k| = \left| \sum_{\substack{j=1 \\ j \ne k}}^{n} a_{kj}v_j \right|$$

$$\le \sum_{\substack{j=1 \\ j \ne k}}^{n} |a_{kj}||v_j|$$

Since $|v_j| \le |v_k|$ for $j = 1, 2, \ldots, n$ we have

$$|\lambda - a_{kk}||v_k| \le \sum_{\substack{j=1 \\ j \ne k}}^{n} |a_{kj}||v_k|$$

Recalling that $v_k \ne 0$ we conclude that

$$|\lambda - a_{kk}| \le \sum_{\substack{j=1 \\ j \ne k}}^{n} |a_{kj}|$$

which shows that λ is contained in D_k. The proof of the second part of the theorem involves concepts beyond the scope of this course and will be omitted. ∎

Example 1.   Consider the matrix

$$A = \begin{bmatrix} 5 & 1 & 1 & 0 \\ 0 & -6 & 2 & 1 \\ 1 & -1 & 4 & 0 \\ 0 & 1 & 0 & 2 \end{bmatrix}$$

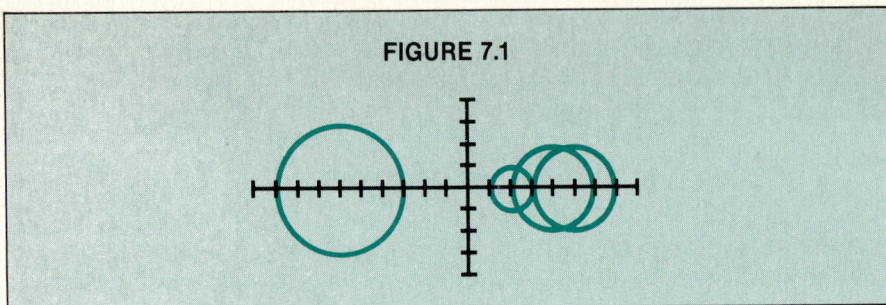

FIGURE 7.1

The disks described in the previous theorem are

$$D_1 = \{z : |z - 5| \le 2\}, \qquad D_3 = \{z : |z - 4| \le 2\}$$

$$D_2 = \{z : |z + 6| \le 3\}, \qquad D_4 = \{z : |z - 2| \le 1\}$$

By the first part of Theorem 9 each eigenvalue of A lies in at least one of the four disks depicted in Figure 7.1. The second part assures us that the disk on the left contains one eigenvalue while the group of three disks on the right contains three eigenvalues. Since the components of A are real numbers, the coefficients of the characteristic polynomial are real numbers. The complex roots of a polynomial with real coefficients occur in conjugate pairs. That is, if $a + bi$, $b \ne 0$, is a root, then $a - bi$ is also a root. Moreover $a + bi$ and $a - bi$ are symmetrically placed with respect to the (real) x-axis. Therefore the disk on the left cannot contain a complex eigenvalue $a + bi$ of A with $b \ne 0$. For if it did then $a - bi$ would also be in the disk and we would have two eigenvalues in the disk, which is impossible. Thus the root in the disk on the left is a real number. A similar argument shows that the group of disks on the right contains at least one real eigenvalue. Hence A has at least two real eigenvalues.

Notice that zero is not contained in any of the disks. Hence 0 is not an eigenvalue of A. By Theorem 2 of Section 7.2 the matrix A is nonsingular.

―――――――――――――――――――――― **Exercises** ――――――――――――――――――――――

In Exercises 1–4 draw the disks described in Theorem 9 and find an upper bound for $|\lambda|$ where λ is an eigenvalue of the given matrix.

1. $\begin{bmatrix} 5 & 2 & 4 \\ -3 & 9 & 1 \\ 1 & 1 & -5 \end{bmatrix}$

   2. $\begin{bmatrix} 7 & -1 & 1 \\ 1 & 2 & 1 \\ 2 & -3 & -8 \end{bmatrix}$

3. $\begin{bmatrix} 6 & 1 & 0 & 1 \\ 2 & 4 & -1 & 0 \\ -3 & 1 & 7 & -2 \\ 0 & -2 & -2 & 5 \end{bmatrix}$

   4. $\begin{bmatrix} 3 & 0 & 1 & 1 \\ 1 & 4 & 1 & 1 \\ 2 & -2 & 6 & 1 \\ 0 & 1 & -1 & 3 \end{bmatrix}$

5. Using Theorem 9, explain why the matrix in Exercise 3 is nonsingular.

6. Using Theorem 9, explain why the matrix in Exercise 4 is nonsingular.
7. Let $A = (a_{ij})$ be an $n \times n$ matrix and let

$$D_i' = \left\{ z : |z - a_{ii}| \leq \sum_{\substack{j=1 \\ j \neq i}}^{n} |a_{ji}| \right\}.$$

That is, D_i' is the disk in the complex plane with center at a_{ii} and radius the sum of the absolute values of all nondiagonal components of the ith column of A. Show that every eigenvalue of A is contained in at least one of these disks. (Hint: Use part 1 of Theorem 4 in Section 7.2.

8. Use Exercise 7 to find an upper bound for $|\lambda|$ where λ is an eigenvalue of the matrix in Exercise 4. Compare this with your answer to Exercise 4.

9. Use Exercise 7 to find an upper bound for $|\lambda|$ where λ is an eigenvalue of the matrix in Exercise 3. Compare this with your answer to Exercise 3.

10. An $n \times n$ matrix $A = (a_{ij})$ is called **diagonally dominant** if the absolute value of each diagonal component is greater than the sum of the absolute values of the remaining components in that row. That is,

$$|a_{ii}| > \sum_{\substack{j=1 \\ j \neq i}}^{n} |a_{ij}|$$

for $i = 1, 2, \ldots, n$. Show that a diagonally dominant matrix is nonsingular.

11. *Strict* inequality is essential in the inequality in Exercise 10. Let

$$A = \begin{bmatrix} 2 & -1 & 1 \\ -1 & 2 & 1 \\ 1 & 1 & 2 \end{bmatrix}$$

Show that $|a_{ii}| \geq \sum_{\substack{j=1 \\ j \neq i}}^{3} |a_{ij}|$ for $i = 1, 2, 3$ and that A is singular.

7.5 EIGENSPACES (OPTIONAL)

In Examples 3 and 4 of Section 7.1 we saw that it is possible for a matrix to have an eigenvalue λ with multiplicity greater than one and that the number of linearly independent eigenvectors associated with λ may or may not be equal to this multiplicity. To more easily discuss the situation when A has multiple eigenvalues we introduce the following definition.

Definition 3.   Let λ be an eigenvalue of an $n \times n$ matrix A. The set

$$A_\lambda = \{ \mathbf{v} : A\mathbf{v} = \lambda\mathbf{v} \}$$

is called the **eigenspace** of A associated with λ.

Notice that the eigenspace of A associated with λ consists of all eigenvectors of A associated with λ and the zero vector.

Example 1.   In Example 4 of Section 7.1 we found that the matrix

$$A = \begin{bmatrix} 7 & 4 & -4 \\ 4 & -8 & -1 \\ -4 & -1 & -8 \end{bmatrix}$$

has -9, -9, and 9 as its eigenvalues. The eigenvectors associated with 9 are all the nonzero scalar multiples of

$$\mathbf{u}_1 = \begin{bmatrix} 4 \\ 1 \\ -1 \end{bmatrix}$$

while the eigenvalues of A associated with -9 are all of the nonzero linear combinations of

$$\mathbf{u}_2 = \begin{bmatrix} 1 \\ 0 \\ 4 \end{bmatrix} \quad \text{and} \quad \mathbf{u}_3 = \begin{bmatrix} 0 \\ 1 \\ 1 \end{bmatrix}$$

Therefore A_9 consists of all scalar multiples of $\mathbf{u}_1$ and A_{-9} consists of all linear combinations of $\mathbf{u}_2$ and $\mathbf{u}_3$. Notice that A_9 and A_{-9} are linear spaces. The following theorem shows us that an eigenspace is always a linear space.

Theorem 10.   *Let λ be an eigenvalue of an $n \times n$ matrix A. Then A_λ is a linear space. Moreover, if $\mathbf{v}$ is any element of A_λ, then $A\mathbf{v}$ is also an element of A_λ.*

Proof: An n-vector $\mathbf{v}$ is an element of A if and only if $(\lambda I - A)\mathbf{v} = 0$. That is, $\mathbf{v}$ is in A_λ if and only if $\mathbf{v}$ is an element of the solution space (see Example 3 of Section 4.3) of $\lambda I - A$. Hence A_λ coincides with the solution space of $\lambda I - A$ and, therefore, is a linear space.

   If $\mathbf{v}$ is any element of A_λ, then $A(A\mathbf{v}) = A(\lambda\mathbf{v}) = \lambda(A\mathbf{v})$. Therefore $A\mathbf{v}$ is also an element of A_λ.   ∎

   Definition 4.   Let λ be an eigenvector of an $n \times n$ matrix A. The dimension of the eigenspace A_λ is called the **geometric multiplicity** of λ.

Theorem 11.   *Let λ be an eigenvalue of an $n \times n$ matrix A. Then the geometric multiplicity of λ is less than or equal to the multiplicity of λ.*

Proof: Let μ be an eigenvalue of A and let $B = \{\mathbf{v}_1, \mathbf{v}_2, \ldots, \mathbf{v}_k\}$ be a basis for A_μ. By Theorem 18 of Section 4.9 we can extend B to a basis B' of R^n. Relative to the basis B', the matrix representation of the linear transformation $T: R^n \to R^n$ defined by $T(\mathbf{x}) = A\mathbf{x}$ has the form

$$A' = \begin{bmatrix} \mu & 0 & 0 & \cdots & 0 & c_{1,k+1} & \cdots & c_{1n} \\ 0 & \mu & 0 & \cdots & 0 & c_{2,k+1} & \cdots & c_{2n} \\ \vdots & \vdots & \vdots & & \vdots & \vdots & & \vdots \\ 0 & 0 & 0 & \cdots & \mu & c_{k+1,k+1} & \cdots & c_{k+1,n} \\ 0 & 0 & 0 & \cdots & 0 & c_{k+2,k+1} & \cdots & c_{k+2,n} \\ \vdots & \vdots & \vdots & & \vdots & \vdots & & \vdots \\ 0 & 0 & 0 & \cdots & 0 & c_{n,k+1} & \cdots & c_{nn} \end{bmatrix}$$

Using cofactor expansions it is easy to show that

$$\det(\lambda I - A') = (\lambda - \mu)^k \det(\lambda I - C)$$

where

$$C = \begin{bmatrix} c_{k+1,k+1} & c_{k+1,k+2} & \cdots & c_{k+1,n} \\ c_{k+2,k+1} & c_{k+2,k+2} & \cdots & c_{k+2,n} \\ \vdots & \vdots & & \vdots \\ c_{n,k+1} & c_{n,k+2} & \cdots & c_{nn} \end{bmatrix}$$

By Theorem 5 of Section 6.6 there is a matrix P such that $A = P^{-1}A'P$. Hence A and A' have the same eigenvalues (see Exercise 17 of Section 7.2). Therefore μ is an eigenvalue of A of multiplicity at least k. This completes the proof. ∎

Example 2.　In Example 1 we found that the eigenvalue -9 of

$$A = \begin{bmatrix} 7 & 4 & -4 \\ 4 & -8 & -1 \\ -4 & -1 & -8 \end{bmatrix}$$

has multiplicity two and that the dimension of A_{-9} is also two. Hence the geometric multiplicity of an eigenvalue may equal the multiplicity of the eigenvalue. The following example shows that this is not always the case.

Example 3.　In Example 3 of Section 7.1 we found that 1 is an eigenvalue of multiplicity two of

$$A = \begin{bmatrix} 3 & -1 \\ 4 & -1 \end{bmatrix}$$

and that the eigenvectors of A associated with 1 are all the nonzero multiples of

$$\begin{bmatrix} 1 \\ 2 \end{bmatrix}$$

Therefore the dimension, 1, of the eigenspace A_1 is less than the multiplicity of the eigenvalue 1.

Exercises

In Exercises 1–10 compute the eigenspace for each eigenvalue of the given matrix. Also give the algebraic and geometric multiplicities of each eigenvalue.

1. $\begin{bmatrix} 6 & 5 \\ 4 & 5 \end{bmatrix}$

2. $\begin{bmatrix} 7 & 9 \\ 1 & -1 \end{bmatrix}$

3. $\begin{bmatrix} 2 & 3 & 0 \\ 2 & 1 & 0 \\ 0 & 0 & 3 \end{bmatrix}$

4. $\begin{bmatrix} 5 & -1 & 1 \\ 0 & 4 & 0 \\ 1 & -1 & 5 \end{bmatrix}$

5. $\begin{bmatrix} 3 & 0 & 1 \\ 0 & 4 & 0 \\ 1 & 0 & 3 \end{bmatrix}$
6. $\begin{bmatrix} 5 & 0 & 1 \\ 0 & -5 & 0 \\ 4 & 0 & 2 \end{bmatrix}$

7. $\begin{bmatrix} -5 & -9 & 9 \\ -9 & -5 & 9 \\ -1 & -1 & 5 \end{bmatrix}$
8. $\begin{bmatrix} 3 & 1 & 0 \\ -1 & 1 & 0 \\ 0 & 0 & 2 \end{bmatrix}$

9. $\begin{bmatrix} 3 & 1 & 0 & 0 \\ -1 & 1 & 0 & 0 \\ 0 & 0 & 3 & 1 \\ 0 & 0 & 1 & 3 \end{bmatrix}$
10. $\begin{bmatrix} 5 & 3 & 0 & 0 \\ 4 & 4 & 0 & 0 \\ 0 & 0 & 6 & 8 \\ 0 & 0 & 5 & 9 \end{bmatrix}$

11. Let λ be an eigenvalue of a matrix A. Is the set of all eigenvectors of A associated with λ a linear space? Explain your answer.

12. Let λ_1 and λ_2 be eigenvalues of a matrix A. Is the set $\{\mathbf{v}: A\mathbf{v} = \lambda_1\mathbf{v} \text{ or } A\mathbf{v} = \lambda_2\mathbf{v}\}$ necessarily a linear space? Explain your answer.

7.6 SYMMETRIC MATRICES (OPTIONAL)*

In applications involving linear transformations of finite dimensional linear spaces into themselves it is frequently possible to represent the transformation by a real matrix with the property that $A^t = A$. Such a matrix is called **symmetric**. That is, an $n \times n$ real matrix $A = (a_{ij})$ is symmetric if and only if $a_{ij} = a_{ji}$ for $i = 1, 2, \ldots, n$ and $j = 1, 2, \ldots, n$. For example, the matrices

$$\begin{bmatrix} 1 & 2 \\ 2 & -7 \end{bmatrix} \quad \text{and} \quad \begin{bmatrix} 1 & 0 & -2 \\ 0 & -4 & 6 \\ -2 & 6 & 7 \end{bmatrix}$$

are symmetric, while the matrices

$$\begin{bmatrix} 1 & 2 \\ 3 & 4 \end{bmatrix} \quad \text{and} \quad \begin{bmatrix} 4 & 0 & 1 \\ 2 & 4 & 3 \\ 0 & 1 & 5 \end{bmatrix}$$

are not symmetric. As the following theorem shows, symmetric matrices have an exceptional relationship with respect to the Euclidean inner product on R^n.

Theorem 12.   *Let A be an $n \times n$ matrix and let $\langle \mathbf{x}, \mathbf{y} \rangle$ denote the Euclidean inner product on R^n. Then*

$$\langle A\mathbf{x}, \mathbf{y} \rangle = \langle \mathbf{x}, A^t\mathbf{y} \rangle$$

for all $\mathbf{x}$ and $\mathbf{y}$ in R^n. In particular, if A is symmetric, then

$$\langle A\mathbf{x}, \mathbf{y} \rangle = \langle \mathbf{x}, A\mathbf{y} \rangle$$

for all $\mathbf{x}$ and $\mathbf{y}$ in R^n.

---

* This section requires material from Section 7.5.

Proof: Recall that $\langle \mathbf{x}, \mathbf{y} \rangle = \mathbf{y}^t \mathbf{x}$ and $\langle \mathbf{x}, \mathbf{y} \rangle = \mathbf{x}^t \mathbf{y}$ (see Section 5.1). Then $\langle A\mathbf{x}, \mathbf{y} \rangle = (A\mathbf{x})^t \mathbf{y} = (\mathbf{x}^t A^t)\mathbf{y} = \mathbf{x}^t(A^t \mathbf{y}) = \langle \mathbf{x}, A^t \mathbf{y} \rangle$. ∎

This theorem enables us to prove a fundamental property of the eigenvectors of a symmetric matrix.

Theorem 13. *Eigenvectors associated with distinct eigenvalues of a symmetric matrix are orthogonal with respect to the Euclidean inner product.*

Proof: Let $\mathbf{u}$ and $\mathbf{v}$ be eigenvectors of a symmetric matrix A associated with distinct eigenvalues λ and μ. Then $A\mathbf{u} = \lambda\mathbf{u}$ and $A\mathbf{v} = \mu\mathbf{v}$ so that

$$\lambda\langle \mathbf{u}, \mathbf{v} \rangle = \langle \lambda\mathbf{u}, \mathbf{v} \rangle$$
$$= \langle A\mathbf{u}, \mathbf{v} \rangle$$
$$= \langle \mathbf{u}, A\mathbf{v} \rangle \qquad \text{(Theorem 12)}$$
$$= \langle \mathbf{u}, \mu\mathbf{v} \rangle$$
$$= \mu\langle \mathbf{u}, \mathbf{v} \rangle$$

Hence $(\lambda - \mu)\langle \mathbf{u}, \mathbf{v} \rangle = 0$. Since $\lambda \neq \mu$ we must have $\langle \mathbf{u}, \mathbf{v} \rangle = 0$ so that $\mathbf{u}$ and $\mathbf{v}$ are orthogonal. ∎

Example 1. In Example 2 of Section 7.1, we found that

$$\mathbf{u} = \begin{bmatrix} 1 \\ -1 \end{bmatrix} \quad \text{and} \quad \mathbf{v} = \begin{bmatrix} 1 \\ 1 \end{bmatrix}$$

are the eigenvectors of

$$A = \begin{bmatrix} 1 & 2 \\ 2 & 1 \end{bmatrix}$$

associated with -1 and 3, respectively. Clearly $\mathbf{u}$ and $\mathbf{v}$ are orthogonal with respect to the Euclidean inner product.

Example 2. In Example 4 of Section 7.1 we found that

$$\mathbf{u} = \begin{bmatrix} 4 \\ 1 \\ -1 \end{bmatrix} \quad \mathbf{v} = \begin{bmatrix} 1 \\ 0 \\ 4 \end{bmatrix} \quad \mathbf{w} = \begin{bmatrix} 0 \\ 1 \\ 1 \end{bmatrix}$$

are eigenvectors of the symmetric matrix

$$A = \begin{bmatrix} 7 & 4 & -4 \\ 4 & -8 & -1 \\ -4 & -1 & -8 \end{bmatrix}$$

associated with the eigenvalues 9, -9, and -9, respectively. Notice that $\langle \mathbf{u}, \mathbf{v} \rangle = 0$ and that $\langle \mathbf{u}, \mathbf{w} \rangle = 0$, as is guaranteed by Theorem 13. In Example 1 of Section 7.5 we found that the eigenspace A_{-9} is spanned by $\mathbf{v}$ and $\mathbf{w}$. By Theorem 7 of Section 5.4 there is an orthonormal basis $\{\mathbf{v}', \mathbf{w}'\}$ for A_{-9}. Such a basis may be found by applying the Gram-Schmidt process (see Section 5.4) to the basis $\{\mathbf{v}, \mathbf{w}\}$ for A_{-9}. Doing this we find that $\mathbf{v}'$ and $\mathbf{w}'$ may be chosen to be

$$v' = \frac{1}{\sqrt{17}} \begin{bmatrix} 1 \\ 0 \\ 4 \end{bmatrix} \qquad w' = \frac{1}{3\sqrt{34}} \begin{bmatrix} -4 \\ 17 \\ 1 \end{bmatrix}$$

Hence $\langle \mathbf{v}', \mathbf{w}' \rangle = 0$. By Theorem 13 we also have $\langle \mathbf{u}, \mathbf{v}' \rangle = 0$ and $\langle \mathbf{u}, \mathbf{w}' \rangle = 0$. Therefore the set $B = \{\mathbf{u}, \mathbf{v}', \mathbf{w}'\}$ is an orthogonal set of eigenvectors of A. For this symmetric matrix we have shown that

1. A has only real eigenvalues.
2. A has three orthogonal eigenvectors with real components.
3. The dimension of A_λ (the geometric multiplicity of λ) equals the multiplicity of λ for every eigenvalue λ of A.

These are specific examples of the following theorems.

Theorem 14.   *An $n \times n$ symmetric matrix has only real eigenvalues.*

Proof: This proof is based on the fact that the product of two $n \times n$ matrices is singular if either matrix is singular (Exercises 17 and 18 of Section 1.9). Let $a + ib$ be any eigenvalue of an $n \times n$ symmetric matrix A. We will prove that $b = 0$ so that the eigenvalue is real. Since $(a + ib)I - A$ is singular, so is the matrix

$$\begin{aligned} B &= [(a + ib)I - A][(a - ib)I - A] \\ &= (a^2 + b^2)I - 2aA + A^2 \\ &= b^2 I + (A - aI)^2 \end{aligned}$$

Since B is $n \times n$, real, and singular, there is a nonzero n-vector $\mathbf{x}$ such that $B\mathbf{x} = \mathbf{0}$ (Theorem 17 of Section 1.9). Using the Euclidean inner product and noticing that $A - aI$ is symmetric, we have

$$\begin{aligned} 0 &= \langle \mathbf{x}, B\mathbf{x} \rangle \\ &= \langle \mathbf{x}, b^2\mathbf{x} + (A - aI)^2\mathbf{x} \rangle \\ &= b^2 \langle \mathbf{x}, \mathbf{x} \rangle + \langle \mathbf{x}, (A - aI)^2\mathbf{x} \rangle \\ &= b^2 \langle \mathbf{x}, \mathbf{x} \rangle + \langle (A - aI)\mathbf{x}, (A - aI)\mathbf{x} \rangle \qquad \text{(Theorem 12)} \\ &= b^2 \langle \mathbf{x}, \mathbf{x} \rangle + \langle \mathbf{y}, \mathbf{y} \rangle \qquad\qquad\qquad\qquad\qquad (1) \end{aligned}$$

where $\mathbf{y} = (A - aI)\mathbf{x}$. Since $\mathbf{x}$ is nonzero we have $\langle \mathbf{x}, \mathbf{x} \rangle > 0$. Moreover $\langle \mathbf{y}, \mathbf{y} \rangle \geq 0$ for any vector $\mathbf{y}$. Hence $b^2 \langle \mathbf{x}, \mathbf{x} \rangle + \langle \mathbf{y}, \mathbf{y} \rangle > 0$ if $b \neq 0$. In light of the identity in (1), we must have $b = 0$. Therefore A has only real eigenvalues. ∎

Theorem 15. *Let A be n × n symmetric matrix. Then*

    1. *For each eigenvalue λ of A the dimension of the eigenspace equals the multiplicity of λ.*
    2. *There is an orthonormal basis for R^n consisting of eigenvectors of A.*

The proof of this theorem is complicated. The interested reader is referred to Appendix 2.

Combining these results, we are able to prove the following fundamental theorem for symmetric matrices.

Theorem 16. *Let A be an n × n matrix. R^n has an orthonormal basis consisting of eigenvectors of A if and only if A is symmetric.*

Proof: Suppose that A is symmetric. Then, by Theorem 15, there is an orthonormal set consisting of n eigenvectors of A. This set must be linearly independent (Theorem 5 of Section 5.4). Therefore this set of eigenvectors of A is an orthonormal basis for R^n.

Now suppose that A has an orthonormal basis consisting of eigenvectors of A. Let P be the $n \times n$ matrix having the vectors in this basis as its columns. In Exercise 11 the reader is asked to show that $P^tP = I_n$ so that $P^{-1} = P^t$. By Theorem 7 of Section 7.3

$$P^{-1}AP = D$$

where D is a diagonal matrix having the eigenvalues of A on its diagonal. Then

$$A = PDP^{-1} = PDP^t$$

and

$$A^t = (PDP^t)^t = (P^t)^t D^t P^t = PDP^t = A$$

so that A is symmetric. ∎

Example 3. In Example 2 we found that

$$\left\{ \begin{bmatrix} 4 \\ 1 \\ -1 \end{bmatrix}, \frac{1}{\sqrt{17}} \begin{bmatrix} 1 \\ 0 \\ 4 \end{bmatrix}, \frac{1}{3\sqrt{34}} \begin{bmatrix} -4 \\ 17 \\ 1 \end{bmatrix} \right\}$$

is an orthogonal set of eigenvectors of

$$A = \begin{bmatrix} 7 & 4 & -4 \\ 4 & -8 & -1 \\ -4 & -1 & -8 \end{bmatrix}$$

Then

$$S = \left\{ \frac{1}{3\sqrt{2}} \begin{bmatrix} 4 \\ 1 \\ -1 \end{bmatrix}, \frac{1}{\sqrt{17}} \begin{bmatrix} 1 \\ 0 \\ 4 \end{bmatrix}, \frac{1}{3\sqrt{34}} \begin{bmatrix} -4 \\ 17 \\ 1 \end{bmatrix} \right\}$$

is an orthonormal basis for R^3.

Example 4.   In Example 2 of Section 7.1, we found that

$$\mathbf{u} = \begin{bmatrix} 1 \\ -1 \end{bmatrix} \quad \text{and} \quad \mathbf{v} = \begin{bmatrix} 1 \\ 1 \end{bmatrix}$$

are eigenvectors of

$$A = \begin{bmatrix} 1 & 2 \\ 2 & 1 \end{bmatrix}$$

associated with -1 and 3, respectively. Then

$$\left\{ \frac{1}{\sqrt{2}} \begin{bmatrix} 1 \\ -1 \end{bmatrix}, \frac{1}{\sqrt{2}} \begin{bmatrix} 1 \\ 1 \end{bmatrix} \right\}$$

is an orthonormal basis for R^2.

In the proof of Theorem 15 we considered an $n \times n$ matrix P whose columns formed an orthonormal set. In Exercise 11 the reader is asked to show that $P^t P = I_n$ so that $P^{-1} = P^t$. A matrix whose inverse equals its transpose is called **orthogonal**. This is a remarkable property for a matrix to have. Usually it takes some effort to compute the inverse of a nonsingular matrix. The inverse of an orthogonal matrix is obtained by merely interchanging its rows and columns.

Example 5.   Consider the matrix

$$P = \begin{bmatrix} \cos\theta & -\sin\theta \\ \sin\theta & \cos\theta \end{bmatrix}$$

that can be used to define the linear transformation R_θ that rotates each point in R^2 counterclockwise about the origin through an angle θ (see Example 2 of Section 6.3). Using the trigonometric identity $\cos^2\theta + \sin^2\theta = 1$, we can easily show that $P^t P = I_2$. Therefore P is an orthogonal matrix.

───────────────── **Exercises** ─────────────────

In Exercises 1–6 find an orthonormal basis of R^n consisting of eigenvectors of the given $n \times n$ matrix A. Also find a matrix P such that $P^t A P$ is a diagonal matrix, and determine $P^t A P$.

1. $\begin{bmatrix} 0 & 1 \\ 1 & 0 \end{bmatrix}$
　　　　　　　　　　　　　　2. $\begin{bmatrix} 3 & 1 \\ 1 & 3 \end{bmatrix}$

3. $\begin{bmatrix} 3 & 0 & 2 \\ 0 & 5 & 0 \\ 2 & 0 & 3 \end{bmatrix}$
　　　　　　4. $\begin{bmatrix} 2 & 1 & 0 \\ 1 & 2 & 0 \\ 0 & 0 & 3 \end{bmatrix}$

5. $\begin{bmatrix} 4 & 1 & -1 \\ 1 & 4 & -1 \\ -1 & -1 & 4 \end{bmatrix}$
　　　　6. $\begin{bmatrix} 1 & -1 & 1 \\ -1 & 1 & -1 \\ 1 & -1 & 1 \end{bmatrix}$

7. Show that the sum of two $n \times n$ symmetric matrices is a symmetric matrix.
8. Find two 2×2 symmetric matrices whose product is not a symmetric matrix.
9. Let A be a nonsingular symmetric matrix. Show that A^{-1} is symmetric.
10. Let A and B be $n \times n$ symmetric matrices such that $AB = BA$. Show that AB is symmetric.
11. Let $\{v_1, v_2, \ldots, v_n\}$ be an orthonormal set of n vectors and let P be the matrix having $v_1, v_2, \ldots, v_n$ as its columns. Show that $P^t P = I_n$.
12. Find a matrix P such that $P^{-1}AP$ is a diagonal matrix where

$$A = \begin{bmatrix} 0 & b \\ b & 0 \end{bmatrix}$$

and $b \neq 0$.
13. Let P be an orthogonal matrix. Show that $\det P = \pm 1$.
14. An $n \times n$ matrix A is called **orthogonally diagonalizable** if there is an orthogonal matrix P such that $P^t AP$ is a diagonal matrix. Show that an $n \times n$ matrix is orthogonally diagonalizable if and only if it is symmetric.
15. Let P be an $n \times n$ orthogonal matrix. Show that multiplication by P does not change an n-vector's length. That is, show that $\|Px\| = \|x\|$ for every x in R^n.

7.7 JORDAN CANONICAL FORM (OPTIONAL)

In Section 7.3 we found that if an $n \times n$ matrix A has n linearly independent eigenvectors $v_1, v_2, \ldots, v_n$ associated with the eigenvalues $\lambda_1, \lambda_2, \ldots, \lambda_n$ and if P is the matrix with $v_1, v_2, \ldots, v_n$ as its columns, then $P^{-1}AP = D$, where D is the diagonal matrix having $\lambda_1, \lambda_2, \ldots, \lambda_n$ on its diagonal.

Even if A does not have n linearly independent eigenvectors it is still possible to find a matrix M so that $M^{-1}AM$ has an exceptionally simple form. Before describing how to obtain the matrix M we describe the form of $M^{-1}AM$.

An $m \times m$ matrix of the form

$$\begin{bmatrix} a & 1 & 0 & 0 & \cdots & 0 & 0 \\ 0 & a & 1 & 0 & \cdots & 0 & 0 \\ \vdots & \vdots & \vdots & \vdots & & \vdots & \vdots \\ 0 & 0 & 0 & 0 & \cdots & a & 1 \\ 0 & 0 & 0 & 0 & \cdots & 0 & a \end{bmatrix}$$

is called a **Jordan block**. An $n \times n$ matrix J is said to be in **Jordan canonical form** if it is a diagonal matrix or has the form

$$\begin{bmatrix} M_1 & \text{all} & \text{all} & & \text{all} \\ & \text{zeros} & \text{zeros} & & \text{zeros} \\ \text{all} & M_2 & \text{all} & & \text{all} \\ \text{zeros} & & \text{zeros} & & \text{zeros} \\ \text{all} & \text{all} & M_3 & & \text{all} \\ \text{zeros} & \text{zeros} & & & \text{zeros} \\ \vdots & \vdots & \vdots & & \vdots \\ \text{all} & \text{all} & \text{all} & & M_k \\ \text{zeros} & \text{zeros} & \text{zeros} & & \end{bmatrix}$$

where $M_1, M_2, \ldots, M_k$ are

1. All Jordan blocks, or
2. M_1 is a diagonal matrix and $M_2, M_3, \ldots, M_k$ are Jordan blocks.

The next theorem, which extends the notion of a matrix being diagonalizable, is one of the fundamental theorems of matrix theory.

Theorem 17. *Every n × n matrix is similar to a matrix in Jordan canonical form.*

A proof of this theorem is beyond the scope of this text, but we will describe how to find a matrix M so that $M^{-1}AM$ is in Jordan canonical form. This description is an outline of the proof of Theorem 17. The interested reader is referred to a text on matrix theory for the details and proofs.

Throughout the following discussion, A will denote an $n \times n$ matrix. If A has n linearly independent eigenvectors, then A is diagonalizable and we can choose M to be a matrix having these eigenvectors as its columns. If A does not have n linearly independent eigenvectors, then A is not diagonalizable. Nonetheless, the matrix M is found in a method similar to that when A is diagonalizable. That is, to each eigenvalue we associate as many linearly independent vectors as the multiplicity of the eigenvalue and form a matrix having these vectors as its columns. We now describe how to find the columns of M.

An n-vector $\mathbf{x}$ is called a **generalized eigenvector of rank r** associated with λ if $(A - \lambda I)^r \mathbf{x} = \mathbf{0}$ and $(A - \lambda I)^{r-1}\mathbf{x} \neq \mathbf{0}$. Notice that a generalized eigenvector of rank 1 is an eigenvector of A associated with λ and that generalized eigenvectors are nonzero.

Theorem 18. *If λ is an eigenvalue of multiplicity m of an $n \times n$ matrix A, then there are exactly m linearly independent generalized eigenvectors associated with λ.*

Example 1. Consider the matrix

$$A = \begin{bmatrix} 4 & 1 & 2 \\ 0 & 4 & 1 \\ 0 & 0 & 4 \end{bmatrix}$$

that has 4 as an eigenvalue of multiplicity 3. A short calculation shows that

$$(A - 4I) = \begin{bmatrix} 0 & 1 & 2 \\ 0 & 0 & 1 \\ 0 & 0 & 0 \end{bmatrix} \qquad (A - 4I)^2 = \begin{bmatrix} 0 & 0 & 1 \\ 0 & 0 & 0 \\ 0 & 0 & 0 \end{bmatrix}$$

$$(A - 4I)^3 = \begin{bmatrix} 0 & 0 & 0 \\ 0 & 0 & 0 \\ 0 & 0 & 0 \end{bmatrix}$$

Evidently the generalized eigenvectors of rank 1 have the form

$$\begin{bmatrix} a \\ 0 \\ 0 \end{bmatrix}, \qquad a \neq 0$$

while the generalized eigenvectors of rank 2 have the form

$$\begin{bmatrix} a \\ b \\ 0 \end{bmatrix}, \qquad b \neq 0$$

We must have $b \neq 0$ because if $b = 0$ then we have a solution of $(A - 3I)\mathbf{x} = \mathbf{0}$. The generalized eigenvectors of rank 3 have the form

$$\begin{bmatrix} a \\ b \\ c \end{bmatrix}, \qquad c \neq 0$$

We must have $c \neq 0$ because if $c = 0$ then we have a solution of $(A - 3I)^2\mathbf{x} = \mathbf{0}$. Notice that

$$\left\{ \begin{bmatrix} 1 \\ 0 \\ 0 \end{bmatrix}, \begin{bmatrix} 0 \\ 1 \\ 0 \end{bmatrix}, \begin{bmatrix} 0 \\ 0 \\ 1 \end{bmatrix} \right\}$$

is a linearly independent set of generalized eigenvectors (one of rank 1, one of rank 2, and one of rank 3) and that the eigenvalue 4 has multiplicity 3.

We will not be interested in just any m linearly independent generalized eigenvectors, but in those that are related in a special fashion.

Let λ be an eigenvalue of A with multiplicity m. Then there is a smallest positive integer r such that the rank of the matrix $(A - \lambda I)^r$ is $n - m$. Let $\mathbf{x}_r$ be a generalized eigenvector of rank r associated with λ. That is, let $\mathbf{x}_r$ be a solution of $(A - \lambda I)^r\mathbf{x} = \mathbf{0}$ that is not a solution of $(A - \lambda I)^{r-1}\mathbf{x} = \mathbf{0}$. The set of vectors $\{\mathbf{x}_1, \mathbf{x}_2, \ldots, \mathbf{x}_r\}$ defined by

$$\mathbf{x}_{r-1} = (A - \lambda I)\mathbf{x}_r$$
$$\mathbf{x}_{r-2} = (A - \lambda I)\mathbf{x}_{r-1} = (A - \lambda I)^2\mathbf{x}_r$$
$$\mathbf{x}_{r-3} = (A - \lambda I)\mathbf{x}_{r-2} = (A - \lambda I)^3\mathbf{x}_r$$
$$\vdots$$
$$\mathbf{x}_1 = (A - \lambda I)\mathbf{x}_2 \quad = (A - \lambda I)^{r-1}\mathbf{x}_r$$

is called the **chain** generated by $\mathbf{x}_r$ and can be shown to be linearly independent.

Notice that $\mathbf{x}_k = (A - \lambda I)^{r-k}\mathbf{x}_r$ and $(A - \lambda I)^k\mathbf{x}_k = (A - \lambda I)^r\mathbf{x}_r = \mathbf{0}$. Thus $\mathbf{x}_k$ is a generalized eigenvector of rank k associated with λ.

Example 2.   In Example 1 we found that

$$\mathbf{x}_3 = \begin{bmatrix} 0 \\ 0 \\ 1 \end{bmatrix}$$

is a solution of $(A - 4I)^3\mathbf{x} = \mathbf{0}$. Then $\{\mathbf{x}_1, \mathbf{x}_2, \mathbf{x}_3\}$ is the chain generated by $\mathbf{x}_3$ where

$$\mathbf{x}_2 = (A - 4I)\mathbf{x}_3 = \begin{bmatrix} 2 \\ 1 \\ 0 \end{bmatrix}$$

$$\mathbf{x}_1 = (A - 4I)\mathbf{x}_2 = \begin{bmatrix} 1 \\ 0 \\ 0 \end{bmatrix}$$

A set of generalized eigenvectors associated with λ is called a **canonical set** of generalized eigenvectors for λ if

1. The number of elements in S is the multiplicity of λ;
2. S is linearly independent; and
3. S consists of chains of generalized eigenvectors associated with λ.

To obtain such a set it is useful to know the number of generalized eigenvectors of each rank that appear in a canonical set. Fortunately there is a simple formula for these numbers. The number

$$r_k = \text{rank}(A - \lambda I)^{k-1} - \text{rank}(A - \lambda I)^k, \quad \text{where} \quad k = 1, 2, \dots, m$$

is the number of linearly independent generalized eigenvectors of rank k that appear in a canonical set of generalized eigenvectors for λ. Note that rank $(A - \lambda I)^0 = $ rank $I = n$.

Example 3.   Consider the 8×8 matrix

$$A = \begin{bmatrix} 2 & 1 & 1 & 0 & 0 & 0 & 0 & 0 \\ 0 & 2 & 1 & 0 & 1 & 0 & 0 & 0 \\ 0 & 0 & 2 & 1 & 0 & 1 & 0 & 0 \\ 0 & 0 & 0 & 2 & 0 & 0 & 0 & 0 \\ 0 & 0 & 0 & 0 & 2 & 1 & 0 & 0 \\ 0 & 0 & 0 & 0 & 0 & 2 & 0 & 0 \\ 0 & 0 & 0 & 0 & 0 & 0 & 3 & 1 \\ 0 & 0 & 0 & 0 & 0 & 0 & 0 & 3 \end{bmatrix}$$

having 2 as an eigenvalue of multiplicity 6 and 3 as an eigenvalue of multiplicity 2. We begin by looking for an integer r such that the rank of $(A - 2I)^r$ is $8 - 6 = 2$.

$$(A - 2I) = \begin{bmatrix} 0 & 1 & 1 & 0 & 0 & 0 & 0 & 0 \\ 0 & 0 & 1 & 0 & 1 & 0 & 0 & 0 \\ 0 & 0 & 0 & 1 & 0 & 1 & 0 & 0 \\ 0 & 0 & 0 & 0 & 0 & 0 & 0 & 0 \\ 0 & 0 & 0 & 0 & 0 & 1 & 0 & 0 \\ 0 & 0 & 0 & 0 & 0 & 0 & 0 & 0 \\ 0 & 0 & 0 & 0 & 0 & 0 & 1 & 1 \\ 0 & 0 & 0 & 0 & 0 & 0 & 0 & 1 \end{bmatrix}$$

$$(A - 2I)^2 = \begin{bmatrix} 0 & 0 & 1 & 1 & 1 & 1 & 0 & 0 \\ 0 & 0 & 0 & 1 & 0 & 2 & 0 & 0 \\ 0 & 0 & 0 & 0 & 0 & 0 & 0 & 0 \\ 0 & 0 & 0 & 0 & 0 & 0 & 0 & 0 \\ 0 & 0 & 0 & 0 & 0 & 0 & 0 & 0 \\ 0 & 0 & 0 & 0 & 0 & 0 & 0 & 0 \\ 0 & 0 & 0 & 0 & 0 & 0 & 1 & 2 \\ 0 & 0 & 0 & 0 & 0 & 0 & 0 & 1 \end{bmatrix}$$

$$(A - 2I)^3 = \begin{bmatrix} 0 & 0 & 0 & 1 & 0 & 2 & 0 & 0 \\ 0 & 0 & 0 & 0 & 0 & 0 & 0 & 0 \\ 0 & 0 & 0 & 0 & 0 & 0 & 0 & 0 \\ 0 & 0 & 0 & 0 & 0 & 0 & 0 & 0 \\ 0 & 0 & 0 & 0 & 0 & 0 & 0 & 0 \\ 0 & 0 & 0 & 0 & 0 & 0 & 0 & 0 \\ 0 & 0 & 0 & 0 & 0 & 0 & 1 & 3 \\ 0 & 0 & 0 & 0 & 0 & 0 & 0 & 1 \end{bmatrix}$$

$$(A - 2I)^4 = \begin{bmatrix} 0 & 0 & 0 & 0 & 0 & 0 & 0 & 0 \\ 0 & 0 & 0 & 0 & 0 & 0 & 0 & 0 \\ 0 & 0 & 0 & 0 & 0 & 0 & 0 & 0 \\ 0 & 0 & 0 & 0 & 0 & 0 & 0 & 0 \\ 0 & 0 & 0 & 0 & 0 & 0 & 0 & 0 \\ 0 & 0 & 0 & 0 & 0 & 0 & 0 & 0 \\ 0 & 0 & 0 & 0 & 0 & 0 & 1 & 4 \\ 0 & 0 & 0 & 0 & 0 & 0 & 0 & 1 \end{bmatrix}$$

Evidently 4 is the least integer r such that $(A - 2I)^r$ has rank 2. Since

$$r_1 = \text{rank}(A - 2I)^0 - \text{rank}(A - 2I)^1 - 8 - 6 - 2$$
$$r_2 = \text{rank}(A - 2I)^1 - \text{rank}(A - 2I)^2 = 6 - 4 = 2$$
$$r_3 = \text{rank}(A - 2I)^2 - \text{rank}(A - 2I)^3 = 4 - 3 = 1$$
$$r_4 = \text{rank}(A - 2I)^3 - \text{rank}(A - 2I)^4 = 3 - 2 = 1$$

a canonical set of generalized eigenvectors for the eigenvalue 2 has

1. One generalized eigenvector of rank 4.
2. One generalized eigenvector of rank 3.
3. Two generalized eigenvectors of rank 2.
4. Two generalized eigenvectors of rank 1.

Notice that

$$\mathbf{x}_4 = \begin{bmatrix} 0 \\ 0 \\ 0 \\ 1 \\ 0 \\ 0 \\ 0 \\ 0 \end{bmatrix}$$

is a generalized eigenvector of rank 4. We now obtain the chain generated by $\mathbf{x}_4$:

$$\mathbf{x}_3 = (A - 2I)\mathbf{x}_4 = \begin{bmatrix} 0 \\ 0 \\ 1 \\ 0 \\ 0 \\ 0 \\ 0 \\ 0 \end{bmatrix} \qquad \mathbf{x}_2 = (A - 2I)\mathbf{x}_3 = \begin{bmatrix} 1 \\ 1 \\ 0 \\ 0 \\ 0 \\ 0 \\ 0 \\ 0 \end{bmatrix}$$

$$\mathbf{x}_1 = (A - 2I)\mathbf{x}_2 = \begin{bmatrix} 1 \\ 0 \\ 0 \\ 0 \\ 0 \\ 0 \\ 0 \\ 0 \end{bmatrix}$$

Hence we have found a chain $\{\mathbf{x}_1, \mathbf{x}_2, \mathbf{x}_3, \mathbf{x}_4\}$ of generalized eigenvectors having ranks 1, 2, 3, and 4. By the preceding list, any canonical set containing this chain will contain another chain with two vectors. Next we find a generalized eigenvector $\mathbf{y}_2$ of rank two that is not a linear combination of the chain we just found:

$$\mathbf{y}_2 = \begin{bmatrix} 0 \\ 0 \\ 1 \\ -2 \\ 0 \\ 1 \\ 0 \\ 0 \end{bmatrix}$$

The other vector in the chain $\{\mathbf{y}_1, \mathbf{y}_2\}$ determined by $\mathbf{y}_2$ is

$$\mathbf{y}_1 = (A - 2I)\mathbf{y}_2 = \begin{bmatrix} 1 \\ 1 \\ -1 \\ 0 \\ 1 \\ 0 \\ 0 \\ 0 \end{bmatrix}$$

Then $\{\mathbf{x}_1, \mathbf{x}_2, \mathbf{x}_3, \mathbf{x}_4, \mathbf{y}_1, \mathbf{y}_2\}$ is a canonical set of generalized eigenvectors for the eigenvalue 2.

An ordered linearly independent set $\{\mathbf{v}_1, \mathbf{v}_2, ..., \mathbf{v}_n\}$ consisting of chains of generalized eigenvectors of the $n \times n$ matrix A is called a **Jordan basis** for R^n relative to A if

1. All chains one vector in length appear before longer chains;
2. All vectors in one chain appear consecutively; and
3. Each chain appears in order of increasing rank. That is, the generalized eigenvector of rank 1 in a chain appears before the generalized eigenvector in the chain of rank 2, which appears before the generalized eigenvector in the chain of rank 3, etc.

Let $\lambda_1, \lambda_2, ..., \lambda_s$ be the distinct eigenvectors of A and let $S_1, S_2, ..., S_s$ be corresponding canonical sets of generalized eigenvectors. After an appropriate reordering, if necessary, the set $S = \bigcup_{i=1}^{s} S_i$ is a Jordan basis for R^n relative to A.

Example 4.   We will find a Jordan basis for R^8 relative to the 8×8 matrix in Example 3. In that example we found a canonical set of generalized eigenvectors for the eigenvalue 2. We now do the same for the eigenvalue 3, which has multiplicity 2. To begin we find the smallest integer r such that $(A - 3I)^r$ has rank $8 - 2 = 6$.

$$(A - 3I) = \begin{bmatrix} -1 & 1 & 1 & 0 & 0 & 0 & 0 & 0 \\ 0 & -1 & 1 & 0 & 1 & 0 & 0 & 0 \\ 0 & 0 & -1 & 1 & 0 & 1 & 0 & 0 \\ 0 & 0 & 0 & -1 & 0 & 0 & 0 & 0 \\ 0 & 0 & 0 & 0 & -1 & 1 & 0 & 0 \\ 0 & 0 & 0 & 0 & 0 & -1 & 0 & 0 \\ 0 & 0 & 0 & 0 & 0 & 0 & 0 & 1 \\ 0 & 0 & 0 & 0 & 0 & 0 & 0 & 0 \end{bmatrix}$$

$$(A - 3I)^2 = \begin{bmatrix} 1 & -2 & -1 & 1 & 1 & 1 & 0 & 0 \\ 0 & 1 & -2 & 1 & -2 & 2 & 0 & 0 \\ 0 & 0 & 1 & -2 & 0 & -2 & 0 & 0 \\ 0 & 0 & 0 & 1 & 0 & 0 & 0 & 0 \\ 0 & 0 & 0 & 0 & 1 & -2 & 0 & 0 \\ 0 & 0 & 0 & 0 & 0 & 1 & 0 & 0 \\ 0 & 0 & 0 & 0 & 0 & 0 & 0 & 0 \\ 0 & 0 & 0 & 0 & 0 & 0 & 0 & 0 \end{bmatrix}$$

Hence $r = 2$. Notice that

$$\mathbf{z}_2 = \begin{bmatrix} 0 \\ 0 \\ 0 \\ 0 \\ 0 \\ 0 \\ 0 \\ 1 \end{bmatrix}$$

is a generalized eigenvector of rank 2 associated with the eigenvalue 3. Then $\{\mathbf{z}_1, \mathbf{z}_2\}$, where

$$\mathbf{z}_1 = (A - 3I)\mathbf{z}_2 = \begin{bmatrix} 0 \\ 0 \\ 0 \\ 0 \\ 0 \\ 0 \\ 1 \\ 0 \end{bmatrix}$$

is the chain generated by $\mathbf{z}_2$. If $\mathbf{x}_1, \mathbf{x}_2, \mathbf{x}_3, \mathbf{x}_4, \mathbf{y}_1$, and $\mathbf{y}_2$ are as in Example 3, then $\{\mathbf{x}_1, \mathbf{x}_2, \mathbf{x}_3, \mathbf{x}_4, \mathbf{y}_1, \mathbf{y}_2, \mathbf{z}_2\}$ is a Jordan basis for R^8 relative to A.

We are now ready to describe a matrix M such that $M^{-1}AM$ is in Jordan canonical form.

Theorem 19. *Let S be a Jordan basis for R^n relative to an $n \times n$ matrix A and let M be the matrix having the vectors in S as its columns. Then $M^{-1}AM$ is in Jordan canonical form.*

This theorem tells us that with respect to the basis S for R^n the linear transformation $T: R^n \to R^n$ defined by $T(\mathbf{x}) = A\mathbf{x}$ has a matrix in Jordan canonical form as its matrix representation.

In fact, we can determine the form of the Jordan canonical form $M^{-1}AM$ from the Jordan basis S. If $\mathbf{v}_k, \mathbf{v}_{k+1}, \ldots, \mathbf{v}_{k+r}$ is a chain in S of generalized eigenvectors associated with an eigenvalue λ of A, then the $r \times r$ Jordan block

$$\begin{bmatrix} \lambda & 1 & 0 & 0 & \cdots & 0 & 0 \\ 0 & \lambda & 1 & 0 & \cdots & 0 & 0 \\ \vdots & \vdots & \vdots & \vdots & & \vdots & \vdots \\ 0 & 0 & 0 & 0 & \cdots & \lambda & 1 \\ 0 & 0 & 0 & 0 & \cdots & 0 & \lambda \end{bmatrix}$$

appears in $M^{-1}AM$ between the kth and $(k + r)$-th columns and between the kth and $(k + r)$-th rows.

$$\begin{array}{cc} \text{column } k & \text{column } k + r \end{array}$$

$$\begin{array}{c} \\ \text{row } r \\ \\ \\ \text{row } k + r \end{array} \begin{bmatrix} \vdots & & & & \vdots & \\ \cdots & \lambda & 1 & 0 & 0 & \cdots & 0 & 0 & \cdots \\ & 0 & \lambda & 1 & 0 & \cdots & 0 & 0 \\ & \vdots & \vdots & \vdots & \vdots & & \vdots & \vdots \\ & 0 & 0 & 0 & 0 & & \lambda & 1 \\ \cdots & 0 & 0 & 0 & 0 & \cdots & 0 & \lambda & \cdots \\ & \vdots & & & & \vdots & \end{bmatrix}$$

Example 5. In Example 4 we found a Jordan basis

$$S = \{\mathbf{x}_1, \mathbf{x}_2, \mathbf{x}_3, \mathbf{x}_4, \mathbf{y}_1, \mathbf{y}_2, \mathbf{z}_1, \mathbf{z}_2\}$$

for the 8×8 matrix given in Example 3. Let M be the matrix having the vectors in S as its columns. Then

$$M^{-1}AM = \begin{bmatrix} 2 & 1 & 0 & 0 & 0 & 0 & 0 & 0 \\ 0 & 2 & 1 & 0 & 0 & 0 & 0 & 0 \\ 0 & 0 & 2 & 1 & 0 & 0 & 0 & 0 \\ 0 & 0 & 0 & 2 & 0 & 0 & 0 & 0 \\ 0 & 0 & 0 & 0 & 2 & 1 & 0 & 0 \\ 0 & 0 & 0 & 0 & 0 & 2 & 0 & 0 \\ 0 & 0 & 0 & 0 & 0 & 0 & 3 & 1 \\ 0 & 0 & 0 & 0 & 0 & 0 & 0 & 3 \end{bmatrix}$$

──────────────────── **Exercises** ────────────────────

Find a Jordan basis and the corresponding Jordan canonical form for each of the following matrices.

1. $\begin{bmatrix} 1 & -1 \\ 1 & 3 \end{bmatrix}$
2. $\begin{bmatrix} 2 & -1 \\ 1 & 4 \end{bmatrix}$

3. $\begin{bmatrix} 4 & 1 & 2 \\ 0 & 4 & 2 \\ 0 & 0 & 4 \end{bmatrix}$
4. $\begin{bmatrix} 4 & 1 & 2 \\ 0 & 4 & 0 \\ 0 & 0 & 4 \end{bmatrix}$

5. $\begin{bmatrix} 1 & 1 & 1 & 0 \\ 0 & 2 & 0 & 0 \\ 0 & 0 & 1 & 1 \\ 0 & 0 & 0 & 2 \end{bmatrix}$
6. $\begin{bmatrix} 1 & 0 & 0 & 0 \\ 0 & 3 & 0 & 0 \\ 2 & 1 & 1 & 0 \\ 0 & 0 & 0 & 3 \end{bmatrix}$

7. $\begin{bmatrix} 1 & -1 & 0 & 0 & 0 & 0 \\ 1 & 3 & 0 & 0 & 0 & 0 \\ 0 & 0 & 2 & -1 & 0 & 0 \\ 0 & 0 & 1 & 4 & 0 & 0 \\ 0 & 0 & 0 & 0 & 4 & 2 \\ 0 & 0 & 0 & 0 & -1 & 1 \end{bmatrix}$
8. $\begin{bmatrix} 1 & 0 & 0 & 0 & 0 & 0 \\ 0 & 3 & 0 & 0 & 0 & 0 \\ 2 & 1 & 1 & 0 & 0 & 0 \\ 0 & 0 & 0 & 3 & 1 & 0 \\ 0 & 0 & 0 & 0 & 3 & 0 \\ 0 & 0 & 0 & 0 & 0 & 1 \end{bmatrix}$

7.8 CAYLEY–HAMILTON THEOREM (OPTIONAL)

In applications it is often necessary to consider functions of a matrix. The simplest of such functions are polynomials. If A is an $n \times n$ matrix, then

$$A^2 = AA$$
$$A^3 = AA^2$$
$$\vdots \qquad \vdots$$
$$A^k = AA^{k-1}$$

for any positive integer k. That is,

$$k \text{ terms}$$
$$A^k = \overbrace{AA \cdots A}$$

Let $f(x) = a_m x^m + a_{m-1} x^{m-1} + \cdots + a_1 x + a_0$ be a polynomial. The matrix polynomial $f(A)$ of an $n \times n$ matrix A is defined by

$$f(A) = a_m A^m + a_{m-1} A^{m-1} + \cdots + a_1 A + a_0 I_n \qquad (1)$$

Notice that the last summand in (1) is $a_0 I_n$ and not simply a_0. The "I_n" term is required so that the last summand in (1) is an $n \times n$ matrix and as such can be added to the previous summands. The sum of a matrix and a scalar is not defined.

Example 1.   Consider the matrix

$$A = \begin{bmatrix} 1 & 2 \\ 2 & 1 \end{bmatrix}$$

If $f(x) = 3x^2 + 4x - 5$, then

$$f(A) = 3A^2 + 4A - 5I_2$$

$$= 3\begin{bmatrix} 5 & 4 \\ 4 & 5 \end{bmatrix} + 4\begin{bmatrix} 1 & 2 \\ 2 & 1 \end{bmatrix} - 5\begin{bmatrix} 1 & 0 \\ 0 & 1 \end{bmatrix}$$

$$= \begin{bmatrix} 14 & 20 \\ 20 & 14 \end{bmatrix}$$

We now consider one of the most useful theorems in matrix theory.

Theorem 20   (Cayley–Hamilton Theorem). *An $n \times n$ matrix satisfies its own characteristic polynomial. That is, if $f(\lambda) = \det(\lambda I - A)$, then $f(A) = 0$.*

Proof: We will prove this theorem only for the case in which A has n linearly independent eigenvectors. If this is not the case, the proof is more complicated and will be omitted. The interested reader can find a proof for the general case in almost any text on matrix theory.

Assume then that A has n linearly independent eigenvectors. There is a matrix P such that

$$P^{-1}AP = \begin{bmatrix} \lambda_1 & 0 & 0 & \cdots & 0 \\ 0 & \lambda_2 & 0 & \cdots & 0 \\ \vdots & \vdots & \vdots & & \vdots \\ 0 & 0 & 0 & \cdots & \lambda_n \end{bmatrix}$$

where $\lambda_1, \lambda_2, \ldots, \lambda_n$ are the eigenvalues of A. In Exercise 14 the reader is asked to show that

(2) $$P^{-1}A^kP = (P^{-1}AP)^k = \begin{bmatrix} \lambda_1^k & 0 & 0 & \cdots & 0 \\ 0 & \lambda_2^k & 0 & \cdots & 0 \\ \vdots & \vdots & \vdots & & \vdots \\ 0 & 0 & 0 & \cdots & \lambda_n^k \end{bmatrix}$$

for every positive integer k. If

$$f(\lambda) = \det(\lambda I - A)$$
$$= \lambda^n + a_{n-1}\lambda^{n-1} + \cdots + a_1\lambda + a_0$$

Then

$$f(A) = A^n + a_{n-1}A^{n-1} + \cdots + a_1A + a_0I$$

and

$$P^{-1}f(A)P = P^{-1}A^nP + a_{n-1}P^{-1}A^{n-1}P + \cdots + a_1P^{-1}AP + a_0P^{-1}IP$$
$$= (P^{-1}AP)^n + a_{n-1}(P^{-1}AP)^{n-1} + \cdots + a_1P^{-1}AP + a_0P^{-1}IP$$

$$= \begin{bmatrix} \lambda_1^n & 0 & 0 & \cdots & 0 \\ 0 & \lambda_2^n & 0 & \cdots & 0 \\ \vdots & \vdots & \vdots & & \vdots \\ 0 & 0 & 0 & \cdots & \lambda_n^n \end{bmatrix} + a_{n-1}\begin{bmatrix} \lambda_1^{n-1} & 0 & 0 & \cdots & 0 \\ 0 & \lambda_2^{n-1} & 0 & \cdots & 0 \\ \vdots & \vdots & \vdots & & \vdots \\ 0 & 0 & 0 & \cdots & \lambda_n^{n-1} \end{bmatrix}$$

$$+ \cdots + a_1\begin{bmatrix} \lambda_1 & 0 & 0 & \cdots & 0 \\ 0 & \lambda_2 & 0 & \cdots & 0 \\ \vdots & \vdots & \vdots & & \vdots \\ 0 & 0 & 0 & \cdots & \lambda_n \end{bmatrix} + a_0\begin{bmatrix} 1 & 0 & 0 & \cdots & 0 \\ 0 & 1 & 0 & \cdots & 0 \\ \vdots & \vdots & \vdots & & \vdots \\ 0 & 0 & 0 & \cdots & 1 \end{bmatrix}$$

$$= \begin{bmatrix} f(\lambda_1) & 0 & 0 & \cdots & 0 \\ 0 & f(\lambda_2) & 0 & \cdots & 0 \\ \vdots & \vdots & \vdots & & \vdots \\ 0 & 0 & 0 & \cdots & f(\lambda_n) \end{bmatrix}$$

Since $\lambda_1, \lambda_2, \ldots, \lambda_n$ are eigenvalues of A we have $f(\lambda_1) = 0, f(\lambda_2) = 0, \ldots, f(\lambda_n) = 0$. Therefore $P^{-1}f(A)P = 0$, so that $f(A) = P0P^{-1} = 0$. Hence, A satisfies its own characteristic polynomial. ∎

Example 2. The characteristic polynomial of

$$A = \begin{bmatrix} 1 & 2 \\ 2 & 1 \end{bmatrix}$$

is $f(\lambda) = \lambda^2 - 2\lambda - 3$. Then

$$f(A) = A^2 - 2A - 3I$$

$$= \begin{bmatrix} 5 & 4 \\ 4 & 5 \end{bmatrix} - 2\begin{bmatrix} 1 & 2 \\ 2 & 1 \end{bmatrix} - 3\begin{bmatrix} 1 & 0 \\ 0 & 1 \end{bmatrix} = \begin{bmatrix} 0 & 0 \\ 0 & 0 \end{bmatrix}$$

Example 3. The characteristic polynomial of

$$A = \begin{bmatrix} 4 & 2 & -2 \\ -5 & 3 & 2 \\ -2 & 4 & 1 \end{bmatrix}$$

is $f(\lambda) = \lambda^3 - 8\lambda^2 + 17\lambda - 10$. Then

$$f(A) = A^3 - 8A^2 + 17A - 10I$$

$$= \begin{bmatrix} 22 & 14 & -14 \\ -227 & 15 & 110 \\ -206 & 28 & 97 \end{bmatrix} - 8 \begin{bmatrix} 10 & 6 & -6 \\ -39 & 7 & 18 \\ -30 & 12 & 13 \end{bmatrix}$$

$$+ 17 \begin{bmatrix} 4 & 2 & -2 \\ -5 & 3 & 2 \\ -2 & 4 & 1 \end{bmatrix} - 10 \begin{bmatrix} 1 & 0 & 0 \\ 0 & 1 & 0 \\ 0 & 0 & 1 \end{bmatrix}$$

$$= \begin{bmatrix} 0 & 0 & 0 \\ 0 & 0 & 0 \\ 0 & 0 & 0 \end{bmatrix}$$

One immediate consequence of the Cayley–Hamilton Theorem is a new method for computing the inverse of a nonsingular matrix.

If

$$f(\lambda) = \lambda^n + a_{n-1}\lambda^{n-1} + \cdots + a_1\lambda + a_0$$

is the characteristic equation of a matrix A, then by the Cayley–Hamilton Theorem

$$0 = f(A) = A^n + a_{n-1}A^{n-1} + \cdots + a_1 A + a_0 I$$

so that

$$(A^{n-1} + a_{n-1}A^{n-2} + \cdots + a_1 I)A = -a_0 I$$

Hence

$$A^{-1} = -\frac{1}{a_0}(A^{n-1} + a_{n-1}A^{n-2} + \cdots + a_1 I)$$

whenever $a_0 \neq 0$. If $a_0 = 0$, then $0 = a_0 = f(0) = \det(0I - A) = \det(-A) = (-1)^n \det A$ so that $\det A = 0$. Therefore, if $a_0 = 0$, then A is singular, so A^{-1} does not exist.

Example 4.   Consider the matrix

$$A = \begin{bmatrix} 4 & 2 & -2 \\ -5 & 3 & 2 \\ -2 & 4 & 1 \end{bmatrix}$$

In Example 3 we showed that $A^3 - 8A^2 + 17A - 10I = 0$. Therefore

$$A^3 - 8A^2 + 17A = 10I$$

and

$$A\left[\frac{1}{10}(A^2 - 8A + 17I)\right] = I$$

Hence

$$A^{-1} = \frac{1}{10}(A^2 - 8A + 17I)$$

$$= \frac{1}{10}\begin{bmatrix} 10 & 6 & -6 \\ -39 & 7 & 18 \\ -30 & 12 & 13 \end{bmatrix} - \frac{8}{10}\begin{bmatrix} 4 & 2 & -2 \\ -5 & 3 & 2 \\ -2 & 4 & 1 \end{bmatrix} + \frac{17}{10}\begin{bmatrix} 1 & 0 & 0 \\ 0 & 1 & 0 \\ 0 & 0 & 1 \end{bmatrix}$$

$$= \frac{1}{10}\begin{bmatrix} -5 & -10 & 10 \\ 1 & 0 & 2 \\ -14 & -20 & 22 \end{bmatrix}$$

The Cayley–Hamilton Theorem also gives a means to simplify the calculation of polynomials of a matrix. Let $p(\lambda)$ be any polynomial and let $f(\lambda)$ be the characteristic polynomial of a square matrix A. A theorem of algebra tells us that there are polynomials $q(\lambda)$ and $r(\lambda)$ such that

$$p(\lambda) = q(\lambda)f(\lambda) + r(\lambda)$$

with the degree of $r(\lambda)$ less than the degree of $f(\lambda)$. Then

$$p(A) = q(A)f(A) + r(A)$$

Since, by the Cayley–Hamilton Theorem, $f(A) = 0$ we have

$$p(A) = r(A)$$

Thus the problem of evaluating a polynomial of an $n \times n$ matrix can be reduced to the problem of evaluating a polynomial of degree less than n.

Example 5.   Consider the matrix

$$A = \begin{bmatrix} 1 & 2 \\ 2 & 1 \end{bmatrix}$$

whose characteristic polynomial is $f(\lambda) = \lambda^2 - 2\lambda - 3$. Let $p(\lambda) = \lambda^4 - 7\lambda^3 - 3\lambda^2 + \lambda + 4$. A straightforward calculation shows that

$$p(\lambda) = (\lambda^2 - 5\lambda - 10)f(\lambda) - 34\lambda - 26$$

Therefore

$$p(A) = (A^2 - 5A + 10)f(A) - 34A - 26I$$
$$= -34A - 26I$$

$$= -34\begin{bmatrix} 1 & 2 \\ 2 & 1 \end{bmatrix} - 26\begin{bmatrix} 1 & 0 \\ 0 & 1 \end{bmatrix}$$

$$= \begin{bmatrix} -60 & -68 \\ -68 & -60 \end{bmatrix}$$

Example 6. In Example 2 of Section 6.5 we found that the Fibonacci numbers p_k satisfy

$$\begin{bmatrix} p_{k-1} \\ p_k \end{bmatrix} = A^{k-2} \begin{bmatrix} 1 \\ 1 \end{bmatrix}$$

where

$$A = \begin{bmatrix} 0 & 1 \\ 1 & 1 \end{bmatrix}$$

We will use the Cayley–Hamilton Theorem to evaluate A^{k-2}. To begin we notice that $f(\lambda) = \lambda^2 - \lambda - 1$ is the characteristic polynomial of A and that there are polynomials $q(x)$ and $r(x) = a + bx$ such that $x^{k-2} = q(x)f(x) + r(x)$. Since

$$\lambda_1 = \frac{1 + \sqrt{5}}{2} \quad \text{and} \quad \lambda_2 = \frac{1 - \sqrt{5}}{2}$$

are the eigenvalues of A we have

$$\lambda_1^{k-2} = q(\lambda_1)f(\lambda_1) + r(\lambda_1) = a + b\lambda_1$$
$$\lambda_2^{k-2} = q(\lambda_2)f(\lambda_2) + r(\lambda_2) = a + b\lambda_2$$

Solving this pair of equations for a and b yields

$$a = \frac{\lambda_1 \lambda_2^{k-2} - \lambda_1^{k-2} \lambda_2}{\lambda_1 - \lambda_2} \qquad b = \frac{\lambda_1^{k-2} - \lambda_2^{k-2}}{\lambda_1 - \lambda_2}$$

Hence

$$A^{k-2} = q(A)f(A) + r(A)$$
$$= aI + bA$$
$$= \frac{\lambda_1 \lambda_2^{k-2} - \lambda_1^{k-2}\lambda_2}{\lambda_1 - \lambda_2} \begin{bmatrix} 1 & 0 \\ 0 & 1 \end{bmatrix} + \frac{\lambda_1^{k-2} - \lambda_2^{k-2}}{\lambda_1 - \lambda_2} \begin{bmatrix} 0 & 1 \\ 1 & 1 \end{bmatrix}$$
$$= \frac{1}{\lambda_1 - \lambda_2} \begin{bmatrix} \lambda_1 \lambda_2^{k-2} - \lambda_1^{k-2}\lambda_2 & \lambda_1^{k-2} - \lambda_2^{k-2} \\ \lambda_1^{k-2} - \lambda_2^{k-2} & \lambda_1 \lambda_2^{k-2} - \lambda_1^{k-2}\lambda_2 + \lambda_1^{k-2} - \lambda_2^{k-2} \end{bmatrix}$$

Therefore

$$\begin{bmatrix} p_{k-1} \\ p_k \end{bmatrix} = A^{k-2} \begin{bmatrix} 1 \\ 1 \end{bmatrix}$$
$$= \frac{1}{\lambda_1 - \lambda_2} \begin{bmatrix} \lambda_1 \lambda_2^{k-2} - \lambda_1^{k-2}\lambda_2 + \lambda_1^{k-2} - \lambda_2^{k-2} \\ \lambda_1 \lambda_2^{k-2} - \lambda_1^{k-2}\lambda_2 + 2(\lambda_1^{k-2} - \lambda_2^{k-2}) \end{bmatrix}$$

We now notice that $-\lambda_1\lambda_2 = 1$ and that

$$\lambda_1\lambda_2^{k-2} - \lambda_1^{k-2}\lambda_2 + 2(\lambda_1^{k-2} - \lambda_2^{k-2}) = \lambda_1^{k-2}(2 - \lambda_2) - \lambda_2^{k-2}(2 - \lambda_1)$$

$$= \lambda_1^{k-2}\left(\frac{3 + \sqrt{5}}{2}\right) - \lambda_2^{k-2}\left(\frac{3 - \sqrt{5}}{2}\right)$$

$$= \lambda_1^{k-2}(1 + \lambda_1) - \lambda_2^{k-2}(1 + \lambda_2)$$

$$= \lambda_1^{k-2} - \lambda_2^{k-2} + \lambda_1^{k-1} - \lambda_2^{k-1}$$

$$= (\lambda_1^{k-2} - \lambda_2^{k-2})(-\lambda_1\lambda_2) + \lambda_1^{k-1} - \lambda_2^{k-1}$$

$$= \lambda_1\lambda_2^{k-1} - \lambda_1^{k-1}\lambda_2 + \lambda_1^{k-1} - \lambda_2^{k-1}$$

so that

$$p_k = \frac{\lambda_1\lambda_2^{k-1} - \lambda_1^{k-1}\lambda_2 + \lambda_1^{k-1} - \lambda_2^{k-1}}{\lambda_1 - \lambda_2}$$

which is equivalent to the expression we found for p_k in Example 5 of Section 7.3. In that example we further simplified this expression to obtain

$$p_k = \frac{(1 + \sqrt{5})^k - (1 - \sqrt{5})^k}{2^k\sqrt{5}}$$

Exercises

In Exercises 1–8 show that the given matrix A satisfies its own characteristic equation and use the Cayley–Hamilton Theorem to compute A^{-1} whenever A^{-1} exists.

1. $\begin{bmatrix} 0 & 1 \\ 4 & 0 \end{bmatrix}$

2. $\begin{bmatrix} 1 & 1 \\ 0 & 1 \end{bmatrix}$

3. $\begin{bmatrix} 1 & 2 \\ 2 & 4 \end{bmatrix}$

4. $\begin{bmatrix} 2 & 3 \\ 4 & 6 \end{bmatrix}$

5. $\begin{bmatrix} 1 & 0 & 1 \\ 0 & 2 & 0 \\ 0 & 0 & 2 \end{bmatrix}$

6. $\begin{bmatrix} 2 & 3 & 1 \\ 0 & 4 & 0 \\ 2 & 5 & 9 \end{bmatrix}$

7. $\begin{bmatrix} 1 & 2 & 3 \\ 4 & 5 & 6 \\ 7 & 8 & 9 \end{bmatrix}$

8. $\begin{bmatrix} 1 & 3 & 6 \\ 4 & 2 & 5 \\ 3 & -1 & -1 \end{bmatrix}$

In Exercises 9–12 use the Cayley–Hamilton Theorem to evaluate the given polynomial.

9. $A^3 + 2A^2 - 3A + 4I$, where A is the matrix in Exercise 1.
10. $A^4 + A^2 + I$, where A is the matrix in Exercise 2.
11. $A^5 + 3A^4 + A^3 - A^2 + 4A + 6I$, where A is the matrix in Exercise 5.
12. $A^4 - 15A^3 + 60A^2 - 64A + 3I$, where A is the matrix in Exercise 6.
13. Show by direct substitution that a 2×2 matrix satisfies its own characteristic polynomial.
14. Establish the identity in equation (2).

8

Applications

8.1 COMPUTER GRAPHICS

We now indicate how several of the concepts and techniques of earlier sections can be used to generate computer graphics. Suppose that we wish to view an object, such as the L-shaped solid in Figure 8.1, from a certain direction and to represent what we

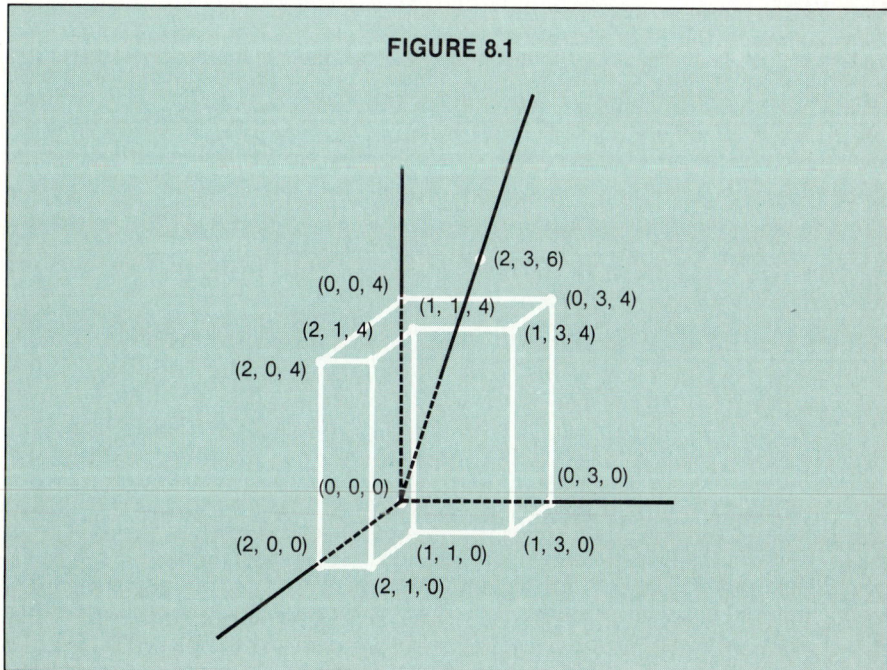

FIGURE 8.1

see by a 2-dimensional picture. To be specific, suppose that we wish to view the L-shaped object in Figure 8.1 from the direction of a ray emanating from the origin in the direction of the vector

$$\mathbf{x} = \begin{bmatrix} 2 \\ 3 \\ 6 \end{bmatrix}$$

In order to obtain a 2-dimensional picture of what we see, we construct a plane perpendicular to the ray and project the object onto this plane along lines that are parallel to the ray (see Figure 8.2). Conceptually this is a rather simple process, but in practice it requires some effort to find this projection.

To begin we extend the set $\{\mathbf{x}\}$ to a basis

$$B' = \left\{ \begin{bmatrix} 2 \\ 3 \\ 6 \end{bmatrix}, \begin{bmatrix} 1 \\ 0 \\ 0 \end{bmatrix}, \begin{bmatrix} 0 \\ 1 \\ 0 \end{bmatrix} \right\}$$

for R^3 by using the technique described before Example 4 of Section 4.11. The Gram–Schmidt process now allows us to obtain an orthonormal basis $B = \{\mathbf{u}, \mathbf{v}, \mathbf{w}\}$

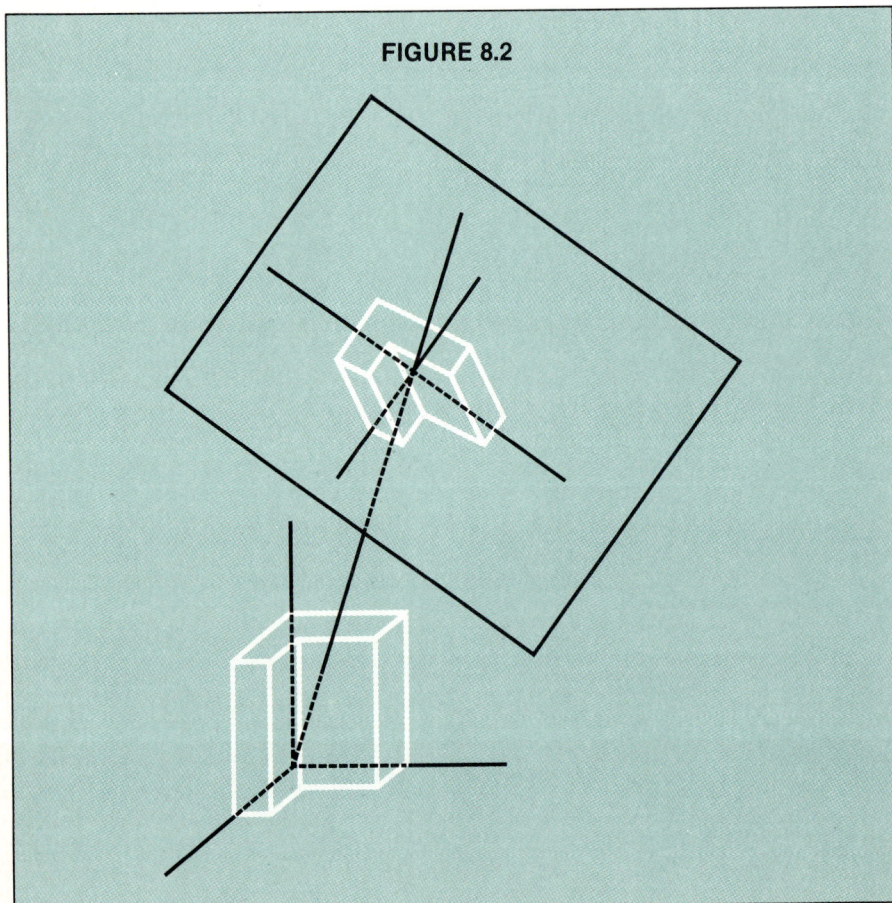

FIGURE 8.2

for R^3 where

$$\mathbf{u} = \frac{1}{7}\begin{bmatrix} 2 \\ 3 \\ 6 \end{bmatrix}, \qquad \mathbf{v} = \frac{1}{7\sqrt{5}}\begin{bmatrix} 15 \\ -2 \\ -4 \end{bmatrix}, \qquad \mathbf{w} = \frac{1}{\sqrt{5}}\begin{bmatrix} 0 \\ 2 \\ -1 \end{bmatrix}$$

(The reader should verify that B' and B are actually obtained as described.) Notice that $\mathbf{u}$ and $\mathbf{x}$ have the same direction. Moreover the orthogonal matrix

$$P = \begin{bmatrix} \dfrac{2}{7} & \dfrac{15}{7\sqrt{5}} & 0 \\[2mm] \dfrac{3}{7} & \dfrac{-2}{7\sqrt{5}} & \dfrac{2}{\sqrt{5}} \\[2mm] \dfrac{6}{7} & \dfrac{-4}{7\sqrt{5}} & \dfrac{-1}{\sqrt{5}} \end{bmatrix}$$

whose columns are $\mathbf{u}, \mathbf{v}$, and $\mathbf{w}$, is the transition matrix from the natural basis N for R^3 to B. Therefore

$$P^{-1} = P^t = \begin{bmatrix} \dfrac{2}{7} & \dfrac{3}{7} & \dfrac{6}{7} \\[2mm] \dfrac{15}{7\sqrt{5}} & \dfrac{-2}{7\sqrt{5}} & \dfrac{-4}{7\sqrt{5}} \\[2mm] 0 & \dfrac{2}{\sqrt{5}} & \dfrac{-1}{\sqrt{5}} \end{bmatrix}$$

is the transition matrix from N to B.

Any plane with $\mathbf{u}$ as a normal vector can be written as

$$\Pi_k = \left\{ \begin{bmatrix} k \\ b \\ c \end{bmatrix}_B : k \text{ a fixed number, } a \text{ and } b \text{ arbitrary} \right\}$$

for some number k. Whenever a vector

$$\mathbf{y} = \begin{bmatrix} a \\ b \\ c \end{bmatrix}_B$$

is projected onto Π_k it is projected along lines perpendicular to Π_k. Hence the projection of y onto Π_k is

$$\begin{bmatrix} k \\ b \\ c \end{bmatrix}_B$$

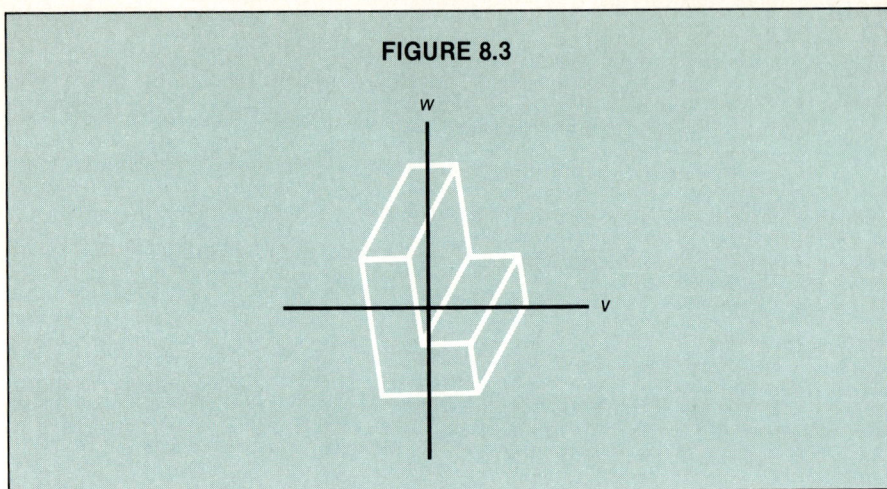

FIGURE 8.3

so that the projection of the object onto Π_k is

$$\left\{ \begin{bmatrix} k \\ b \\ c \end{bmatrix}_B : \begin{bmatrix} a \\ b \\ c \end{bmatrix}_B \text{ is the coordinate vector for a point in the object} \right\}$$

Since the first component of each vector in this set is always k, we can graph the projection in a vw-plane by merely plotting the last two components of each vector.

We now notice that the projection onto the plane Π_k takes a line segment in R^3 onto either a line segment or a point in Π_k. This observation allows us to compute the projection of the L-shaped object by merely projecting its vertices and then connecting the images of adjacent vertices with line segments. To do this we first compute the coordinate vector with respect to B of each of the vertices of the object. This is done by multiplying each of the natural coordinate vectors by P^t. Doing this and then plotting in a vw-plane the points formed by the last two components of each vector yields the picture in Figure 8.3.

Exercises

In Exercises 1–4 find the projection of the L-shaped region given in Figure 8.1 onto a plane perpendicular to a ray emanating from the origin in the direction of the given vector.

1. $\begin{bmatrix} 0 \\ 0 \\ 1 \end{bmatrix}$ 2. $\begin{bmatrix} 0 \\ 1 \\ 0 \end{bmatrix}$ 3. $\begin{bmatrix} 1 \\ 0 \\ 1 \end{bmatrix}$ 4. $\begin{bmatrix} 1 \\ -1 \\ 0 \end{bmatrix}$

In Exercises 5–8 find the projection of the cube with vertices $(\pm 1, \pm 1, \pm 1)$ onto a plane perpendicular to a ray emanating from the origin in the direction of the given vector.

5. $\begin{bmatrix} 0 \\ 0 \\ 1 \end{bmatrix}$ 6. $\begin{bmatrix} 0 \\ 1 \\ 0 \end{bmatrix}$ 7. $\begin{bmatrix} 1 \\ 0 \\ 1 \end{bmatrix}$ 8. $\begin{bmatrix} 1 \\ -1 \\ 0 \end{bmatrix}$

8.2 QUADRATIC FORMS AND POSITIVE DEFINITE MATRICES

A function of the form

$$F(x, y) = ax^2 + bxy + cy^2$$

is called a **quadratic form** in x and y. More generally, a function of the form

$$F(x_1, x_2, \cdots, x_n) = \sum_{i=1}^{n} \left(\sum_{j=1}^{n} a_{ij} x_j \right) x_i$$

$$= a_{11} x_1^2 + a_{12} x_1 x_2 + \cdots + a_{1n} x_1 x_n$$

$$+ a_{21} x_1 x_2 + a_{22} x_2^2 + \cdots + a_{2n} x_2 x_n$$

(1)

$$+ a_{n1} x_1 x_n + a_{n2} x_2 x_n + \cdots + a_{nn} x_n^2$$

is called a **quadratic form** in $x_1, x_2, \ldots, x_n$.

Quadratic forms arise in numerous areas: economics, geometry, mechanics, relativity, and statistics. It is frequently convenient to represent a quadratic form in terms of the Euclidean inner product as $\langle B\mathbf{x}, \mathbf{x} \rangle$ where B is a symmetric $n \times n$ matrix and $\mathbf{x}$ is the n-vector having $x_1, x_2, \ldots, x_n$ as its components. Fortunately this is easy to do. We begin by noting that in (1) the product of x_i and x_j occurs twice: once as $a_{ij} x_i x_j$ and again as $a_{ji} x_j x_i$. We now let

$$b_{ij} = b_{ji} = \frac{1}{2}(a_{ij} + a_{ji})$$

and notice that

$$a_{ij} x_i x_j + a_{ji} x_j x_i = (a_{ij} + a_{ji}) x_i x_j$$

$$= (b_{ij} + b_{ji}) x_i x_j$$

$$= b_{ij} x_i x_j + b_{ji} x_j x_i$$

Hence, the quadratic form in (1) can be rewritten as

(2)

$$F(x_1, x_2, \ldots, x_n) = \sum_{i=1}^{n} \left(\sum_{j=1}^{n} b_{ij} x_j \right) x_i$$

Notice that

$$\sum_{j=1}^{n} b_{ij} x_j = b_{i1} x_1 + b_{i2} x_2 + \cdots + b_{in} x_n$$

is the ith component of $B\mathbf{x}$ where

$$B = \begin{bmatrix} b_{11} & b_{12} & \cdots & b_{1n} \\ b_{21} & b_{22} & \cdots & b_{2n} \\ \vdots & \vdots & & \vdots \\ b_{n1} & b_{n2} & \cdots & b_{nn} \end{bmatrix} \quad \text{and} \quad \mathbf{x} = \begin{bmatrix} x_1 \\ x_2 \\ \vdots \\ x_n \end{bmatrix}$$

With this observation we see that the quadratic form as expressed in (2) is the sum of the products of the corresponding components of $\mathbf{x}$ and $B\mathbf{x}$. Therefore, the quadratic form is the Euclidean inner product of $\mathbf{x}$ and $B\mathbf{x}$:

$$F(x_1, x_2, \ldots, x_n) = \langle B\mathbf{x}, \mathbf{x} \rangle$$
$$= \mathbf{x}^t B \mathbf{x}$$

Moreover, we constructed the matrix B in such a way that it is symmetric. This is especially important because in many applications it is advantageous to diagonalize B. Quadratic forms are commonly written as $\mathbf{x}^t B \mathbf{x}$ where B is a symmetric $n \times n$ matrix and $\mathbf{x}$ is an n-vector.

Example 1. Consider the quadratic form

$$F(x, y) = 3x^2 + 4xy + 5y^2 \qquad (3)$$

Then

$$F(x, y) = 3x^2 + 2xy + 2yx + 5y^2$$

$$= [x \ y] \begin{bmatrix} 3 & 2 \\ 2 & 5 \end{bmatrix} \begin{bmatrix} x \\ y \end{bmatrix}$$

Example 2. Consider the quadratic form $F(x_1, x_2) = \mathbf{x}^t B \mathbf{x}$ where

$$B = \begin{bmatrix} 3 & 2 \\ 2 & 5 \end{bmatrix} \qquad \mathbf{x} = \begin{bmatrix} x_1 \\ x_2 \end{bmatrix}$$

Since B is symmetric there is an orthogonal matrix P such that (see the discussion before Example 4 of Section 7.6)

$$P^t B P = \begin{bmatrix} \lambda_1 & 0 \\ 0 & \lambda_2 \end{bmatrix}$$

where λ_1 and λ_2 are the eigenvalues of B. Hence

$$B = P \begin{bmatrix} \lambda_1 & 0 \\ 0 & \lambda_2 \end{bmatrix} P^{-1}$$

$$= P \begin{bmatrix} \lambda_1 & 0 \\ 0 & \lambda_2 \end{bmatrix} P^t$$

A straightforward calculation shows that the eigenvalues of B are $4 + \sqrt{5}$ and $4 - \sqrt{5}$. Thus, we may choose P so that $\lambda_1 = 4 + \sqrt{5}$ and $\lambda_2 = 4 - \sqrt{5}$. Let

$$\begin{bmatrix} x_1' \\ x_2' \end{bmatrix} = P^t \begin{bmatrix} x_1 \\ x_2 \end{bmatrix}$$

In this new notation we have

$$F(x_1, x_2) = [x_1 x_2] B \begin{bmatrix} x_1 \\ x_2 \end{bmatrix}$$

$$= [x_1 x_2] P \begin{bmatrix} \lambda_1 & 0 \\ 0 & \lambda_2 \end{bmatrix} P^t \begin{bmatrix} x_1 \\ x_2 \end{bmatrix}$$

$$F(x_1, x_2) = \left(P^t \begin{bmatrix} x_1 \\ x_2 \end{bmatrix} \right)^t \begin{bmatrix} \lambda_1 & 0 \\ 0 & \lambda_2 \end{bmatrix} \left(P^t \begin{bmatrix} x_1 \\ x_2 \end{bmatrix} \right)$$

$$= \begin{bmatrix} x_1' \\ x_2' \end{bmatrix}^t \begin{bmatrix} \lambda_1 & 0 \\ 0 & \lambda_2 \end{bmatrix} \begin{bmatrix} x_1' \\ x_2' \end{bmatrix}$$

$$= \lambda_1(x_1')^2 + \lambda_2(x_2')^2$$

We have shown that if B is the basis for R^2 consisting of the columns of P, then

$$F(x_1, x_2) = \lambda_1(x_1')^2 + \lambda_2(x_2')^2$$

where λ_1, λ_2 are the eigenvalues of B and x_1', x_2' are the components of $[\mathbf{x}]_B$. This is an example of a general property of quadratic forms that is given in the following theorem.

Theorem 1. *Let $F(x_1, x_2, \ldots, x_n) = \mathbf{x}^t A\mathbf{x}$ be a quadratic form where A is a symmetric $n \times n$ matrix. Then there is an orthonormal basis B for R^n consisting of eigenvalues of A so that*

$$F(x_1, x_2, \ldots, x_n) = \lambda_1(x_1')^2 + \lambda_2(x_2')^2 + \cdots + \lambda_n(x_n')^2$$

where $\lambda_1, \lambda_2, \ldots, \lambda_n$ are the eigenvalues of A,

$$\mathbf{x} = \begin{bmatrix} x_1 \\ x_2 \\ \vdots \\ x_n \end{bmatrix}, \quad and \quad [\mathbf{x}]_B = \begin{bmatrix} x_1' \\ x_2' \\ \vdots \\ x_n' \end{bmatrix}_B$$

Proof: By Theorem 16 of Section 7.6 there is an orthonormal basis B for R^n consisting of eigenvectors of A. Let P be the orthogonal matrix having the vectors in B as its columns (see Exercise 11 of Section 7.6). Then $P^{-1} = P^t$ and $P^{-1}AP$ is a diagonal matrix D having the eigenvalues $\lambda_1, \lambda_2, \ldots, \lambda_n$ on its diagonal. Notice that P is the transition matrix from the basis B to the natural basis N for R^n. Therefore $P^{-1} = P^t$ is the transition matrix from N to B. Since $A = PDP^t$ we have

$$F(x_1, x_2, \ldots, x_n) = \mathbf{x}^t A\mathbf{x}$$

$$= \mathbf{x}^t PDP^t\mathbf{x}$$

$$= (P^t\mathbf{x})^t D(P^t\mathbf{x})$$

$$= [\mathbf{x}]_B^t D[\mathbf{x}]_B$$

$$= \lambda_1(x_1')^2 + \lambda_2(x_2')^2 + \cdots + \lambda_n(x_n')^2 \quad \blacksquare$$

One of the properties that is commonly desired of a quadratic form, or equivalently of its associated symmetric matrix, is described in the following definition.

Definition 1. A symmetric $n \times n$ matrix A is called **positive definite** if $\mathbf{x}^t A\mathbf{x} > 0$ for every nonzero n-vector $\mathbf{x}$.

Example 3.   In Example 1 we found that there is an orthogonal matrix P, whose columns form a basis B for R^2, such that

$$F(x_1, x_2) = \mathbf{x}^t \begin{bmatrix} 3 & 2 \\ 2 & 5 \end{bmatrix} \mathbf{x}$$

$$= \lambda_1(x_1')^2 + \lambda_2(x_2')^2$$

where $\lambda_1 = 4 + \sqrt{5}$, $\lambda_2 = 4 - \sqrt{5}$, and

$$P\mathbf{x} = [\mathbf{x}]_B = \begin{bmatrix} x_1 \\ x_2 \end{bmatrix}_B$$

If $\begin{bmatrix} x_1 \\ x_2 \end{bmatrix} \neq \begin{bmatrix} 0 \\ 0 \end{bmatrix}$, then $\begin{bmatrix} x_1' \\ x_2' \end{bmatrix} \neq \begin{bmatrix} 0 \\ 0 \end{bmatrix}$ since P^t is nonsingular. Moreover, if $\begin{bmatrix} x_1 \\ x_2 \end{bmatrix} \neq \begin{bmatrix} 0 \\ 0 \end{bmatrix}$, then $F(x_1, x_2) > 0$ because $\lambda_1 > 0$, $\lambda_2 > 0$, and either $x_1' \neq 0$ or $x_2' \neq 0$. Hence the matrix

$$\begin{bmatrix} 3 & 2 \\ 2 & 5 \end{bmatrix}$$

is positive definite. Notice that if the eigenvalues were not positive, then the matrix would not be positive definite.

The following theorem shows that the signs of the eigenvalues determine whether a symmetric matrix is positive definite. Its proof follows directly from Theorem 1 and is left as an exercise.

Theorem 2.   *A symmetric matrix is positive definite if and only if all its eigenvalues are positive.*

Example 4.   In Example 4 of Section 7.1 we found that the eigenvalues of

$$A = \begin{bmatrix} 7 & 4 & -4 \\ 4 & -8 & -1 \\ -4 & -1 & -8 \end{bmatrix}$$

and 9, -9, and -9. By Theorem 2 the matrix A is not positive definite.

Example 5.   Consider the matrix

$$A = \begin{bmatrix} 7 & 1 & 1 & 2 \\ 1 & 4 & 0 & 1 \\ 1 & 0 & 3 & 1 \\ 2 & 1 & 1 & 5 \end{bmatrix}$$

By Gershgorin's Theorem (Theorem 9 of Section 7.4), the eigenvalues of A lie in the disks indicated in Figure 8.4. Since symmetric matrices have only real eigenvalues (Theorem 14 of Section 7.6), the eigenvalues of A must lie between 1 and 11. Hence all the eigenvalues of A are positive. By Theorem 2 the matrix A is positive definite.

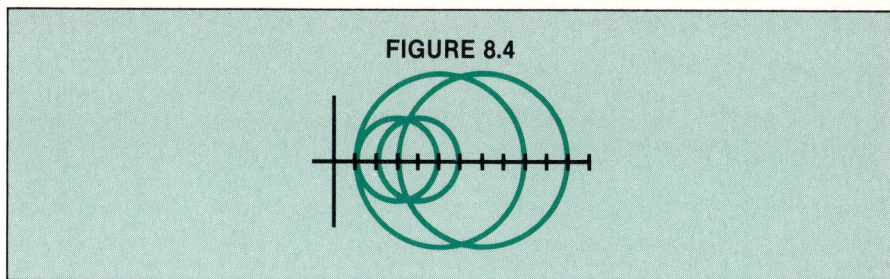

FIGURE 8.4

Let A be an $n \times n$ symmetric matrix. If n is greater than 3 it may not be a simple matter to determine the eigenvalues of A. If we are interested in determining whether A is positive definite, we do not need to know its eigenvalues. We only need to know the signs of the eigenvalues. Example 5 shows how Gershgorin's Theorem can be used to show that a matrix is positive definite without computing its eigenvalues. The following theorem, whose proof is beyond the scope of this course, shows how determinants may be used in a similar fashion.

Theorem 3.   *A symmetric matrix*

$$A = \begin{bmatrix} a_{11} & a_{12} & \cdots & a_{1n} \\ a_{21} & a_{22} & \cdots & a_{2n} \\ \vdots & \vdots & & \vdots \\ a_{n1} & a_{n2} & \cdots & a_{nn} \end{bmatrix}$$

is positive definite if and only if

$$a_{11} > 0, \det \begin{bmatrix} a_{11} & a_{12} \\ a_{21} & a_{22} \end{bmatrix} > 0, \quad \det \begin{bmatrix} a_{11} & a_{12} & a_{13} \\ a_{21} & a_{22} & a_{23} \\ a_{31} & a_{32} & a_{33} \end{bmatrix} > 0, \ldots,$$

$$\det \begin{bmatrix} a_{11} & a_{12} & \cdots & a_{1n} \\ a_{21} & a_{22} & \cdots & a_{2n} \\ \vdots & \vdots & & \vdots \\ a_{n1} & a_{n2} & \cdots & a_{nn} \end{bmatrix} > 0$$

Example 6.   Consider the matrix

$$A = \begin{bmatrix} 7 & 4 & -4 \\ 4 & -8 & -1 \\ -4 & -1 & -8 \end{bmatrix}$$

Since

$$\det \begin{bmatrix} 7 & 4 \\ 4 & -8 \end{bmatrix} = -72 < 0$$

The matrix is not positive definite. This agrees with our findings in Example 4.

Example 7. Consider the matrix

$$A = \begin{bmatrix} 7 & 1 & 1 & 2 \\ 1 & 4 & 0 & 1 \\ 1 & 0 & 3 & 1 \\ 2 & 1 & 1 & 5 \end{bmatrix}$$

Since

$$7 > 0, \qquad \det \begin{bmatrix} 7 & 1 \\ 1 & 4 \end{bmatrix} = 27 > 0,$$

$$\det \begin{bmatrix} 7 & 1 & 1 \\ 1 & 4 & 0 \\ 1 & 0 & 3 \end{bmatrix} = 77 > 0, \qquad \text{and} \qquad \det A = 316 > 0$$

the matrix A is positive definite. This agrees with our findings in **Example 5**.

Exercises

In Exercises 1–6 write the quadratic form as $\mathbf{x}^t A \mathbf{x}$, where A is a symmetric matrix.

1. $3x^2 + 6xy + 6y^2$
2. $5x^2 - 10xy - 6y^2$
3. $2x_1^2 + 3x_2^2 + 7x_3^2 + 4x_1x_2 + 6x_1x_3 + 2x_2x_3$
4. $-3x_1^2 + x_2^2 + 5x_3^2 + x_1x_2 + x_2x_3$
5. $x_1^2 - x_2^2 + x_3^2 - 2x_1x_2 + 4x_1x_3 - 5x_2x_3$
6. $(x_1 + x_2)^2 + (x_1 - x_3)^2 + 2(x_2 - x_3)^2$

In Exercises 7–12 determine whether the given matrix is positive definite by using Theorem 2 and then by using Theorem 3.

7. $\begin{bmatrix} 2 & 1 \\ 1 & 4 \end{bmatrix}$

8. $\begin{bmatrix} 1 & 2 \\ 2 & 5 \end{bmatrix}$

9. $\begin{bmatrix} 5 & 2 & 0 \\ 2 & 3 & 0 \\ 0 & 0 & 4 \end{bmatrix}$

10. $\begin{bmatrix} 5 & 0 & 0 \\ 0 & 1 & 3 \\ 0 & 3 & 2 \end{bmatrix}$

11. $\begin{bmatrix} 1 & 0 & 2 \\ 0 & 7 & 0 \\ 2 & 0 & 1 \end{bmatrix}$

12. $\begin{bmatrix} 3 & 0 & 1 \\ 0 & 7 & 0 \\ 1 & 0 & 6 \end{bmatrix}$

In Exercises 13–16 use Theorem 9 of Section 7.4, to determine whether the given matrix is positive definite.

13. $\begin{bmatrix} 3 & 1 & 1 \\ 1 & 5 & 2 \\ 1 & 2 & 7 \end{bmatrix}$

14. $\begin{bmatrix} 3 & 1 & 1 \\ 1 & -5 & 2 \\ 1 & 2 & 7 \end{bmatrix}$

15. $\begin{bmatrix} 5 & 3 & 1 \\ 3 & 7 & 2 \\ 1 & 2 & -6 \end{bmatrix}$ **16.** $\begin{bmatrix} 5 & 3 & 1 \\ 3 & 7 & 2 \\ 1 & 2 & 6 \end{bmatrix}$

17. Show that a symmetric diagonally dominant matrix (see Exercise 10 of Section 7.4) with positive diagonal components is positive definite.

18. Prove Theorem 2.

8.3 AN APPLICATION TO ANALYTIC GEOMETRY

In this section we will apply the results on diagonalizing symmetric matrices to the study of conic sections. Before doing this we give a brief summary of the basic equations of conic sections.

The graph of

$$\frac{(x-h)^2}{r^2} + \frac{(y-k)^2}{s^2} = 1 \qquad\qquad r > 0, \quad s > 0$$

is a circle if $r = s$ and an ellipse if $r \neq s$. The shape of the graph is determined by the relative sizes of r and s while its position in the xy-plane is determined by h and k as shown in Figure 8.5.

The graphs of

$$\frac{(x-h)^2}{r^2} - \frac{(y-k)^2}{s^2} = 1 \qquad\qquad r > 0, \quad s > 0$$

and

$$\frac{(y-k)^2}{s^2} - \frac{(x-h)^2}{r^2} = 1 \qquad\qquad r > 0, \quad s > 0$$

are hyperbolas whose positions and orientations in the xy-plane are indicated in Figure 8.6.

The graphs of

$$(y-k)^2 = c(x-h)$$

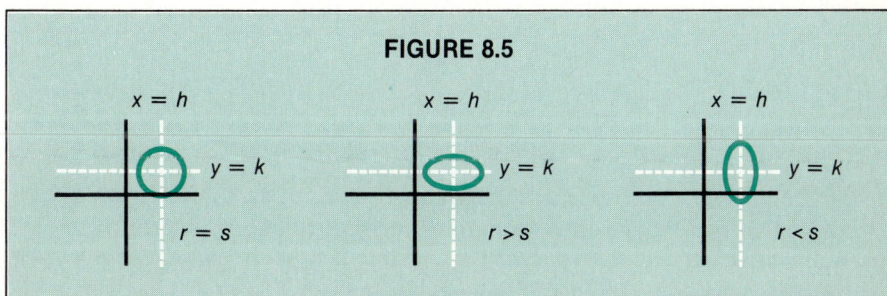

FIGURE 8.5

FIGURE 8.6

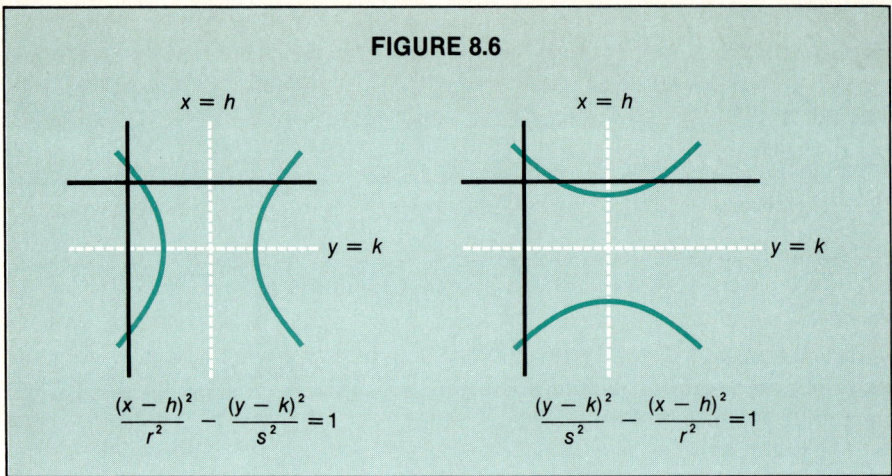

$$\frac{(x-h)^2}{r^2} - \frac{(y-k)^2}{s^2} = 1 \qquad\qquad \frac{(y-k)^2}{s^2} - \frac{(x-h)^2}{r^2} = 1$$

FIGURE 8.7

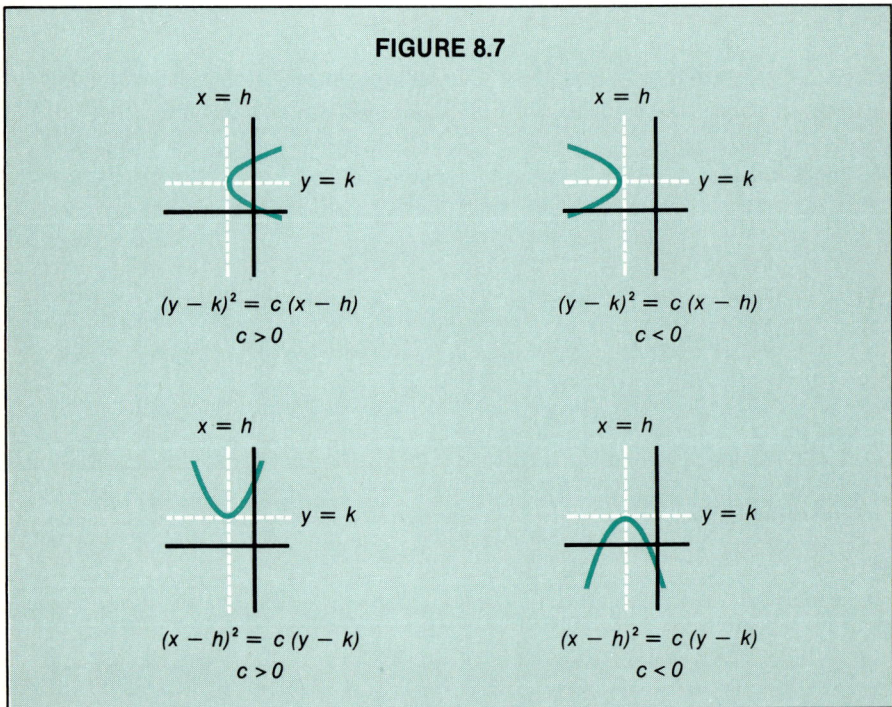

$(y-k)^2 = c\,(x-h)$
$c > 0$

$(y-k)^2 = c\,(x-h)$
$c < 0$

$(x-h)^2 = c\,(y-k)$
$c > 0$

$(x-h)^2 = c\,(y-k)$
$c < 0$

and

$$(x-h)^2 = c(y-k)$$

are parabolas whose positions and orientations are as indicated in Figure 8.7.

Notice that each of the above equations can be written in the form

$$ax^2 + cy^2 + dx + ey + f = 0 \qquad\qquad (1)$$

for some choice of the constants a, c, d, e, and f with either $a \neq 0$ or $c \neq 0$. Moreover, the graph of any equation of the form in (1) with $a \neq 0$ or $c \neq 0$ is a conic section. This can be proved by grouping terms in an appropriate fashion. A complete proof is tiresome and will be omitted. We will indicate the nature of the proof by determining the type of graph of a specific equation of the form in (1).

Example 1.   Consider the equation

(2)
$$16x^2 - 32x + 9y^2 + 36y - 92 = 0$$

which can be written in the equivalent form

$$16(x^2 - 2x) + 9(y^2 + 4y) = 92$$

By completing the squares of the two terms in parentheses, we obtain

$$16(x^2 - 2x + 1 - 1) + 9(y^2 + 4y + 4 - 4) = 92$$

or

$$16(x^2 - 2x + 1) - 16 + 9(y^2 + 4y + 4) - 36 = 92$$

or

$$16(x - 1)^2 + 9(y + 2)^2 = 144$$

or

$$\frac{(x - 1)^2}{3^2} + \frac{(y + 2)^2}{4^2} = 1$$

so that the graph of equation (2) is an ellipse.

    Notice that equation (1) has no cross-product term xy. The presence of such a term indicates that the conic section has been rotated. We will now show how to identify the graph of an equation having the form

(3)
$$ax^2 + 2bxy + cy^2 + dx + ey + f = 0$$

with a, b, or c nonzero. The expression

(4)
$$ax^2 + 2bxy + cy^2$$

is called the **quadratic form** associated with equation (3).

    Notice that the quadratic form in (4) can be represented as

$$ax^2 + 2bxy + cy^2 = [x \ \ y] \begin{bmatrix} a & b \\ b & c \end{bmatrix} \begin{bmatrix} x \\ y \end{bmatrix}$$

$$= \mathbf{u}^t A \mathbf{u}$$

where

$$\mathbf{u} = \begin{bmatrix} x \\ y \end{bmatrix} \quad \text{and} \quad A = \begin{bmatrix} a & b \\ b & c \end{bmatrix}$$

Let B be an orthonormal basis for R^2 consisting of eigenvectors of A and let P be the orthogonal matrix having the vectors in B as its columns. Such a basis exists because the matrix A is symmetric (Theorem 16 of Section 7.6). Then

$$P^t A P = D = \begin{bmatrix} a' & 0 \\ 0 & c' \end{bmatrix}$$

where a' and c' are the eigenvalues of A (Theorem 1 of Section 8.2). Notice that P is the transition matrix from the basis B to the natural basis N for R^2. Therefore $P^{-1} = P^t$ is the transition matrix from N to B. The coordinate vectors relative to B of

$$\begin{bmatrix} x \\ y \end{bmatrix} \quad \text{and} \quad \begin{bmatrix} d \\ e \end{bmatrix}$$

are

$$\begin{bmatrix} x' \\ y' \end{bmatrix} = P^t \begin{bmatrix} x \\ y \end{bmatrix} \quad \text{and} \quad \begin{bmatrix} d' \\ e' \end{bmatrix} = P^t \begin{bmatrix} d \\ e \end{bmatrix}$$

respectively. Hence

$$\begin{bmatrix} x \\ y \end{bmatrix} = P \begin{bmatrix} x' \\ y' \end{bmatrix} \quad \text{and} \quad \begin{bmatrix} d \\ e \end{bmatrix} = P \begin{bmatrix} d' \\ e' \end{bmatrix}$$

Notice that

$$dx + ey = [d \ e] \begin{bmatrix} x \\ y \end{bmatrix}$$

and that

$$[d \ e] = \begin{bmatrix} d \\ e \end{bmatrix}^t = \left(P \begin{bmatrix} d' \\ e' \end{bmatrix} \right)^t = [d' \ e'] P^t$$

With the above notation , equation (3) can be rewritten as

$$[x \ y] \begin{bmatrix} a & b \\ b & c \end{bmatrix} \begin{bmatrix} x \\ y \end{bmatrix} + [d \ e] \begin{bmatrix} x \\ y \end{bmatrix} + f = 0$$

or

$$[x' \ y'] P^t \begin{bmatrix} a & b \\ b & c \end{bmatrix} P \begin{bmatrix} x' \\ y' \end{bmatrix} + [d' \ e'] P^t P \begin{bmatrix} x' \\ y' \end{bmatrix} + f = 0$$

or

$$[x' \ y'] \begin{bmatrix} a' & 0 \\ 0 & c' \end{bmatrix} \begin{bmatrix} x' \\ y' \end{bmatrix} + [d' \ e'] \begin{bmatrix} x' \\ y' \end{bmatrix} + f = 0$$

Performing the matrix multiplications yields equation (3) written with respect to the basis B rather than with respect to the natural basis N for R^2. This equation is

$$a'(x')^2 + c'(y')^2 + d'x' + e'y' + f = 0$$

which is an equation of the form in (1).

We have proved the following theorem.

Theorem 4. *Let*

$$ax^2 + 2bxy + cy^2 + dx + ey + f = 0$$

be the equation of a conic C and let

$$\mathbf{u}^t A\mathbf{u} = ax^2 + 2bxy + cy^2$$

be its associated quadratic form. Then there is an orthonormal basis B for $\mathbf{R}^2$ *consisting of eigenvectors of the matrix*

$$A = \begin{bmatrix} a & b \\ b & c \end{bmatrix}$$

such that relative to B the conic C has an equation of the form

$$a'(x')^2 + c'(y')^2 + d'x' + e'y' + f = 0$$

where a' and c' are eigenvalues of A and $\begin{bmatrix} d' \\ e' \end{bmatrix}$ *is the coordinate vector of* $\begin{bmatrix} d \\ e \end{bmatrix}$ *with respect to B.*

Example 2.   Consider the conic C described by the equation

$$x^2 + 4xy + y^2 + \sqrt{2}x - \sqrt{2}y + 3 = 0$$

The associated quadratic form is

$$[x \; y]\begin{bmatrix} 1 & 2 \\ 2 & 1 \end{bmatrix}\begin{bmatrix} x \\ y \end{bmatrix} = x^2 + 4xy + y^2$$

In Example 3 of Section 7.6 we found that

$$B = \left\{ \begin{bmatrix} \frac{1}{\sqrt{2}} \\ -\frac{1}{\sqrt{2}} \end{bmatrix}, \begin{bmatrix} \frac{1}{\sqrt{2}} \\ \frac{1}{\sqrt{2}} \end{bmatrix} \right\}$$

is an orthonormal basis for $\mathbf{R}^2$ consisting of eigenvectors of

$$\begin{bmatrix} 1 & 2 \\ 2 & 1 \end{bmatrix}$$

associated with the eigenvalues -1 and 3, respectively.

Let P be the matrix having the vectors in B as its columns. Then

$$\begin{bmatrix} d' \\ e' \end{bmatrix} = P^t\begin{bmatrix} \sqrt{2} \\ -\sqrt{2} \end{bmatrix} = \begin{bmatrix} 2 \\ 0 \end{bmatrix}$$

The equation of the conic C relative to the basis B is

$$(-1)(x')^2 + 3(y')^2 + 2x' + 3 = 0$$

which can be rewritten as

$$-(x' - 1)^2 + 3(y')^2 = -4$$

FIGURE 8.8

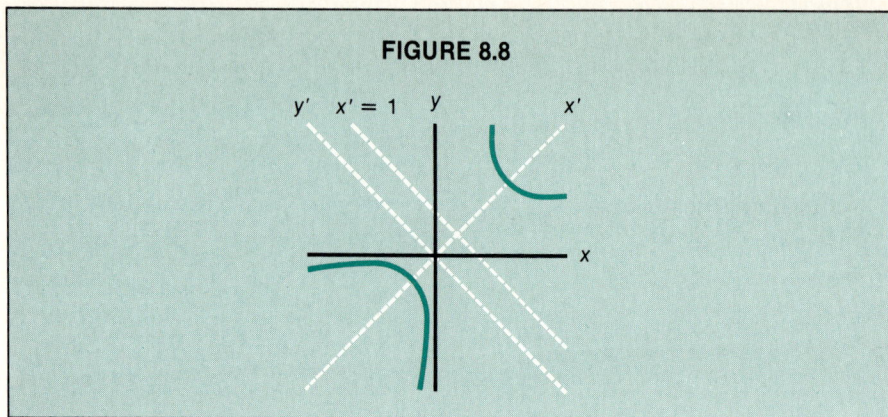

or equivalently

$$\frac{(x'-1)^2}{2^2} - \frac{(y')^2}{(2/\sqrt{3})^2} = 1$$

Hence C is a hyperbola.

Notice that the x' and y' axes are determined by the equations $y' = 0$ and $x' = 0$ respectively. Since

$$\begin{bmatrix} x' \\ y' \end{bmatrix} = P \begin{bmatrix} x \\ y \end{bmatrix} = \begin{bmatrix} \frac{1}{\sqrt{2}} & \frac{1}{\sqrt{2}} \\ -\frac{1}{\sqrt{2}} & \frac{1}{\sqrt{2}} \end{bmatrix} \begin{bmatrix} x \\ y \end{bmatrix}$$

$$= \begin{bmatrix} \frac{1}{\sqrt{2}}x + \frac{1}{\sqrt{2}}y \\ -\frac{1}{\sqrt{2}}x + \frac{1}{\sqrt{2}}y \end{bmatrix}$$

the x' and y' axes coincide with the lines $x + y = 0$ and $-x + y = 0$ (see Figure 8.8).

Exercises

In Exercises 1–6 determine the quadratic form associated with the given equation and determine the symmetric matrix such that the quadratic form can be written as

$$[x \ y] \ A \begin{bmatrix} x \\ y \end{bmatrix}$$

1. $3x^2 - 4xy + 2y^2 + 6x - 4y + 6 = 0$
2. $7x^2 + 6xy - 5y^2 + 5x + 4y - 7 = 0$
3. $2x^2 + 8xy + 7y - 2 = 0$
4. $6xy - 2y^2 + 5x - 8y + 2 = 0$
5. $x^2 + xy + y^2 - x = 0$
6. $5x^2 + 3xy - y^2 = 0$

In Exercises 7–12 name the graph of the given equation.

7. $x^2 - 3y^2 - 2 = 0$

8. $x^2 + 3y^2 - 2 = 0$

9. $y^2 - 2y + x = 0$

10. $x^2 - y^2 + 2 = 0$

11. $2x^2 + 5y^2 + 4x = 0$

12. $x^2 - 4x + y = 0$

In Exercises 13–22 name the graph of the given equation and give its equation in a coordinate system in which the equation has the form in (1).

13. $2x^2 + 2xy + 2y^2 - 1 = 0$

14. $4x^2 + 2xy + 4y^2 - 1 = 0$

15. $2x^2 + 6xy + 2y^2 - 1 = 0$

16. $x^2 - 2xy + y^2 - 1 = 0$

17. $x^2 + 4xy + 4y^2 - 4 = 0$

18. $4x^2 + 10xy + 4y^2 - 1 = 0$

19. $2x^2 + 2xy + 2y^2 + \sqrt{2}x - \sqrt{2}y = 0$

20. $4x^2 + 2xy + 4y^2 + 8\sqrt{2}x + 2\sqrt{2}y = 0$

21. $2x^2 + 6xy + 2y^2 + 4\sqrt{2}x + 6\sqrt{2}y - 1 = 0$

22. $x^2 - 2xy + y^2 + x + y = 0$

8.4 QUADRIC SURFACES

When we move from R^2 to R^3 and consider quadric surfaces (to be defined subsequently) instead of conics, we can still use transition matrices to rotate coordinates in an appropriate manner.

An equation of the form

(1) $ax^2 + by^2 + cz^2 + 2dxy + 2exz + 2fyz + gx + hy + iz + j = 0$

where at least one of a, b, c, d, e, or f is nonzero, is called a **quadratic equation** in x, y, z. The term

$$ax^2 + by^2 + cz^2 + 2dxy + 2exz + 2fyz$$

is called the **quadratic form** associated with the quadratic equation in (1).

The quadratic equation in equation (1) can be rewritten by using matrices as

$$[x\ y\ z]\begin{bmatrix} a & d & e \\ d & b & f \\ e & f & c \end{bmatrix}\begin{bmatrix} x \\ y \\ z \end{bmatrix} + [g\ h\ i]\begin{bmatrix} x \\ y \\ z \end{bmatrix} + j = 0$$

Setting

$$\mathbf{x} = \begin{bmatrix} x \\ y \\ z \end{bmatrix}, \quad A = \begin{bmatrix} a & d & e \\ d & b & f \\ e & f & c \end{bmatrix}, \quad B = [g\ h\ i]$$

enables us to write the quadratic equation and its associated quadratic form as

$$\mathbf{x}^t A \mathbf{x} + B\mathbf{x} + j = 0$$

and

$$\mathbf{x}^t A \mathbf{x}$$

respectively. The symmetric matrix A is called the matrix associated with the quadratic form

$$\mathbf{x}^t A \mathbf{x} = ax^2 + by^2 + cz^2 + 2dxy + 2exz + 2fyz$$

Example 1. The quadratic equation

$$7x^2 - 8y^2 - 5z^2 + 8xy - 6xz - 2yz + 5x + 9y - 4z = 0$$

has

$$7x^2 - 8y^2 - 5z^2 + 8xy - 6xz - 2yz$$

as its associated quadratic form, which in turn has

$$\begin{bmatrix} 7 & 4 & -3 \\ 4 & -8 & -1 \\ -3 & -1 & -5 \end{bmatrix}$$

as its associated matrix.

Graphs of quadratic equations in x, y, and z are called **quadric surfaces** (sometimes simply **quadrics**). Examples of the basic eleven quadric surfaces are depicted in Figure 8.9. A quadric surface whose equation has one of the forms listed in Figure 8.9 is said to be in standard position.

To begin, we consider the case in which the coefficients d, e, and f of the cross-product terms are zero. The presence of both x^2 and x terms, y^2 and y terms, or z^2 and z terms in a quadric equation indicates that the quadric surface is translated out of standard position. In such a case a change of coordinates of the form $x' = x - x_0$, $y' = y - y_0$, $z' = z - z_0$, possibly accompanied by a renaming of the coordinates, places the surface in standard position.

Example 2. Consider the quadratic equation

$$-16x^2 + 3y^2 + 4z^2 + 32x + 6y - 4z - 13 = 0$$

Rearranging terms and completing the squares yields

$$-16(x^2 - 2x) + 3(y^2 + 2y) + 4(z^2 - z) = 13$$

$$-16(x^2 - 2x + 1) + 3(y^2 + 2y + 1) + 4\left(z^2 - z + \frac{1}{4}\right) = 13 - 16 + 3 + 1$$

$$-16(x - 1)^2 + 3(y + 1)^2 + 4\left(z - \frac{1}{2}\right)^2 = 1$$

Setting $x' = x - 1$, $y' = y + 1$, and $z' = z - \frac{1}{2}$, we obtain $-16(x')^2 + 3(y')^2 + 4(z')^2 = 1$. This is not the equation for a quadric surface in standard position. However if we change coordinates again according to

$$x'' = z', \quad y'' = y', \quad z'' = x'$$

we obtain

$$4(x'')^2 + 3(y'')^2 - 16(z'')^2 = 1$$

This equation is equivalent to

$$\frac{(x'')^2}{(1/2)^2} + \frac{(y'')^2}{(1/\sqrt{3})^2} - \frac{(z'')^2}{(1/4)^2} = 1$$

which is the equation of a quadric surface in standard position.

FIGURE 8.9

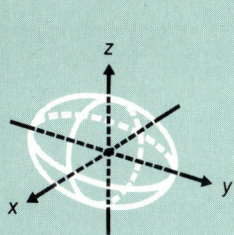

ellipsoid

$$\frac{x^2}{a^2} + \frac{y^2}{b^2} + \frac{z^2}{c^2} = 1$$

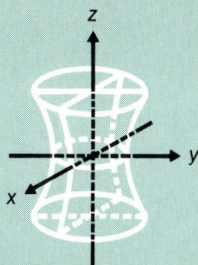

hyperboloid of one sheet

$$\frac{x^2}{a^2} + \frac{y^2}{b^2} - \frac{z^2}{c^2} = 1$$

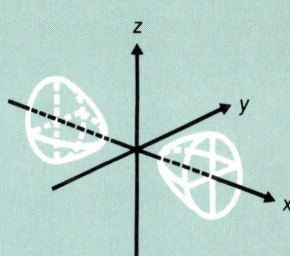

hyperboloid of two sheets

$$\frac{x^2}{a^2} - \frac{y^2}{b^2} - \frac{z^2}{c^2} = 1$$

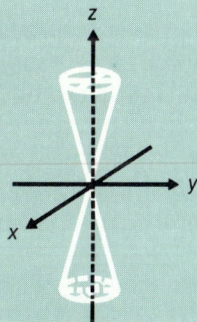

cone

$$\frac{x^2}{a^2} + \frac{y^2}{b^2} - \frac{z^2}{c^2} = 0$$

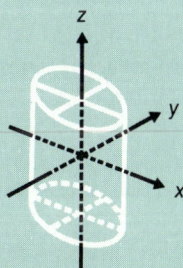

elliptic cylinder

$$\frac{x^2}{a^2} + \frac{y^2}{b^2} = 1$$

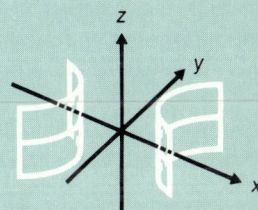

hyperbolic cylinder

$$\frac{x^2}{a^2} - \frac{y^2}{h^2} = 1$$

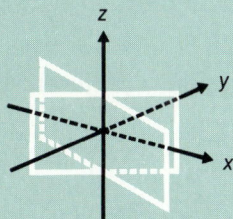

two intersecting planes

$$\frac{x^2}{a^2} - \frac{y^2}{b^2} = 0$$

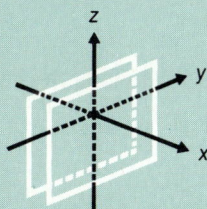

two parallel planes

$$x^2 = d^2$$

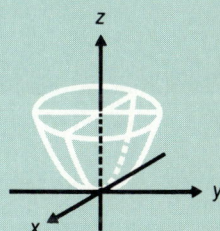

elliptic paraboloid

$$\frac{x^2}{a^2} + \frac{y^2}{b^2} = z$$

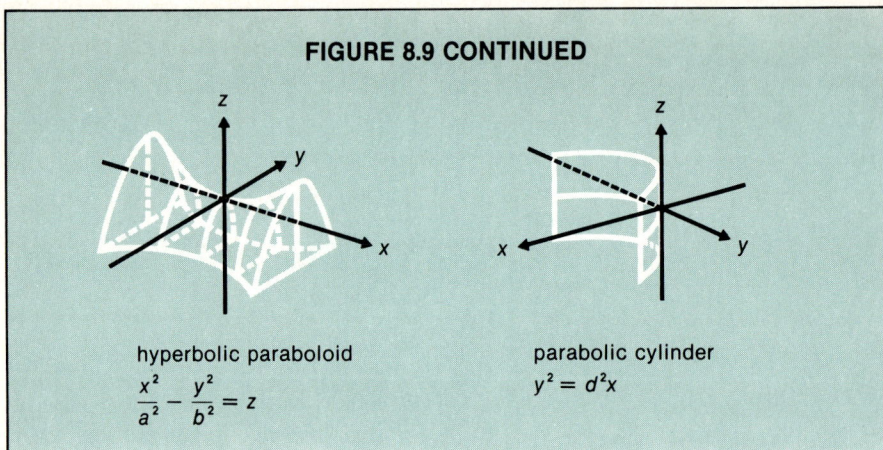

FIGURE 8.9 CONTINUED

hyperbolic paraboloid
$$\frac{x^2}{a^2} - \frac{y^2}{b^2} = z$$

parabolic cylinder
$$y^2 = d^2 x$$

The following theorem shows that it is possible to remove the cross-product terms from the equation of a quadric surface by rotating the coordinates. Once this has been done, an equation in standard form for the surface may be obtained by the method illustrated in the previous example.

Theorem 5. *Let*

$$ax^2 + by^2 + cz^2 + dxy + exz + fyz + gx + hy + iz + j = 0$$

be the equation of a quadric surface and let

$$\mathbf{u}^t A \mathbf{u} = ax^2 + by^2 + cxz^2 + dxy + exz + fyz$$

be its associated quadratic form. Then there is an orthonormal basis B for R^3 consisting of eigenvectors of the matrix A such that the quadric surface has an equation relative to B of the form

$$a'(x')^2 + b'(y')^2 + c'(z')^2 + g'x' + h'y' + i'z' + j = 0$$

where a', b', c' are eigenvalues of A and $[g'\ h'\ i']^t$ is the coordinate vector with respect to B of $[g\ h\ i]^t$.

The proof of this theorem is analogous to that of Theorem 4 of the previous section and is left as an exercise.

Example 3.   Consider the quadric surface described by the equation

$$7x^2 - 8y^2 - 8z^2 + 8xy - 8xz - 2yz + \tfrac{1}{\sqrt{17}}x + \tfrac{4}{\sqrt{17}}z - \tfrac{325}{36} = 0$$

The associated quadratic form is

$$[x\ y\ z]\begin{bmatrix} 7 & 4 & -4 \\ 4 & -8 & -1 \\ -4 & -1 & -8 \end{bmatrix}\begin{bmatrix} x \\ y \\ z \end{bmatrix} = 7x^2 - 8y^2 - 8z^2 + 8xy - 8xz - 2yz$$

From Example 2 of Section 7.6 we find that

$$\frac{1}{3\sqrt{2}}\begin{bmatrix} 4 \\ 1 \\ -1 \end{bmatrix}, \quad \frac{1}{\sqrt{17}}\begin{bmatrix} 1 \\ 0 \\ 4 \end{bmatrix}, \quad \frac{1}{3\sqrt{34}}\begin{bmatrix} -4 \\ 17 \\ 1 \end{bmatrix}$$

are orthonormal eigenvectors of

$$\begin{bmatrix} 7 & 4 & -4 \\ 4 & -8 & -1 \\ -4 & -1 & -8 \end{bmatrix}$$

associated with the eigenvalues 9, -9, and -9, respectively. Letting

$$P = \begin{bmatrix} \dfrac{4}{3\sqrt{2}} & \dfrac{1}{\sqrt{17}} & \dfrac{-4}{3\sqrt{34}} \\[2mm] \dfrac{1}{3\sqrt{2}} & 0 & \dfrac{17}{3\sqrt{34}} \\[2mm] \dfrac{-1}{3\sqrt{2}} & \dfrac{4}{\sqrt{17}} & \dfrac{1}{3\sqrt{34}} \end{bmatrix}$$

we have

$$P^{-1}\begin{bmatrix} 7 & 4 & -4 \\ 4 & -8 & -1 \\ -4 & -1 & -8 \end{bmatrix}P = \begin{bmatrix} 9 & 0 & 0 \\ 0 & -9 & 0 \\ 0 & 0 & -9 \end{bmatrix}$$

Let B be the orthonormal basis for R^3 consisting of the columns of P. Then P is the transition matrix from B to the natural basis N for R^3 and $P^{-1} = P^t$ is the transition matrix from N to B. Since

$$\begin{bmatrix} g' \\ h' \\ i' \end{bmatrix} = P^t \begin{bmatrix} \frac{1}{\sqrt{17}} \\ 0 \\ \frac{4}{\sqrt{17}} \end{bmatrix} = \begin{bmatrix} 0 \\ 1 \\ 0 \end{bmatrix}$$

we can write the equation for the quadric surface as

$$9(x')^2 - 9(y')^2 - 9(z')^2 + y' - \tfrac{325}{36} = 0$$

or equivalently as

$$(x')^2 - (y' - \tfrac{1}{18})^2 - (z')^2 = 1$$

Hence the quadric surface is a hyperboloid of two sheets that is translated and rotated out of standard position.

--------- **Exercises** ---------

In Exercises 1–6 determine the quadratic form associated with the given equation and determine the symmetric matrix associated with this quadratic form.

1. $3x^2 + 4y^2 + 5z^2 - 8xy + 10xz + 4yz - 9x + 6y - 4z + 17 = 0$
2. $-7x^2 + 5y^2 - 9z^2 + 6xy - 12xz - 18yz + 6x - 9y - 6z - 2 = 0$
3. $x^2 - y^2 + 2z^2 - 3xy + 4xz - 16yz + 5x - 4z + 1 = 0$
4. $4x^2 - 3y^2 - 5z^2 - 6xy + 5xz - 7yz + 2y - z + 3 = 0$
5. $x^2 - z^2 + 8xy - 4xz + 18x - 20z = 0$
6. $7x^2 - 9y^2 - xy + 3xz + yz - 6 = 0$

In Exercises 7–14 name the surface determined by the given equation and give its equation in a coordinate system in which the surface is in standard position.

7. $4x^2 - y^2 + 2z^2 + 8x - 4y - 1 = 0$
8. $-x^2 + 3y^2 - z^2 - 6x - 18y - 16z - 48 = 0$
9. $x^2 + 2y^2 + 4z^2 + 4xz - 1 = 0$
10. $-2x^2 + y^2 + 4z^2 - 4yz - 1 = 0$
11. $x^2 + y^2 - z^2 + 2xy - 1 = 0$
12. $2y^2 + 2z^2 - 4yz - 1 = 0$
13. $z^2 + 2xy + 2\sqrt{2}x + 2\sqrt{2}y + 2 = 0$
14. $2xy - z = 0$

8.5 DIFFERENTIAL EQUATIONS*

In this section we consider systems of first-order linear differential equations of the form

$$\frac{dx_1}{dt} = a_{11}x_1 + a_{12}x_2 + \cdots + a_{1n}x_n$$

$$\frac{dx_2}{dt} = a_{21}x_1 + a_{22}x_2 + \cdots + a_{2n}x_n \tag{1}$$

$$\vdots \qquad \vdots \qquad \vdots \qquad \qquad \vdots$$

$$\frac{dx_n}{dt} = a_{n1}x_1 + a_{n2}x_2 + \cdots + a_{nn}x_n$$

where the a_{ij} are real numbers and each x_i is an unknown function of t.

Such systems of differential equations arise frequently in applications. For example, consider the following sequence of chemical reactions in a well-stirred tank:

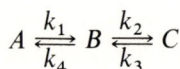

$$A \underset{k_4}{\overset{k_1}{\rightleftharpoons}} B \underset{k_3}{\overset{k_2}{\rightleftharpoons}} C$$

We assume that the rate at which each reaction occurs is proportional to the concentration of the chemical. The constants of proportionality are k_1, k_2, k_3, and

---

* This section is intended for those who have taken a course in differential equations.

k_4. If $C_1(t)$, $C_2(t)$, and $C_3(t)$ denote the concentrations at time t of A, B, and C, respectively, then

$$\frac{dC_1}{dt} = -k_1 C_1 + k_4 C_2$$

(2)

$$\frac{dC_2}{dt} = k_1 C_1 - (k_2 + k_4)C_2 + k_3 C_3$$

$$\frac{dC_3}{dt} = k_2 C_2 - k_3 C_3$$

In order to simplify the notation we introduce the concept of a vector function.

Definition 2.   If each component of a vector

$$\mathbf{x}(t) = \begin{bmatrix} x_1(t) \\ x_2(t) \\ \vdots \\ x_n(t) \end{bmatrix}$$

is a function on an interval I, then $\mathbf{x}$ is called a **vector function** on I. If, in addition, each component of $\mathbf{x}$ is differentiable, then $\mathbf{x}$ is called a **differentiable vector function** and its derivative is given by

$$\frac{d\mathbf{x}}{dt} = \begin{bmatrix} \dfrac{dx_1}{dt} \\ \dfrac{dx_2}{dt} \\ \vdots \\ \dfrac{dx_n}{dt} \end{bmatrix}$$

Example 1.   If

$$\mathbf{x}(t) = \begin{bmatrix} e^{4t} + 5t \\ e^{-2t} + 4 \\ \cos t \end{bmatrix}$$

Then

$$\frac{d\mathbf{x}}{dt} = \begin{bmatrix} 4e^{4t} + 5 \\ -2e^{-2t} \\ -\sin t \end{bmatrix}$$

With this notation the system of differential equations in (1) can be rewritten as

$$\frac{d\mathbf{x}}{dt} = A\mathbf{x}$$

where

$$A = \begin{bmatrix} a_{11} & a_{12} & \cdots & a_{1n} \\ a_{21} & a_{22} & \cdots & a_{2n} \\ \cdots & \cdots & \cdots & \cdots \\ a_{n1} & a_{n2} & \cdots & a_{nn} \end{bmatrix} \quad \text{and} \quad \mathbf{x}(t) = \begin{bmatrix} x_1(t) \\ x_2(t) \\ \cdots \\ x_n(t) \end{bmatrix}$$

Example 2.   The system of differential equations in (2) can be rewritten as

$$\frac{d\mathbf{x}}{dt} = \begin{bmatrix} -k_1 & k_4 & 0 \\ k_1 & -(k_2 + k_4) & k_3 \\ 0 & k_2 & -k_3 \end{bmatrix} \mathbf{x}$$

where

$$\mathbf{x} = \begin{bmatrix} C_1(t) \\ C_2(t) \\ C_3(t) \end{bmatrix}$$

Definition 3.   Let A be an $n \times n$ matrix. A solution of

$$\frac{d\mathbf{x}}{dt} = A\mathbf{x}$$

is an n-vector function $\mathbf{u}(t)$ defined on $(-\infty, \infty)$ such that

$$\frac{d\mathbf{u}}{dt} = A\mathbf{u}(t)$$

for every t.

Example 3.   Consider the equation

$$\frac{d\mathbf{x}}{dt} = \begin{bmatrix} 1 & 2 \\ 2 & 1 \end{bmatrix} \mathbf{x} \tag{3}$$

and the function

$$\mathbf{u}(t) = \begin{bmatrix} e^{-t} \\ -e^{-t} \end{bmatrix}$$

Then

$$\frac{d\mathbf{u}}{dt} = \begin{bmatrix} -e^{-t} \\ e^{-t} \end{bmatrix}$$

and

$$\begin{bmatrix} 1 & 2 \\ 2 & 1 \end{bmatrix} \mathbf{u}(t) = \begin{bmatrix} 1 & 2 \\ 2 & 1 \end{bmatrix} \begin{bmatrix} e^{-t} \\ -e^{-t} \end{bmatrix} = \begin{bmatrix} -e^{-t} \\ e^{-t} \end{bmatrix}$$

so that $\mathbf{u}$ is a solution of the differential equation in (3).

The equation

(4)
$$\frac{d\mathbf{x}}{dt} = A\mathbf{x}$$

in which A is an $n \times n$ matrix and $\mathbf{x}$ is an n-vector function is reminiscent of the equation

(5)
$$\frac{dx}{dt} = ax$$

in which a is a real number and x is a real valued function. It is easily verified that

$$x(t) = e^{at}c$$

is a solution of equation (5) for every real number c. (In fact every solution of equation (5) has this form.) We will now show that equation (4) has analogous functions for solutions. That is, we will show that equation (4) has solutions of the form

$$\mathbf{x}(t) = e^{\lambda t}\mathbf{v}$$

where λ is a number (possibly a complex number) and $\mathbf{v}$ is a fixed vector. To begin we note that if

$$\mathbf{v} = \begin{bmatrix} v_1 \\ v_2 \\ \vdots \\ v_n \end{bmatrix}$$

then

$$\frac{d}{dt}(e^{\lambda t}\mathbf{v}) = \frac{d}{dt}\begin{bmatrix} e^{\lambda t}v_1 \\ e^{\lambda t}v_2 \\ \vdots \\ e^{\lambda t}v_n \end{bmatrix}$$

$$= \begin{bmatrix} \lambda e^{\lambda t}v_1 \\ \lambda e^{\lambda t}v_2 \\ \vdots \\ \lambda e^{\lambda t}v_n \end{bmatrix}$$

$$= \lambda e^{\lambda t}\begin{bmatrix} v_1 \\ v_2 \\ \vdots \\ v_n \end{bmatrix} = \lambda e^{\lambda t}\mathbf{v}$$

Thus $\mathbf{x}(t) = e^{\lambda t}\mathbf{v}$ is a solution of $d\mathbf{x}/dt = A\mathbf{x}$ if and only if

$$\lambda e^{\lambda t}\mathbf{v} = e^{\lambda t}A\mathbf{v}$$

or, equivalently, if and only if

$$A\mathbf{v} = \lambda\mathbf{v}$$

Hence if $\mathbf{x}(t) \neq 0$, then λ is an eigenvalue of A with $\mathbf{v}$ as an associated eigenvector. We have proved the following theorem.

Theorem 6. *If λ is an eigenvalue of the $n \times n$ matrix A and $\mathbf{v}$ is an associated eigenvector, then $\mathbf{x}(t) = e^{\lambda t}\mathbf{v}$ is a solution of $d\mathbf{x}/dt = A\mathbf{x}$.*

Example 4.   In Example 2 of Section 7.1, we found that -1 and 3 are eigenvalues of

$$A = \begin{bmatrix} 1 & 2 \\ 2 & 1 \end{bmatrix}$$

with

$$\mathbf{v} = \begin{bmatrix} 1 \\ -1 \end{bmatrix} \quad \text{and} \quad \mathbf{w} = \begin{bmatrix} 1 \\ 1 \end{bmatrix}$$

as associated eigenvectors, respectively. Therefore

$$\mathbf{v}(t) = e^{-t}\begin{bmatrix} 1 \\ -1 \end{bmatrix} = \begin{bmatrix} e^{-t} \\ -e^{-t} \end{bmatrix}$$

and

$$\mathbf{w}(t) = e^{3t}\begin{bmatrix} 1 \\ 1 \end{bmatrix} = \begin{bmatrix} e^{3t} \\ e^{3t} \end{bmatrix}$$

are solutions of

$$\frac{d\mathbf{x}}{dt} = A\mathbf{x}$$

Important properties of differential equations of the form $d\mathbf{x}/dt = A\mathbf{x}$ are listed in the following theorem.

Theorem 7.   *Let A be an $n \times n$ matrix and $\mathbf{x}_0$ be any n-vector. Then*

   1. *There is exactly one solution of $d\mathbf{x}/dt = A\mathbf{x}$ satisfying the initial conditional $\mathbf{x}(0) = \mathbf{x}_0$.*
   2. *The set of all solutions of $d\mathbf{x}/dt = A\mathbf{x}$ is an n-dimensional linear space.*

The second part of the theorem tells us that if we can find n linearly independent solutions $\mathbf{x}_1, \mathbf{x}_2, \ldots, \mathbf{x}_n$ of $d\mathbf{x}/dt = A\mathbf{x}$ then any solution of $d\mathbf{x}/dt = A\mathbf{x}$ can be written as a linear combination of these n solutions. This motivates the following definition.

Definition 4. Let $\{x_1, x_2, \ldots, x_n\}$ be a linearly independent set of solutions of $dx/dt = Ax$ where A is an $n \times n$ matrix. The expression

$$x(t) = c_1 x_1(t) + c_2 x_2(t) + \cdots + c_n x_n(t)$$

where $c_1, c_2, \ldots, c_n$ are constants, is called a **general solution** of $dx/dt = Ax$.

Theorem 6 gives us a means to find solutions of $dx/dt = Ax$. The following theorem tells us when these solutions are linearly independent.

Theorem 8. *Let $v_1, v_2, \ldots, v_k$ be eigenvectors of an $n \times n$ matrix A associated with the distinct eigenvalues $\lambda_1, \lambda_2, \ldots, \lambda_n$. Then $x_1(t) = e^{\lambda_1 t} v_1, x_2(t) = e^{\lambda_2 t} v_2, \ldots, x_k(t) = e^{\lambda_k t} v_k$ are linearly independent solutions of $dx/dt = Ax$. Moreover, if $x = n$ and if $P(t)$ is the matrix having $x_1(t), x_2(t), \ldots, x_n(t)$ as its columns, then for any n-vector x_0*

$$x(t) = P(t)P^{-1}(0)x_0$$

is the only solution of $dx/dt = Ax$, satisfying $x(0) = x_0$.

Proof: Let $c_1, c_2, \ldots, c_k$ be scalars such that

$$c_1 e^{\lambda_1 t} v_1 + c_2 e^{\lambda_2 t} v_2 + \cdots + c_k e^{\lambda_k t} v_k = 0$$

for all t. In particular, when $t = 0$ we have

$$c_1 v_1 + c_2 v_2 + \cdots + c_k v_k = 0$$

Since $\lambda_1, \lambda_2, \ldots, \lambda_k$ are distinct, their associated eigenvectors are linearly independent (Theorem 5 of Section 7.3). Hence $c_1 = 0, c_2 = 0, \ldots, c_k = 0$. Therefore $\{x_1(t), x_2(t), \ldots, x_k(t)\}$ is linearly independent.

A tedious calculation, which is omitted, shows that $x(t) = P(t)P^{-1}(0)x_0$. ∎

Example 5. In Example 4 we found that

$$v(t) = e^{-t} \begin{bmatrix} 1 \\ -1 \end{bmatrix} \quad \text{and} \quad w(t) = e^{3t} \begin{bmatrix} 1 \\ 1 \end{bmatrix}$$

are solutions of

$$(6) \qquad \frac{dx}{dt} = \begin{bmatrix} 1 & 2 \\ 2 & 1 \end{bmatrix} x$$

By Theorem 8, v and w are linearly independent solutions of equation (6). Therefore

$$x(t) = c_1 e^{-1} \begin{bmatrix} 1 \\ -1 \end{bmatrix} + c_2 e^{3t} \begin{bmatrix} 1 \\ 1 \end{bmatrix}$$

is a general solution of equation (6). We can find the solution of equation (6) satisfying the initial condition

$$(7) \qquad x(0) = \begin{bmatrix} 1 \\ -2 \end{bmatrix}$$

in two ways. First, we can choose the coefficients c_1 and c_2 so that

$$\begin{bmatrix} 1 \\ -2 \end{bmatrix} = \mathbf{x}(0) = c_1 \begin{bmatrix} 1 \\ -1 \end{bmatrix} + c_2 \begin{bmatrix} 1 \\ 1 \end{bmatrix}$$

That is, we must choose c_1 and c_2 so that

$$c_1 + c_2 = 1$$
$$-c_1 + c_2 = -2$$

Evidently

$$c_1 = \frac{3}{2} \qquad c_2 = -\frac{1}{2}$$

Thus the solution of equation (6) satisfying the initial condition in (7) is

$$\mathbf{x}(t) = \frac{3}{2} e^{-t} \begin{bmatrix} 1 \\ -1 \end{bmatrix} - \frac{1}{2} e^{3t} \begin{bmatrix} 1 \\ 1 \end{bmatrix}$$

$$= \frac{1}{2} \begin{bmatrix} 3e^{-t} - e^{3t} \\ -3e^{-t} - e^{3t} \end{bmatrix}$$

Second, we can use Theorem 8 with

$$P(t) = \begin{bmatrix} e^{-t} & e^{3t} \\ -e^{-t} & e^{3t} \end{bmatrix}$$

A short calculation shows that

$$P(0)^{-1} = \begin{bmatrix} 1 & 1 \\ -1 & 1 \end{bmatrix}^{-1} = \begin{bmatrix} \frac{1}{2} & -\frac{1}{2} \\ \frac{1}{2} & \frac{1}{2} \end{bmatrix}$$

Therefore

$$\mathbf{x}(t) = P(t)P(0)^{-1} \begin{bmatrix} 1 \\ -2 \end{bmatrix}$$

$$= \begin{bmatrix} e^{-t} & e^{3t} \\ -e^{-t} & e^{3t} \end{bmatrix} \begin{bmatrix} \frac{1}{2} & -\frac{1}{2} \\ \frac{1}{2} & \frac{1}{2} \end{bmatrix} \begin{bmatrix} 1 \\ -2 \end{bmatrix}$$

$$= \frac{1}{2} \begin{bmatrix} 3e^{-t} - e^{3t} \\ -3e^{-t} - e^{3t} \end{bmatrix}$$

is the solution of equation (6) satisfying the initial condition in (7). This is precisely the same solution we found using the other method.

If the $n \times n$ matrix A does not have n linearly independent eigenvectors, then it is more difficult to find a general solution of $d\mathbf{x}/dt = A\mathbf{x}$. However, a generalization of the above method allows this to be done. In order to describe why this method works we need the exponential of a matrix which we have not discussed. In light of this we omit a discussion of this case and refer the interested reader to an intermediate or advanced text on differential equations.

Exercises

In Exercises 1–4 show that the given vector function is a solution of $d\mathbf{x}/dt = A\mathbf{x}$.

1. $\begin{bmatrix} e^{-4t} \\ -e^{-4t} \end{bmatrix}$ $\quad A = \begin{bmatrix} 1 & 5 \\ 5 & 1 \end{bmatrix}$

2. $\begin{bmatrix} e^{6t} \\ e^{6t} \end{bmatrix}$ $\quad A = \begin{bmatrix} 1 & 5 \\ 5 & 1 \end{bmatrix}$

3. $\begin{bmatrix} e^{2t} \\ 0 \\ e^{2t} \end{bmatrix}$ $\quad A = \begin{bmatrix} 3 & 1 & -1 \\ -3 & 7 & 3 \\ -4 & 4 & 6 \end{bmatrix}$

4. $\begin{bmatrix} e^{4t} \\ e^{4t} \\ 0 \end{bmatrix}$ $\quad A = \begin{bmatrix} 3 & 1 & -1 \\ -3 & 7 & 2 \\ -4 & 4 & 6 \end{bmatrix}$

In Exercises 5–10 find a general solution of the given differential equation.

5. $\dfrac{d\mathbf{x}}{dt} = \begin{bmatrix} 5 & -3 \\ -2 & 4 \end{bmatrix} \mathbf{x}$

6. $\dfrac{d\mathbf{x}}{dt} = \begin{bmatrix} 0 & 1 \\ 1 & 0 \end{bmatrix} \mathbf{x}$

7. $\dfrac{d\mathbf{x}}{dt} = \begin{bmatrix} 1 & 1 & 0 \\ 2 & 2 & 0 \\ 0 & 0 & 4 \end{bmatrix} \mathbf{x}$

8. $\dfrac{d\mathbf{x}}{dt} = \begin{bmatrix} 2 & 0 & 1 \\ 0 & -5 & 0 \\ 1 & 0 & 2 \end{bmatrix} \mathbf{x}$

9. $\dfrac{d\mathbf{x}}{dt} = \begin{bmatrix} -1 & 2 & 2 \\ 2 & 2 & 2 \\ -3 & -6 & -6 \end{bmatrix} \mathbf{x}$

10. $\dfrac{d\mathbf{x}}{dt} = \begin{bmatrix} 1 & 2 & -4 \\ 0 & 0 & 1 \\ 0 & 4 & 0 \end{bmatrix} \mathbf{x}$

11. Find the solution to the differential equation in Exercise 5 satisfying

$$\mathbf{x}(0) = \begin{bmatrix} 1 \\ 0 \end{bmatrix}.$$

12. Find the solution to the differential equation in Exercise 6 satisfying

$$\mathbf{x}(0) = \begin{bmatrix} 0 \\ 1 \end{bmatrix}.$$

13. Find the solution to the differential equation in Exercise 7 satisfying

$$\mathbf{x}(0) = \begin{bmatrix} 1 \\ 0 \\ 1 \end{bmatrix}.$$

14. Find the solution to the differential equation in Exercise 8 satisfying

$$\mathbf{x}(0) = \begin{bmatrix} 1 \\ 0 \\ 1 \end{bmatrix}.$$

9

Introduction to Numerical Linear Algebra

9.1 ITERATIVE METHODS FOR SOLVING $A\mathbf{x} = \mathbf{b}$

In Section 1.2 we discussed a method, Gaussian elimination with backward substitution, for computing the solution of a system of n linear equations in n unknowns. For most such systems this is a highly efficient method of solution, but not always. In Section 1.3 we found that it may be desirable to modify the method slightly to cut down on round-off errors. If n is very large, then the amount of storage or time required to use Gaussian elimination may be prohibitive. Such systems, involving thousands of unknowns, arise in various mathematical models, such as modeling the flow of air about an airfoil. Fortunately, there are relatively efficient methods for solving most large systems of equations that arise in such models. In this section we describe the two simplest methods.

Suppose that we wish to solve

$$
\begin{aligned}
a_{11}x_1 + a_{12}x_2 + \cdots + a_{1n}x_n &= b_1 \\
a_{21}x_1 + a_{22}x_2 + \cdots + a_{2n}x_n &= b_2 \\
\vdots \qquad \vdots \qquad\qquad \vdots \qquad \vdots & \\
a_{n1}x_1 + a_{n2}x_2 + \cdots + a_{nn}x_n &= b_n
\end{aligned}
\tag{1}
$$

We will suppose that this system has a unique solution and that the diagonal coefficients $a_{11}, a_{22}, \ldots, a_{nn}$ are nonzero. Rather than determine the solution $x_1 = \bar{x}_1, x_2 = \bar{x}_2, \ldots, x_n = \bar{x}_n$ of this system, we will construct sequences

$$\{x_1^{(k)}\}_{k=0}^{\infty}, \quad \{x_2^{(k)}\}_{k=0}^{\infty}, \quad \ldots, \quad \{x_n^{(k)}\}_{k=0}^{\infty}$$

such that, ideally

$$\lim_{k \to \infty} x_1^{(k)} = \bar{x}_1, \lim_{k \to \infty} x_2^{(k)} = \bar{x}_2, \ldots, \lim_{k \to \infty} x_n^{(k)} = \bar{x}_n$$

In terms of vectors this means that if we set

$$\mathbf{x}^{(k)} = \begin{bmatrix} x_1^{(k)} \\ x_2^{(k)} \\ \vdots \\ x_n^{(k)} \end{bmatrix} \quad \text{and} \quad \bar{\mathbf{x}} = \begin{bmatrix} \bar{x}_1 \\ \bar{x}_2 \\ \vdots \\ \bar{x}_n \end{bmatrix}$$

then

$$\lim_{k \to \infty} \mathbf{x}^{(k)} = \bar{\mathbf{x}}$$

That is, we will determine an approximation to the solution that is as accurate as we wish.

Before describing how to obtain such approximating sequences we will consider an iterative method for approximating $\sqrt{2}$ so that the reader can see the basic idea in a simple setting. Let a_0 be any positive number, which will be our initial approximation to $\sqrt{2}$, and note that $(a_0)(2/a_0) = 2$. It follows easily from this identity that

1. If $a_0 \geq \sqrt{2}$, then $2/a_0 \leq \sqrt{2}$ and
2. If $a_0 \leq \sqrt{2}$, then $2/a_0 \geq \sqrt{2}$

Thus either a_0 or $2/a_0$ is greater than or equal to $\sqrt{2}$ when the other is less than or equal to $\sqrt{2}$. For our next approximation, a_1 to $\sqrt{2}$, we take the average of these two numbers:

$$a_1 = \frac{1}{2}\left(a_0 + \frac{2}{a_0}\right)$$

The next approximation, a_2, is obtained by repeating the above argument with a_0 replaced by a_1:

$$a_2 = \frac{1}{2}\left(a_1 + \frac{2}{a_1}\right)$$

Continuing in this fashion we define

$$a_{k+1} = \frac{1}{2}\left(a_k + \frac{2}{a_k}\right)$$

for $k = 2, 3, \ldots$.

It is possible that the Babylonians used the first two steps of this method to obtain approximations to $\sqrt{2}$. If we take $a_0 = \frac{3}{2}$, then

$$a_1 = \frac{1}{2}\left(\frac{3}{2} + \frac{2}{(3/2)}\right) = \frac{17}{12} \approx 1.4166667$$

and

$$a_2 = \frac{1}{2}\left(\frac{17}{12} + \frac{2}{(17/12)}\right)$$

$$= \frac{577}{408} \approx 1.4142157$$

where the decimal approximations are rounded to eight digits. The numbers a_1 and a_2 correspond to approximations to $\sqrt{2}$ given in Babylonian texts. These values approximate $\sqrt{2} \approx 1.4142136$ to an accuracy that is acceptable for many calculations. In fact, a_2 approximates $\sqrt{2}$ with an error of less than 2.2×10^{-6}.

Readers interested in reading more about Babylonian mathematics in general or their approximations to $\sqrt{2}$ in particular are referred to the second chapter of *The Exact Sciences in Antiquity* by O. Neugenbauer or the second and third chapters of *Science Awakening* by B. L. van der Waerden.

In this section we present two iterative methods for approximating the solution of the system of equations in (1). The basic idea behind both of these methods is simple. To begin, we rewrite the system of equations so that it has the form

$$\mathbf{x} = B\mathbf{x} + \mathbf{c}$$

where B is an $n \times n$ matrix and $\mathbf{c}$ is an n-vector. How B and $\mathbf{c}$ are chosen will be described later. We now take an initial guess $\mathbf{x}^{(0)}$ to the solution. If $\mathbf{x}^{(0)} = B\mathbf{x}^{(0)} + \mathbf{c}$ then we have guessed the solution. Otherwise we set $\mathbf{x}^{(1)} = B\mathbf{x}^{(0)} + \mathbf{c}$ and hope that $\mathbf{x}^{(1)}$ is "closer" to the solution than is $\mathbf{x}^{(0)}$. Continuing in this fashion, we form $\mathbf{x}^{(2)} = B\mathbf{x}^{(1)} + \mathbf{c}$, $\mathbf{x}^{(3)} = B\mathbf{x}^{(2)} + \mathbf{c}, \ldots, \mathbf{x}^{(k+1)} = B\mathbf{x}^{(k)} + \mathbf{c}$, ... and hope that the sequence $\{\mathbf{x}^{(k)}\}_{k=1}^{\infty}$ converges to some vector $\mathbf{y}$. If $\lim_{k\to\infty} \mathbf{x}^{(k)} = \mathbf{y}$, then $\lim_{k\to\infty} \mathbf{x}^{(k+1)} = \mathbf{y}$ and $\lim_{k\to\infty} (B\mathbf{x}^{(k)} + \mathbf{c}) = B\mathbf{y} + \mathbf{c}$. Since $\mathbf{x}^{(k+1)} = B\mathbf{x}^{(k)} + \mathbf{c}$ we conclude that $\mathbf{y} = B\mathbf{y} + \mathbf{c}$ so that $\mathbf{y}$ is the solution of the system of equations. Thus we would like to choose B and $\mathbf{c}$ so that the sequence $\{\mathbf{x}^{(k)}\}_{k=1}^{\infty}$ converges. Unfortunately there is no choice that causes the sequence to converge for every system of equations. However, there are two choices that work for many systems of equations that arise in practice.

The first iterative method to approximate the solution of the system of equations in (1) is called **Jacobi iteration**. To begin (assuming that $a_{11}, a_{22}, \ldots, a_{nn}$ are nonzero) we rewrite the system of equations in (1) as

$$x_1 = \frac{1}{a_{11}}(b_1 - a_{12}x_2 - a_{13}x_3 - \cdots - a_{1n}x_n)$$

$$x_2 = \frac{1}{a_{22}}(b_2 - a_{21}x_1 - a_{23}x_3 - \cdots - a_{2n}x_n)$$

$$\vdots \qquad \vdots \quad \vdots \qquad \vdots \qquad \vdots \qquad \qquad \vdots$$

$$x_n = \frac{1}{a_{nn}}(b_n - a_{n1}x_1 - a_{n2}x_2 - \cdots - a_{n,n-1}x_{n-1})$$

For example, the system of equations

$$4x_1 + 2x_2 + x_3 = 2$$

$$2x_1 + 5x_2 + x_3 = 0 \qquad\qquad (2)$$

$$2x_1 - x_2 + 4x_3 = 9$$

is rewritten as

$$x_1 = \frac{1}{4}(2 - 2x_2 - x_3)$$

$$x_2 = \frac{1}{5}(-2x_1 - x_3)$$

$$x_3 = \frac{1}{4}(9 - 2x_1 + x_2)$$

or, equivalently,

$$x_1 = .5 - .5x_2 - .25x_3$$

(3) $$x_2 = -.4x_1 - .2x_3$$

$$x_3 = 2.25 - .5x_1 + .25x_2$$

Next we take an initial approximation to the solution. If no a priori information about the solution is available, the initial approximation is frequently taken to be $x_1^{(0)} = 0$, $x_2^{(0)} = 0, \ldots, x_n^{(0)} = 0$. We then obtain approximations to the solution by using the iterative formula.

$$x_1^{(k+1)} = \frac{1}{a_{11}}(b_1 - a_{12}x_2^{(k)} - a_{13}x_3^{(k)} - \cdots - a_{1n}x_n^{(k)})$$

(4) $$x_2^{(k+1)} = \frac{1}{a_{22}}(b_2 - a_{21}x_1^{(k)} - a_{23}x_3^{(k)} - \cdots - a_{2n}x_n^{(k)})$$

$$\vdots \quad \vdots \quad \vdots \qquad \vdots \qquad \vdots \qquad \qquad \vdots$$

$$x_n^{(k+1)} = \frac{1}{a_{nn}}(b_n - a_{n1}x_1^{(k)} - a_{n2}x_2^{(k)} - \cdots - a_{n,n-1}x_{n-1}^{(k)})$$

For example, the equations in (3) give the iterative formula

$$x_1^{(k+1)} = .5 - .5x_2^{(k)} - .25x_3^{(k)}$$

(5) $$x_2^{(k+1)} = -.4x_1^{(k)} - .2x_3^{(k)}$$

$$x_3^{(k+1)} = 2.25 - .5x_1^{(k)} + .25x_2^{(k)}$$

Beginning with $x_1^{(0)} = 0$, $x_2^{(0)} = 0$, $x_3^{(0)} = 0$ and setting $k = 0$ in (5), we have

$$x_1^{(1)} = .5 - 0 - 0 = .5$$

$$x_2^{(1)} = 0 - 0 = 0$$

$$x_3^{(1)} = 2.25 - 0 - 0 = 2.25$$

Using these values with $k = 1$ in (5), we obtain

$$x_1^{(2)} = .5 - .5(0) - (.25)(2.25) = -.0625$$

$$x_2^{(2)} = -.4(.5) - .2(2.25) = -.65$$

$$x_3^{(2)} = 2.25 - .5(.5) + .25(0) = 2.0$$

Continuing in this manner we obtain the data (rounded to 5 digits) in Table 9.1.

TABLE 9.1

k	$x_1^{(k)}$	$x_2^{(k)}$	$x_3^{(k)}$
0	0	0	0
1	.5000	.00000	2.2500
2	.06250	−.65000	2.0000
3	.32500	−.37500	2.1188
4	.15781	−.55375	1.9938
5	.27844	−.46188	2.0327
6	.22277	−.51791	1.9953
7	.26013	−.48817	2.0091
8	.24180	−.50588	1.9979
9	.25347	−.49630	2.0026
10	.24749	−.50191	1.9942
11	.25116	−.49883	2.0008
12	.24922	−.50062	1.9997
13	.25038	−.49963	2.0002
14	.24976	−.50020	1.9999
15	.25012	−.49488	2.0001
16	.24992	−.50006	2.0000
17	.25004	−.49996	2.0000
18	.24998	−.50002	2.0000
19	.25001	−.49998	2.0000
20	.24999	−.50001	2.0000
21	.25000	−.50000	2.0000

It is easily verified that $x_1 = .25, x_2 = -.5$, and $x_3 = 2$ is the solution of the system of equations in (2). Thus it takes 21 iterations to obtain the first five digits (rounded) of the solution.

Notice that each new approximation $x_1^{(k+1)}, x_2^{(k+1)}, \ldots, x_n^{(k+1)}$ in (4) is determined exclusively by the previous approximation $x_1^{(k)}, x_2^{(k)}, \ldots, x_n^{(k)}$. Since we compute $x_1^{(k+1)}$ before $x_2^{(k+1)}, x_3^{(k+1)}, \ldots, x_n^{(k+1)}$, we could use this value, which we hope is more accurate than $x_1^{(k)}$, in place of $x_1^{(k)}$. Similarly, we could use $x_2^{(k+1)}$ in place of $x_2^{(k)}$ when we compute $x_3^{(k+1)}, x_4^{(k+1)}, \ldots, x_n^{(k+1)}$. Continuing in this manner we obtain the iterative formula

$$x_1^{(k+1)} = \frac{1}{a_{11}}(b_1 - a_{12}x_2^{(k)} - a_{13}x_3^{(k)} - \cdots - a_{1n}x_n^{(k)})$$

$$x_2^{(k+1)} = \frac{1}{a_{22}}(b_2 - a_{21}x_1^{(k+1)} - a_{23}x_3^{(k)} - \cdots - a_{2n}x_n^{(k)})$$

$$x_3^{(k+1)} = \frac{1}{a_{33}}(b_3 - a_{31}x_1^{(k+1)} - a_{32}x_2^{(k+1)} - a_{34}x_4^{(k)} - \cdots - a_{3n}x_n^{(k)}) \qquad (6)$$

$$\vdots \qquad \vdots \quad \vdots \quad \vdots \qquad \vdots$$

$$x_n^{(k+1)} = \frac{1}{a_{nn}}(b_n - a_{n1}x_1^{(k+1)} - a_{n2}x_2^{(k+1)} - \cdots - a_{n,n-1}x_{n-1}^{(k+1)})$$

which is known as the **Gauss–Seidel iteration.**

For example, the Gauss–Seidel iterative formula for the system of equations in (2) is

(7)
$$x_1^{(k+1)} = .5 - .5x_2^{(k)} - .25x_3^{(k)}$$
$$x_2^{(k+1)} = -.4x_1^{(k+1)} - .2x_3^{(k)}$$
$$x_3^{(k+1)} = 2.25 - .5x_1^{(k+1)} + .25x_2^{(k+1)}$$

As before, we begin the iteration with the initial approximation $x_1^{(0)} = 0, x_2^{(0)} = 0$, $x_3^{(0)} = 0$. Setting $k = 0$ in (7) we obtain

$$x_1^{(1)} = .5 - 0 - 0 = .5$$
$$x_2^{(1)} = -(.4)(.5) - 0 = -.2$$
$$x_3^{(1)} = 2.25 - (.5)(.5) + (.25)(-.2) = 1.95$$

Using these values with $k = 1$ in (6) we obtain

$$x_1^{(2)} = .5 - (.5)(-.2) - (.25)(1.95) = .1125$$
$$x_2^{(2)} = -(.4)(.1125) - (.2)(1.95) = -.435$$
$$x_3^{(2)} = 2.25 - (.5)(.1125) + (.25)(-.435) = 2.085$$

Continuing in this manner we obtain the data (rounded to five digits) in Table 9.2.

Thus we see that for the system of equations in (2) both Jacobi iteration and Gauss–Seidel iteration yield sequences that converge to the solution. From Tables 1 and 2, Gauss–Seidel iteration may seem to require only half as many iterations as does Jacobi iteration to obtain approximations with the same degree of accuracy. While this frequently happens, it does not always happen. There are systems of equations for which neither method of iteration yields convergent sequences. There are other systems of equations for which one method, but not the other, yields convergent sequences.

There are theoretically simple means for determining whether these methods yield convergent sequences. To describe how to determine this we define the

TABLE 9.2

k	$x_1^{(k)}$	$x_2^{(k)}$	$x_3^{(k)}$
0	0	0	0
1	.50000	−.20000	1.9500
2	.11250	−.43500	2.0850
3	.19625	−.49550	2.0280
4	.24075	.50190	2.0042
5	.24991	−.50080	1.9998
6	.25044	−.50014	1.9997
7	.25014	−.50000	1.9999
8	.25002	−.50000	2.0000
9	.25000	−.50000	2.0000

following matrices. Let

$$A = \begin{bmatrix} a_{11} & a_{12} & \cdots & a_{1n} \\ a_{21} & a_{22} & \cdots & a_{2n} \\ \vdots & \vdots & & \vdots \\ a_{n1} & a_{n2} & \cdots & a_{nn} \end{bmatrix}$$

$$L = \begin{bmatrix} 0 & 0 & 0 & \cdots & 0 \\ -a_{21} & 0 & 0 & \cdots & 0 \\ -a_{31} & -a_{32} & 0 & \cdots & 0 \\ \vdots & \vdots & \vdots & & \vdots \\ -a_{n1} & -a_{n2} & -a_{n3} & \cdots & 0 \end{bmatrix}$$

$$U = \begin{bmatrix} 0 & -a_{12} & -a_{13} & \cdots & -a_{1n} \\ 0 & 0 & -a_{23} & \cdots & -a_{2n} \\ \vdots & \vdots & \vdots & & \vdots \\ 0 & 0 & 0 & \cdots & -a_{n-1,n} \\ 0 & 0 & 0 & \cdots & 0 \end{bmatrix}$$

$$D = \begin{bmatrix} a_{11} & 0 & 0 & \cdots & 0 \\ 0 & a_{22} & 0 & \cdots & 0 \\ \vdots & \vdots & \vdots & & \vdots \\ 0 & 0 & 0 & \cdots & a_{nn} \end{bmatrix}$$

$$\mathbf{x}^{(k)} = \begin{bmatrix} x_1^{(k)} \\ x_2^{(k)} \\ \vdots \\ x_n^{(k)} \end{bmatrix} \qquad \mathbf{b} = \begin{bmatrix} b_1 \\ b_2 \\ \vdots \\ b_n \end{bmatrix}$$

With this notation the systems of equations in (4) and (6) can be written as

$$\mathbf{x}^{(k+1)} = D^{-1}(L + U)\mathbf{x}^{(k)} + D^{-1}\mathbf{b}$$

and

$$\mathbf{x}^{(k+1)} = (D - L)^{-1}U\mathbf{x}^{(k)} + (D - L)^{-1}\mathbf{b}$$

respectively. It can be shown that the sequence generated by Jacobi iteration and the sequence generated by Gauss–Seidel iteration converge for any choice of the initial approximation if and only if all of the eigenvalues of $D^{-1}(L + U)$ and $(D - L)^{-1}U$, respectively, have absolute values less than one. In general these are not easy conditions to verify. However, there is one type of matrix that arises frequently in applications for which it is easy to verify that these conditions hold. An $n \times n$ matrix $A = (a_{ij})$ is called **diagonally dominant** if

$$|a_{11}| > |a_{12}| + |a_{13}| + \cdots + |a_{1n}|$$
$$|a_{22}| > |a_{21}| + |a_{23}| + \cdots + |a_{2n}|$$
$$\vdots \qquad \vdots \qquad \vdots \qquad \qquad \vdots$$
$$|a_{nn}| > |a_{n1}| + |a_{n2}| + \cdots + |a_{n,n-1}|$$

That is, A is strictly diagonally dominant if the absolute value of each diagonal element is greater than the sum of the absolute values of the remaining components in the same row. For example, the matrix

$$\begin{bmatrix} 4 & 2 & 1 \\ 2 & 5 & 1 \\ 2 & -1 & 4 \end{bmatrix}$$

associated with the system of equations in (2) is diagonally dominant.

It is possible to use Gershgorin's theorem (Theorem 9 of Section 7.4) to show that if A is diagonally dominant, then all of the eigenvalues of $D^{-1}(L + U)$ and $(D - L)^{-1}U$ have absolute values less than one. (In the exercises, the reader is asked to show this for $D^{-1}(L + U)$.) Thus if A is diagonally dominant, then both Jacobi iteration and Gauss–Seidel iteration yield sequences that converge to the solution of the linear system of equations in (1).

A difficult part of any iterative method is deciding when to stop. There are ways of estimating how accurate an approximation is, but these estimations are usually very pessimistic. In practice, an iterative method such as those described above is run for a fixed number of iterations or until the absolute value of the difference between two successive approximations is less than some predetermined number.

In practice, iterative methods are not usually used to solve small systems of linear equations because the time required to obtain a sufficiently accurate solution is greater than the time required to solve the system by using Gaussian elimination. For large systems (e.g., a hundred equations in a hundred unknowns) with a high percentage of zero coefficients, iterative techniques require less computer storage and less time to obtain a sufficiently accurate solution than does Gaussian elimination. Systems of this type arise frequently in the numerical solution of partial differential equations.

--- **Exercises** ---

In Exercises 1–8, beginning with $\mathbf{x}_0 = \mathbf{0}$, find the first ten iterations of the methods of Jacobi iteration and Gauss–Seidel iteration. Compare the results to the exact solutions.

1. $\begin{bmatrix} 3 & 1 \\ 2 & 5 \end{bmatrix} \mathbf{x} = \begin{bmatrix} 1 \\ 4 \end{bmatrix}$

2. $\begin{bmatrix} 2 & 1 \\ 3 & 4 \end{bmatrix} \mathbf{x} = \begin{bmatrix} -2 \\ 3 \end{bmatrix}$

3. $\begin{bmatrix} 4 & 2 \\ 1 & 5 \end{bmatrix} \mathbf{x} = \begin{bmatrix} 0 \\ 1 \end{bmatrix}$

4. $\begin{bmatrix} -2 & 1 \\ -3 & 6 \end{bmatrix} \mathbf{x} = \begin{bmatrix} 7 \\ 5 \end{bmatrix}$

5. $\begin{bmatrix} 5 & 3 & 1 \\ 2 & 4 & 1 \\ 3 & 2 & 6 \end{bmatrix} \mathbf{x} = \begin{bmatrix} 1 \\ 0 \\ 2 \end{bmatrix}$

6. $\begin{bmatrix} -8 & 2 & 4 \\ 2 & 6 & -3 \\ 1 & 3 & -6 \end{bmatrix} \mathbf{x} = \begin{bmatrix} 2 \\ 4 \\ 7 \end{bmatrix}$

7. $\begin{bmatrix} 3 & 0 & 2 \\ 2 & 5 & 1 \\ 6 & 2 & 9 \end{bmatrix} \mathbf{x} = \begin{bmatrix} 1 \\ 1 \\ 1 \end{bmatrix}$

8. $\begin{bmatrix} 9 & 5 & 1 \\ 2 & -9 & 6 \\ 3 & 2 & 7 \end{bmatrix} \mathbf{x} = \begin{bmatrix} 1 \\ -9 \\ 6 \end{bmatrix}$

9. Show that the methods of Jacobi iteration and Gauss–Seidel iteration yield sequences that converge to the solution of

$$\begin{bmatrix} 1 & 2 \\ 2 & 6 \end{bmatrix} \mathbf{x} = \begin{bmatrix} 5 \\ 7 \end{bmatrix}$$

even though the matrix is not diagonally dominant.

10. (a) Show that the method of Jacobi iteration beginning with $x_0 = \begin{bmatrix} 0 \\ 0 \end{bmatrix}$ does not yield a sequence that converges to the solution of

$$\begin{bmatrix} 2 & 3 \\ 2 & 1 \end{bmatrix} \begin{bmatrix} x_1 \\ x_2 \end{bmatrix} = \begin{bmatrix} 2 \\ -4 \end{bmatrix}$$

(b) Rename the variables according to $y_1 = x_2$ and $y_2 = x_1$. Show that the equation in (a) can be rewritten as

$$\begin{bmatrix} 3 & 2 \\ 1 & 2 \end{bmatrix} \begin{bmatrix} y_1 \\ y_2 \end{bmatrix} = \begin{bmatrix} 2 \\ -4 \end{bmatrix}$$

(c) Show that beginning with x_0, the method of Jacobi iteration yields a sequence that converges to the solution of the system of equations in (b).

11. Let A be a diagonal matrix with $\det A \neq 0$. Show that 0 is the only eigenvalue of $D^{-1}(L + U)$ and $(D - L)^{-1}U$.

12. Let A be a diagonally dominant matrix. Use Gershgorin's theorem (Theorem 9 of Section 7.4) to show that the eigenvalues of $D^{-1}(L + U)$ are less than one in absolute value.

9.2 APPROXIMATION OF EIGENVALUES AND EIGENVECTORS

In many problems it is useful to know the size of the absolute values of the eigenvalues of a matrix. For example, in the previous section we found that if all the eigenvalues of a certain matrix have absolute values less than one, then the method of Gauss–Seidel iteration can be used to approximate the solution of an associated system of linear equations as accurately as we desire (if errors in calculation such as round-off and truncation errors are ignored). In this section we will discuss a method for approximating the eigenvalue with the largest absolute value of a matrix A. This method will also yield an approximation to an eigenvector associated with that eigenvalue. The reader may wonder why we do not merely find the characteristic polynomial $\det(\lambda I - A)$ of A, calculate its roots, and determine directly an eigenvalue with the largest absolute value. There are three reasons why this approach is not practical if the dimension of A is not very small. First, it is a time consuming and tedious process to compute $\det(\lambda I - A)$. Second, it is not necessarily an easy matter to find the roots of a polynomial of large degree. And third, very small errors in the coefficients of $\det(\lambda I - A)$ may lead to very large errors in its roots.

Definition 1.   Let $\lambda_1, \lambda_2, \ldots, \lambda_n$ be the eigenvalues of an $n \times n$ matrix A. The eigenvalue λ_1 is called the **dominant eigenvalue** of A if $|\lambda_1| > |\lambda_i|$ for $i = 2, 3, \ldots, n$. An eigenvector of A associated with the dominant eigenvalue is called a **dominant eigenvector**.

Example 1. If the eigenvalues of A are -5, -3, 0, 1, and 4, then -5 is the dominant eigenvalue.

Example 2. If the eigenvalues of A are -2, -1, $\frac{1}{2}$, and 2, then A has no dominant eigenvalue, because $|-2| = |2|$.

We now describe a method, called the **power method**, for approximating the dominant eigenvalue of a matrix, if it has one.

Let A be an $n \times n$ matrix with a dominant eigenvalue λ_1 and having n linearly independent eigenvectors $\mathbf{u}_1, \mathbf{u}_2, \ldots, \mathbf{u}_n$ associated with the eigenvalues $\lambda_1, \lambda_2, \ldots, \lambda_n$, respectively. For simplicity of exposition we will assume that these eigenvectors form an orthogonal set. This assumption is not necessary, but it simplifies the calculations.

To begin, we choose any nonzero n-vector $\mathbf{x}_0$ to be an approximation to an eigenvector associated with the dominant eigenvalue. Since the eigenvectors are linearly independent, there are scalars $c_1, c_2, \ldots, c_n$ such that

$$\mathbf{x}_0 = c_1 \mathbf{u}_1 + c_2 \mathbf{u}_2 + \cdots + c_n \mathbf{u}_n$$

We now assume that $\mathbf{x}_0$ was chosen so that $c_1 \neq 0$ and consider what happens when we repeatedly multiply $\mathbf{x}_0$ by A. If we set $\mathbf{x}_k = A^k \mathbf{x}_0$, then

$$\mathbf{x}_k = c_1 A^k \mathbf{u}_1 + c_2 A^k \mathbf{u}_2 + \cdots + c_n A^k \mathbf{u}_n$$

$$= c_1 \lambda_1^k \mathbf{u}_1 + c_2 \lambda_2^k \mathbf{u}_2 + \cdots + c_n \lambda_n^k \mathbf{u}_n$$

(1)
$$= \lambda_1^k \left(c_1 \mathbf{u}_1 + c_2 \left(\frac{\lambda_2}{\lambda_1} \right)^k \mathbf{u}_2 + \cdots + c_n \left(\frac{\lambda_n}{\lambda_1} \right)^k \mathbf{u}_n \right)$$

Recalling that $|\lambda_i/\lambda_1| < 1$ for $i = 1, 2, \ldots, n$ and that $\lim_{k \to \infty} r^k = 0$ whenever $|r| < 1$, we conclude that $\mathbf{x}_k \approx \lambda_1^k c_1 \mathbf{u}_1$ whenever k is sufficiently large.

Now using the identity in (1) and the basic properties of inner product, while recalling that the eigenvectors are orthogonal, we have

$$\langle \mathbf{x}_{k-1}, A\mathbf{x}_{k-1} \rangle = \langle \mathbf{x}_{k-1}, \mathbf{x}_k \rangle = \lambda_1^{2k-1} \left(c_1 \langle \mathbf{u}_1, \mathbf{u}_1 \rangle + c_2 \left(\frac{\lambda_2}{\lambda_1} \right)^{2k-1} \langle \mathbf{u}_2, \mathbf{u}_2 \rangle \right.$$
$$\left. + \cdots + c_n \left(\frac{\lambda_n}{\lambda_1} \right)^{2k-1} \langle \mathbf{u}_n, \mathbf{u}_n \rangle \right)$$

and

$$\langle \mathbf{x}_{k-1}, \mathbf{x}_{k-1} \rangle = \lambda_1^{2k-2} \left(c_1 \langle \mathbf{u}_1, \mathbf{u}_1 \rangle + c_2 \left(\frac{\lambda_2}{\lambda_1} \right)^{2k-2} \langle \mathbf{u}_2, \mathbf{u}_2 \rangle \right.$$
$$\left. + \cdots + c_n \left(\frac{\lambda_n}{\lambda_1} \right)^{2k-2} \langle \mathbf{u}_n, \mathbf{u}_n \rangle \right)$$

Thus whenever k is relatively large we have

$$\frac{\langle \mathbf{x}_{k-1}, A\mathbf{x}_{k-1} \rangle}{\langle \mathbf{x}_{k-1}, \mathbf{x}_{k-1} \rangle} \approx \lambda_1$$

and this approximation becomes better as k increases.

Example 3.   We will illustrate the above calculations by using the matrix

$$A = \begin{bmatrix} 1 & 2 \\ 2 & 1 \end{bmatrix}$$

To begin, we take $\mathbf{x}_0$ to be any nonzero vector, say

$$\mathbf{x}_0 = \begin{bmatrix} 3 \\ 2 \end{bmatrix}$$

Straightforward calculations show that

$$\mathbf{x}_1 = A\mathbf{x}_0 = \begin{bmatrix} 7 \\ 8 \end{bmatrix}$$

$$\mathbf{x}_2 = A^2\mathbf{x}_0 = A\mathbf{x}_1 = \begin{bmatrix} 23 \\ 22 \end{bmatrix}$$

$$\mathbf{x}_3 = A^3\mathbf{x}_0 = A\mathbf{x}_2 = \begin{bmatrix} 67 \\ 68 \end{bmatrix}$$

$$\mathbf{x}_4 = A^4\mathbf{x}_0 = A\mathbf{x}_3 = \begin{bmatrix} 203 \\ 202 \end{bmatrix}$$

$$\mathbf{x}_5 = A^4\mathbf{x}_0 = A\mathbf{x}_4 = \begin{bmatrix} 607 \\ 608 \end{bmatrix}$$

and

$$\frac{\langle \mathbf{x}_1, A\mathbf{x}_1 \rangle}{\langle \mathbf{x}_1, \mathbf{x}_1 \rangle} = \frac{\langle \mathbf{x}_1, \mathbf{x}_2 \rangle}{\langle \mathbf{x}_1, \mathbf{x}_1 \rangle} = \frac{337}{113} \approx 2.98230$$

$$\frac{\langle \mathbf{x}_2, A\mathbf{x}_2 \rangle}{\langle \mathbf{x}_2, \mathbf{x}_2 \rangle} = \frac{\langle \mathbf{x}_2, \mathbf{x}_3 \rangle}{\langle \mathbf{x}_2, \mathbf{x}_2 \rangle} = \frac{3037}{1013} \approx 2.99803$$

$$\frac{\langle \mathbf{x}_3, A\mathbf{x}_3 \rangle}{\langle \mathbf{x}_3, \mathbf{x}_3 \rangle} = \frac{\langle \mathbf{x}_3, \mathbf{x}_4 \rangle}{\langle \mathbf{x}_3, \mathbf{x}_3 \rangle} = \frac{27{,}337}{9{,}113} \approx 2.99978$$

$$\frac{\langle \mathbf{x}_4, A\mathbf{x}_4 \rangle}{\langle \mathbf{x}_4, \mathbf{x}_4 \rangle} = \frac{\langle \mathbf{x}_4, \mathbf{x}_5 \rangle}{\langle \mathbf{x}_4, \mathbf{x}_4 \rangle} = \frac{246{,}037}{82{,}013} \approx 2.99998$$

It is easily verified that the eigenvalues of A are -1 and 3. Thus the power method is giving us approximations to the dominant eigenvalue.

Notice that in Example 3 the numbers $\langle \mathbf{x}_{k-1}, \mathbf{x}_k \rangle$ and $\langle \mathbf{x}_k, \mathbf{x}_k \rangle$ increased by roughly a multiple of 10 at each step. If convergence to the dominant eigenvalue had not been so rapid we would have had to deal with very large numbers. These can be troublesome when using a computer, because we may lose accuracy in our calculations from round-off or truncation errors. To avoid this we will modify the power method slightly so that the components of each x_k lie between -1 and 1. This can be achieved by multiplying the vector by the reciprocal of its component with the largest absolute value. Instead of having $x_k = A^k x_0$ (as in the power method) we now have

$$y_1 = Ax_0 \qquad x_1 = \frac{1}{\alpha_1} y_1$$

$$y_2 = Ax_1 \qquad x_2 = \frac{1}{\alpha_2} y_2$$

$$\vdots \qquad \vdots \qquad \vdots \qquad \vdots$$

$$y_k = Ax_{k-1} \qquad x_k = \frac{1}{\alpha_k} y_k$$

where α_j is the component of y_j with the largest absolute value.

Just as in the power method, the dominant eigenvalue is now approximated by $\langle x_{k-1}, Ax_{k-1} \rangle / \langle x_{k-1}, x_{k-1} \rangle$. In fact, it can be shown that we obtain the same approximations as in the power method. An advantage of this modification of the power method, called the **power method with scaling**, is that the sequence of vectors x_k approximate a dominant eigenvector.

Example 4. Consider the matrix

$$A = \begin{bmatrix} 1 & 2 \\ 2 & 1 \end{bmatrix}$$

from Example 3 and let

$$x_0 = \begin{bmatrix} 3 \\ 2 \end{bmatrix}$$

Then

$$y_1 = Ax_0 = \begin{bmatrix} 7 \\ 8 \end{bmatrix} \qquad x_1 = \frac{1}{8} y_1 = \begin{bmatrix} .87500 \\ 1.00000 \end{bmatrix}$$

$$y_2 = Ax_1 = \begin{bmatrix} 2.875 \\ 2.75 \end{bmatrix} \qquad x_2 \approx \frac{1}{2.875} y_2 \approx \begin{bmatrix} 1.00000 \\ .95652 \end{bmatrix}$$

$$y_3 = Ax_2 \approx \begin{bmatrix} 2.91304 \\ 2.95652 \end{bmatrix} \qquad x_3 \approx \frac{1}{2.95652} y_3 \approx \begin{bmatrix} .98529 \\ 1.00000 \end{bmatrix}$$

$$y_4 = Ax_3 \approx \begin{bmatrix} 2.98529 \\ 2.97058 \end{bmatrix} \qquad x_4 \approx \frac{1}{2.98529} y_4 \approx \begin{bmatrix} 1.00000 \\ .99507 \end{bmatrix}$$

$$y_5 = Ax_4 \approx \begin{bmatrix} 2.99014 \\ 2.99507 \end{bmatrix} \qquad x_5 \approx \frac{1}{2.99507} y_5 \approx \begin{bmatrix} .99835 \\ 1.00000 \end{bmatrix}$$

and

$$\frac{\langle \mathbf{x}_1, A\mathbf{x}_1 \rangle}{\langle \mathbf{x}_1, \mathbf{x}_1 \rangle} \approx \frac{5.26563}{1.76563} \approx 2.98230$$

$$\frac{\langle \mathbf{x}_2, A\mathbf{x}_2 \rangle}{\langle \mathbf{x}_2, \mathbf{x}_2 \rangle} \approx \frac{5.74101}{1.91493} \approx 2.99803$$

$$\frac{\langle \mathbf{x}_3, A\mathbf{x}_3 \rangle}{\langle \mathbf{x}_3, \mathbf{x}_3 \rangle} \approx \frac{5.91196}{1.97080} \approx 2.99978$$

$$\frac{\langle \mathbf{x}_4, A\mathbf{x}_4 \rangle}{\langle \mathbf{x}_4, \mathbf{x}_4 \rangle} \approx \frac{5.97044}{1.99016} \approx 2.99998$$

Notice that the approximations $\langle \mathbf{x}_k, A\mathbf{x}_k \rangle / \langle \mathbf{x}_k, \mathbf{x}_k \rangle$ to the dominant eigenvalue are the same as in Example 3. Also notice that the sequence $\{\mathbf{x}_k\}_{k=0}^{\infty}$ appears to be converging to

$$\begin{bmatrix} 1 \\ 1 \end{bmatrix}$$

which is an eigenvector associated with the dominant eigenvalue 3.

There is not any good rule for determining the accuracy of any iterate $\lambda_1(k) = \langle \mathbf{x}_{k-1}, A\mathbf{x}_k \rangle / \langle \mathbf{x}_{k-1}, \mathbf{x}_{k-1} \rangle$ from either of the methods described above. There are two commonly used ways of measuring the error in this approximation. The first

$$|\lambda_1 - \lambda_1(k)|$$

called the **absolute error** measures how close the approximation $\lambda_1(k)$ is to the dominant eigenvalue λ_1. The second, if $\lambda_1 \neq 0$,

$$\left| \frac{\lambda_1 - \lambda_1(k)}{\lambda_1} \right|$$

called the **relative error**, measures how close $\lambda_1(k)$ is to λ_1 relative to the size of λ_1.

Before starting the power method, we decide which of these two measures of error is the more appropriate for the use to which we will put our approximation to λ_1. Having done this, we next decide how small an error, E, we can tolerate. We then use the power method with the intention of stopping as soon as the error is less than E. Since λ_1 is not known we cannot compute either the absolute or relative error. However, we can approximate the absolute and relative errors by

$$|\lambda_1(k+1) - \lambda_1(k)|$$

and

$$\left| \frac{\lambda_1(k+1) - \lambda_1(k)}{\lambda_1(k+1)} \right|$$

respectively. Then we stop iterating as soon as the approximation to the appropriate error is less than E.

Recalling that the characteristic equation of a matrix with real components has real coefficients and that the roots of a polynomial with real coefficients occur in conjugate pairs, we conclude that the dominant eigenvalue, if one exists, of a matrix with real coefficients is a real number. In general, it is not possible to tell a priori whether a given matrix has a dominant eigenvalue. This is the main drawback of the power method, with or without scaling. However, the following theorem (due to O. Perron) guarantees the existence of a dominant eigenvalue. The proof of this theorem is beyond the scope of this course.

Theorem 1. *Let A be an $n \times n$ with positive components. Then A has a dominant positive eigenvalue, and there is a dominant eigenvector of A having positive components.*

Example 5. We have seen above that the power method with scaling gives sequences which converge to the dominant eigenvalue, 3, and dominant eigenvector, $\begin{bmatrix} 1 \\ 1 \end{bmatrix}$, of

$$A = \begin{bmatrix} 1 & 2 \\ 2 & 1 \end{bmatrix}$$

This matrix, which has positive components, has a dominant positive eigenvalue and a dominant eigenvector with positive components, as assured by Theorem 1.

Exercises

In Exercises 1–6, beginning with an initial vector, each of whose components are one, use the power method with scaling to approximate the dominant eigenvalue and an associated eigenvector of the given matrix. Continue the iterations until the approximate absolute error is less than the given number E.

1. $\begin{bmatrix} 1 & 2 \\ 3 & 4 \end{bmatrix}$ $E = .01$ **2.** $\begin{bmatrix} 2 & 6 \\ 8 & 4 \end{bmatrix}$ $E = .01$

3. $\begin{bmatrix} 4 & 2 \\ 2 & 1 \end{bmatrix}$ $E = .001$ **4.** $\begin{bmatrix} 5 & 1 \\ 6 & 2 \end{bmatrix}$ $E = .001$

5. $\begin{bmatrix} 1 & 2 & 3 \\ 4 & 5 & 6 \\ 7 & 8 & 9 \end{bmatrix}$ $E = .001$ **6.** $\begin{bmatrix} 1 & 3 & 2 \\ 6 & 5 & 4 \\ 4 & 4 & 1 \end{bmatrix}$ $E = .001$

7. Show that the power method beginning with $\mathbf{x}_0 = \begin{bmatrix} 1 \\ 1 \end{bmatrix}$ does not work for the matrix

$$A = \begin{bmatrix} 0 & 1 \\ 1 & 0 \end{bmatrix}$$

Explain why it does not work.

9.3 NEWTON'S METHOD*

It is well known that there is a formula, the quadratic formula, for finding the roots of the quadratic equation $a_2 x^2 + a_1 x + a_0 = 0$; $a_2 \neq 0$. However, there is not always a formula for finding the roots of an equation. For example, if $n \geq 5$ there is no formula for finding the roots of a general nth degree polynomial equation $a_n x^n + a_{n-1} x^{n-1} + \cdots + a_1 x + a_0 = 0$.

We begin this section by describing a method, called **Newton's method**, for "finding" a root of $f(x) = 0$, where f is a twice differentiable function on an open interval containing the root. By "finding" a root, we mean approximating a root as closely as desired. Then we generalize this method to be used on systems of nonlinear equations.

Newton's method can be derived in several ways. One of the simplest is to take an approximation p_1 to a root p of $f(x) = 0$ and approximate the graph of f by the tangent line to the graph at the point $(p_1, f(p_1))$ (see Figure 9.1).

Since

$$y = f(p_1) + f'(p_1)(x - p_1)$$

is an equation for this tangent line, we have

$$f(x) \approx f(p_1) + f'(p_1)(x - p_1)$$

In particular, when $x = p$ we have

$$0 = f(p) \approx f(p_1) + f'(p_1)(p - p_1)$$

Solving for p yields

$$p \approx p_1 - \frac{f(p_1)}{f'(p_1)}$$

We now set

$$p_2 = p_1 - \frac{f(p_1)}{f'(p_1)}$$

and repeat the above processes to obtain a new approximation p_3 to p:

$$p_3 = p_2 - \frac{f(p_2)}{f'(p_2)}$$

Continuing in this fashion we generate a sequence $\{p_k\}_{k=1}^{\infty}$ of approximations to the root p, where

$$p_k = p_{k-1} - \frac{f(p_{k-1})}{f'(p_{k-1})} \tag{1}$$

It can be shown that if $f'(p) \neq 0$ and if p_1 is sufficiently close to p, then this sequence converges to p.

For example, consider the function $f(x) = \cos x - x$. From Figure 9.2 we see that the graphs of $g(x) = \cos x$ and $h(x) = x$ intersect at precisely one point and the

---

* This section is intended for readers who have studied partial differentiation.

FIGURE 9.1

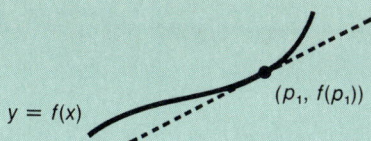

$y = f(x)$

$(p_1, f(p_1))$

FIGURE 9.2

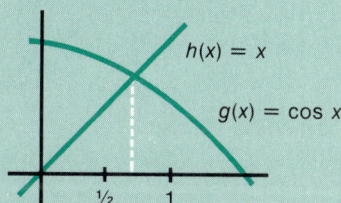

$h(x) = x$

$g(x) = \cos x$

$\frac{1}{2}$ 1

first coordinate, p, of this point lies between .5 and 1. The number p is the only root of $f(x) = 0$. Newton's method for $f(x) = \cos x - x$ yields the sequence $\{p_k\}_{k=1}^{\infty}$ where

$$p_k = p_{k-1} + \frac{\cos p_{k-1} - p_{k-1}}{\sin p_{k-1} + 1}, \qquad 2 \leq k$$

Beginning with $p_1 = .5$, we obtain a sequence whose first four terms (to six decimal places) are

$$p_1 = .5$$
$$p_2 = .755222$$
$$p_3 = .739141$$
$$p_4 = .739085$$

and, to six decimal places, $p_k = p_4$ for $k \geq 4$.

Newton's method can be generalized to systems of nonlinear equations of the form

(2)
$$f_1(x_1, x_2, \ldots, x_n) = 0$$
$$f_2(x_1, x_2, \ldots, x_n) = 0$$
$$f_n(x_1, x_2, \ldots, x_n) = 0$$

The equations

(3)
$$x_1^2 + 16x_2^2 - 1 = 0$$
$$8x_1^3 + x_2^3 - 1 = 0$$

form such a system.

If we let

$$\mathbf{x} = \begin{bmatrix} x_1 \\ x_2 \\ \vdots \\ x_n \end{bmatrix} \quad \text{and} \quad F(\mathbf{x}) = \begin{bmatrix} f_1(x_1, x_2, \ldots, x_n) \\ f_2(x_1, x_2, \ldots, x_n) \\ \vdots \\ f_n(x_1, x_2, \ldots, x_n) \end{bmatrix}$$

then finding a solution of the system of equations in (2) is equivalent to finding a root of the vector equation

$$F(\mathbf{x}) = \mathbf{0}$$

The matrix, called the **Jacobian** of F at $\mathbf{x}$,

$$J(\mathbf{x}) = \begin{bmatrix} \dfrac{\partial f_1}{\partial x_1}(\mathbf{x}) & \dfrac{\partial f_1}{\partial x_2}(\mathbf{x}) & \cdots & \dfrac{\partial f_1}{\partial x_n}(\mathbf{x}) \\ \dfrac{\partial f_2}{\partial x_1}(\mathbf{x}) & \dfrac{\partial f_2}{\partial x_2}(\mathbf{x}) & \cdots & \dfrac{\partial f_2}{\partial x_n}(\mathbf{x}) \\ \vdots & \vdots & & \vdots \\ \dfrac{\partial f_n}{\partial x_n}(\mathbf{x}) & \dfrac{\partial f_n}{\partial x_n}(\mathbf{x}) & \cdots & \dfrac{\partial f_n}{\partial x_n}(\mathbf{x}) \end{bmatrix}$$

plays the same role with respect to the vector function $F(x)$ as the derivative $f'(x)$ does to the scalar function $f(x)$. When we accept this as a fact, the vector analogy to equation (1) is

$$\mathbf{p}_k = \mathbf{p}_{k-1} - J(\mathbf{p}_{k-1})^{-1} F(\mathbf{p}_{k-1}) \tag{4}$$

where $\mathbf{p}_k$ and $\mathbf{p}_{k-1}$ are n-vectors approximating a root $\mathbf{p}$ of $F(\mathbf{x}) = \mathbf{0}$. If $n = 2$ there is a simple formula for the inverse of $J(\mathbf{p}_{k-1})$ (see Theorem 12 in Section 1.8). In practice, the inverse of $J(\mathbf{p}_{k-1})$ is not computed at each step if $n > 2$. Instead we solve the equation

$$J(\mathbf{p}_{k-1})\mathbf{q}_{k-1} = -F(\mathbf{p}_{k-1}) \tag{5}$$

for $\mathbf{q}_{k-1}$ and let

$$\mathbf{p}_k = \mathbf{p}_{k-1} + \mathbf{q}_{k-1} \tag{6}$$

This method yields the same approximations as does the method described by (4), but it requires fewer arithmetic operations than does computing the inverse of $J(\mathbf{p}_{k-1})$ at each step. As for the scalar case, there are theorems that guarantee the convergence of the sequence $\{\mathbf{p}_k\}_{k=1}^{\infty}$ to a root $\mathbf{p}$ of $F(\mathbf{x}) = \mathbf{0}$ whenever the initial approximation is sufficiently close to the root $\mathbf{p}$.

For example, the system of equations in (3) can be rewritten as

$$\begin{bmatrix} x_1^2 + 16x_2^2 - 1 \\ 8x_1^3 + x_2^3 - 1 \end{bmatrix} = \begin{bmatrix} 0 \\ 0 \end{bmatrix}$$

Then

$$J(\mathbf{x}) = \begin{bmatrix} 2x_1 & 32x_2 \\ 24x_1^2 & 3x_2^2 \end{bmatrix}$$

Even though $n = 2$, we will illustrate the use of the formulas in (5) and (6) on this equation. We can begin with

$$\mathbf{p}_1 = \begin{bmatrix} 1 \\ 1 \end{bmatrix}$$

and compute a sequence $\{\mathbf{p}_k\}_{k=1}^{\infty}$ of 2-vectors by solving

$$\begin{bmatrix} 2p_1^{(k-1)} & 32p_2^{(k-1)} \\ 24(p_1^{(k-1)})^2 & 3(p_2^{(k-1)})^2 \end{bmatrix} \mathbf{q}_{k-1} = - \begin{bmatrix} (p_1^{(k-1)})^2 + 16(p_2^{(k-1)})^2 - 1 \\ 8(p_1^{(k-1)})^3 + (p_2^{(k-1)})^3 - 1 \end{bmatrix}$$

for $\mathbf{q}_{k-1}$ and then setting

$$\mathbf{p}_k = \mathbf{p}_{k-1} + \mathbf{q}_{k-1}$$

where

$$\mathbf{p}_j = \begin{bmatrix} p_1^{(j)} \\ p_2^{(j)} \end{bmatrix}$$

for $j = 1, 2, 3, \ldots$. The first six terms of the sequence of $\{\mathbf{p}_k\}_{k=1}^{\infty}$ (to six decimal places) are

$$\mathbf{p}_1 = \begin{bmatrix} 1.00000 \\ 1.00000 \end{bmatrix} \qquad \mathbf{p}_2 = \begin{bmatrix} .727034 \\ .517060 \end{bmatrix}$$

$$\mathbf{p}_3 = \begin{bmatrix} .566271 \\ .301149 \end{bmatrix} \qquad \mathbf{p}_4 = \begin{bmatrix} .506487 \\ .228094 \end{bmatrix}$$

$$\mathbf{p}_5 = \begin{bmatrix} .498435 \\ .217023 \end{bmatrix} \qquad \mathbf{p}_6 = \begin{bmatrix} .498297 \\ .216751 \end{bmatrix}$$

while, to six decimal places, $\mathbf{p}_k = \mathbf{p}_6$ for $k \geq 6$. Thus, to six decimal places,

$$\mathbf{x}_1 = .498297 \qquad \mathbf{x}_2 = .216751$$

is a solution of the system of equations in (3).

--- **Exercises** ---

Use Newton's method to approximate solutions of the following systems of equations. Begin the method with the given vectors.

1. $\begin{aligned} x_1^2 + x_2^2 - 5 &= 0 \\ x_1^3 - 2x_1 + x_2^2 - 3 &= 0 \end{aligned}$ $\begin{bmatrix} 1 \\ 1 \end{bmatrix}$ $\begin{bmatrix} 1.5 \\ 1.5 \end{bmatrix}$

2. $\begin{aligned} x_1^2 - x_2^2 + 2 &= 0 \\ x_1^4 - x_1^2 x_2^2 + x_1^2 + 5 &= 0 \end{aligned}$ $\begin{bmatrix} 2 \\ 2 \end{bmatrix}$ $\begin{bmatrix} -2 \\ 2 \end{bmatrix}$

3. $\begin{aligned} x_1^2 - x_2^2 + x_1 x_2 + 2 &= 0 \\ x_1^3 - x_2^3 - 3x_1 + x_2 &= 0 \end{aligned}$ $\begin{bmatrix} 0 \\ 1 \end{bmatrix}$ $\begin{bmatrix} 1 \\ 1 \end{bmatrix}$

4. $\begin{aligned} x_1 - x_2 + \sin x_1 &= 0 \\ x_1^2 - x_1 x_2 + \cos x_2 + 1 &= 0 \end{aligned}$ $\begin{bmatrix} 1 \\ 1 \end{bmatrix}$ $\begin{bmatrix} -3 \\ 3 \end{bmatrix}$

5. $\begin{aligned} \cos x_1 + 2\cos x_2 + 2x_3 &= 0 \\ \sin x_1 + 2\sin x_2 - x_3 &= 0 \\ \sin(x_2 - x_1) - x_3 &= 0 \end{aligned}$ $\begin{bmatrix} 1.5 \\ 3 \\ 0 \end{bmatrix}$

6. $\begin{aligned} x^2 + y^2 + z - 3 &= 0 \\ x^3 + z^4 - 1 &= 0 \\ y^4 + z^3 - 4 &= 0 \end{aligned}$ $\begin{bmatrix} 2 \\ 1 \\ 0 \end{bmatrix}$

Appendix 1

In this appendix we prove the second part of Theorem 23 of Section 4.11, which we restate for convenience.

Theorem.   *Let A be an m $\times$ n matrix that is row equivalent to a matrix B that is in row-echelon form. If $B_{i_1}, B_{i_2}, \ldots, B_{i_k}$ are the columns of B containing a nonzero leading entry of some row of B, then the corresponding columns $A_{i_1}, A_{i_2}, \ldots, A_{i_k}$ of A form a basis for the column space of A.*

Before furnishing a proof for the theorem, we need a basis for the column space of B. The following lemma and its proof provide such a basis.

Lemma.   *The set $S = \{B_{i_1}, B_{i_2}, \ldots, B_{i_k}\}$ is a basis for the column space of B.*

Proof: We first show that S is linearly independent. Let $c_1, c_2, \ldots, c_k$ be numbers such that

$$c_1 B_{i_1} + c_2 B_{i_2} + \cdots + c_k B_{i_k} = 0 \tag{1}$$

Let j be the number of the component of B_{i_k} that is a nonzero leading entry b of some row of B. Then the jth component of any of the other elements of S is zero. Hence the jth component of $c_1 B_{i_1} + c_2 B_{i_2} + \cdots + c_k B_{i_k}$ is $c_k b$. From equation (1) we see that this component must be zero. Since $b \neq 0$, we have $c_k = 0$. Repeating this argument we find that $c_{k-1} = 0$, $c_{k-2} = 0, \ldots, c_1 = 0$. Therefore S is linearly independent.

Since S has k elements and each element of S contains exactly one nonzero leading entry of some row of B, the last $m - k$ components must consist entirely of zeros. Hence, the column space of B is a subspace of the linear space V consisting of all m-vectors with the last $m - k$ components equal to zero. Evidently the following vectors form a basis for V:

$$\mathbf{v}_1 = \begin{bmatrix} 1 \\ 0 \\ 0 \\ 0 \\ \vdots \\ 0 \end{bmatrix}, \quad \mathbf{v}_2 = \begin{bmatrix} 0 \\ 1 \\ 0 \\ 0 \\ \vdots \\ 0 \end{bmatrix}, \quad \dots, \quad \mathbf{v}_k = \begin{bmatrix} 0 \\ \vdots \\ 0 \\ 1 \\ 0 \\ \vdots \\ 0 \end{bmatrix} \quad \leftarrow k\text{th component}$$

Thus dim $V = k$. Since S is a linearly independent subset of V containing k elements, S is a basis for V. Therefore the column space of B equals V and has S as a basis. ∎

Proof of Theorem: Since A is row equivalent to B, there is a nonsingular matrix P such that $PA = B$. From this identity it is easy to show that $\mathbf{x}$ is an n-vector such that $A\mathbf{x} = \mathbf{0}$ if and only if $B\mathbf{x} = \mathbf{0}$. Using Theorem 4 of Section 4.4 we find that this statement is equivalent to the following statement: $x_1, x_2, \dots, x_n$ are numbers such that $x_1 A_1 + x_2 A_2 + \cdots + x_n A_n = \mathbf{0}$ if and only if $x_1 B_1 + x_2 B_2 + \cdots + x_n B_n = \mathbf{0}$. Therefore a set of columns of A is linearly independent if and only if the set of corresponding columns of B is linearly independent. By the lemma, $S = \{B_{i_1}, B_{i_2}, \dots, B_{i_k}\}$ is a basis for the column space of B. That is, S is linearly independent, and if we add any other column of B to S we obtain a linearly dependent set. Hence, $T = \{A_{i_1}, A_{i_2}, \dots, A_{i_k}\}$ is linearly independent, and if we add any other column of A to T we obtain a linearly dependent set. Therefore T is a basis for the column space of A. This completes the proof. ∎

Appendix 2

In this appendix we prove Theorem 15 of Section 7.6, which we restate for convenience.

Theorem. *Let A be an n × n symmetric matrix. Then*

    1. *For each eigenvalue λ of A the dimension of the eigenspace equals the multiplicity of λ.*

    2. *There is an orthonormal basis for R^n consisting of eigenvectors of A.*

To prove this theorem we use the following lemma.

Lemma. *Let A be an n × n symmetric matrix and let V be a subspace of R^n having dimension k, k ≥ 1. If Av is an element of V for every v in V, then V contains an eigenvector of A.*

Proof: The basic idea of this proof is simple. By the discussion in Section 4.8 we can identify R^k with V. Since the matrix A determines a linear transformation from V into V it also determines a linear transformation T from R^k into R^k. An eigenvector of the matrix representation for T will be shown to determine an eigenvector of A that is in V.

Let $U = \{\mathbf{v}_1, \mathbf{v}_2, \ldots, \mathbf{v}_k\}$ be a basis for V. Then each $A\mathbf{v}_i$ can be written as

$$(1) \qquad\qquad A\mathbf{v}_i = c_{i1}\mathbf{v}_1 + c_{i2}\mathbf{v}_2 + \cdots + c_{ik}\mathbf{v}_k$$

Let B be the $n \times k$ matrix having $\mathbf{v}_1, \mathbf{v}_2, \ldots, \mathbf{v}_k$ as its columns and let C be the $k \times k$ matrix having c_{ij} as its ij-component. Notice that B is the matrix representation of a linear transformation identifying R^k with V and that C is the matrix representation of the linear transformation induced on R^k by the linear transformation $T: V \to V$ defined by $T(\mathbf{v}) = A\mathbf{v}$. By Theorem 8 of Section 1.7 we have that $A\mathbf{v}_i$ is the ith column of AB. From equation (1) we see that $A\mathbf{v}_i$ is also the ith column of BC. Therefore $AB = BC$.

Let λ be an eigenvalue (possibly complex) of C and let $\mathbf{x}$ be an associated eigenvector. Then

$$(2) \qquad A(B\mathbf{x}) = (AB)\mathbf{x} = (BC)\mathbf{x} = B(C\mathbf{x}) = B(\lambda\mathbf{x}) = \lambda(B\mathbf{x})$$

Notice that

$$Bx = x_1 v_1 + x_2 v_2 + \cdots + x_k v_k \tag{3}$$

where $x_1, x_2, \ldots, x_k$ are the components of x. Since the set U is linearly independent and x is not the zero vector, we conclude from equation (3) that $Bx \neq 0$. Therefore, by equation (2), Bx is an eigenvector of A associated with λ, which must be real since A is symmetric. Since λ is real the matrix C must have a nonzero element of R^k as an associated eigenvector. We may assume that x is such an element. Then Bx is in R^n. Equation (3) shows that Bx is an element of the subspace V of R^n. This completes the proof. ∎

Proof of Theorem.   For each eigenvalue λ of A let n_λ denote the multiplicity of λ. By Theorem 16 of Section 4.9 and Theorem 11 of Section 7.5 the dimension, k_λ, of the eigenspace A_λ is less than or equal to n_λ. Theorem 7 of Section 5.4 assures us that each A_λ has an orthonormal basis B_λ consisting of k_λ eigenvectors of A associated with λ. Since eigenvectors associated with distinct eigenvalues are orthogonal, the set

$$B = \bigcup \{B_\lambda : \lambda \text{ an eigenvalue of } A\}$$

is orthonormal.

Suppose that $n_\lambda \neq k_\lambda$ for some eigenvalue λ of A. Then B has fewer than n elements. Let $\{v_1, v_2, \ldots, v_k\}$ be the elements of B. Extend B to an orthonormal basis $\{v_1, \ldots, v_k, v_{k+1}, \ldots, v_n\}$ for R^n. This can be done by using the procedure in Example 4 of Section 4.11 and the Gram–Schmidt process.

Let U be the subspace of R^n spanned by B and let V be the subspace of R^n spanned by $B' = \{v_{k+1}, \ldots, v_n\}$. Notice that U contains all the eigenvectors of A. Since each element of B is orthogonal to each element of B', it follows that V is the set of all n-vectors that are orthogonal to every element of U. In Exercises 1 and 2 the reader is asked to show that

1. Au is an element of U whenever u is an element of U.
2. $U \cap V = \{0\}$.

Let v be any element of V. Since Au is an element of U whenever u is an element of U, we have

$$0 = \langle v, Au \rangle = \langle Av, u \rangle$$

for every element u of U. Thus Av is orthogonal to every element of U and, hence, Av is an element of V. By the lemma the subspace V contains an eigenvector of A. This is impossible because U contains all the eigenvalues of A and $U \cap V = \{0\}$. From this contradiction we conclude that the dimension of A_λ must equal the multiplicity of λ for every eigenvalue λ of A. Therefore B contains n elements and is a basis for R^n. This completes the proof. ∎

――――――――――――――――― Exercises ―――――――――――――――――

1. Let U be the linear space in the proof of the theorem. Show that Au is an element of U whenever u is an element of U.
2. Let U and V be the linear spaces in the proof of the theorem. Show that $U \cap V = \{0\}$.

Answers to Odd-Numbered Exercises

SECTION 1.1

1. Linear   **3.** Nonlinear   **5.** Linear   **7.** Nonlinear   **9.** Linear   **11.** No
13. No   **15.** Yes   **17.** Yes   **19.** Yes   **21.** No   **23.** One solution
25. No solution   **27.** Infinitely many solutions   **29.** One solution

31. $\begin{bmatrix} 1 \\ -1 \end{bmatrix}$ **33.** $\begin{bmatrix} -2 \\ 3 \\ -5 \end{bmatrix}$ **35.** $\begin{bmatrix} -.25 \\ .25 \\ .5 \end{bmatrix}$

SECTION 1.2

1. $\left[\begin{array}{cc|c} 1 & 3 & 1 \\ 3 & 1 & 4 \end{array}\right]$ **3.** $\left[\begin{array}{ccc|c} 5 & 3 & 2 & 0 \\ -2 & 3 & -5 & 1 \\ 1 & -1 & 1 & 6 \end{array}\right]$ **5.** $\left[\begin{array}{cccc|c} 2 & 1 & 3 & -4 & 2 \\ -1 & 2 & -1 & 5 & 3 \\ 1 & -1 & 4 & -3 & -1 \end{array}\right]$

7. $\left[\begin{array}{cccc|c} 6 & 4 & -5 & -2 & 1 \\ 2 & -3 & 2 & -5 & -1 \end{array}\right]$ **9.** $\begin{aligned} x_1 + 2x_2 &= 3 \\ 4x_1 + 5x_2 &= 6 \end{aligned}$

11. $\begin{aligned} x_1 + 2x_2 + 3x_3 &= 5 \\ 6x_1 + 2x_2 - x_3 &= 7 \\ -6x_1 - 4x_2 + 5x_3 &= 3 \end{aligned}$ **13.** $\begin{aligned} 2x_1 + 3x_2 + x_3 - 2x_4 &= 9 \\ -3x_1 + 4x_2 - x_3 + 5x_4 &= 5 \end{aligned}$

15. $\begin{aligned} 3x_1 + 2x_2 + 4x_3 + 8x_5 &= 5 \\ 9x_1 + 3x_2 + 2x_3 + 4x_4 &= 3 \\ 2x_1 + 2x_2 + 3x_3 + 2x_4 + 5x_5 &= 3 \\ 4x_1 + x_2 + 4x_3 + 6x_4 &= 1 \\ 5x_1 + 4x_2 + 3x_3 + 2x_4 + x_5 &= 0 \end{aligned}$ **17.** Yes   **19.** No   **21.** Yes   **23.** No

25. $\begin{bmatrix} .5 \\ .5 \end{bmatrix}$ 27. $\begin{bmatrix} c \\ c \\ -c \end{bmatrix}$ 29. $\begin{bmatrix} .5 \\ .75 \\ .25 \end{bmatrix}$ 31. $\begin{bmatrix} -\dfrac{2c}{3} \\ \dfrac{c}{3} \\ c \end{bmatrix}$ 33. $\begin{bmatrix} -\dfrac{2c}{3} \\ \dfrac{c}{3} \\ c \end{bmatrix}$

35. $\begin{bmatrix} -\dfrac{3c}{2} - \dfrac{d}{2} + \dfrac{1}{2} \\ c \\ d \end{bmatrix}$ 37. $\begin{bmatrix} \dfrac{3}{2} \\ \dfrac{1}{2} \\ -\dfrac{1}{3} \\ -\dfrac{5}{3} \end{bmatrix}$ 39. $\begin{bmatrix} -.6c + .4 \\ -.2c - d + .8 \\ c \\ d \end{bmatrix}$

41. No solution   43. $\begin{bmatrix} \dfrac{(a+b)}{2} \\ \dfrac{(a-b)}{2} \end{bmatrix}$ 45. $\begin{bmatrix} \dfrac{(2b+c)}{4} \\ \dfrac{(a-b)}{2} \\ \dfrac{(2a-c)}{4} \end{bmatrix}$

47. $c + 3b - 2a = 0$ 49. $.5x^2 + x + 5$ 51. $\begin{bmatrix} -.2c \\ -.8c \\ c \end{bmatrix}$

SECTION 1.3

There are no exercises in this section.

SECTION 1.4

1. $\begin{bmatrix} .5 \\ .5 \end{bmatrix}$ 3. $\begin{bmatrix} c \\ c \\ -c \end{bmatrix}$ 5. $\begin{bmatrix} .5 \\ .75 \\ .25 \end{bmatrix}$ 7. $\begin{bmatrix} -\dfrac{2c}{3} \\ \dfrac{c}{3} \\ c \end{bmatrix}$ 9. $\begin{bmatrix} -\dfrac{2c}{3} \\ \dfrac{c}{3} \\ 3c \end{bmatrix}$

11. $\begin{bmatrix} \dfrac{(-3c - d + 1)}{2} \\ c \\ d \end{bmatrix}$ 13. $\begin{bmatrix} \dfrac{3}{2} \\ \dfrac{1}{2} \\ -\dfrac{1}{3} \\ -\dfrac{5}{3} \end{bmatrix}$ 15. $\begin{bmatrix} -.6c + .4 \\ -.2c - d + .8 \\ c \\ d \end{bmatrix}$

17. No solution　　**19.** $\begin{bmatrix} \dfrac{(a+b)}{2} \\[2mm] \dfrac{(a-b)}{2} \end{bmatrix}$　　**21.** $\begin{bmatrix} \dfrac{(2b+c)}{4} \\[2mm] \dfrac{(a-b)}{2} \\[2mm] \dfrac{(2a-c)}{4} \end{bmatrix}$

SECTION 1.5

1. The unique solution is $\begin{bmatrix} 3 \\ -2 \end{bmatrix}$.　　**3.** No solution

5. The unique solution is $\dfrac{1}{10}\begin{bmatrix} 4b_1 + 3b_2 \\ 2b_1 - b_2 \end{bmatrix}$.

7. The unique solution is $\dfrac{1}{2}\begin{bmatrix} b_2 + b_3 \\ b_1 - b_2 \\ b_1 - b_3 \end{bmatrix}$.

9. There is no solution if $b_2 - 2b_1 \neq 0$.
11. There is no solution if $b_1 - b_2 - b_3 \neq 0$.
13. By Theorem 2 there is exactly one solution.

SECTION 1.6

1. $\begin{bmatrix} 1 & 5 \\ 1 & 1 \\ 3 & 3 \end{bmatrix}$　**3.** $\begin{bmatrix} 1 & 12 \\ 4 & 3 \\ 5 & 8 \end{bmatrix}$　**5.** $\begin{bmatrix} 1 & 3 \\ 4 & 2 \\ 5 & 0 \end{bmatrix}$　**7.** $\begin{bmatrix} 0 & 42 \\ 18 & 12 \\ 12 & 30 \end{bmatrix}$　**9.** Not defined

11. $\begin{bmatrix} -7 & -18 \\ 8 & 1 \\ -11 & -12 \end{bmatrix}$　**13.** $\begin{bmatrix} -1 & -5 \\ -1 & -1 \\ -3 & -3 \end{bmatrix}$

15. The ij-components of $(a+b)A$, aA, and bA are $(a+b)a_{ij}$, aa_{ij}, ba_{ij} respectively, so that the ij-component of $(a+b)A$ is the sum of the ij-components of aA and bA. Therefore $(a+b)A = aA + bA$.

17. $\begin{bmatrix} 6+12i & 9+9i \\ & 3i & -6 \end{bmatrix}$　**19.** $\begin{bmatrix} 9-5i & 15-11i \\ -4+7i & 15-11i \end{bmatrix}$　**21.** $\begin{bmatrix} 2-17i & -2-18i \\ 11i & 8-i \end{bmatrix}$

SECTION 1.7

1. $\begin{bmatrix} 2 & -1 \\ 8 & 7 \end{bmatrix}$　**3.** $\begin{bmatrix} 9 & 5 \\ 5 & 13 \end{bmatrix}$　**5.** $\begin{bmatrix} 25 & 16 \\ 7 & 32 \end{bmatrix}$　**7.** Not defined　**9.** $\begin{bmatrix} 9 & 4 & 7 \\ 1 & 4 & 3 \\ 3 & 12 & 9 \end{bmatrix}$

11. $\begin{bmatrix} 21 & 7 & 8 \\ 5 & 7 & 4 \\ 15 & 21 & 12 \end{bmatrix}$　**13.** Not defined　**15.** $\begin{bmatrix} 6 \\ 9 \\ 7 \end{bmatrix}$　**17.** $\begin{bmatrix} 6 & -1 & 4 \\ 2 & 8 & -1 \\ 7 & 9 & 3 \end{bmatrix}$

19. $\begin{bmatrix} 1 & 0 & 0 \\ 0 & 0 & 2 \\ 2 & 0 & 3 \end{bmatrix}$ **21.** $\begin{bmatrix} a+b & 2a+b & a+2b \\ a+2b & a+b & 2a+b \\ a+2b & 2a+b & a+b \end{bmatrix}$ **23.** $\dfrac{1}{3}\begin{bmatrix} -1 & 2 \\ 2 & -1 \end{bmatrix}$ **25.** $\begin{bmatrix} \frac{1}{2}c+d & 0 \\ c & d \end{bmatrix}$

27. $(AB)C = \left(\begin{bmatrix} a_{11} & a_{12} \\ a_{21} & a_{22} \end{bmatrix}\begin{bmatrix} b_{11} & b_{12} \\ b_{21} & b_{22} \end{bmatrix}\right)\begin{bmatrix} c_{11} & c_{12} \\ c_{21} & c_{22} \end{bmatrix}$

$= \begin{bmatrix} a_{11}b_{11}+a_{12}b_{21} & a_{11}b_{12}+a_{12}b_{22} \\ a_{21}b_{11}+a_{22}b_{21} & a_{21}b_{12}+a_{22}b_{22} \end{bmatrix}\begin{bmatrix} c_{11} & c_{12} \\ c_{21} & c_{22} \end{bmatrix}$

$= \begin{bmatrix} (a_{11}b_{11}+a_{12}b_{21})c_{11}+(a_{11}b_{12}+a_{12}b_{22})c_{21} \\ (a_{21}b_{11}+a_{22}b_{21})c_{11}+(a_{21}b_{12}+a_{22}b_{22})c_{21} \end{bmatrix}$

$\begin{bmatrix} (a_{11}b_{11}+a_{12}b_{21})c_{12}+(a_{11}b_{12}+a_{12}b_{22})c_{22} \\ (a_{21}b_{11}+a_{22}b_{21})c_{12}+(a_{21}b_{12}+a_{22}b_{22})c_{22} \end{bmatrix}$

$= \begin{bmatrix} a_{11}(b_{11}c_{11}+b_{12}c_{21})+a_{12}(b_{21}c_{11}+b_{22}c_{21}) \\ a_{21}(b_{11}c_{11}+b_{12}c_{21})+a_{22}(b_{21}c_{11}+b_{22}c_{21}) \end{bmatrix}$

$\begin{bmatrix} a_{11}(b_{11}c_{12}+b_{12}c_{22})+a_{12}(b_{21}c_{12}+b_{22}c_{22}) \\ a_{21}(b_{11}c_{12}+b_{12}c_{22})+a_{22}(b_{21}c_{12}+b_{22}c_{22}) \end{bmatrix}$

$= \begin{bmatrix} a_{11} & a_{12} \\ a_{21} & a_{22} \end{bmatrix}\begin{bmatrix} b_{11}c_{11}+b_{12}c_{21} & b_{11}c_{12}+b_{12}c_{22} \\ b_{21}c_{11}+b_{22}c_{21} & b_{21}c_{12}+b_{22}c_{22} \end{bmatrix}$

$= \begin{bmatrix} a_{11} & a_{12} \\ a_{21} & a_{22} \end{bmatrix}\left(\begin{bmatrix} b_{11} & b_{12} \\ b_{21} & b_{22} \end{bmatrix}\begin{bmatrix} c_{11} & c_{12} \\ c_{21} & c_{22} \end{bmatrix}\right)$

$= A(BC)$

29. There are infinitely many such matrices. One such matrix is

$$\begin{bmatrix} 2 & 3 \\ -1 & -2 \end{bmatrix}$$

31. There are infinitely many such matrices. One such pair of matrices is

$$A = \begin{bmatrix} 1 & 1 \\ 0 & 0 \end{bmatrix} \qquad B = \begin{bmatrix} 1 & 0 \\ -1 & 0 \end{bmatrix}$$

33. $C(A+B) = \begin{bmatrix} c_{11} & c_{12} \\ c_{21} & c_{22} \end{bmatrix}\left(\begin{bmatrix} a_{11} & a_{12} \\ a_{21} & a_{22} \end{bmatrix}+\begin{bmatrix} b_{11} & b_{12} \\ b_{21} & b_{22} \end{bmatrix}\right)$

$= \begin{bmatrix} c_{11} & c_{12} \\ c_{21} & c_{22} \end{bmatrix}\begin{bmatrix} a_{11}+b_{11} & a_{12}+b_{12} \\ a_{21}+b_{21} & a_{22}+b_{22} \end{bmatrix}$

$= \begin{bmatrix} c_{11}(a_{11}+b_{11})+c_{12}(a_{21}+b_{21}) & c_{11}(a_{12}+b_{12})+c_{12}(a_{22}+b_{22}) \\ c_{21}(a_{11}+b_{11})+c_{22}(a_{22}+b_{22}) & c_{21}(a_{12}+b_{12})+c_{22}(a_{22}+b_{22}) \end{bmatrix}$

$= \begin{bmatrix} c_{11}a_{11}+c_{12}a_{21} & c_{11}a_{12}+c_{12}a_{22} \\ c_{21}a_{11}+c_{22}a_{22} & c_{21}a_{12}+c_{22}a_{22} \end{bmatrix}+\begin{bmatrix} c_{11}b_{11}+c_{12}b_{21} & c_{11}b_{12}+c_{12}b_{22} \\ c_{21}b_{11}+c_{22}b_{22} & c_{21}b_{12}+c_{22}b_{22} \end{bmatrix}$

$= \begin{bmatrix} c_{11} & c_{12} \\ c_{21} & c_{22} \end{bmatrix}\begin{bmatrix} a_{11} & a_{12} \\ a_{21} & a_{22} \end{bmatrix}+\begin{bmatrix} c_{11} & c_{12} \\ c_{21} & c_{22} \end{bmatrix}\begin{bmatrix} b_{11} & b_{12} \\ b_{21} & b_{22} \end{bmatrix}$

35. The *ij*-component of AB and the *i*th component of A times the *j*th column of B are both equal to $\sum_{k=1}^{n} a_{ik}b_{kj}$.

37. $\begin{bmatrix} 0 & 12-6i \\ 0 & 18i \\ 0 & 6+6i \end{bmatrix}$ **39.** $\begin{bmatrix} -36-40i & 4-20i \\ -46-26i & -6+22i \\ -38- & 4i & 2+10i \end{bmatrix}$

SECTION 1.8

1. $\begin{bmatrix} -2 & 1 \\ 1.5 & -.5 \end{bmatrix}$ **3.** No inverse

5. $\begin{bmatrix} 1 & -1 & \frac{1}{3} \\ 0 & \frac{1}{2} & -\frac{2}{3} \\ 0 & 0 & \frac{1}{3} \end{bmatrix}$ **7.** $\begin{bmatrix} \frac{4}{3} & -1 & \frac{1}{3} \\ \frac{5}{6} & \frac{1}{2} & -\frac{2}{3} \\ -\frac{2}{3} & 0 & \frac{1}{3} \end{bmatrix}$ **9.** No inverse **11.** $\frac{1}{2}\begin{bmatrix} -1 & 1 & 1 \\ 1 & -1 & 1 \\ 1 & 1 & -1 \end{bmatrix}$

13. $\begin{bmatrix} 1 & 1 & -1 & 0 \\ -1 & 0 & 0 & 1 \\ 0 & -1 & 1 & 0 \\ 1 & 1 & 0 & -1 \end{bmatrix}$ **15.** $a=0, c^{-1}=b\neq 0$ **17.** $\begin{bmatrix} a_{11}^{-1} & 0 & \cdots & 0 \\ 0 & a_{22}^{-1} & \cdots & 0 \\ \vdots & \vdots & \ddots & \vdots \\ 0 & 0 & \cdots & a_{nn}^{-1} \end{bmatrix}$

19. The matrix is nonsingular for all values of θ.

21. Suppose that $a \neq 0$. Then $d - cb/a = a^{-1}(ad - bc) = 0$ and

$$\begin{bmatrix} a & b & | & 1 & 0 \\ c & d & | & 0 & 1 \end{bmatrix} \sim \begin{bmatrix} a & b & | & 1 & 0 \\ 0 & d-\dfrac{cb}{a} & | & -\dfrac{c}{a} & 1 \end{bmatrix} \quad \left(-\dfrac{c}{a}\right)R_1 + R_2 \to R_2$$

$$\sim \begin{bmatrix} a & b & | & 1 & 0 \\ 0 & 0 & | & -\dfrac{c}{a} & 1 \end{bmatrix}$$

Thus the left side of $[A|I_2]$ cannot be changed into I_2 by using elementary row operations. Therefore A has no inverse. If $c \neq 0$, a similar argument shows that the left side of $[A|I_2]$ cannot be changed into I_2. If $a = c = 0$, it is clear that the left side of $[A|I_2]$ cannot be changed into I_2. Hence, in any case A has no inverse.

23. $\begin{bmatrix} -i & 0 \\ 1 & -i \end{bmatrix}$

SECTION 1.9

1. $\dfrac{1}{5}\begin{bmatrix} -1 & 3 \\ 2 & -1 \end{bmatrix}$ **3.** $\dfrac{1}{12}\begin{bmatrix} 2 & 8 & -4 \\ -1 & 2 & 2 \\ 2 & -4 & 2 \end{bmatrix}$

9. $(c^{-1}A^{-1})(cA) = (c^{-1}c)(A^{-1}A) = (1)I = I$ **11.** $C^{-1}B^{-1}A^{-1}$

13. Since A is nonsingular, A^{-1} exists and we have $B = IB = A^{-1}AB = A^{-1}AC = IC = C$.

15. There are infinitely many such matrices. For example, the matrices

$$A = \begin{bmatrix} 1 & 0 \\ 0 & 1 \end{bmatrix}, \qquad B = \begin{bmatrix} 1 & 2 \\ 1 & 0 \end{bmatrix},$$

and

$$C = \begin{bmatrix} 0 & 1 \\ 1 & 0 \end{bmatrix}$$

are nonsingular, but $A + B$ is nonsingular and $A + C$ is singular.

17. Since B is singular, there is a solution $\mathbf{y}$, $\mathbf{y} \neq \mathbf{0}$, of $B\mathbf{x} = \mathbf{0}$. Then $(AB)\mathbf{y} = A(B\mathbf{y}) = A\mathbf{0} = \mathbf{0}$ so that $AB\mathbf{x} = \mathbf{0}$ has a solution other than $\mathbf{x} = \mathbf{0}$. Therefore AB is singular.

19. $I = -A^2 - 3A = A(-A - 3I)$. By part 2 of Theorem 18 we have $A^{-1} = -A - 3I$.

SECTION 1.10

1. $\mathbf{0}$ is the only solution.

3. Each scalar multiple of $\begin{bmatrix} 1 \\ -2 \\ 1 \end{bmatrix}$ is a solution.

5. Each scalar multiple of $\begin{bmatrix} -5 \\ 7 \\ 3 \end{bmatrix}$ is a solution.

7. (b) $3a - 2b + c = 0$

(c) If the system of equations is consistent, then $\begin{bmatrix} b - a \\ 2a - b \end{bmatrix}$ is the only solution.

9. (a) $\begin{bmatrix} 2s - 3t \\ s \\ t \end{bmatrix}$ is a solution for every choice of the scalars s and t.

(b) $b = 2a, c = -3a$

(c) $\begin{bmatrix} a \\ 0 \\ 0 \end{bmatrix} + \begin{bmatrix} 2s - 3t \\ s \\ t \end{bmatrix}$ is a solution for every choice of the scalars s and t.

11. $A(\mathbf{y} - \mathbf{z}) = A\mathbf{y} - A\mathbf{z} = \mathbf{b} - \mathbf{b} = \mathbf{0}$. Therefore $\mathbf{y} - \mathbf{z}$ is a solution of $A\mathbf{x} = \mathbf{0}$.

SUPPLEMENTARY EXERCISES FOR CHAPTER 1

1. True. For example, $\begin{bmatrix} 1 & 2 \\ 2 & 4 \end{bmatrix} \mathbf{x} = \begin{bmatrix} 1 \\ 1 \end{bmatrix}$.

3. True. $\mathbf{x} = \mathbf{0}$ is a solution.

5. True. Theorem 2 of Section 1.5.

7. True. Theorem 16 of Section 1.9.

9. False. $A\mathbf{x} = \mathbf{0}$ always has $\mathbf{x} = \mathbf{0}$ as a solution.

11. False. Definition 4 of Section 1.7; p must equal n; q can be any positive integer.

13. True. Theorem 20 of Section 1.10.   **15.** False. See Exercise 1.

17. True. Definition 7 of Section 1.9.

19. True. By Theorem 6 of Section 1.7 and Theorem 4 of Section 1.6, $A(B + C) = AB + AC = AC + AB$.

21. True. Using Exercise 10 of Section 1.9 twice, we have $(ABC)^{-1} = ((AB)C)^{-1} = C^{-1}(AB)^{-1} = C^{-1}B^{-1}A^{-1}$.

23. False. See Exercise 15 of Section 1.9.

25. True. By Exercise 10 of Section 1.9, AB has $B^{-1}A^{-1}$ as an inverse. By Theorem 16 of Section 1.9, AB is nonsingular.

SECTION 2.1

3. -12 **5.** 51   **7.** $\mathbf{x} = \begin{bmatrix} .7 \\ .2 \end{bmatrix}$ **9.** $\mathbf{x} = \begin{bmatrix} -\dfrac{5}{32} \\[2mm] -\dfrac{73}{32} \\[2mm] \dfrac{9}{4} \end{bmatrix}$ **11.** Minus

SECTION 2.2

There are no exercises in this section.

SECTION 2.3

1. -2 **3.** 7   **5.** 0   **7.** 83   **9.** 0   **11.** 360

13. $\begin{bmatrix} 2 & 5 & 3 \\ 4 & -1 & 0 \end{bmatrix}$ **15.** $\begin{bmatrix} 1 & 0 & 6 \\ 2 & 4 & 5 \\ 2 & 1 & 4 \\ -2 & 3 & -1 \end{bmatrix}$ **17.** $\begin{bmatrix} 1 \\ 2 \\ 3 \\ 4 \end{bmatrix}$

19. The ij-components of $(A \pm B)^t$, A^t, and B^t are $a_{ji} + b_{ji}, a_{ji}$, and b_{ji}, respectively, so that the ij-component of $(A \pm B)^t$ is the sum of the ij-components of A^t and B^t. Therefore $(A \pm B)^t = A^t \pm B^t$.

21. The ij-components of $(cA)^t$ and cA^t are ca_{ji} and $c(a_{ji})$, respectively, so that the ij-component of $(cA)^t$ equals the ij-component of cA^t. Therefore $(cA)^t = cA^t$.

23. $(A^{-1}B^{-1})^t = (B^{-1})^t(A^{-1})^t = (B^t)^{-1}(A^t)^{-1} = (A^tB^t)^{-1}$

25. $(B^tB)^t = B^t(B^t)^t = B^tB, \quad (B + B^t)^t = B^t + (B^t)^t = B^t + B = B + B^t$

SECTION 2.4

1. -7 **3.** 0   **5.** 0   **7.** -72 **9.** -3

11. Interchanging the two identical rows of A leaves A unchanged, but by part 3 of Theorem 4 the sign of the determinant changes. Hence $\det A = -\det A$. This is possible only if $\det A = 0$.

13. det $A = \det A^t = \det((-1)A) = (-1)^n \det A$ (Exercise 14 of Section 2.1). If n is odd, then det $A = -\det A$ so that det $A = 0$.

15. Using elementary row operations and part 2 of Theorem 4, we can obtain a matrix B with one row consisting entirely of zeros such that det $A = \det B$. By Exercise 15 of Section 2.1 we have det $A = \det B = 0$.

17. Suppose that the matrix B is obtained by multiplying the kth row of A by c. Then the ij-component of B is ca_{ij} if $i = k$ and a_{ij} if $i \neq k$.

$$
\begin{aligned}
\det B &= \sum s(j_1, j_2, \ldots, j_n) b_{1j_1} b_{2j_2} \cdots b_{kj_k} \cdots b_{nj_n} \\
&= \sum s(j_1, j_2, \ldots, j_n) a_{1j_1} a_{2j_2} \cdots (ca_{kj_k}) \cdots a_{nj_n} \\
&= c \sum s(j_1, j_2, \ldots, j_n) a_{1j_1} a_{2j_2} \ldots, a_{nj_n} \\
&= c \det A
\end{aligned}
$$

19. If $A = \begin{bmatrix} a_{11} & a_{12} \\ a_{21} & a_{22} \end{bmatrix}$ then $B = \begin{bmatrix} a_{21} & a_{22} \\ a_{11} & a_{12} \end{bmatrix}$ so that

$$
\det B = a_{21}a_{21} - a_{22}a_{11} = -(a_{11}a_{22} - a_{12}a_{21}) = -\det A.
$$

SECTION 2.5

1. Nonsingular   **3.** Nonsingular   **5.** Singular   **7.** Unique solution
9. No unique solution   **11.** $1 = \det I = \det(AA^{-1}) = (\det A)(\det A^{-1})$
13. There are infinitely many such matrices. One such pair is

$$
A = \begin{bmatrix} 1 & 1 \\ 2 & 1 \end{bmatrix} \qquad B = \begin{bmatrix} 0 & 1 \\ 0 & 0 \end{bmatrix}
$$

SECTION 2.6

1. 266   **3.** 332   **5.** 348   **7.** 7   **9.** 82   **11.** -40 **13.** -770 **15.** -80 **17.** 16

SECTION 2.7

1. $\begin{bmatrix} -.6 & .4 \\ .8 & -.2 \end{bmatrix}$ **3.** $\dfrac{1}{11}\begin{bmatrix} -11 & 11 & -11 \\ 10 & -5 & 6 \\ -2 & 1 & 1 \end{bmatrix}$ **5.** $\begin{bmatrix} 4 & -1 & -1 & -1 \\ -1 & 1 & 0 & 0 \\ -1 & 0 & 1 & 0 \\ -1 & 0 & 0 & 1 \end{bmatrix}$

7. $\begin{bmatrix} -\dfrac{9}{71} \\ \dfrac{2}{71} \end{bmatrix}$ **9.** $\begin{bmatrix} \dfrac{3}{43} \\ -\dfrac{14}{43} \\ \dfrac{5}{43} \end{bmatrix}$

11. If the components of A are integers then each cofactor is an integer. The desired result follows directly from Theorem 12.

SECTION 3.1

1. $\dfrac{\pi}{4}, 3\sqrt{2}, \begin{bmatrix} 3 \\ 3 \end{bmatrix}$ **3.** $\dfrac{\pi}{4}, \sqrt{2}, \begin{bmatrix} 1 \\ 1 \end{bmatrix}$ **5.** $\dfrac{\pi}{2}, 3, \begin{bmatrix} 0 \\ 3 \end{bmatrix}$

7. $\dfrac{3\pi}{2}, 3, \begin{bmatrix} 0 \\ -3 \end{bmatrix}$ **9.** $\tan^{-1}(1.5), 3\sqrt{13}, \begin{bmatrix} 6 \\ 9 \end{bmatrix}$ **11.** $\dfrac{7\pi}{4}, 2\sqrt{2}$

13. $\dfrac{5\pi}{4}, \sqrt{14}$ **15.** $\dfrac{5\pi}{14}, 1$ **17.** $\dfrac{\pi}{2} - .5, 1$

19. $\tan^{-1} \dfrac{4}{3}, 5$ **21.** $\pi + \tan^{-1} \dfrac{2}{3}, \sqrt{13}$

SECTION 3.2

1. $2\sqrt{5}$ **3.** $\sqrt{34}$ **5.** $\sqrt{170}$ **7.** $\begin{bmatrix} -5 \\ 12 \end{bmatrix}$ **9.** $\dfrac{1}{5}\begin{bmatrix} \sqrt{5} - 4 \\ 2\sqrt{5} + 3 \end{bmatrix}$

11. $\dfrac{1}{\sqrt{34}}\begin{bmatrix} -3 \\ 5 \end{bmatrix}$ **13.** $\begin{bmatrix} -9 \\ 2 \end{bmatrix}$ **15.** $\dfrac{1}{2}\begin{bmatrix} \sqrt{3} \\ 1 \end{bmatrix}$ **17.** $\dfrac{1}{\sqrt{2}}\begin{bmatrix} 3 \\ -3 \end{bmatrix}$ **19.** $\begin{bmatrix} 0 \\ -2 \end{bmatrix}$

21. $\dfrac{1}{5}\begin{bmatrix} 3 \\ -4 \end{bmatrix}$ **23.** $\left\| \dfrac{1}{\|\mathbf{u}\|}\mathbf{u} \right\| = \dfrac{1}{\|\mathbf{u}\|}\|\mathbf{u}\| = 1$

25. $\|\mathbf{u}\| = \|(\mathbf{u} + \mathbf{v}) + (-\mathbf{v})\| \leq \|\mathbf{u} + \mathbf{v}\| + \|-\mathbf{v}\| = \|\mathbf{u} + \mathbf{v}\| + \|\mathbf{v}\|$
Therefore $\|\mathbf{u}\| - \|\mathbf{v}\| \leq \|\mathbf{u} + \mathbf{v}\|$

27. $k\left(\begin{bmatrix} a \\ b \end{bmatrix} + \begin{bmatrix} c \\ d \end{bmatrix}\right) = k\begin{bmatrix} a + c \\ b + d \end{bmatrix} = \begin{bmatrix} ka + kc \\ kb + kd \end{bmatrix} = \begin{bmatrix} ka \\ kb \end{bmatrix} + \begin{bmatrix} kc \\ kd \end{bmatrix} = k\begin{bmatrix} a \\ b \end{bmatrix} + k\begin{bmatrix} c \\ d \end{bmatrix}$

29. $\begin{bmatrix} a \\ b \end{bmatrix} + \begin{bmatrix} 0 \\ 0 \end{bmatrix} = \begin{bmatrix} a + 0 \\ b + 0 \end{bmatrix} = \begin{bmatrix} a \\ b \end{bmatrix}$

SECTION 3.3

1. 7   **3.** -6 **5.** 0   **7.** $\dfrac{\pi}{4}$ **9.** π **11.** $\cos^{-1}\dfrac{9\sqrt{3} - 4}{\sqrt{403}}$

13. $\cos^{-1}\dfrac{1}{\sqrt{26}}$ **15.** $\dfrac{1}{5}\begin{bmatrix} 2 \\ 1 \end{bmatrix}, \dfrac{1}{5}\begin{bmatrix} 3 \\ -6 \end{bmatrix}$ **17.** $\dfrac{5}{13}\begin{bmatrix} 3 \\ -2 \end{bmatrix}, \dfrac{1}{13}\begin{bmatrix} 24 \\ 36 \end{bmatrix}$

19. 40   **21.** $-\dfrac{8}{5\sqrt{2}}\begin{bmatrix} 4 \\ 5 \end{bmatrix}$ **23.** $-\dfrac{6}{5}$

25. (a) and (b) The dot product is defined only for vectors.
(c) A number cannot be subtracted from a vector.

27. $\|u + v\|^2 + \|u - v\|^2 = (u + v)\cdot(u + v) + (u - v)\cdot(u - v)$
$$= (u\cdot u + 2u\cdot v + v\cdot v) + (u\cdot u - 2u\cdot v + v\cdot v)$$
$$= 2(u\cdot u + v\cdot v)$$
$$= 2(\|u\|^2 + \|v\|^2)$$

29. $u\cdot u = u_1 u_1 + u_2 u_2 = (\sqrt{u_1^2 + u_2^2})^2 = \|u\|^2$ **31.** $u\cdot 0 = u_1 0 + u_2 0 = 0$

33. $k(u\cdot v) = k(u_1 v_1 + u_2 v_2) = (ku_1)v_1 + (ku_2)v_2 = (ku)\cdot v$
$k(u\cdot v) = k(u_1 v_1 + u_2 v_2) = u_1(kv_1) + u_2(kv_2) = u\cdot(kv)$

SECTION 3.4

1. $\sqrt{3}$ **3.** $3\sqrt{2}$ **5.** $\sqrt{14}$ **7.** $\sqrt{14}$ **9.** $\begin{bmatrix} 7 \\ 6 \\ -7 \end{bmatrix}$

11. $\begin{bmatrix} \dfrac{1}{\sqrt{14}} - \dfrac{4}{\sqrt{101}} + \dfrac{2}{\sqrt{5}} \\ \dfrac{-2}{\sqrt{14}} - \dfrac{6}{\sqrt{101}} \\ \dfrac{3}{\sqrt{14}} - \dfrac{7}{\sqrt{101}} + \dfrac{1}{\sqrt{5}} \end{bmatrix}$ **13.** $\dfrac{1}{\sqrt{74}}\begin{bmatrix} 7 \\ 4 \\ -3 \end{bmatrix}$ **15.** $\sqrt{\dfrac{101}{5}}$

17. $\|u\| = \|(u + v) + (-v)\| \le \|u + v\| + \|-v\| = \|u + v\| + \|v\|$
Therefore $\|u\| - \|v\| \le \|u + v\|$

SECTION 3.5

1. -4 **3.** -15 **5.** 8 **7.** $\cos^{-1}\dfrac{-2}{\sqrt{154}}$ **9.** $\cos^{-1}\dfrac{-15}{2\sqrt{105}}$

11. $14i + 10k$ **13.** 1 **15.** $2\sqrt{14}$ **17.** $\dfrac{1}{\sqrt{14}}(5i + j + 2k)$

19. -2 **21.** $\begin{bmatrix} a \\ b \\ c \end{bmatrix}$ with $a + 2b + 3c = 0$

23. See the answer to Exercise 27 of Section 3.3.

SECTION 3.6

1. $-2i + 7j - 4k$ **3.** $4i + 40j - 17k$ **5.** $23i + 9j - k$

7. $41i + 11j - k$ **9.** $-10i + 10j + 10k$ **11.** 0 **13.** $5\sqrt{3} + 3$

15. $-2j$ **17.** $\sqrt{2}$ **19.** $(1 + c)i + (2 - 2c)j + ck$

21. If $v = ai + bj + ck$, then $u\cdot v = 0$ if and only if $a + 2b + 3c = 0$. In particular, $v = i + j - k$ is orthogonal to u. Then $w = u \times v = -5i + 4j - k$ is orthogonal to both u and v.

SECTION 3.7

1. $-(x - 3) + 3(y - 4) + 2(z - 1) = 0, -x + 3y + 2z - 11 = 0$
3. $5(x + 2) + 3(y + 3) + 2z = 0, 5x + 3y + 2z + 19 = 0$
5. $51(x - 1) + 21y + 19(z - 5) = 0,$ **7.** $6(x + \frac{1}{3}) + 4y + 5z = 0$

9. $\dfrac{11}{\sqrt{14}}$ **11.** $\dfrac{4}{\sqrt{38}}$ **13.** $\dfrac{20}{3}$ **15.** $3x - 7y + 2z + 25 = 0$

17. $ax + by + cz + d = 0$, where $2a + 3b + c = 0$ and d is arbitrary.

SECTION 3.8

1. $x = 2 - 5t, y = 1 - t, z = 3 + t; \dfrac{x - 2}{-5} = \dfrac{y - 1}{-1} = \dfrac{z - 3}{1}$

3. $x = 5 - 2t, y = 3 + 3t, z = -1 + 9t; \dfrac{x - 5}{-2} = \dfrac{y - 3}{3} = \dfrac{z + 1}{9}$

5. $x = -2 + 6t, y = -1 + 8t, z = -4 + t; \dfrac{x + 2}{6} = \dfrac{y + 1}{8} = \dfrac{z + 4}{1}$

7. $x = 2 + 3t, y = 3 - 2t, z = 1 + t; \dfrac{x - 2}{3} = \dfrac{y - 3}{-2} = \dfrac{z - 1}{1}$

9. $x = -17 - \dfrac{43}{2}t, y = t, z = 6 + 8t$

11. The lines do not intersect.

13. $\dfrac{x - 1}{1} = \dfrac{y - 2}{2} = \dfrac{z - 3}{3}$ **15.** $\dfrac{\sqrt{333}}{3}$

17. $x = 7 + at, y = 1 + bt, z = 5 + ct$, where $a + b + 2c = 0$.

SECTION 4.1

1. $\begin{bmatrix} 15 \\ -26 \\ 21 \\ -3 \\ -43 \end{bmatrix}$ **3.** $\begin{bmatrix} 8 \\ 16 \\ 17 \\ -11 \\ -33 \end{bmatrix}$ **5.** $\begin{bmatrix} -3 \\ 76 \\ -18 \\ 48 \\ 116 \end{bmatrix}$ **7.** $\begin{bmatrix} -1 \\ -10 \\ 3 \\ -13 \\ -25 \end{bmatrix}$ **9.** No

11. $(\mathbf{u} + \mathbf{v}) + (-\mathbf{u} - \mathbf{v}) = (\mathbf{u} + (-\mathbf{u})) + (\mathbf{v} + (-\mathbf{v})) = \mathbf{0} + \mathbf{0} = \mathbf{0}.$
Therefore $-(\mathbf{u} + \mathbf{v}) = -\mathbf{u} - \mathbf{v}.$

13. Let $\mathbf{u} = \begin{bmatrix} a_{11} \\ a_{21} \end{bmatrix}$, $\mathbf{v} = \begin{bmatrix} a_{12} \\ a_{22} \end{bmatrix}$, and $\mathbf{w} = \begin{bmatrix} a_{13} \\ a_{23} \end{bmatrix}$.

Then $\mathbf{0} = a\mathbf{u} + b\mathbf{v} + c\mathbf{w}$ if and only if $a_{11}a + a_{12}b + a_{13}c = 0$ and $a_{21}a + a_{22}b + a_{23}c = 0$. A homogeneous system of linear equations with more unknowns than equations has infinitely many solutions (Theorem 20 of Section 1.10). Hence, there are numbers $a, b,$ and c other than $a = 0, b = 0,$ and $c = 0$ such that $a\mathbf{u} + b\mathbf{v} + c\mathbf{w} = \mathbf{0}$.

15. $\begin{bmatrix} u_1 \\ u_2 \\ \vdots \\ u_n \end{bmatrix} + \begin{bmatrix} v_1 \\ v_2 \\ \vdots \\ v_n \end{bmatrix} = \begin{bmatrix} u_1 + v_1 \\ u_2 + v_2 \\ \vdots \\ u_n + v_n \end{bmatrix}$

17. $\left(\begin{bmatrix} u_1 \\ u_2 \\ \vdots \\ u_n \end{bmatrix} + \begin{bmatrix} v_1 \\ v_2 \\ \vdots \\ v_n \end{bmatrix} \right) + \begin{bmatrix} w_1 \\ w_2 \\ \vdots \\ w_n \end{bmatrix} = \begin{bmatrix} u_1 + v_1 \\ u_2 + v_2 \\ \vdots \\ u_n + v_n \end{bmatrix} + \begin{bmatrix} w_1 \\ w_2 \\ \vdots \\ w_n \end{bmatrix}$

$$= \begin{bmatrix} u_1 + v_1 + w_1 \\ u_2 + v_2 + w_2 \\ \vdots \\ u_n + v_n + w_n \end{bmatrix}$$

$$= \begin{bmatrix} u_1 \\ u_2 \\ \vdots \\ u_n \end{bmatrix} + \begin{bmatrix} v_1 + w_1 \\ v_2 + w_2 \\ \vdots \\ v_n + w_n \end{bmatrix}$$

$$= \begin{bmatrix} u_1 \\ u_2 \\ \vdots \\ u_n \end{bmatrix} + \left(\begin{bmatrix} v_1 \\ v_2 \\ \vdots \\ v_n \end{bmatrix} + \begin{bmatrix} w_1 \\ w_2 \\ \vdots \\ w_n \end{bmatrix} \right)$$

19. $-\begin{bmatrix} u_1 \\ u_2 \\ \vdots \\ u_n \end{bmatrix} = \begin{bmatrix} -u_1 \\ -u_2 \\ \vdots \\ -u_n \end{bmatrix}$

21. $(ab)\begin{bmatrix} u_1 \\ u_2 \\ \vdots \\ u_n \end{bmatrix} = \begin{bmatrix} abu_1 \\ abu_2 \\ \vdots \\ abu_n \end{bmatrix} = a\begin{bmatrix} bu_1 \\ bu_2 \\ \vdots \\ bu_n \end{bmatrix} = a\left(b\begin{bmatrix} u_1 \\ u_2 \\ \vdots \\ u_n \end{bmatrix} \right)$

23. $a\left(\begin{bmatrix} u_1 \\ u_2 \\ \vdots \\ u_n \end{bmatrix} + \begin{bmatrix} v_1 \\ v_2 \\ \vdots \\ v_n \end{bmatrix} \right) = a\begin{bmatrix} u_1 + v_1 \\ u_2 + v_2 \\ \vdots \\ u_n + v_n \end{bmatrix} = \begin{bmatrix} au_1 + av_1 \\ au_2 + av_2 \\ \vdots \\ au_n + av_n \end{bmatrix} = a\begin{bmatrix} u_1 \\ u_2 \\ \vdots \\ u_n \end{bmatrix} + a\begin{bmatrix} v_1 \\ v_2 \\ \vdots \\ v_n \end{bmatrix}$

SECTION 4.2

1. Yes **3.** Yes **5.** No **7.** Yes **9.** Yes **11.** Yes **13.** Yes **15.** No **17.** No
19. Suppose that $\mathbf{w}'$ is an element of V such that $\mathbf{u} + \mathbf{w}' = \mathbf{0}$. Then $\mathbf{w} = \mathbf{w} + \mathbf{0} = \mathbf{w} + (\mathbf{u} + \mathbf{w}') = (\mathbf{w} + \mathbf{u}) + \mathbf{w}' = (\mathbf{u} + \mathbf{w}) + \mathbf{w}' = \mathbf{0} + \mathbf{w}' = \mathbf{w}'$.
21. $c\mathbf{x} = c(\mathbf{x} + \mathbf{0}) = c\mathbf{x} + c\mathbf{0}$. By part 2 of Theorem 2 we have $c\mathbf{0} = \mathbf{0}$.
23. $\mathbf{0} = 0\mathbf{v} = (1 + (-1))\mathbf{v} = \mathbf{v} + (-1)\mathbf{v}$. By part 7 of Theorem 2 we have $(-1)\mathbf{v} = -\mathbf{v}$.

SECTION 4.3

1. Elements of V have the form $\begin{bmatrix} a \\ 2a \end{bmatrix}$. If $\begin{bmatrix} a \\ 2a \end{bmatrix}$ and $\begin{bmatrix} b \\ 2b \end{bmatrix}$ are any two elements of V, then

$\begin{bmatrix} a \\ 2a \end{bmatrix} + \begin{bmatrix} b \\ 2b \end{bmatrix} = \begin{bmatrix} a + b \\ 2(a + b) \end{bmatrix}$ and $c\begin{bmatrix} a \\ 2a \end{bmatrix} = \begin{bmatrix} ca \\ 2(ca) \end{bmatrix}$ so that V is closed under both addition and scalar multiplication. Hence V is a subspace of R^2. Suppose that W is a subspace of V that contains a nonzero element $\begin{bmatrix} a \\ 2a \end{bmatrix}$. If c is any scalar, then $c\begin{bmatrix} a \\ 2a \end{bmatrix} = (ca)\begin{bmatrix} 1 \\ 2 \end{bmatrix}$ is an element of W. Thus W contains all of the scalar multiples of $\begin{bmatrix} 1 \\ 2 \end{bmatrix}$. That is, $W = V$.

3. $m = n$ **5.** No **7.** Yes **9.** Yes **11.** No **13.** No **15.** Yes **17.** Yes

19. If $A\mathbf{x} \neq \mathbf{0}$, then $\mathbf{0} = 0(A\mathbf{x}) = A(0\mathbf{x})$. Thus the set is not closed under scalar multiplication. It is not closed under addition either.

21. Yes

23. V_k is closed under both addition and scalar multiplication. Hence it is a subspace of V. Evidently $\mathbf{v}_1 + k\mathbf{v}_2$ is an element of V_k. If it were also an element of $V_{k'}$, $k \neq k'$, then $\mathbf{v}_1 + k\mathbf{v}_2 = c(\mathbf{v}_1 + k'\mathbf{v}_2)$ for some number c. Hence $(1 - c)\mathbf{v}_1 = (ck' - k)\mathbf{v}_2$, from which it follows that $\mathbf{v}_1$ is a scalar multiple of $\mathbf{v}_2$ if $c \neq 1$. If $c = 1$, then $\mathbf{v}_2 = \mathbf{0}$, which is impossible. Thus regardless of the value of c we cannot have $\mathbf{v}_1 + k\mathbf{v}_2 = c(\mathbf{v}_1 + k'\mathbf{v}_2)$. Therefore $\mathbf{v}_1 + k\mathbf{v}_2$ is not an element of $V_{k'}$ if $k \neq k'$. The subspaces V_k and $V_{k'}$ are different if k and k' are different.

25. Let $\mathbf{u}$ and $\mathbf{v}$ be elements of $U \cap W$. Then $\mathbf{u}$ and $\mathbf{v}$ are elements of both U and W. Hence $\mathbf{u} + \mathbf{v}$ and $c\mathbf{u}$ are elements of both U and W for any scalar c. Therefore $\mathbf{u} + \mathbf{v}$ and $c\mathbf{u}$ are elements of $U \cap W$ so that $U \cap W$ is closed under both addition and scalar multiplication. $U \cap W$ is a subspace of V.

27. No, the scalars are different.

SECTION 4.4

1. $\begin{bmatrix} 1 \\ 2 \\ 3 \end{bmatrix} = \frac{1}{2}\begin{bmatrix} 2 \\ 4 \\ 6 \end{bmatrix}$ **3.** $\begin{bmatrix} 1 \\ 2 \end{bmatrix} = \begin{bmatrix} 1 \\ 0 \end{bmatrix} + 2\begin{bmatrix} 0 \\ 1 \end{bmatrix}$ **5.** $x^2 + x = \frac{1}{2}(2x) + \frac{1}{3}(3x^2)$

7. $\begin{bmatrix} 0 & 1 \\ 2 & 0 \end{bmatrix} = \frac{1}{3}\begin{bmatrix} 2 & 1 \\ 0 & 2 \end{bmatrix} + \frac{2}{3}\begin{bmatrix} -1 & 1 \\ 3 & -1 \end{bmatrix}$ **9.** $\left\{ \begin{bmatrix} a \\ b \\ c \end{bmatrix} : c - a + b = 0 \right\}$

11. $\left\{ \begin{bmatrix} a \\ b \\ 0 \end{bmatrix} : a, b \text{ arbitrary} \right\}$ **13.** $\left\{ \begin{bmatrix} a \\ b \\ c \\ d \end{bmatrix} : c + a - 2b = 0, \ d + 2a - 3b = 0 \right\}$

15. $\{ax^2 + bx + 2a : a, b \text{ arbitrary}\}$ **17.** P_2 **19.** P_2

21. Let $a_1\mathbf{v}_1 + a_2\mathbf{v}_2$ and $b_1\mathbf{v}_1 + b_2\mathbf{v}_2$ by any two linear combinations of $\mathbf{v}_1$ and $\mathbf{v}_2$. Then $(a_1\mathbf{v}_1 + a_2\mathbf{v}_2) + (b_1\mathbf{v}_1 + b_2\mathbf{v}_2) = (a_1 + b_1)\mathbf{v}_1 + (a_2 + b_2)\mathbf{v}_2$ and $c(a_1\mathbf{v}_1 + a_2\mathbf{v}_2) = (ca_1)\mathbf{v}_1 + (ca_2)\mathbf{v}_2$. Thus the set of all linear combinations of $\mathbf{v}_1$ and $\mathbf{v}_2$ is closed under both addition and scalar multiplication. Therefore this set is a subspace of V.

23. $\left\{ \begin{bmatrix} -2 \\ 1 \\ 0 \end{bmatrix}, \begin{bmatrix} -3 \\ 0 \\ 1 \end{bmatrix} \right\}$

25. Let $\mathbf{e}_i$ denote the n-vector with 1 in the ith component and zeros elsewhere. By Theorem 4 we have that $A\mathbf{e}_i$ and $B\mathbf{e}_i$ are the ith columns of A and B, respectively. Since $A\mathbf{e}_i = B\mathbf{e}_i$ we conclude that the ith columns of A and B are identical for $i = 1, 2, \ldots, n$. Therefore $A = B$.

SECTION 4.5

1. No   **3.** Yes   **5.** Yes   **7.** No   **9.** No

11. Let $\mathbf{u}$ be any element of V. Since $\{\mathbf{v}_1, \mathbf{v}_2, \cdots \mathbf{v}_n\}$ spans V, there are scalars $c_1, c_2, \ldots, c_n$ such that $\mathbf{u} = c_1\mathbf{v}_1 + c_2\mathbf{v}_2 + \cdots + c_n\mathbf{v}_n$. Then $\mathbf{u} = c_1\mathbf{v}_1 + c_2\mathbf{v}_2 + \cdots + c_n\mathbf{v}_n + 0\mathbf{v}$ so $\{\mathbf{v}_1, \mathbf{v}_2, \ldots, \mathbf{v}_n, \mathbf{v}\}$ spans V.

13.

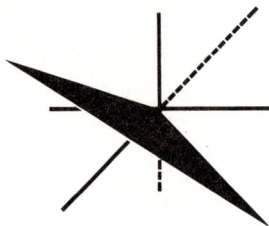

A plane passing through the origin.

15.

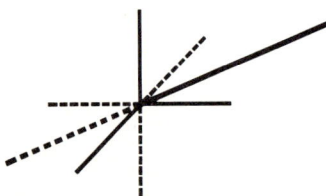

A line passing through the origin.

17. $\left\{ \begin{bmatrix} -2 \\ 1 \\ 0 \\ 0 \end{bmatrix}, \begin{bmatrix} -1 \\ 0 \\ 1 \\ 0 \end{bmatrix} \right\}$ **19.** $\left\{ \begin{bmatrix} 0 \\ 1 \\ 1 \\ 0 \end{bmatrix}, \begin{bmatrix} -1 \\ -2 \\ 0 \\ 3 \end{bmatrix} \right\}$

21. By Theorem 7 the set S spans R^n if and only if $A\mathbf{x} = \mathbf{b}$ is consistent for every n-vector $\mathbf{b}$. By definition, A is nonsingular if and only if $A\mathbf{x} = \mathbf{b}$ is consistent for every n-vector $\mathbf{b}$.

23. Yes

SECTION 4.6

1. No   **3.** No   **5.** No   **7.** Yes   **9.** No   **11.** Yes   **13.** $c \neq 0$

15. Since A is nonsingular the equation $A\mathbf{x} = \mathbf{0}$ has $\mathbf{0}$ as its only solution. Hence there are scalars $c_1, c_2, \cdots c_k$ such that $c_1\mathbf{v}_1 + c_2\mathbf{v}_2 + \cdots + c_k\mathbf{v}_k = \mathbf{0}$ if and only if $\mathbf{0} = A(c_1\mathbf{v}_1 + c_2\mathbf{v}_2 + \cdots + c_k\mathbf{v}_k)$. However $A(c_1\mathbf{v}_1 + c_2\mathbf{v}_2 + \cdots + c_k\mathbf{v}_k) = c_1(A\mathbf{v}_1) + c_2(A\mathbf{v}_2) + \cdots + c_k(A\mathbf{v}_k)$. Hence $\{\mathbf{v}_1, \mathbf{v}_2, \ldots, \mathbf{v}_k\}$ is linearly independent if and only if $\{A\mathbf{v}_1, A\mathbf{v}_2, \ldots, A\mathbf{v}_k\}$ is linearly independent.

17. Let $c_1, c_2, \ldots, c_m$ be scalars such that $c_1\mathbf{v}_1 + c_2\mathbf{v}_2 + \cdots + c_m\mathbf{v}_m = \mathbf{0}$. Then $c_1\mathbf{v}_1 + c_2\mathbf{v}_2 + \cdots + c_m\mathbf{v}_m + 0\mathbf{v}_{m+1} + \cdots + 0\mathbf{v}_n = \mathbf{0}$. Since $\mathbf{v}_1, \mathbf{v}_2, \ldots, \mathbf{v}_n$ is linearly independent, each coefficient in this last identity must be zero. Therefore $c_1 = 0, c_2 = 0, \ldots, c_n = 0$. The set $\{\mathbf{v}_1, \mathbf{v}_2, \ldots, \mathbf{v}_m\}$ is linearly independent.

19. Let p_1, p_2, p_3, p_4 be any four elements of P_2. Suppose that c_1, c_2, c_3, c_4 are scalars such that $c_1 p_1(x) + c_2 p_2(x) + c_3 p_3(x) + c_4 p_4(x) = 0$ for all x. Rearranging the terms on the left side, we obtain a polynomial each of whose coefficients must be 0. Thus we have three linear equations (one for each coefficient) in four unknowns (c_1, c_2, c_3, c_4). This system of equations has infinitely many solutions. Therefore any four elements of P_2 form a linearly dependent set.

21. If S is linearly dependent, then there are scalars, not both zero, such that $c_1\mathbf{v}_1 + c_2\mathbf{v}_2 = \mathbf{0}$. If $c_1 \neq 0$, then $\mathbf{v}_1 = -(c_1^{-1}c_2)\mathbf{v}_2$. If $c_2 \neq 0$, then $\mathbf{v}_2 = -(c_2^{-1}c_1)\mathbf{v}_1$.
23. No. The scalars are different.

SECTION 4.7

1. $\left\{ \begin{bmatrix} 2 \\ 1 \\ 3 \\ 0 \end{bmatrix}, \begin{bmatrix} 1 \\ 0 \\ 1 \\ 1 \end{bmatrix} \right\}$ **3.** $\{x + 1, x - 1, x^3 + x\}$

5. Yes   **7.** Yes   **9.** Yes   **11.** No   **13.** Yes

15. $\left\{ \begin{bmatrix} -1 \\ 1 \\ -1 \end{bmatrix} \right\}$ **17.** $\left\{ \begin{bmatrix} -1 \\ 1 \\ -1 \\ 0 \end{bmatrix}, \begin{bmatrix} 0 \\ 0 \\ -1 \\ 1 \end{bmatrix} \right\}$ **19.** $\{x - x^2, 1\}$

21. $\left\{ \begin{bmatrix} -1 & 1 \\ 0 & 0 \end{bmatrix}, \begin{bmatrix} -1 & 0 \\ 1 & 0 \end{bmatrix}, \begin{bmatrix} -1 & 0 \\ 0 & 1 \end{bmatrix} \right\}$ **23.** $\left\{ \begin{bmatrix} 1 \\ -2 \\ 1 \end{bmatrix} \right\}$ **25.** $\left\{ \begin{bmatrix} 1 \\ 3 \\ -2 \\ 0 \end{bmatrix}, \begin{bmatrix} 1 \\ 5 \\ 0 \\ -2 \end{bmatrix} \right\}$

27. $\left\{ \begin{bmatrix} -1 \\ 1 \\ 0 \\ 0 \\ 0 \end{bmatrix}, \begin{bmatrix} 0 \\ 0 \\ -1 \\ 1 \\ 0 \end{bmatrix}, \begin{bmatrix} -1 \\ 0 \\ 0 \\ 0 \\ 1 \end{bmatrix} \right\}$

SECTION 4.8

1. $\begin{bmatrix} 1 \\ 0 \\ 2 \end{bmatrix}$ **3.** $\begin{bmatrix} 0 \\ 1 \\ 0 \end{bmatrix}$ **5.** $\begin{bmatrix} 5 \\ 2 \\ 3 \end{bmatrix}$ **7.** $\begin{bmatrix} 3.25 \\ 1.25 \\ .25 \end{bmatrix}$ **9.** $\begin{bmatrix} -1 \\ 4 \\ -6 \end{bmatrix}$

11. Let $\mathbf{u} = a_1\mathbf{u}_1 + a_2\mathbf{u}_2 + \cdots + a_n\mathbf{u}_n$ and $\mathbf{v} = b_1\mathbf{u}_1 + b_2\mathbf{u}_2 + \cdots + b_n\mathbf{u}_n$. Then $\mathbf{u} + \mathbf{v} = (a_1 + b_1)\mathbf{u}_1 + (a_2 + b_2)\mathbf{u}_2 + \cdots + (a_n + b_n)\mathbf{u}_n$ and $c\mathbf{u} = (ca_1)\mathbf{u}_1 + (ca_2)\mathbf{u}_2 + \cdots + (ca_n)\mathbf{u}_n$ so that

$$[\mathbf{u} + \mathbf{v}]_B = \begin{bmatrix} a_1 + b_1 \\ a_2 + b_2 \\ \vdots \\ a_n + b_n \end{bmatrix} = \begin{bmatrix} a_1 \\ a_2 \\ \vdots \\ a_n \end{bmatrix} + \begin{bmatrix} b_1 \\ b_2 \\ \vdots \\ b_n \end{bmatrix} = [\mathbf{u}]_B + [\mathbf{v}]_B$$

$$[c\mathbf{u}]_B = \begin{bmatrix} ca_1 \\ ca_2 \\ \vdots \\ ca_n \end{bmatrix} = c\begin{bmatrix} a_1 \\ a_2 \\ \vdots \\ a_n \end{bmatrix} = c[\mathbf{u}]_B$$

SECTION 4.9

1. 2 **3.** 3 **5.** 2 **7.** 2

9. Evidently $U \cap W \subset U$ and $U \cap W \subset W$. By Theorem 16 we have $\dim(U \cap W) \leq \dim U$ and $\dim(U \cap W) \leq \dim/W$.

11. mn **13.** In Section 4.7 we found that every spanning set contains a basis.

15. If S is linearly independent, then by Theorem 18 it is a subset of a basis for V. This is impossible because every basis for V contains n elements. Therefore S must be linearly dependent.

17. $\left\{ \begin{bmatrix} 1 \\ 0 \\ 0 \end{bmatrix}, \begin{bmatrix} i \\ 0 \\ 0 \end{bmatrix}, \begin{bmatrix} 0 \\ 1 \\ 0 \end{bmatrix}, \begin{bmatrix} 0 \\ i \\ 0 \end{bmatrix}, \begin{bmatrix} 0 \\ 0 \\ 1 \end{bmatrix}, \begin{bmatrix} 0 \\ 0 \\ i \end{bmatrix} \right\}; \ 6$

19. Let $\mathbf{e}_j$ be the n-vector that has a 1 as its jth component and zeros elsewhere. A basis for D^n is $\{\mathbf{e}_1, i\mathbf{e}_1, \mathbf{e}_2, i\mathbf{e}_2, \ldots, \mathbf{e}_n, i\mathbf{e}_n\}$. The dimension of D^n is $2n$.

SECTION 4.10

1. $\begin{bmatrix} \frac{1}{3} & 1 \\ \frac{1}{3} & -1 \end{bmatrix}, \begin{bmatrix} \frac{3}{2} & \frac{3}{2} \\ \frac{1}{2} & -\frac{3}{2} \end{bmatrix}$ **3.** $\begin{bmatrix} \frac{1}{3} & 1 \\ \frac{1}{3} & -1 \end{bmatrix}, \begin{bmatrix} \frac{3}{2} & \frac{3}{2} \\ \frac{1}{2} & -\frac{3}{2} \end{bmatrix}$

5. $\begin{bmatrix} 1 & 0 & 1 \\ -1 & 1 & -1 \\ 1 & 0 & 0 \end{bmatrix}, \begin{bmatrix} 0 & 0 & 1 \\ 1 & 1 & 0 \\ 1 & 0 & -1 \end{bmatrix}$ **7.** $\begin{bmatrix} 2 & 1 & 2 \\ 1 & 0 & 2 \\ 0 & 0 & -1 \end{bmatrix}, \begin{bmatrix} 0 & 1 & 2 \\ 1 & -2 & -2 \\ 0 & 0 & -1 \end{bmatrix}$

11. $(_C P_B)(_D P_B)^{-1}$

SECTION 4.11

1. Rows: $[1 \quad 2 \quad 3 \quad 4], [0 \ -1 \quad 0 \quad 2], [2 \quad 1 \quad 2 \quad -3]$,

Columns: $\begin{bmatrix} 1 \\ 0 \\ 2 \end{bmatrix}, \begin{bmatrix} 2 \\ -1 \\ 1 \end{bmatrix}, \begin{bmatrix} 3 \\ 0 \\ 2 \end{bmatrix}, \begin{bmatrix} 4 \\ 2 \\ -3 \end{bmatrix}$

3. $\{[1 \ 2 \ 3], [0 \ 1 \ 2]\}, \left\{ \begin{bmatrix} 1 \\ 4 \\ 2 \end{bmatrix}, \begin{bmatrix} 2 \\ 5 \\ 1 \end{bmatrix} \right\}, 2$

5. $\{[1 \ -1 \quad 2 \ -3], [0 \quad 0 \quad 1 \ -1]\}, \left\{ \begin{bmatrix} 1 \\ 3 \\ 2 \\ 0 \\ -1 \end{bmatrix}, \begin{bmatrix} 2 \\ 1 \\ 0 \\ 4 \\ 2 \end{bmatrix} \right\}, 2$

7. The rows are linearly independent and form a basis for the row space.

$$\left\{ \begin{bmatrix} 1 \\ 2 \\ 1 \end{bmatrix}, \begin{bmatrix} 2 \\ 1 \\ -1 \end{bmatrix}, \begin{bmatrix} 4 \\ 3 \\ 0 \end{bmatrix} \right\}, 3$$

9. $\left\{ \begin{bmatrix} 1 \\ 1 \\ 0 \\ 1 \end{bmatrix}, \begin{bmatrix} 4 \\ 5 \\ 1 \\ 5 \end{bmatrix}, \begin{bmatrix} 1 \\ 2 \\ 1 \\ 1 \end{bmatrix} \right\}$ **11.** $\left\{ \begin{bmatrix} 1 \\ 2 \\ 1 \\ 1 \end{bmatrix}, \begin{bmatrix} 0 \\ 1 \\ 0 \\ -1 \end{bmatrix}, \begin{bmatrix} 0 \\ 0 \\ 3 \\ 1 \end{bmatrix} \right\}$

13. $\left\{ \begin{bmatrix} 1 \\ 1 \\ 1 \\ 1 \\ 1 \end{bmatrix}, \begin{bmatrix} 0 \\ 1 \\ 1 \\ 1 \\ 1 \end{bmatrix}, \begin{bmatrix} 0 \\ 1 \\ 0 \\ 0 \\ 0 \end{bmatrix}, \begin{bmatrix} 0 \\ 0 \\ 1 \\ 0 \\ 0 \end{bmatrix}, \begin{bmatrix} 0 \\ 0 \\ 0 \\ 1 \\ 0 \end{bmatrix} \right\}$ **15.** $\left\{ \begin{bmatrix} 0 \\ 1 \\ 1 \\ 1 \\ 0 \end{bmatrix}, \begin{bmatrix} 2 \\ 0 \\ 0 \\ 0 \\ 1 \end{bmatrix}, \begin{bmatrix} 1 \\ 1 \\ 1 \\ 1 \\ 1 \end{bmatrix}, \begin{bmatrix} 0 \\ 1 \\ 0 \\ 0 \\ 0 \end{bmatrix}, \begin{bmatrix} 0 \\ 0 \\ 1 \\ 1 \\ 0 \end{bmatrix} \right\}$

17. The row rank of A is the dimension of the row space of A. Therefore the row rank of A is at most 3. The column rank of A is the dimension of the column rank of A. Therefore the column rank of A is at most 4. Since the row rank and the column rank are equal, the dimension of the column space is at most 3. Therefore the columns of A are linearly dependent.

19. The smaller of m and n.

21. The column rank of A equals the column rank of $[A \mid \mathbf{b}]$ if and only if b is a linear combination of the columns of A. By Theorem 22, rank A = rank $[A \mid \mathbf{b}]$ if and only if $A\mathbf{x} = \mathbf{b}$ is consistent for every n-vector $\mathbf{b}$.

SECTION 4.12

1. Rank $= 1$, nullity $= 1$ **3.** Rank $= 2$, nullity $= 1$ **5.** Rank $= 2$, nullity $= 0$
7. Rank $= 4$, nullity $= 1$ **9.** 6, yes

SUPPLEMENTARY EXERCISES FOR CHAPTER 4

1. True. Theorem 6 of Section 4.5.

3. False. $S = \left\{ \begin{bmatrix} 1 \\ 0 \end{bmatrix}, \begin{bmatrix} 0 \\ 1 \end{bmatrix} \right\}$ is a linearly independent subset of R^2, but $\left\{ \begin{bmatrix} 1 \\ 0 \end{bmatrix}, \begin{bmatrix} 0 \\ 1 \end{bmatrix}, \begin{bmatrix} 2 \\ 3 \end{bmatrix} \right\}$ is not.

5. True. Exercise 17 of Section 4.6.

7. False. $S = \left\{ \begin{bmatrix} 1 \\ 0 \end{bmatrix}, \begin{bmatrix} 2 \\ 0 \end{bmatrix} \right\}$ is a linearly dependent subset of R^2, but $\left\{ \begin{bmatrix} 1 \\ 0 \end{bmatrix} \right\}$ is not.

9. False. $S = \left\{ \begin{bmatrix} 1 \\ 0 \end{bmatrix}, \begin{bmatrix} 0 \\ 1 \end{bmatrix}, \begin{bmatrix} 2 \\ 3 \end{bmatrix} \right\}$ spans R^2, but is not a basis for R^2.

11. True. Theorem 17 of Section 4.9.
13. True. dim V is the number of elements in a basis for V.
15. True. Theorem 19 of Section 4.9.
17. True. Theorem 19 of Section 4.9.
19. False. S must be linearly dependent (Theorem 19 of Section 4.9).
21. True. dim V is the number of elements in a basis for V.

23. True. Theorem 22 of Section 4.11.

25. False. Theorem 22 of Section 4.11.

27. True. Theorem 22 of Section 4.11.

29. False. It is a subspace of R^n, not of R^m (Example 3 of Section 4.3).

31. True. If z is a solution of $Ax = b$, then $2z$ is a solution of $Ax = 2b$, so the set is not closed under scalar multiplication. It is not closed under addition either.

33. True. Part 3 of Theorem 21 of Section 1.10.

35. False. The rank of A is the dimension of the column space of A, which is at most the number of columns of A.

37. True. By Theorem 13 of Section 4.7, A is nonsingular if and only if the columns of A are linearly independent. Therefore A is nonsingular if and only if the rank of A equals n. Since the rank of A is the dimension of the row space of A, we have that A is nonsingular if and only if the dimension of the row space is n.

39. True. By Exercise 37, A is nonsingular if and only if the dimension of the row space of A is n. Therefore A is nonsingular if and only if the rows of A are linearly independent.

SECTION 5.1

1. 8   3. 48   5. 15   7. Inner product   9. Not an inner product

11. Since $\langle p, q \rangle = p(-1)q(-1) + p(0)q(0) + p(1)q(1)$

$$= q(-1)p(-1) + q(0)p(0) + q(1)p(1)$$
$$= \langle q, p \rangle$$

the first axiom is satisfied. If r is any polynomial in P_2, then

$$\langle p + q, r \rangle = [p(-1) + q(-1)]r(-1) + [p(0) + q(0)]r(0) + [p(1) + q(1)]r(1)$$
$$= p(-1)r(-1) + p(0)r(0) + p(1)r(1) + q(-1)r(-1) + q(0)r(0) + q(1)r(1)$$
$$= \langle p, r \rangle + \langle q, r \rangle$$

The second axiom is satisfied. If c is any scalar, then $\langle cp, q \rangle = cp(-1)q(-1) + cp(0)q(0) + cp(1)q(1) = c\langle p, q \rangle$, so the third axiom is satisfied. Since $\langle p, p \rangle = [p(-1)]^2 + [p(0)]^2 + [p(1)]^2$ we conclude that $\langle p, p \rangle \geq 0$ and that $\langle p, p \rangle = 0$ if and only if $p(-1) = p(0) = p(1) = 0$. If $p(x) = a + bx + cx^2$, then $\langle p, p \rangle = 0$ if and only if

$$a - b + c = p(-1) = 0$$
$$a = p(0) \quad = 0$$
$$a + b + c = p(1) \quad = 0$$

It follows that $\langle p, p \rangle = 0$ if and only if $a = b = c = 0$, i.e., p is the zero polynomial. Thus the fourth and fifth axioms are satisfied.

13. No. $\langle p, p \rangle = a_0^2 - a_1^2 + a_2^2$ is not necessarily nonnegative. If $p(x) = 1 + 5x + x^2$, then $\langle p, p \rangle = -23$.

15. $\langle u, cv \rangle = \langle cv, u \rangle = c\langle v, u \rangle = c\langle u, v \rangle$

17. $\langle u, v \rangle = \sum_{i=1}^{n} u_i v_i = \sum_{i=1}^{n} v_i u_i = \langle v, u \rangle$

$$\langle u + v, w \rangle = \sum_{i=1}^{n} (u_i + v_i)w_i = \sum_{i=1}^{n} u_i w_i + \sum_{i=1}^{n} v_i w_i = \langle u, w \rangle + \langle v, w \rangle$$

$$\langle cu, v \rangle = \sum_{i=1}^{n} cu_i v_i = c\sum_{i=1}^{n} u_i v_i = c\langle u, v \rangle$$

$$\langle u, u \rangle = \sum_{i=1}^{n} u_i^2 \geq 0 \text{ and equals } 0 \text{ if and only if } u_i = 0 \text{ for every } i. \text{ Thus } \langle u, u \rangle = 0 \text{ if and only if } u = 0.$$

19. $\overline{\langle \mathbf{u}, \mathbf{v} \rangle} = \overline{\sum_{i=1}^{n} u_i \bar{v}_i} = \sum_{i=1}^{n} \bar{u}_i v_i = \sum_{i=1}^{n} v_i \bar{u}_i = \langle \mathbf{v}, \mathbf{u} \rangle$

$\langle \mathbf{u} + \mathbf{v}, \mathbf{w} \rangle = \sum_{i=1}^{n} (u_i + v_i) \bar{w}_i = \sum_{i=1}^{n} u_i \bar{w}_i + \sum_{i=1}^{n} v_i \bar{w}_i = \langle \mathbf{u}, \mathbf{w} \rangle + \langle \mathbf{v}, \mathbf{w} \rangle$

$\langle c\mathbf{u}, \mathbf{v} \rangle = \sum_{i=1}^{n} c u_i \bar{v}_i = c \sum_{i=1}^{n} u_i \bar{v}_i = c \langle \mathbf{u}, \mathbf{v} \rangle$

$\langle \mathbf{u}, \mathbf{u} \rangle = \sum_{i=1}^{n} |u_i|^2 \geq 0$

and equals zero if and only if $u_i = 0$ for every i. Thus $\langle \mathbf{u}, \mathbf{u} \rangle = 0$ if and only if $\mathbf{u} = 0$.

SECTION 5.2

1. $\sqrt{14}$ **3.** $\sqrt{30}$

5. $\langle t\mathbf{u} + \mathbf{v}, t\mathbf{u} + \mathbf{v} \rangle = \langle t\mathbf{u} + \mathbf{v}, t\mathbf{u} \rangle + \langle t\mathbf{u} + \mathbf{v}, \mathbf{v} \rangle$
$$= \langle t\mathbf{u}, t\mathbf{u} \rangle + \langle \mathbf{v}, t\mathbf{u} \rangle + \langle t\mathbf{u}, \mathbf{v} \rangle + \langle \mathbf{v}, \mathbf{v} \rangle$$
$$= t^2 \langle \mathbf{u}, \mathbf{u} \rangle + t \langle \mathbf{v}, \mathbf{u} \rangle + t \langle \mathbf{u}, \mathbf{v} \rangle + \langle \mathbf{v}, \mathbf{v} \rangle$$
$$= \langle \mathbf{u}, \mathbf{u} \rangle t^2 + 2 \langle \mathbf{u}, \mathbf{v} \rangle t + \langle \mathbf{v}, \mathbf{v} \rangle$$

7. $\|\mathbf{u}\| = \sqrt{\langle \mathbf{u}, \mathbf{u} \rangle} \geq 0$ since $\langle \mathbf{u}, \mathbf{u} \rangle \geq 0$

9. $\|c\mathbf{u}\| = \sqrt{\langle c\mathbf{u}, c\mathbf{u} \rangle} = \sqrt{c^2 \langle \mathbf{u}, \mathbf{u} \rangle} = |c| \sqrt{\langle \mathbf{u}, \mathbf{u} \rangle} = |c| \|\mathbf{u}\|$

11. $\|\mathbf{u} + \mathbf{v}\|^2 + \|\mathbf{u} - \mathbf{v}\|^2 = \langle \mathbf{u} + \mathbf{v}, \mathbf{u} + \mathbf{v} \rangle + \langle \mathbf{u} - \mathbf{v}, \mathbf{u} - \mathbf{v} \rangle$
$$= (\langle \mathbf{u}, \mathbf{u} \rangle + 2\langle \mathbf{u}, \mathbf{v} \rangle + \langle \mathbf{v}, \mathbf{v} \rangle) + (\langle \mathbf{u}, \mathbf{u} \rangle - 2\langle \mathbf{u}, \mathbf{v} \rangle + \langle \mathbf{v}, \mathbf{v} \rangle)$$
$$= 2(\langle \mathbf{u}, \mathbf{u} \rangle + \langle \mathbf{v}, \mathbf{v} \rangle)$$
$$= 2(\|\mathbf{u}\|^2 + \|\mathbf{v}\|^2)$$

13. Clearly $\|f\| \geq 0$ and $\|f\| = 0$ if and only if $f(x) = 0$ for every x in $[0, 1]$. Therefore
$$\|f\| = 0 \text{ if and only if } f \text{ is the zero function.}$$

$$\|cf\| = \max_{0 \leq x \leq 1} |cf(x)| = |c| \max_{0 \leq x \leq 1} |f(x)| = |c| \|f\|.$$

$$\|f + g\| = \max_{0 \leq x \leq 1} |f(x) + g(x)| \leq \max_{0 \leq x \leq 1} (|f(x)| + |g(x)|)$$

$$\leq \max_{0 \leq x \leq 1} |f(x)| + \max_{0 \leq x \leq 1} |g(x)| = \|f + g\|$$

SECTION 5.3

1. $\cos^{-1} \dfrac{1}{\sqrt{3}} \approx .955$ radians **3.** $\cos^{-1} \dfrac{-46}{\sqrt{55}\sqrt{157}} \approx 2.089$ radians **5.** 3

7. $\cos^{-1} \dfrac{19}{21} \approx .440$ radians **9.** $\cos^{-1} \dfrac{10}{\sqrt{30}\sqrt{39}} \approx 1.274$ radians

11. $\langle a\mathbf{v} + b\mathbf{w}, \mathbf{u} \rangle = \langle a\mathbf{v}, \mathbf{u} \rangle + \langle b\mathbf{w}, \mathbf{u} \rangle = a\langle \mathbf{v}, \mathbf{u} \rangle + b\langle \mathbf{w}, \mathbf{u} \rangle = a(0) + b(0) = 0$

13. (a) Integrating by parts twice, we have

$$\langle f_n, g_m \rangle = \int_0^\pi \sin nx \cos mx \, dx$$

$$= \frac{m \sin nx \sin mx + n \cos nx \cos mx}{n^2 + n^2} \Big|_0^\pi = 0$$

(b) $\langle f_n, f_m \rangle = \int_0^\pi \sin nx \sin mx \, dx$

$$= \left[\frac{\sin(n-m)x}{2(n-m)} - \frac{\sin(n+m)x}{2(n+m)} \right] \Big|_0^\pi = 0$$

(c) $\langle g_n, g_m \rangle = \int_0^\pi \cos nx \cos mx \, dx$

$$= \left[\frac{\sin(n-m)x}{2(n-m)} - \frac{\sin(n+m)x}{2(n+m)} \right] \Big|_0^\pi = 0$$

15. Let $\mathbf{u}$ and $\mathbf{w}$ be any elements of W, and let c be any scalar. Then for every i

$$\langle \mathbf{u} + \mathbf{w}, \mathbf{v}_i \rangle = \langle \mathbf{u}, \mathbf{v}_i \rangle + \langle \mathbf{w}, \mathbf{v}_i \rangle = 0 + 0 = 0$$

$$\langle c\mathbf{u}, \mathbf{v}_i \rangle = c\langle \mathbf{u}, \mathbf{v}_i \rangle = c(0) = 0$$

Thus W is closed under addition and scalar multiplication. Therefore W is a subspace of V.

SECTION 5.4

1. Orthogonal   **3.** Neither   **5.** Orthonormal   **7.** Orthogonal   **9.** Neither
11. Let $\mathbf{v}_1, \mathbf{v}_2,$ and $\mathbf{v}_3$ denote the vectors in Exercise 5. Then

$$\mathbf{v} = \frac{6}{\sqrt{3}} \mathbf{v}_2 - \frac{2}{\sqrt{2}} \mathbf{v}_3.$$

13. $\left\{ \dfrac{1}{\sqrt{2}} \begin{bmatrix} 1 \\ 0 \\ 1 \end{bmatrix}, \dfrac{1}{\sqrt{6}} \begin{bmatrix} -1 \\ 2 \\ 1 \end{bmatrix} \right\}$

15. $\left\{ \dfrac{1}{\sqrt{2}} \begin{bmatrix} 1 \\ 0 \\ 1 \\ 0 \end{bmatrix}, \dfrac{1}{\sqrt{6}} \begin{bmatrix} 1 \\ 2 \\ -1 \\ 0 \end{bmatrix}, \dfrac{1}{2\sqrt{3}} \begin{bmatrix} 1 \\ -1 \\ -1 \\ 3 \end{bmatrix} \right\}$ **17.** $\left\{ \dfrac{1}{\sqrt{5}} \begin{bmatrix} 1 \\ 0 \end{bmatrix}, \dfrac{1}{2} \begin{bmatrix} 0 \\ 1 \end{bmatrix} \right\}$

19. $\left\{ \dfrac{1}{\sqrt{2}}, \sqrt{\dfrac{3}{2}} x, \dfrac{3}{2} \sqrt{\dfrac{2}{5}} \left(x^2 - \dfrac{1}{3} \right) \right\}$

21. Let $a = \langle \mathbf{w}, \mathbf{v}_1 \rangle, b = \langle \mathbf{w}, \mathbf{v}_2 \rangle, c = \langle \mathbf{w}, \mathbf{v}_3 \rangle$. Then $\mathbf{w} = a\mathbf{v}_1 + b\mathbf{v}_2 + c\mathbf{v}_3$ so that

$$\|\mathbf{w}\|^2 = \langle \mathbf{w}, \mathbf{w} \rangle = \langle a\mathbf{v}_1 + b\mathbf{v}_2 + c\mathbf{v}_3, a\mathbf{v}_1 + b\mathbf{v}_2 + c\mathbf{v}_3 \rangle$$

$$= a^2 \langle \mathbf{v}_1, \mathbf{v}_1 \rangle + b^2 \langle \mathbf{v}_2, \mathbf{v}_2 \rangle + c^2 \langle \mathbf{v}_3, \mathbf{v}_3 \rangle$$

$$+ 2ab \langle \mathbf{v}_1, \mathbf{v}_2 \rangle + 2ac \langle \mathbf{v}_1, \mathbf{v}_3 \rangle + 2bc \langle \mathbf{v}_2, \mathbf{v}_3 \rangle$$

$$= a^2 + b^2 + c^2 = \langle \mathbf{w}, \mathbf{v}_1 \rangle^2 + \langle \mathbf{w}, \mathbf{v}_2 \rangle^2 + \langle \mathbf{w}, \mathbf{v}_3 \rangle^2$$

SECTION 5.5

1. $\begin{bmatrix} 2 \\ 2 \\ 2 \end{bmatrix}$ **3.** $\begin{bmatrix} \dfrac{8}{3} \\ \dfrac{14}{3} \\ \dfrac{2}{3} \end{bmatrix}$ **5.** $\begin{bmatrix} \dfrac{46}{45} \\ \dfrac{115}{45} \\ -\dfrac{92}{45} \end{bmatrix}$ **7.** $\begin{bmatrix} \dfrac{5}{3} \\ \dfrac{16}{3} \\ \dfrac{11}{3} \end{bmatrix}$

9. $x + x^2$ **11.** $\dfrac{1}{2}x + \dfrac{1}{2}x^2$ **13.** 1 **15.** $-\dfrac{1}{3} + \dfrac{1}{3}x + \dfrac{2}{3}x^2$

17. $3 + 2\sum\limits_{j=1}^{\infty} (-1)^{j+1}\dfrac{1}{j}\sin jx$

SECTION 5.6

1. $\begin{bmatrix} -2 \\ 1.2 \end{bmatrix}$ **3.** $\begin{bmatrix} .2 \\ 0 \end{bmatrix}$ **5.** $\begin{bmatrix} -\dfrac{208}{641} \\ -\dfrac{38}{641} \\ \dfrac{55}{641} \end{bmatrix}$ **7.** $107.18e^{.01138x}$

9. Suppose that $\mathbf{y}$ is the solution of $A\mathbf{x} = \mathbf{b}$ so that $A\mathbf{y} = \mathbf{b}$. Then $A^t A\mathbf{y} = A^t(A\mathbf{y}) = A^t\mathbf{b}$ so that $\mathbf{y}$ is a least squares solution.

SECTION 6.1

1. Linear transformation   **3.** Not a linear transformation
5. Linear transformation   **7.** Not a linear transformation
9. Linear transformation   **11.** Not a linear transformation
13. Linear transformation   **15.** Linear transformation

17. $T\left(\begin{bmatrix} a \\ b \end{bmatrix} + \begin{bmatrix} c \\ d \end{bmatrix}\right) = T\left(\begin{bmatrix} a+c \\ b+d \end{bmatrix}\right) = \begin{bmatrix} a+c \\ -b-d \end{bmatrix} = \begin{bmatrix} a \\ -b \end{bmatrix} + \begin{bmatrix} c \\ -d \end{bmatrix}$

$= T\left(\begin{bmatrix} a \\ b \end{bmatrix}\right) + T\left(\begin{bmatrix} c \\ d \end{bmatrix}\right)$

$T\left(c\begin{bmatrix} a \\ b \end{bmatrix}\right) = T\left(\begin{bmatrix} ca \\ cb \end{bmatrix}\right) = \begin{bmatrix} ca \\ -cb \end{bmatrix} = c\begin{bmatrix} a \\ -b \end{bmatrix} = cT\left(\begin{bmatrix} a \\ b \end{bmatrix}\right)$

T gives a reflection through the horizontal axis.

SECTION 6.2

There are no exercises in this section.

SECTION 6.3

1. (a) $\begin{bmatrix} -10 \\ 10 \end{bmatrix}$ **(b)** $\begin{bmatrix} x - 2y \\ 3x + 4y \end{bmatrix}$ **(c)** $\begin{bmatrix} 1 & -2 \\ 3 & 4 \end{bmatrix}$

3. (a) $\begin{bmatrix} 1 \\ \dfrac{57}{2} \\ 20 \end{bmatrix}$ **(c)** $\begin{bmatrix} x - y \\ \dfrac{9}{2}x - \dfrac{1}{2}y \\ 2x + y \end{bmatrix}$ **(d)** $\begin{bmatrix} 1 & -1 \\ \dfrac{9}{2} & -\dfrac{1}{2} \\ 2 & 1 \end{bmatrix}$

5. (a) $3x^2 + 25x - 36$ **(b)** $bx^2 + (2a - 3c)x + (4a - 3b + 5c)$

7. Let $\mathbf{w}_1$ and $\mathbf{w}_2$ be any elements of W. Then there are $\mathbf{v}_1$ and $\mathbf{v}_2$ in V such that $\mathbf{w}_1 = T(\mathbf{v}_1)$ and $\mathbf{w}_2 = T(\mathbf{v}_2)$ so that $\mathbf{w}_1 + \mathbf{w}_2 = T(\mathbf{v}_1) + T(\mathbf{v}_2) = T(\mathbf{v}_1 + \mathbf{v}_2)$. Therefore the set is closed under addition. Moreover, $c\mathbf{w}_1 = cT(\mathbf{v}_1) = T(c\mathbf{v}_1)$ so the set is also closed under scalar multiplication. The set is a subspace of W.

9. Let $c_1, c_2, \ldots, c_k$ be scalars such that $c_1\mathbf{v}_1 + c_2\mathbf{v}_2 + \cdots + c_k\mathbf{v}_k = \mathbf{0}$. Then $\mathbf{0} = T(\mathbf{0}) = T(c_1\mathbf{v}_1 + c_2\mathbf{v}_2 + \cdots + c_k\mathbf{v}_k) = c_1T(\mathbf{v}_1) + c_2T(\mathbf{v}_2) + \cdots + c_kT(\mathbf{v}_k)$.

11. Let $\mathbf{v}$ be any element of V. Then there are scalars $c_1, c_2, \ldots, c_k$ such that $\mathbf{v} = c_1\mathbf{v}_1 + c_2\mathbf{v}_2 + \cdots + c_k\mathbf{v}_k$. Therefore $T(\mathbf{v}) = T(c_1\mathbf{v}_1 + c_2\mathbf{v}_2 + \cdots + c_k\mathbf{v}_k) = c_1T(\mathbf{v}_1) + c_2T(\mathbf{v}_2) + \cdots + c_kT(\mathbf{v}_k) = c_1\mathbf{0} + c_2\mathbf{0} + \cdots + c_k\mathbf{0} = \mathbf{0}$.

SECTION 6.4

1. $\begin{bmatrix} 3 & 2 \\ -5 & 4 \end{bmatrix}$ **3.** $\begin{bmatrix} 2 & 0 & 3 \\ 0 & 3 & 2 \\ 2 & 5 & 0 \end{bmatrix}$ **5.** $\begin{bmatrix} 1 & 7 \\ 3 & -2 \\ 4 & 5 \end{bmatrix}$

7. $\begin{bmatrix} 1 & -1 & 0 & 0 \\ 0 & 0 & 1 & -1 \\ 0 & 1 & 1 & 0 \end{bmatrix}$ **9.** $\begin{bmatrix} 9 & -\dfrac{5}{2} & -\dfrac{3}{2} \\ 2 & 2 & 2 \\ -1 & 2 & -1 \end{bmatrix}, \begin{bmatrix} -5 \\ 0 \end{bmatrix}$ **11.** $\begin{bmatrix} 0 & 0 & 1 \\ 0 & 1 & 0 \\ 1 & 0 & 0 \end{bmatrix}$

13. $\begin{bmatrix} 0 & 0 & 2 \\ 1 & 0 & 0 \\ 0 & 1 & 0 \end{bmatrix}$ **15.** $\begin{bmatrix} -1 & -1 & 1 \\ 1 & 1 & 0 \\ 1 & 2 & 0 \end{bmatrix}$ **17.** $\begin{bmatrix} 0 & -1 & 1 \\ 0 & 1 & 1 \\ 1 & 2 & -1 \end{bmatrix}$ **19.** $\begin{bmatrix} \dfrac{1}{2} & 3 & 2 \\ \dfrac{1}{2} & 1 & 1 \end{bmatrix}$

21. $\begin{bmatrix} 0 & 1 & 0 \\ 0 & 0 & 2 \\ 0 & 0 & 0 \\ 0 & 0 & 0 \end{bmatrix}$ **23.** $\begin{bmatrix} 3 & 3 & 0 \\ 1 & -1 & 0 \\ -1 & 1 & 2 \\ 1 & -1 & -2 \end{bmatrix}$ **25.** $\begin{bmatrix} 5 & -1 \\ -2 & 0 \end{bmatrix}$ **27.** $\begin{bmatrix} -2 & -4 \\ 3 & 7 \end{bmatrix}$

SECTION 6.5

1. $(T \circ R_{\pi/4})(\mathbf{x}) = \begin{bmatrix} \dfrac{\sqrt{2}}{2} & \dfrac{\sqrt{2}}{2} \\ \dfrac{\sqrt{2}}{2} & -\dfrac{\sqrt{2}}{2} \end{bmatrix}\mathbf{x}, \quad (R_{\pi/4} \circ T)(\mathbf{x}) = \begin{bmatrix} -\dfrac{\sqrt{2}}{2} & \dfrac{\sqrt{2}}{2} \\ \dfrac{\sqrt{2}}{2} & \dfrac{\sqrt{2}}{2} \end{bmatrix}\mathbf{x}$

3. Let $\{\mathbf{u}_1, \mathbf{u}_2, \dots \mathbf{u}_n\}$ be an orthonormal basis for U. Then

$$T^2(\mathbf{u}) = T(T(\mathbf{u})) = \text{proj}_U T(\mathbf{u})$$

$$= \sum_{i=1}^{n} \langle T(\mathbf{u}), \mathbf{u}_i \rangle \mathbf{u}_i$$

$$= \sum_{i=1}^{n} \langle \sum_{j=1}^{n} \langle \mathbf{u}, \mathbf{u}_j \rangle \mathbf{u}_j, \mathbf{u}_i \rangle \mathbf{u}_i$$

$$= \sum_{i=1}^{n} \sum_{j=1}^{n} \langle \langle \mathbf{u}, \mathbf{u}_j \rangle \mathbf{u}_j, \mathbf{u}_i \rangle \mathbf{u}_i$$

$$= \sum_{i=n}^{n} \sum_{j=1}^{n} \langle \mathbf{u}, \mathbf{u}_j \rangle \langle \mathbf{u}_j, \mathbf{u}_i \rangle \mathbf{u}_i$$

$$= \sum_{i=1}^{n} \langle \mathbf{u}, \mathbf{u}_i \rangle \mathbf{u}_i$$

$$= \text{proj}_U \mathbf{u} = T(\mathbf{u})$$

5. $\begin{bmatrix} \dfrac{\sqrt{3}}{2} & \dfrac{1}{2} \\ \dfrac{1}{2} & -\dfrac{\sqrt{3}}{2} \end{bmatrix}$ **7.** $\begin{bmatrix} 0 & 1 \\ 1 & 0 \end{bmatrix}$ **9.** $\begin{bmatrix} \dfrac{1}{2} & \dfrac{\sqrt{3}}{2} \\ -\dfrac{\sqrt{3}}{2} & \dfrac{1}{2} \end{bmatrix}$

SECTION 6.6

1. $\begin{bmatrix} 3 & 0 \\ 0 & -1 \end{bmatrix}$ **3.** $\begin{bmatrix} 1 & 1 & 0 \\ -1 & -1 & 0 \\ 0 & 0 & -2 \end{bmatrix}$ **5.** $\begin{bmatrix} 1 & 3 & -5 \\ \dfrac{1}{2} & 1 & 3 \\ -\dfrac{1}{2} & 1 & 1 \end{bmatrix}$ **7.** $\begin{bmatrix} 8 & 13 & 8 \\ -3 & -4 & -4 \\ -1 & -2 & 1 \end{bmatrix}$

9. If $A = P^{-1}BP$, then $PAP^{-1} = B$ so that $B = Q^{-1}AQ$ where $Q = P^{-1}$.

11. If $A = P^{-1}BP$, then by Theorem 10 of Section 2.5 and Exercise 11 of Section 2.5 we have
$\det A = \det(P^{-1}BP) = (\det A^{-1})(\det B)(\det P) = \det B$.

SECTION 6.7

1. $\left\{ \begin{bmatrix} 24 \\ -12 \\ -4 \\ 1 \end{bmatrix} \right\}$, $\left\{ \begin{bmatrix} 1 \\ 0 \\ 0 \end{bmatrix}, \begin{bmatrix} 0 \\ 1 \\ 0 \end{bmatrix}, \begin{bmatrix} 0 \\ 0 \\ 1 \end{bmatrix} \right\}$ **3.** $\left\{ \begin{bmatrix} -1 \\ 0 \\ 1 \\ 1 \end{bmatrix}, \begin{bmatrix} 0 \\ 1 \\ 0 \\ 0 \end{bmatrix} \right\}$, $\left\{ \begin{bmatrix} 2 \\ 0 \\ 1 \end{bmatrix}, \begin{bmatrix} -1 \\ 1 \\ 0 \end{bmatrix} \right\}$

5. $\left\{ \begin{bmatrix} 1 \\ -2 \\ 1 \\ 0 \end{bmatrix}, \begin{bmatrix} 1 \\ -1 \\ 0 \\ 1 \end{bmatrix} \right\}$, $\left\{ \begin{bmatrix} 1 \\ 0 \\ 0 \end{bmatrix}, \begin{bmatrix} 0 \\ 0 \\ 1 \end{bmatrix} \right\}$ **7.** $\{1\}, \{1, x\}$ **9.** $\{1, x\}, \{x^2\}$

SECTION 7.1

1. 1; all nonzero scalar multiples of $\begin{bmatrix} 1 \\ 0 \end{bmatrix}$.

　3; all nonzero scalar multiples of $\begin{bmatrix} 0 \\ 1 \end{bmatrix}$.

3. 1; all nonzero scalar multiples of $\begin{bmatrix} 3 \\ -1 \end{bmatrix}$.

　5; all nonzero scalar multiples of $\begin{bmatrix} 1 \\ 1 \end{bmatrix}$.

5. 3; all nonzero scalar multiples of $\begin{bmatrix} 1 \\ -1 \end{bmatrix}$.

7. 4; all nonzero linear combinations of $\begin{bmatrix} 1 \\ 0 \\ -1 \end{bmatrix}$ and $\begin{bmatrix} 0 \\ 1 \\ 0 \end{bmatrix}$.

9. 1; all nonzero linear combinations of $\begin{bmatrix} -1 \\ 2 \\ 1 \end{bmatrix}$ and $\begin{bmatrix} 1 \\ 0 \\ -2 \end{bmatrix}$.

　2; all nonzero scalar multiples of $\begin{bmatrix} -1 \\ 1 \\ 1 \end{bmatrix}$.

11. 2; all nonzero scalar multiples of $\begin{bmatrix} 1 \\ 1 \\ 0 \end{bmatrix}$.

13. 1; all nonzero linear combinations of $\begin{bmatrix} 1 \\ -1 \\ 0 \end{bmatrix}$ and $\begin{bmatrix} 1 \\ 0 \\ -1 \end{bmatrix}$.

　4; all nonzero scalar multiples of $\begin{bmatrix} 1 \\ 1 \\ 1 \end{bmatrix}$.

15. 1; all nonzero scalar multiples of $\begin{bmatrix} 3 \\ -1 \\ 0 \\ 0 \end{bmatrix}$.

　2; all nonzero scalar multiples of $\begin{bmatrix} 0 \\ 0 \\ 1 \\ 1 \end{bmatrix}$.

　-2; all nonzero scalar multiples of $\begin{bmatrix} 0 \\ 0 \\ 1 \\ -1 \end{bmatrix}$.

　5; all nonzero scalar multiples of $\begin{bmatrix} 1 \\ 1 \\ 0 \\ 0 \end{bmatrix}$.

17. Let $A = [a_{ij}]$ be an $n \times n$ diagonal matrix. Then $\lambda I - A$ is an $n \times n$ diagonal matrix with $\lambda - a_{11}, \lambda - a_{22}, \dots, \lambda - a_{nn}$ on the diagonal. Since the determinant of a diagonal matrix is the product of the diagonal components, $\det(\lambda I - A) = (\lambda - a_{11})(\lambda - a_{22})\dots(\lambda - a_{nn})$. Therefore $a_{11}, a_{22}, \dots, a_{nn}$ are the eigenvalues of A.

19. 1; all nonzero scalar multiples of $\begin{bmatrix} -3 \\ 1 \end{bmatrix}$.

     8; all nonzero scalar multiples of $\begin{bmatrix} 1 \\ 2 \end{bmatrix}$.

21. 0; all nonzero scalar multiples of $\begin{bmatrix} 1 \\ -1 \end{bmatrix}$.

     2; all nonzero scalar multiples of $\begin{bmatrix} 1 \\ 1 \end{bmatrix}$.

23. 1; all nonzero linear combinations of x and $1 + x^2$.

25. $2 + 2i$; all nonzero scalar multiples of $\begin{bmatrix} 1 \\ i \end{bmatrix}$.

     $2 - 2i$; all nonzero scalar multiples of $\begin{bmatrix} 1 \\ -i \end{bmatrix}$.

27. 2; all nonzero scalar multiples of $\begin{bmatrix} 0 \\ 1 \\ 0 \end{bmatrix}$.

     i; all nonzero scalar multiples of $\begin{bmatrix} 1 \\ 0 \\ i \end{bmatrix}$.

     $-i$; all nonzero scalar multiples of $\begin{bmatrix} 1 \\ 0 \\ -i \end{bmatrix}$.

SECTION 7.2

1. The eigenvalues are 1 and 4.

3. The eigenvalues are $\dfrac{1 \pm \sqrt{37}}{2}$.

5. The eigenvalues are 1 and $\dfrac{3 \pm \sqrt{37}}{2}$.

7. The eigenvalues are $-5, 9,$ and $\dfrac{5 \pm \sqrt{65}}{2}$.

9. Let λ be an eigenvalue of A and let $\mathbf{x}$ be an associated eigenvector. Since A is nonsingular $\lambda \neq 0$. Therefore $\mathbf{x} = \lambda^{-1} A \mathbf{x}$ and $A^{-1}\mathbf{x} = A^{-1}(\lambda^{-1} A \mathbf{x}) - \lambda^{-1} A^{-1} A \mathbf{x} = \lambda^{-1}\mathbf{x}$. Thus λ^{-1} is an eigenvalue of A^{-1}.

11. Let $\mathbf{x}$ be an eigenvector of A associated with λ. Then $A\mathbf{x} = \lambda\mathbf{x}$ so that $A^k\mathbf{x} = \lambda^k\mathbf{x}$ when $k = 1$. Suppose that $A^{k-1}\mathbf{x} = \lambda^{k-1}\mathbf{x}$. Then $A^k\mathbf{x} = A(A^{k-1}\mathbf{x}) = A(\lambda^{k-1}\mathbf{x}) = \lambda^{k-1}A\mathbf{x} = \lambda^{k-1}\lambda\mathbf{x} = \lambda^k\mathbf{x}$. By induction, $A^k\mathbf{x} = \lambda^k\mathbf{x}$ for every positive integer k. Therefore λ^k is an eigenvalue of A^k for every positive integer k.

13. $A\mathbf{x} = \lambda\mathbf{x}$ if and only if $(cA)\mathbf{x} = (c\lambda)\mathbf{x}$. Therefore $\mathbf{x}$ is an eigenvector of A associated with λ if and only if $\mathbf{x}$ is an eigenvector of cA associated with $c\lambda$.

15. No. For example, 1 is the only eigenvalue of I_2, so $2 = 1 + 1$ is not an eigenvalue of I_2.

17. $\det(\lambda I - A) = \det(\lambda P^{-1}IP - P^{-1}BP) = \det P^{-1}(\lambda I - B)P = \det P^{-1}\det(\lambda I - B)$
$\det P = \det(\lambda I - B)$. Hence A and B have the same characteristic polynomials and, therefore, the same eigenvalues.

SECTION 7.3

1. $\begin{bmatrix} 1 & 1 \\ 1 & -1 \end{bmatrix}, \begin{bmatrix} 1 & 0 \\ 0 & -1 \end{bmatrix}$ **3.** Not diagonalizable

5. $\begin{bmatrix} 3 & 1 & 0 \\ -8 & 0 & 1 \\ 1 & 1 & 0 \end{bmatrix}, \begin{bmatrix} 3 & 0 & 0 \\ 0 & 5 & 0 \\ 0 & 0 & 5 \end{bmatrix}$ **7.** Not diagonalizable

9. $\left\{ \begin{bmatrix} 2 \\ 1 \end{bmatrix}, \begin{bmatrix} 2 \\ -3 \end{bmatrix} \right\}$ **11.** $\{x^2, x + x^2, 2x + x^2\}$

13. The characteristic polynomial for A is $\lambda^2 - (a + d)\lambda + (ad - bc) = 0$, which has roots

$$\frac{(a + d) \pm \sqrt{(a + d)^2 - 4(ad - bc)}}{2} = \frac{(a + d) \pm \sqrt{(a - d)^2 + 4bc}}{2}.$$

Therefore if $(a - d)^2 + 4bc > 0$ the eigenvalues are unequal real numbers. Combining Theorems 6 and 8, the matrix A is diagonalizable.

SECTION 7.4

1. 13   **3.** 13   **5.** 0 is not in any disk.

7. The disks D_i' are the disks given by Gershgorin's Theorem for the matrix A^t. Since A and A^t have the same eigenvalues (part 1 of Theorem 4 of Section 7.2), every eigenvalue of A lies in at least one of the D_i.

9. 11

SECTION 7.5

1. $A_1 = \left\{ c\begin{bmatrix} 1 \\ -1 \end{bmatrix} : c, \text{a real number} \right\}$. The algebraic and geometric multiplicities of 1 are 1.

$A_{10} = \left\{ c\begin{bmatrix} 5 \\ 4 \end{bmatrix} : c, \text{a real number} \right\}$. The algebraic and geometric multiplicities of 10 are 1.

3. $A_{-1} = \left\{ c\begin{bmatrix} 1 \\ -1 \\ 0 \end{bmatrix} : c, \text{a real number} \right\}$. The algebraic and geometric multiplicities of -1 are 1.

$A_3 = \left\{ c\begin{bmatrix} 0 \\ 0 \\ 1 \end{bmatrix} : c, \text{a real number} \right\}$. The algebraic and geometric multiplicities of 3 are 1.

$A_4 = \left\{ c\begin{bmatrix} 3 \\ 2 \\ 0 \end{bmatrix} : c, \text{a real number} \right\}$. The algebraic and geometric multiplicities of 4 are 1.

5. $A_{-5} = \left\{ c \begin{bmatrix} 0 \\ 1 \\ 0 \end{bmatrix} : c, \text{ a real number} \right\}$. The algebraic and geometric multiplicities of -5 are 1.

$A_1 = \left\{ c \begin{bmatrix} 1 \\ 0 \\ -4 \end{bmatrix} : c, \text{ a real number} \right\}$. The algebraic and geometric multiplicities of 1 are 1.

$A_6 = \left\{ c \begin{bmatrix} 1 \\ 0 \\ 1 \end{bmatrix} : c, \text{ a real number} \right\}$. The algebraic and geometric multiplicities of 6 are 1.

7. $A_4 = \left\{ c_1 \begin{bmatrix} 1 \\ 1 \\ 2 \end{bmatrix} + c_2 \begin{bmatrix} 1 \\ 0 \\ 1 \end{bmatrix} : c_1 \text{ and } c_2, \text{ real numbers} \right\}$. The algebraic and geometric

multiplicities of 4 are 2.

$A_{-13} = \left\{ c \begin{bmatrix} 9 \\ 9 \\ 1 \end{bmatrix} : c, \text{ a real number} \right\}$. The algebraic and geometric multiplicities of -13

are 1.

9. $A_2 = \left\{ c_1 \begin{bmatrix} 1 \\ -1 \\ 0 \\ 0 \end{bmatrix} + c_2 \begin{bmatrix} 0 \\ 0 \\ 1 \\ -1 \end{bmatrix} : c_1 \text{ and } c_2, \text{ real numbers} \right\}$. The algebraic multiplicity of 2

is 3 and the geometric multiplicity is 2.

$A_4 = \left\{ c \begin{bmatrix} 0 \\ 0 \\ 1 \\ 1 \end{bmatrix} : c, \text{ a real number} \right\}$. The algebraic and geometric multiplicities of 4 are 1.

11. No. The zero vector is not in the set.

SECTION 7.6

1. $\left\{ \dfrac{1}{\sqrt{2}} \begin{bmatrix} 1 \\ 1 \end{bmatrix}, \dfrac{1}{\sqrt{2}} \begin{bmatrix} 1 \\ -1 \end{bmatrix} \right\}, \dfrac{1}{\sqrt{2}} \begin{bmatrix} 1 & 1 \\ 1 & -1 \end{bmatrix}, \begin{bmatrix} 1 & 0 \\ 0 & -1 \end{bmatrix}$

3. $\left\{ \dfrac{1}{\sqrt{2}} \begin{bmatrix} 1 \\ 0 \\ -1 \end{bmatrix}, \begin{bmatrix} 0 \\ 1 \\ 0 \end{bmatrix}, \dfrac{1}{\sqrt{2}} \begin{bmatrix} 1 \\ 0 \\ 1 \end{bmatrix} \right\}, \dfrac{1}{\sqrt{2}} \begin{bmatrix} 1 & 0 & 1 \\ 0 & \sqrt{2} & 0 \\ -1 & 0 & 1 \end{bmatrix}, \begin{bmatrix} 1 & 0 & 0 \\ 0 & 5 & 0 \\ 0 & 0 & 5 \end{bmatrix}$

5. $\left\{ \dfrac{1}{\sqrt{2}} \begin{bmatrix} 1 \\ -1 \\ 0 \end{bmatrix}, \dfrac{1}{\sqrt{6}} \begin{bmatrix} 1 \\ 1 \\ 2 \end{bmatrix}, \dfrac{1}{\sqrt{3}} \begin{bmatrix} 1 \\ 1 \\ -1 \end{bmatrix} \right\}, \dfrac{1}{\sqrt{6}} \begin{bmatrix} \sqrt{3} & 1 & \sqrt{2} \\ -\sqrt{3} & 1 & \sqrt{2} \\ 0 & 2 & -\sqrt{2} \end{bmatrix}, \begin{bmatrix} 3 & 0 & 0 \\ 0 & 3 & 0 \\ 0 & 0 & 6 \end{bmatrix}$

7. If $A = [a_{ij}]$ and $B = [b_{ij}]$, then $A + B = [a_{ij} + b_{ij}]$. Since $a_{ij} = a_{ji}$ and $b_{ij} = b_{ji}$, we have $a_{ij} + b_{ij} = a_{ji} + b_{ji}$. Therefore $A + B$ is symmetric.

9. Since $I_n = I_n^t = (A^{-1}A)^t = A^t(A^{-1})^t$, we have $(A^t)^{-1} = (A^{-1})^t$. But $A^t = A$. Therefore $A^{-1} = (A^{-1})^t$ so A^{-1} is symmetric.

11. The ij-component of P^tP is $\langle \mathbf{v}_i, \mathbf{v}_j \rangle$. Therefore $P^tP = I_n$ since $\{\mathbf{v}_1, \mathbf{v}_2, \dots, \mathbf{v}_n\}$ is orthonormal.

13. $1 = \det I_n = \det P^{-1}P = \det P^t P = (\det P^t)(\det P) = (\det P)^2$.
Therefore $\det P = \pm 1$.

15. $\|\mathbf{x}\|^2 = \langle \mathbf{x}, \mathbf{x} \rangle = \langle I_n \mathbf{x}, \mathbf{x} \rangle = \langle P^t P \mathbf{x}, \mathbf{x} \rangle = \langle P\mathbf{x}, P\mathbf{x} \rangle = \|P\mathbf{x}\|^2$.
Therefore $\|\mathbf{x}\| = \|P\mathbf{x}\|$.

SECTION 7.7

1. $\left\{ \begin{bmatrix} -1 \\ 1 \end{bmatrix}, \begin{bmatrix} 1 \\ 0 \end{bmatrix} \right\}, \begin{bmatrix} 2 & 1 \\ 0 & 2 \end{bmatrix}$
3. $\left\{ \begin{bmatrix} 2 \\ 0 \\ 0 \end{bmatrix}, \begin{bmatrix} 2 \\ 2 \\ 0 \end{bmatrix}, \begin{bmatrix} 0 \\ 0 \\ 1 \end{bmatrix} \right\}, \begin{bmatrix} 4 & 1 & 0 \\ 0 & 4 & 1 \\ 0 & 0 & 4 \end{bmatrix}$

5. $\left\{ \begin{bmatrix} 1 \\ 1 \\ 0 \\ 0 \end{bmatrix}, \begin{bmatrix} 0 \\ 1 \\ -1 \\ -1 \end{bmatrix}, \begin{bmatrix} 1 \\ 0 \\ 0 \\ 0 \end{bmatrix}, \begin{bmatrix} 0 \\ 0 \\ 1 \\ 0 \end{bmatrix} \right\}, \begin{bmatrix} 2 & 0 & 0 & 0 \\ 0 & 2 & 0 & 0 \\ 0 & 0 & 1 & 1 \\ 0 & 0 & 0 & 1 \end{bmatrix}$

7. $\left\{ \begin{bmatrix} 0 \\ 0 \\ 0 \\ 0 \\ 1 \\ -1 \end{bmatrix}, \begin{bmatrix} 0 \\ 0 \\ 0 \\ 0 \\ 2 \\ -1 \end{bmatrix}, \begin{bmatrix} -1 \\ 1 \\ 0 \\ 0 \\ 0 \\ 0 \end{bmatrix}, \begin{bmatrix} 1 \\ 0 \\ 0 \\ 0 \\ 0 \\ 0 \end{bmatrix}, \begin{bmatrix} 0 \\ 0 \\ -1 \\ 1 \\ 0 \\ 0 \end{bmatrix}, \begin{bmatrix} 0 \\ 0 \\ 1 \\ 0 \\ 0 \\ 0 \end{bmatrix} \right\}, \begin{bmatrix} 2 & 0 & 0 & 0 & 0 & 0 \\ 0 & 3 & 0 & 0 & 0 & 0 \\ 0 & 0 & 2 & 1 & 0 & 0 \\ 0 & 0 & 0 & 2 & 0 & 0 \\ 0 & 0 & 0 & 0 & 3 & 1 \\ 0 & 0 & 0 & 0 & 0 & 3 \end{bmatrix}$

SECTION 7.8

1. $\begin{bmatrix} 0 & .25 \\ 1 & 0 \end{bmatrix}$ **3.** A is singular. **5.** $\begin{bmatrix} 1 & 0 & -.5 \\ 0 & .5 & 0 \\ 0 & 0 & .5 \end{bmatrix}$

7. A is singular. **9.** $\begin{bmatrix} 12 & 1 \\ 4 & 12 \end{bmatrix}$ **11.** $\begin{bmatrix} 54 & 0 & 46 \\ 0 & 98 & 0 \\ 84 & 0 & 136 \end{bmatrix}$

13. If $A = \begin{bmatrix} a & b \\ c & d \end{bmatrix}$, then $f(\lambda) = \det(\lambda I - A) = \lambda^2 - (a + d)\lambda + ad - bc$ so that $f(A) = A^2$
$- (a + d)A + (ad - bc)I = 0$.

SECTION 8.1

1. 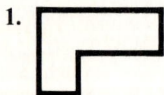 **3.** 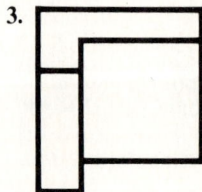 **5.**  **7.**

SECTION 8.2

1. $[x \ y] \begin{bmatrix} 3 & 3 \\ 3 & 6 \end{bmatrix} \begin{bmatrix} x \\ y \end{bmatrix}$ **3.** $[x_1 \ x_2 \ x_3] \begin{bmatrix} 2 & 2 & 3 \\ 2 & 3 & 1 \\ 3 & 1 & 7 \end{bmatrix} \begin{bmatrix} x_1 \\ x_2 \\ x_3 \end{bmatrix}$

5. $[x_1 \ x_2 \ x_3] \begin{bmatrix} 1 & -1 & 2 \\ -1 & -1 & -2.5 \\ 2 & -2.5 & 1 \end{bmatrix} \begin{bmatrix} x_1 \\ x_2 \\ x_3 \end{bmatrix}$

7. Positive definite   **9.** Positive definite   **11.** Not positive definite
13. Positive definite   **15.** Not positive definite
17. Symmetric matrices have real eigenvalues. If A is diagonally dominant with positive diagonal components, then the nonpositive real axis does not intersect any of the disks in Theorem 9 of Section 7.4. Therefore A has positive eigenvalues. By Theorem 2 the matrix A is positive definite.

SECTION 8.3

1. $3x^2 - 4xy + 2y^2, \begin{bmatrix} 3 & -2 \\ -2 & 2 \end{bmatrix}$ **3.** $2x^2 + 8xy, \begin{bmatrix} 2 & 4 \\ 4 & 0 \end{bmatrix}$

5. $x^2 + xy + y^2, \begin{bmatrix} 1 & .5 \\ .5 & 1 \end{bmatrix}$ **7.** Hyperbola   **9.** Parabola

11. Ellipse   **13.** $(x')^2 + 3(y')^2 - 1 = 0$, ellipse   **15.** $(x')^2 - 5(y')^2 - 1 = 0$, hyperbola
17. $5(x')^2 - 4 = 0$, two lines   **19.** $(x')^2 + 3(y')^2 + 2x' = 0$, ellipse
21. $5(x')^2 - (y')^2 + 10x' - 2y' - 1 = 0$, hyperbola

SECTION 8.4

1. $3x^2 + 4y^2 + 5z^2 - 8xy + 10xz + 4yz, \begin{bmatrix} 3 & -4 & 5 \\ -4 & 4 & 2 \\ 5 & 2 & 5 \end{bmatrix}$

3. $x^2 - y^2 + 2z^2 - 3xy + 4xz - 16yz, \begin{bmatrix} 1 & -1.5 & 2 \\ -1.5 & -1 & -8 \\ 2 & -8 & 2 \end{bmatrix}$

5. $x^2 - z^2 + 8xy - 4xz, \begin{bmatrix} 1 & 4 & -2 \\ 4 & 0 & 0 \\ -2 & 0 & -1 \end{bmatrix}$

7. $4(x')^2 + 2(y')^2 - (z')^2 = 1$, hyperboloid of one sheet
9. $5(x')^2 + 2(y')^2 = 1$, elliptic cylinder
11. $(x')^2 - (y')^2 = 1$, hyperbolic cylinder
13. $(x')^2 + (y')^2 - (z')^2 = 1$, hyperboloid of one sheet

SECTION 8.5

5. $c_1 e^{2t} \begin{bmatrix} 1 \\ 1 \end{bmatrix} + c_2 e^{7t} \begin{bmatrix} 3 \\ -2 \end{bmatrix}$ **7.** $c_1 \begin{bmatrix} 1 \\ -1 \\ 0 \end{bmatrix} + c_2 e^{3t} \begin{bmatrix} 1 \\ 2 \\ 0 \end{bmatrix} + c_3 e^{4t} \begin{bmatrix} 0 \\ 0 \\ 1 \end{bmatrix}$

9. $c_1 \begin{bmatrix} 0 \\ 1 \\ -1 \end{bmatrix} + c_2 e^{-2t} \begin{bmatrix} 2 \\ -1 \\ 0 \end{bmatrix} + c_3 e^{-3t} \begin{bmatrix} 1 \\ 0 \\ -1 \end{bmatrix}$ **11.** $\frac{1}{2} e^{2t} \begin{bmatrix} 1 \\ 1 \end{bmatrix} + \frac{1}{2} e^{7t} \begin{bmatrix} 1 \\ -1 \end{bmatrix}$

13. $\frac{2}{3} \begin{bmatrix} 1 \\ -1 \\ 0 \end{bmatrix} + \frac{1}{3} e^{3t} \begin{bmatrix} 1 \\ 2 \\ 0 \end{bmatrix} + e^{4t} \begin{bmatrix} 0 \\ 0 \\ 1 \end{bmatrix}$

SECTION 9.1

1. $\begin{bmatrix} .076919 \\ .769198 \end{bmatrix}, \begin{bmatrix} .076923 \\ .769231 \end{bmatrix}, \begin{bmatrix} .076923 \\ .769231 \end{bmatrix}$ **3.** $\begin{bmatrix} -.111111 \\ .222222 \end{bmatrix}, \begin{bmatrix} -.111111 \\ .222222 \end{bmatrix}, \begin{bmatrix} -.111111 \\ .222222 \end{bmatrix}$

5. $\begin{bmatrix} .258274 \\ -.207202 \\ .258740 \end{bmatrix}, \begin{bmatrix} .266668 \\ -.200000 \\ .266666 \end{bmatrix}, \begin{bmatrix} .266667 \\ -.200000 \\ .266667 \end{bmatrix}$

7. $\begin{bmatrix} .469637 \\ .047218 \\ -.225423 \end{bmatrix}, \begin{bmatrix} .480446 \\ .051955 \\ -.220732 \end{bmatrix}, \begin{bmatrix} .480520 \\ .051948 \\ -.220779 \end{bmatrix}$

9. $\begin{bmatrix} 6.94650 \\ -1.30247 \end{bmatrix}, \begin{bmatrix} 7.92196 \\ -1.47399 \end{bmatrix}, \begin{bmatrix} 8 \\ -1.5 \end{bmatrix}$

11. $L + U$ and U are both the zero matrix. Therefore $D^{-1}(L + U)$ and $(D - 1)^{-1}U$ are also the zero matrix, which has 0 as its only eigenvalue.

SECTION 9.2

1. $\lambda(4) = 5.37217, \quad \mathbf{x}_4 = \begin{bmatrix} .457437 \\ 1 \end{bmatrix}$

3. $\lambda(4) = 5.37235, \quad \mathbf{x}_4 = \begin{bmatrix} 1 \\ .686147 \end{bmatrix}$

5. $\lambda(5) = 16.1169, \quad \mathbf{x}_5 = \begin{bmatrix} .283349 \\ .641674 \\ 1 \end{bmatrix}$

7. The matrix does not have a dominant eigenvalue.

SECTION 9.3

1. $\begin{bmatrix} 1 \\ 2 \end{bmatrix}, \begin{bmatrix} 1.41421 \\ 1.73205 \end{bmatrix}$ **3.** $\begin{bmatrix} -1.41421 \\ 1.41421 \end{bmatrix}$, does not converge **5.** $\begin{bmatrix} 1.57080 \\ 3.14159 \\ 1 \end{bmatrix}$

APPENDIX 2

1. Let $\lambda_1, \lambda_2, \ldots, \lambda_k$ be eigenvalues of A such that $A\mathbf{v}_i = \lambda_i \mathbf{v}_i$ for $i = 1, 2, \ldots k$. Since $\mathbf{u}$ is an element of U there are scalars $c_1, c_2, \ldots, c_k$ such that $\mathbf{u} = c_1 \mathbf{v}_1 + c_2 \mathbf{v}_2 + \cdots + c_k \mathbf{v}_k$. Then $A\mathbf{u} = c_1 A\mathbf{v}_1 + c_2 A\mathbf{v}_2 + \cdots + c_k A\mathbf{v}_k = (c_1 \lambda_1)\mathbf{v}_1 + (c_2 \lambda_2)\mathbf{v}_2 + \cdots + (c_k \lambda_k)\mathbf{v}_k$. Thus $A\mathbf{u}$ is a linear combination of the elements of B and, therefore, is an element of U.

Index